The Biology of Cephalopods

FRONTISPIECE: Professor J. Z. Young, MA, DSc, FRS, in whose honour this Symposium was held.

(*Photograph P. N. Dilly*)

SYMPOSIA OF THE ZOOLOGICAL SOCIETY OF LONDON
NUMBER 38

The Biology of Cephalopods

(The Proceedings of a Symposium held at The Zoological Society of London on 10 and 11 April 1975)

Edited by

MARION NIXON
Department of Anatomy, University College London, England

and

J. B. MESSENGER
Department of Zoology, University of Sheffield, England

Published for
THE ZOOLOGICAL SOCIETY OF LONDON
BY
ACADEMIC PRESS

1977

ACADEMIC PRESS INC. (LONDON) LTD
24/28 Oval Road
London NW1

U.S. Edition published by
ACADEMIC PRESS INC.
111 Fifth Avenue,
New York, New York 10003

Library of Congress Catalog Card Number: 74-5683

ISBN: 0-12-613338-7

PRINTED IN GREAT BRITAIN BY
J. W. ARROWSMITH LTD, BRISTOL

CONTRIBUTORS

BAKER, P. F., *Department of Physiology, King's College London, The Strand, London WC2, England* (p. 243)

BARLOW, J. J., *Institute of Neurology, Queen's Square, London WC1, England* (p. 325)

BOLETZKY, S. v., *Laboratoire Arago, 66650 Banyuls-sur-Mer, France* (p. 557)

BUDELMANN, B. U., *Universität Regensburg, Fachbereich Biologie, 8400 Regensburg, Universitätsstrasse 31, Germany* (p. 309)

CLARKE, M. R., *Marine Biological Association of the United Kingdom, The Laboratory, Citadel Hill, Plymouth PL1 2PB, England* (p. 89)

DILLY, P. N., *Department of Anatomy, University College London, Gower Street, London WC1 6BT, England* (p. 447)

DONOVAN, D. T., *Department of Geology, University College London, Gower Street, London WC1 6BT, England* (p. 15)

FROESCH, D., *Laboratoire Arago, 66650 Banyuls-sur-Mer, France* (p. 541)

GHIRETTI, F., *Istituto di Biologia Animale, Fisiologia Generale, Università di Padova, 35100 Padova, Italy* (p. 513)

GHIRETTI-MAGALDI, ANNA, *Istituto di Biologia Animale, Fisiologia Generale, Università di Padova, 35100 Padova, Italy* (p. 513)

GILPIN-BROWN, J. B., *Marine Biological Association of the United Kingdom, The Laboratory, Citadel Hill, Plymouth PL1 2PB, England* (p. 233)

HERRING, P. J., *Institute of Oceanographic Sciences, Wormley, Surrey, England* (p. 127)

HOCHBERG, F. G., *Museum of Natural History, Santa Barbara, California, USA* (p. 191)

MANGOLD, KATHARINA, *Laboratoire Arago, 66650 Banyuls-sur-Mer, France* (p. 541)

MARTIN, R., *Stazione Zoologica, Villa Communale, Napoli 80121, Italy* (p. 261)

MAURO, A., *The Rockefeller University, New York, NY 10021, USA* (p. 287)

MESSENGER, J. B., *Department of Zoology, University of Sheffield, Sheffield S10 2TN, England* (pp. 1, 347)

MUNTZ, W. R. A., *Department of Biology, Stirling University, Stirling FK9 4LA, Scotland* (p. 277)

NIXON, MARION, *Department of Anatomy, University College London, Gower Street, London WC1 6BT, England* (pp. 1, 447)

PACKARD, A., *Department of Physiology, University Medical School, Teviot Place, Edinburgh EB8 9AG, Scotland* (p. 191)

ROPER, C. F. E., *Department of Invertebrate Zoology (Mollusca), National Museum of Natural History, Smithsonian Institution, Washington, DC 20560, USA* (p. 61)

SALVATO, B., *Istituto di Biologia Animale, Fisiologia Generale, Università di Padova, 35100 Padova, Italy* (p. 513)

SANDERS, G. D., *Department of Psychology, City of London Polytechnic, Calcutta House, Old Castle Street, London E1, England* (p. 435)

VOSS, G. L., *School of Marine and Atmospheric Science, University of Miami, 10 Rickenbacker Causeway, Miami, Florida 33149, USA* (p. 49)

WELLS, J., *Department of Zoology, University of Cambridge, Downing Street, Cambridge, England* (p. 525)

WELLS, M. J., *Department of Zoology, University of Cambridge, Downing Street, Cambridge, England* (p. 525)

YOUNG, J. Z., *The Wellcome Institute for the History of Medicine, 183 Euston Road, London NW1 2BP, England* (p. 377)

YOUNG, R. E., *Department of Oceanography, University of Hawaii, Honolulu, Hawaii 96822* (p. 161)

ORGANIZERS AND CHAIRMEN

ORGANIZERS

MARION NIXON and J. B. MESSENGER, on behalf of The Zoological Society of London

CHAIRMEN OF SESSIONS

ANNA M. BIDDER, *Department of Zoology, University of Cambridge, Downing Street, Cambridge, England*

E. J. DENTON, *Marine Biological Association of the United Kingdom, The Laboratory, Citadel Hill, Plymouth PL1 2PB, England*

BERNARD KATZ, *Department of Biophysics, University College London, Gower Street, London WC1 6BT, England*

M. J. WELLS, *Department of Zoology, University of Cambridge, Downing Street, Cambridge, England*

CONTENTS

Introduction

JOHN MESSENGER and MARION NIXON

Evolution of the Dibranchiate Cephalopoda

D. T. DONOVAN

Present Status and New Trends in Cephalopod Systematics

G. L. VOSS

Comparative Captures of Pelagic Cephalopods by Midwater Trawls

C. F. E. ROPER

Beaks, Nets and Numbers

M. R. CLARKE

Luminescence in Cephalopods and Fish

PETER J. HERRING

Ventral Bioluminescent Countershading in Midwater Cephalopods

R. E. YOUNG

Skin Patterning in *Octopus* and Other Genera

A. PACKARD and F. G. HOCHBERG

The Squid and its Giant Nerve Fibre

J. B. GILPIN-BROWN

The Squid Giant Axon: A Model for Studies of Neuronal Calcium Metabolism

P. F. BAKER

The Giant Nerve Fibre System of Cephalopods. Recent Structural Findings

R. MARTIN

Pupillary Response of Cephalopods

W. R. A. MUNTZ

Extra-ocular Photoreceptors in Cephalopods

A. MAURO

Structure and Function of the Angular Acceleration Receptor Systems in the Statocysts of Cephalopods

B. U. BUDELMANN

Comparative Biochemistry of the Central Nervous System

J. J. BARLOW

Prey-capture and Learning in the Cuttlefish, *Sepia*

J. B. MESSENGER

Brain, Behaviour and Evolution of Cephalopods

J. Z. YOUNG

Multiphasic Retention Performance Curves: Fear or Memory?

G. D. SANDERS

Sucker Surfaces and Prey Capture

MARION NIXON and P. N. DILLY

The Evolution of Haemocyanin

ANNA GHIRETTI-MAGALDI, F. GHIRETTI and B. SALVATO

Optic Glands and the Endocrinology of Reproduction

M. J. WELLS and J. WELLS

A Reconsideration of Factors Associated with Sexual Maturation

KATHARINA MANGOLD and D. FROESCH

Post-hatching Behaviour and Mode of Life in Cephalopods

S. v. BOLETZKY

Appendix I

Studies on Cephalopods by J. Z. Young

Appendix II

Classification of Recent Cephalopods

Symp. zool. Soc. Lond. (1977) No. 38, 1–13.

INTRODUCTION

JOHN MESSENGER and MARION NIXON*

Department of Zoology, University of Sheffield, England and *Department of Anatomy, University College London, England*

Although this Symposium was originally conceived simply as a tribute to Professor J. Z. Young, it soon became apparent that there had never been, so far as we could ascertain, a full-scale meeting devoted exclusively to cephalopods. In a world burgeoning with symposia, large and small, national and international, on a broad canvas or devoted to the latest, and often ephemeral, problems of a particular research field, this seemed to us a quite extraordinary oversight. For not only are the cephalopods fascinating in their own right, but for the last 40 years they have provided material for fundamental research in neurobiology. We therefore set out to produce a meeting that would both honour Professor Young and reflect current research into different aspects of the life of cephalopods.

The result will inevitably be seen to be a compromise and our title over-ambitious. Although there is as strong a neurobiological bias as would be expected (Professor Young having worked on the nervous system throughout his long and active career) this volume could not provide a comprehensive review of the many experiments on learning in *Octopus* and *Sepia* carried out by him and his colleagues: readers seeking this will consult the useful review by Sanders (1975). And we do not pretend that this volume is a definitive Biology of Cephalopods: it is a record of a conference into which we fitted as many speakers as possible. Naturally we had to make some selection, partly because the Zoological Society's Symposia are traditionally limited to two days and partly because, in these times of financial stringency, it was impossible to invite every worker in the field. As it was, several contributors had to finance themselves, despite the very generous financial help given by the Royal Society, by the Wellcome Trustees, and above all by the Zoological Society of London. To all of these bodies we would like to express our warmest thanks for making the Symposium possible. We would like, too, to thank the Zoological Society of

* Present address: *The Wellcome Institute for the History of Medicine, London.*

London for its help in so many other ways, particularly for its guiding hand, in the person of Dr H. G. Vevers (who was also kind enough to introduce the Symposium) and for the executive skill of Ms Unity McDonnell. May we, as organizers, congratulate the Officers of the Society on the style and efficiency of their arrangements, and, on behalf of all the participants, thank them heartily for having us. It is a pleasure too to thank Professor E. J. Denton, Dr A. M. Bidder, Professor Sir Bernard Katz and Dr M. J. Wells for acting as our Chairmen.

Compromise or not, the outcome of our selection is here for all to read, but to forestall criticism about the omissions of this volume and, more important, to help the interested student find his way into the extensive literature on these wonderful animals, we would like to digress for a moment and catalogue some of the topics that could not be discussed at the Symposium.

There is nothing here about digestion, for example, despite the work of Bidder (reviewed 1966) and more recently of Boucaud-Camou (1974; Boucaud-Camou & Péquignat, 1973), predation (Wodinsky, 1971), feeding (Nixon, 1969) or the function of the buccal mass (Arnold & Arnold, 1969; Wodinsky, 1969; Altman & Nixon, 1970). There is nothing about the cardiovascular or respiratory systems although a number of experimenters have considered these problems recently, notably Johansen and his colleagues (e.g. Johansen, 1965; Johansen & Lenfant, 1966; Johansen & Martin, 1962; see also Ghiretti, 1966). There is nothing about excretion, despite the elegant work of A. W. Martin and his collaborators (see Martin & Harrison, 1966; Wittmer & Martin, 1973; Potts, 1967) on *Octopus dofleini*. There are two papers on reproductive physiology but no mention of the detailed studies by Mann and his colleagues on the spermatophores of *O. dofleini* (Mann, Martin & Thiersch, 1970; Hanson, Mann & Martin, 1973) or Maxwell (1974) on *Eledone cirrosa*. There is no chapter on embryology (although fortunately there are two recent reviews available: Arnold, 1971; Arnold & Williams-Arnold, 1976). Those at the conference heard a paper by Dr. J. M. Arnold, who at very short notice kindly came over from Hawaii and spoke about his views on the role of the egg cortex in development (Arnold & Williams-Arnold, 1974); these views are not uncontested, however, and for an alternative interpretation see Marthy (1973, 1975).

Those interested in the growth, development and life history of these animals will also be disappointed, although there is the work of Mangold-Wirz (1963), Mangold & Boletzky (1973), Nixon

(1969, 1973), Packard & Albergoni (1970), Wodinsky (1972), van Heukelem (1973), Aldred (1974) and Dilly & Nixon (1976a). And the literature on the hatching of eggs and rearing of young is also largely ignored (Itami *et al.*, 1963; Choe, 1966; La Roe, 1971; Boletzky, 1974; Bradley, 1974; Opresko & Thomas, 1975).

One of the most remarkable series of experiments on cephalopods recently has been the examination by Denton and Gilpin-Brown of their various buoyancy mechanisms. Dr Gilpin-Brown presented a film about buoyancy at the conference but there is no paper here dealing with this fascinating problem, and we can only urge anyone unfamiliar with this work to read these authors' review papers (e.g. Denton & Gilpin-Brown, 1973). Other omissions are locomotion—studied by such workers as Nesis, Williamson, Zuev and Bradbury and Aldrich (see Packard, 1972) as well as by Packard & Trueman (1974)—and the physiology and structure of muscle (Gray, 1960; Wilson, 1960; Ward & Wainwright, 1972). This is most regrettable but Trueman (1975) gives a most useful introduction to these topics.

Even within the broad area of neurobiology and sensory physiology the coverage has not been truly comprehensive. The sense organs, for example, are only represented here by papers on the statocyst, on the pupillary response and on extra-ocular photoreception. There is no mention of the many other aspects of vision that have been investigated experimentally; there have been studies on the organization and ultrastructure of the retina including its electrophysiology, on the visual pigments, work on acuity, colour-vision, polarized-light detection and an especially extensive consideration of shape discrimination by Sutherland (his paper of 1969 is a useful introduction). Bullock & Horridge (1965) review much of this work and a more extended treatment will be given in Messenger (in press).

The other senses are hardly mentioned, although chemoreception has been examined experimentally by Wells (e.g. Wells, Freeman & Ashburner, 1965), and morphologically by Graziadei (1965), by Woodhams & Messenger (1974), and by Emery (1975). The tactile sense and proprioception, extensively investigated by Wells & Wells (see Wells' review, 1966) is also omitted. Neither is there any consideration of mechanoreceptors, perhaps because we are still very ignorant about them. Are pressure receptors necessary to keep those cephalopods that show extensive diel migration at the appropriate depth (see the chapters by J. Z. Young and R. E. Young, this volume)?

Finally we must regret the short shrift given to purely psychological studies, which have produced some very interesting findings indeed. The justification for a comparative study of learning was made by Boycott & Young in their classic paper of 1950 and need not be restated here; but it is extraordinary, as Mackintosh (1965) has emphasized, how in some of its higher operations the cephalopod brain appears to work in a very similar way to the vertebrate brain. For example cephalopods show such phenomena as transfer along a continuum, learning reversals faster after overtraining on the original problem, and learning a non-reversal shift more slowly after previous overtraining; similarly they do not exhibit cue-additivity (Sutherland & Mackintosh, 1964; Messenger & Sanders, 1972).

The foregoing account should convince the interested student that there is a wealth of information available on many aspects of the life of cephalopods. Much of this has yet to find its way into a single volume, although it is hoped that the forthcoming chapter by Mangold & Portmann for Grassé's *Traité de Zoologie*, volume V, fascicule 4 and M. J. Wells' book, *The physiology of* Octopus, will soon be available. Meanwhile there could be no better general introduction to their biology than Packard's provocative review (1972) nor a more succinct account of their sense organs and nervous systems than that given in Bullock & Horridge (1965).

It is nearly 50 years now since Professor Young, or JZ as he is always known, first became involved with the animals that he has made so very much his own. After graduating from Oxford he was awarded that University's esteemed Naples scholarship and in 1927 went to the world-famous Stazione Zoologica. He was already interested in the nervous system and in 1929 his first paper on the nerves of cephalopods appeared, in the *Bolletino della società italiana di biologia sperimentale.* Appropriately, in view of his lifelong association with the Stazione and his love for Naples, it was written in Italian, and it was to be the first of no fewer than 102 publications on cephalopods published up to the present time [including what is, to date, the most comprehensive study of any invertebrate brain, *The anatomy of the nervous system of* Octopus vulgaris (1971)]. We have taken this opportunity to list these papers for the first time (see Appendix I). We would also remind the reader that JZ has published numerous other papers on various aspects of the nervous system as well and that, difficult though it is to believe, this is the same J. Z. Young who has found time to write

The life of vertebrates (1950); *Doubt and certainty in science* (1951); *The life of mammals* (1957); *A model of the brain* (1964); *The memory system of the brain* (1966) and *An introduction to the study of man* (1971).

In the thirties in Naples (and at Plymouth and at Woods Hole, see Young, 1975) Young continued to study cephalopods, especially the peripheral nervous system of *Loligo*. It is now past history how he rediscovered the giant nerve fibres that were to revolutionize the study of the mechanisms of nervous conduction and, in the hands of Hodgkin, Huxley, Katz and others, provide the basis of modern neurophysiology (see Hodgkin, 1964). It is easy to say that JZ left others to do the really crucial experiments but this does not recognize the fact that biological systems are organized on a number of levels and that different kinds of scientists concern themselves with different levels. Although Young's great virtue as a biologist stems from his desire to relate findings at all levels of analysis he has always been fascinated by the final "operational" level, that of the whole animal. Of course it is important to understand the ionic basis of the action potential but what is more important for biologists of Young's persuasion is the communication system that it subserves. The brain is a complex multi-channel system that must *a priori* be examined at a level higher that that of a single cell if we are to begin to understand the way it operates. Neurons may have properties in common—they may look alike or have similar electrical characteristics—but it is becoming increasingly clear that they are connected in a highly specific way and that it is their *connections* that give them their unique property (Gaze, 1970; Hubel & Wiesel, 1962, 1968). A full understanding of the biophysics of a single cell could never explain how particular sets of neurons act together to bring about a particular motor act (e.g. Eccles, Ito & Szentágothai, 1967), let alone one that is more likely to lead to the animal surviving. For it is characteristic of the homeostats we call animals that the activity of some neurons may affect others in such a way that their owner's behaviour will be modified adaptively as a result of changing information about the external world. Concern for problems of this kind took Young away from isolated nerve preparations into the central nervous system and particularly into the very difficult area of learning and memory. It is typical of him, however, that before abandoning the giant axons for "higher things" he clarified their function in the living squid; in particular he demonstrated the elegant morphological arrangement whereby a set of axons, graded in size and therefore in conduction velocity, would permit synchrony of muscular activity along the mantle and hence bring about the rapid propulsive thrust of the escape jetting (Young, 1938; Pumphrey & Young, 1938).

How often in biology do advances await the discovery of the right preparation! This one has proved to be outstandingly useful: the beautiful film on the squid giant axon by Gilpin-Brown (see his chapter, this volume) brought out this point most strikingly at the conference, but a glance at any *Current Contents* will make the same point. In this volume a biophysical consideration of the squid axon is given by Baker; and the giant fibre system itself is re-examined, with modern techniques, by Martin.

It was during his investigations on the giant fibres that JZ discovered the epistellar body in the stellate ganglion of octopods (Young, 1936). As Young himself admits (p. 387) his interpretation of this structure was never entirely satisfactory but he had, nevertheless, discovered another fascinating organ for, in 1962, Nishioka, Hagadorn & Bern established that it had the ultrastructure and biochemistry of a photoreceptor. We may wonder why cephalopods with their enormous eyes have had to evolve extra-ocular photoreceptors as well, yet it seems they are widespread among living coleoids (J. Z. Young, see his chapter, this volume) and, as Mauro makes clear (see his chapter, this volume), wherever they have been examined physiologically they have been shown unequivocally to be light-sensitive.

At about this time JZ influenced Holmes (1940) to examine the extraordinary chromatophore system of cephalopods, a system that has since been most elegantly studied at the ultrastructural and physiological level by Florey and his colleagues (Florey, 1969; Florey & Kriebel, 1969) and, at the level of the whole animal, by Packard and his co-workers. In this volume Packard & Hochberg (see their chapter) give some idea of the complexity of the skin of *Octopus*, and the way the chromatophores are complemented by other elements. Their work, incidentally, is unique in this volume in that it involves field observations. We would like to draw the attention of all workers, but especially young biologists skilled in skin-diving, to the fact that field studies of littoral species of cephalopods are sadly lacking (though see McGowan, 1954: Arnold, 1962, 1965; Williamson, 1965). This is a virtually untapped area and detailed, quantitative observations are urgently required (Altman, 1967; Kayes, 1974).

By the beginning of the Second World War, Young, in collaboration with F. K. Sanders, had published his first paper on cephalopod learning (Sanders & Young, 1940) using the beautiful

Sepia attack that Messenger elaborates on in this volume (see his chapter). After the war JZ returned to Naples, with B. B. Boycott as his assistant, and began the detailed study of the octopus central nervous system that has continued up to the present. The first of many papers on the brain of *Octopus vulgaris* appeared in 1950; others followed throughout the fifties, and in the sixties the rate of publication increased steadily. Anyone who has ever been associated with JZ knows that his energy is prodigious but, even so, a glance at his output between 1960 and 1970 (Appendix I) is still impressive beyond belief.

Another of JZ's remarkable qualities is the range of his interests; he is at home with quite different kinds of problems, and many of these he has encouraged others to take up and develop. His extensive experimental work on learning, for example, first with Boycott and later with Wells, attracted several other investigators among whom we can mention Maldonado (1963, 1968, 1969), and G. D. Sanders who here considers long-term and short-term memory systems (see his chapter). At the same time JZ has found time to write detailed anatomical papers on the retina (Young, 1962, 1963), some of whose findings are most pertinent to the discussion of optics presented here by Muntz (see his chapter), and on the statocyst: his 1960 paper on the anatomy of this organ is already something of a classic and, again, his descriptions have had the merit of prompting others to carry out ultrastructural studies (Barber, 1966) and experiments. These have been made by a number of workers in recent years, notably by Vinnikov and by Budelmann. Professor Vinnikov from Leningrad was unfortunately unable to accept our invitation to the Symposium. He was sadly missed and those interested in his particular approach to the problems of equilibrium control in cephalopods should consult Vinnikov, 1974. In this volume Budelmann (see his chapter) summarizes his own work on an organ that continues to fascinate JZ.

Young has always been quick to exploit new techniques. The way he encouraged electron microscopy from its earliest days is well known, and several of his colleagues have examined *Octopus* with the transmission electron microscope, among whom we might note Graziadei (1965); Jones (1970); Barber (1966); Barber & Wright (1969); Woodhams (1975; in prep.); and above all Gray (1969, 1970), who has made a number of very detailed studies. More recently the scanning electron microscope has begun to make its mark in morphological studies and again JZ has been quick to encourage others to apply this method to cephalopods (Dilly &

Nixon, 1976b). In this work Nixon & Dilly (see their chapter) examine the curious variety of cephalopod sucker surfaces with this powerful technique.

The nervous system is so difficult to understand that we need to examine it with a whole variety of techniques. Amongst other approaches that JZ has encouraged have been the cybernetic (Maldonado, 1964), the histochemical (Matus, 1973), the pharmacological (Juorio, 1971; Juorio & Philips, 1975) and the biochemical, examined here by Barlow (see his chapter).

We have said before that JZ's view of biology is essentially large scale: although an anatomist he "thinks big". It is characteristic of him that at this stage in his career his studies on the cephalopod CNS are moving from the base he has created in *Octopus vulgaris* out into the wider world of the squids to examine their nervous systems in detail and relate this to their mode of life and phylogeny. He is, indeed, preparing a book on the nervous system of squids; those of us who cannot wait to see this in print will have to content themselves for the moment with the intriguing account he gives here in his chapter. Brilliant in style, extensive in scope, this chapter is typical of JZ; the ease with which he handles facts and ideas, reducing them to order, is remarkable.

It needs emphasizing that Young's overriding interest in the nervous system has never made him myopic. He is fascinated by all aspects of the life of cephalopods and extremely knowledgeable about them. One of the organizers still remembers that it was an aside of JZ's that gave us the clue to the function of the branchial gland for example (Dilly & Messenger, 1972; Messenger, Muzii *et al.*, 1974). The extent of his knowledge on most of the topics discussed here—on phylogeny and systematics: Donovan (pp. 15–48) and Voss (pp. 49–60); on reproduction and development: Wells & Wells (pp. 525–540), Mangold & Froesch (pp. 541–555), Boletzky (pp. 557–567); on various aspects of the ecology of oceanic squids: Clarke (pp. 89–126), Roper (pp. 61–87); on the light organs: R. E. Young (pp. 161–190) and Herring (pp. 127–159); and on haemocyanin: Ghiretti-Magaldi, Ghiretti & Salvato (pp. 513–524)—must have surprised some of the contributors themselves!

We are sure that all the participants of this symposium will join us in thanking JZ for all he has taught us about these beautiful and remarkable animals, and in hoping that he will teach us much more. We hope that he will enjoy these essays, whatever their shortcomings, and that he will find some things here to stimulate him in his future research career.

References

Aldred, R. G. (1974). Structure, growth and distribution of the squid *Bathothauma lyromma* Chun. *J. mar. biol. Ass. U.K.* **54**: 995–1006.

Altman, J. S. (1967). The behaviour of *Octopus vulgaris* Lam. in its natural habitat: a pilot study. *Underwat. Ass. Rep.* **1966–67**: 77–83.

Altman, J. S. & Nixon, M. (1970). Use of the beaks and radula by *Octopus vulgaris* in feeding. *J. Zool., Lond.* **161**: 25–38.

Arnold, J. M. (1962). Mating behaviour and social structure in *Loligo pealeii*. *Biol. Bull. mar. biol. Lab. Woods Hole* **123**: 53–57.

Arnold, J. M. (1965). Observations on the mating behaviour of the squid *Sepioteuthis sepioidea*. *Bull. mar. Sci.* **15**: 216–222.

Arnold, J. M. (1971). Cephalopods. In *Experimental embryology of marine and fresh-water invertebrates*: 265–311. Reverberi, G. (ed.). Amsterdam: North Holland Publishing Co.

Arnold, J. M. & Arnold, K. O. (1969). Some aspects of hole-boring predation by *Octopus vulgaris*. *Am. Zool.* **9**: 991–996.

Arnold, J. M. & Williams-Arnold, L. D. (1974). Cortical-nuclear interactions in cephalopod development: cytochalasin B effects on the informational pattern in the cell surface. *J. Embryol. exp. Morph.* **31**: 1–25.

Arnold, J. M. & Williams-Arnold, L. D. (1976). The egg cortex problem as seen through the squid eye. *Am. Zool.* **16**: 421–446.

Barber, V. C. (1966). The fine structure of the statocyst of *Octopus vulgaris*. *Z. Zellforsch. mikrosk. anat.* **70**: 91–107.

Barber, V. C. & Wright, D. E. (1969). The fine structure of the sense organs of the cephalopod mollusc *Nautilus*. *Z. Zellforsch. mikrosk. anat.* **102**: 293–312.

Bidder, A. M. (1966). Feeding and digestion in cephalopods. In *Physiology of Mollusca* **2**: 97–124. Wilbur, K. M. & Yonge, C. M. (eds). London and New York: Academic Press.

Boletzky, S. v. (1974). Élevage de céphalopodes en aquarium. *Vie Milieu* **24**: 309–340.

Boucaud-Camou, E. (1974). Localisation d'activités enzymatiques impliquées dans la digestion chez *Sepia officinalis* L. *Archs Zool. exp. gén.* **115**: 5–27.

Boucaud-Camou, E. & Péquignat, E. (1973). Étude expérimentale de l'absorption digestive chez *Sepia officinalis* L. *Forma et functio* **6**: 93–111.

Boycott, B. B. & Young, J. Z. (1950). The comparative study of learning. *Symp. Soc. exp. Biol.* No. 4: 432–453.

Bradley, E. A. (1974). Some observations of *Octopus joubini* reared in an inland aquarium. *J. Zool., Lond.* **173**: 355–368.

Bullock, T. H. & Horridge, G. A. (1965). *Structure and function in the nervous systems of invertebrates* **2**. Freeman: San Francisco.

Choe, S. (1966). On the eggs, rearing, habits of the fry, and growth of some cephalopods. *Bull. mar. Sci.* **16**: 330–338.

Denton, E. J. & Gilpin-Brown, J. B. (1973). Flotation mechanisms in modern and fossil cephalopods. *Adv. mar. Biol.* **11**: 197–268.

Dilly, P. N. & Messenger, J. B. (1972). The branchial gland: a site of haemocyanin synthesis in *Octopus*. *Z. Zellforsch. mikrosk. anat.* **132**: 193–201.

Dilly, P. N. & Nixon, M. (1976). Growth of *Taonius megalops* (Mollusca, Cephalopoda). *J. Zool., Lond.* **179**: 19–83.

Dilly, P. N. & Nixon, M. (1976). The dermal tubercles of *Cranchia scabra* (Mollusca, Cephalopoda); surface structure and development. *J. Zool., Lond.* **179**: 291–295.

Eccles, J. C., Ito, M. & Szentágothai, J. (1967). *The cerebellum as a neuronal machine.* Heidelberg: Springer Verlag.

Emery, D. G. (1975). The histology and fine structure of the olfactory organ of the squid *Lolliguncula brevis* Blainville. *Tiss. Cell* **7**: 357–367.

Florey, E. (1969). Ultrastructure and function of cephalopod chromatophores. *Am. Zool.* **9**: 429–442.

Florey, E. & Kriebel, M. E. (1969). Electrical and mechanical responses of chromatophore muscle fibers of the squid, *Loligo opalescens*, to nerve stimulation and drugs. *Z. vergl. Physiol.* **65**: 98–130.

Gaze, M. (1970). *The formation of nerve connections.* London and New York: Academic Press.

Ghiretti, F. (1966). Respiration. In *Physiology of Mollusca* **2**: 175–208. Wilbur, K. M. & Yonge, C. M. (eds). London and New York: Academic Press.

Gray, E. G. (1969). Electron-microscopy of the glio-vascular organisation of the brain of *Octopus. Phil. Trans. R. Soc.* (B) **255**: 13–32.

Gray, E. G. (1970). The fine structure of the vertical lobe of *Octopus* brain. *Phil. Trans. R. Soc.* (B) **258**: 379–395.

Gray, J. A. B. (1960). Mechanically excitable receptor units in the mantle of the octopus and their connexions. *J. Physiol., Lond.* **153**: 573–582.

Graziadei, P. P. C. (1965). Sensory receptor cells and related neurons in cephalopods. *Cold Spring Harb. Symp. quant. Biol.* **30**: 45–57.

Hanson, D., Mann, T. & Martin, A. W. (1973). Mechanism of the spermatophoric reaction in the giant octopus of the North Pacific, *Octopus dofleini martini. J. exp. Biol.* **58**: 711–723.

Hodgkin, A. L. (1964). *The Sherrington Lectures VII. The conduction of the nerve impulse.* Liverpool: University Press.

Holmes, W. (1940). The colour changes and colour patterns of *Sepia officinalis* L. *Proc. zool. Soc. Lond.* (A.) **110**: 17–36.

Hubel, D. H. & Wiesel, T. N. (1962). Receptive fields, binocular interactions, and functional architecture in the cat's visual cortex. *J. Physiol., Lond.* **160**: 106–154.

Hubel, D. H. & Wiesel, T. N. (1968). Receptive fields and functional architecture of monkey striate cortex. *J. Physiol., Lond.* **195**: 215–243.

Itami, K., Izawa, Y., Maeda, S. & Nakai, K. (1963). Notes on the laboratory culture of the octopus larvae. *Bull. Jap. Soc. scient. Fish.* **29**: 514–520.

Johansen, K. (1965). Cardiac output in the large cephalopod *Octopus dofleini. J. exp. Biol.* **42**: 475–480.

Johansen, K. & Lenfant, C. (1966). Gas exchange in the cephalopod *Octopus dofleini. Am. J. Physiol.* **210**: 910–918.

Johansen, K. & Martin, A. W. (1962). Circulation in the cephalopod *Octopus dofleini. Comp. Biochem. Physiol.* **5**: 161–176.

Jones, D. G. (1970). A study of the presynaptic network of *Octopus* synaptosomes. *Brain Res.* **20**: 145–158.

Juorio, A. V. (1971). Catecholamines and 5-hydroxytryptamine in nervous tissue of cephalopods. *J. Physiol., Lond.* **216**: 213–226.

Juorio, A. V. & Philips, S. R. (1975). Tyramines in *Octopus* nerves. *Brain Res.* **83**: 180–184.

Kayes, R. J. (1974). The daily activity pattern of *Octopus vulgaris* in a natural habitat. *Mar. Behav. Physiol.* **2**: 337–343.

La Roe, E. T. (1971). The culture and maintenance of the loliginid squids *Sepioteuthis sepioidea* and *Doryteuthis plei*. *Mar. Biol.* **9**: 9–25.
Mackintosh, N. J. (1965). Discrimination learning in the octopus. *Anim. Behav.* (*Suppl.*) **1**: 129–134.
Maldonado, H. (1963). The general amplification function of the vertical lobe in *Octopus vulgaris*. *Z. vergl. Physiol.* **47**: 215–229.
Maldonado, H. (1964). The control of attack by *Octopus*. *Z. vergl. Physiol.* **47**: 656–674.
Maldonado, H. (1968). Effects of electroconvulsive shock on memory in *Octopus vulgaris* Lamarck. *Z. vergl. Physiol.* **59**: 25–37.
Maldonado, H. (1969). Further experiments on the effect of electroconvulsive shock (ECS) on memory in *Octopus vulgaris*. *Z. vergl. Physiol.* **63**: 113–118.
Mangold-Wirz, K. (1963). Biologie des céphalopodes benthiques et nectoniques de la mer Catalane. *Vie Milieu* (Suppl. No. 13): 1–285.
Mangold, K. & Boletzky, S. v. (1973). New data on reproductive biology and growth of *Octopus vulgaris*. *Mar. Biol.* **19**: 7–12.
Mann, T., Martin, A. W. & Thiersch, J. B. (1970). Male reproductive tract, spermatophores and spermatophoric reaction in the giant octopus of the North Pacific, *Octopus dofleini martini*. *Proc. R. Soc.* (B.) **175**: 31–61.
Marthy, H.-J. (1973). An experimental study of eye development in the cephalopod *Loligo vulgaris*: determination and regulation during formation of the primary optic vesicles. *J. Embryol. exp. Morph.* **29**: 347–361.
Marthy, H.-J. (1975). Organogenesis in Cephalopoda: further evidence of blastodisc-bound developmental information. *J. Embryol. exp. Morph.* **33**: 75–83.
Martin, A. W. & Harrison, F. M. (1966). Excretion. In *Physiology of Mollusca* **2**: 353–386. Wilbur, K. M. & Yonge, C. M. (eds). London and New York: Academic Press.
Matus, A. (1973). Histochemical localisation of biogenic monoamines in the cephalic ganglion of *Octopus vulgaris*. *Tiss. Cell* **5**: 591–601.
Maxwell, W. L. (1974). Spermiogenesis of *Eledone cirrhosa* Lamarck (Cephalopoda, Octopoda). *Proc. R. Soc. Lond.* (B.) **186**: 181–190.
McGowan, J. A. (1954). Observations on the sexual behaviour and spawning of the squid *Loligo opalescens* at La Jolla, California. *Calif. Fish. Game* **40**: 47–54.
Messenger, J. B. (in press). Comparative physiology of vision in molluscs. In *Handbook of sensory physiology* **VII/6B**. Autrum, H. (ed.) Heidelberg: Springer Verlag.
Messenger, J. B., Muzii, E. O., Nardi, G. & Steinberg, H. (1974). Haemocyanin synthesis and the branchial gland of *Octopus*. *Nature, Lond.* **250**: 154–155.
Messenger, J. B. & Sanders, G. D. (1972). Visual preference and two-cue discrimination learning in octopus. *Anim. Behav.* **20**: 580–585.
Nishioka, R. S., Hagadorn, I. R. & Bern, H. A. (1962). The ultrastructure of the epistellar body of the octopus. *Z. Zellforsch. mikrosk. anat.* **27**: 406–421.
Nixon, M. (1969). The time and frequency of responses by *Octopus vulgaris* to an automatic food dispenser. *J. Zool., Lond.* **158**: 475–483.
Nixon, M. (1973). Beak and radula growth in *Octopus vulgaris*. *J. Zool., Lond.* **170**: 451–462.
Opresko, L. & Thomas, R. (1975). Observations on *Octopus joubini*. Some aspects of reproductive biology and growth. *Mar. Biol.* **31**: 51–63.

Packard, A. (1972). Cephalopods and fish: the limits of convergence. *Biol. Rev.* **47**: 241–307.
Packard, A. & Albergoni, V. (1970). Relative growth, nucleic acid content and cell numbers of the brain in *Octopus vulgaris*. *J. exp. Biol.* **52**: 539–553.
Packard, A. & Trueman, E. R. (1974). Muscular activity of the mantle of *Sepia* and *Loligo* (Cephalopoda) during respiratory movements and jetting, and its physiological interpretation. *J. exp. Biol.* **61**: 411–419.
Potts, W. T. W. (1967). Excretion in the molluscs. *Biol. Rev.* **42**: 1–41.
Pumphrey, R. J. & Young, J. Z. (1938). The rates of conduction of nerve fibres of various diameters in cephalopods. *J. exp. Biol.* **15**: 453–466.
Sanders, F. K. & Young, J. Z. (1940). Learning and other functions of the higher nervous centres of *Sepia*. *J. Neurophysiol.* **3**: 501–526.
Sanders, G. D. (1975). The cephalopods. In *Invertebrate learning*: 1–101. Corning, W. C., Dyal, J. A. & Willows, A. O. D. (eds.). New York: Plenum.
Sutherland, N. S. (1969). Shape discrimination in rat, octopus and goldfish: a comparative study. *J. comp. physiol. Psychol.* **67**: 160–176.
Sutherland, N. S. & Mackintosh, J. (1964). Discrimination learning: non-additivity of cues. *Nature, Lond.* **201**: 528–530.
Trueman, E. R. (1975). *The locomotion of soft-bodied animals*. London: Arnold.
van Heukelem, W. F. (1973). Growth and life-span of *Octopus cyanea* (Mollusca: Cephalopoda). *J. Zool., Lond.* **169**: 299–315.
Vinnikov, Ya. A. (1974). *Sensory reception: cytology, molecular mechanisms and evolution*. Berlin: Springer Verlag.
Ward, D. V. & Wainwright, S. A. (1972). Locomotory aspects of squid mantle structure. *J. Zool., Lond.* **167**: 437–450.
Wells, M. J. (1966). Learning in the octopus. *Symp. Soc. exp. Biol.* **20**: 477–507.
Wells, M. J., Freeman, N. H. & Ashburner, M. (1965). Some experiments on the chemotactile sense of octopuses. *J. exp. Biol.* **43**: 553–563.
Williamson, G. R. (1965). Underwater observations of the squid *Illex illecebrosus* Lesueur in Newfoundland waters. *Can. Fld Nat.* **79**: 239–247.
Wilson, D. M. (1960). Nervous control of movements in cephalopods. *J. exp. Biol.* **37**: 57–72.
Wittmer, A. & Martin, A. W. (1973). The fine structure of the branchial heart appendages of the cephalopod *Octopus dofleini martini*. *Z. Zellforsch. mikrosk. anat.* **136**: 545–568.
Wodinsky, J. (1969). Penetration of the shell and feeding on gastropods by *Octopus*. *Am. Zool.* **9**: 997–1010.
Wodinsky, J. (1971). Movement as a necessary stimulus of *Octopus* predation. *Nature, Lond.* **299**: 493–494.
Wodinsky, J. (1972). Breeding season of *Octopus vulgaris*. *Mar. Biol.* **16**: 59–63.
Woodhams, P. L. (1975). *The ultrastructure of a visuomotor centre in* Octopus. Ph.D. thesis: University of Sheffield.
Woodhams, P. L. (in prep.). *The ultrastructure of a cerebellar analogue in* Octopus.
Woodhams, P. L. & Messenger, J. B. (1974). A note on the ultrastructure of the *Octopus* olfactory organ. *Cell Tiss. Res.* **152**: 253–258.
Young, J. Z. (1936). The giant nerve fibres and epistellar body of cephalopods. *Q. Jl microsc. Sci.* **78**: 367–386.
Young, J. Z. (1938). The functioning of the giant nerve fibres of the squid. *J. exp. Biol.* **15**: 170–185.

Young, J. Z. (1960). The statocysts of *Octopus vulgaris*. *Proc. R. Soc.* (B) **152**: 3–29.
Young, J. Z. (1962). The retina of cephalopods and its degeneration after optic nerve section. *Phil. Trans. R. Soc.* (B) **245**: 1–18
Young, J. Z. (1963). Light- and dark-adaptation in the eyes of some cephalopods. *Proc. zool. Soc. Lond.* **140**: 255–272.
Young, J. Z. (1975). Sources of discovery in neuroscience. In *The neurosciences: paths of discovery*: 15–46. Worden, F. G., Swazey, J. P. & Adelman, G. (eds). London: MIT Press.

Symp. zool. Soc. Lond. (1977) No. 38, 15–48.

EVOLUTION OF THE DIBRANCHIATE CEPHALOPODA

D. T. DONOVAN

Department of Geology, University College London, London, England

SYNOPSIS

Upper Palaeozoic and Mesozoic coleoids possessing normal, straight, phragmocones are surveyed. They comprise: Aulacocerida (Lower Carboniferous—Upper Jurassic), with telum ("guard") and body-chamber; Belemnitida (Jurassic and Cretaceous), with calcite guard and simple pro-ostracum; Phragmoteuthida (Upper Permian—Lower Jurassic), with guard inconspicuous or absent and three-lobed pro-ostracum; and a series of rare fossils with ten, hooked arms. Groenlandibelidae have been rejected from Belemnitida but do not at present find a place in any other group.

Origin of living dibranchiates appears to lie in the Phragmoteuthida in the early Jurassic. One lineage led, by loss of phragmocone, to Loligosepiina with three-lobed gladius, and thence to living *Vampyroteuthis*. The Octopoda may have been a specialization from this line but there is no fossil evidence.

Another lineage of Phragmoteuthida appears to have quickly acquired the arrangement of eight arms and two tentacles characteristic of the Decapoda, while still possessing a phragmocone. One group, by the Upper Jurassic at the latest, lost the phragmocone and evolved a gladius (*Plesioteuthis*) almost indistinguishable from that of some living squids (e.g. *Ommastrephes*). This group led to the modern oegopsids. In another group the phragmocone became specialized into a cuttlebone, typical cuttlebones retaining the lateral "wings" of the phragmoteuthid pro-ostracum being known from the Upper Jurassic (*Trachyteuthis*).

The origin of *Spirula* is obscure but it may be connected with the late Cretaceous Groenlandibelidae, which share the same kind of protoconch, and with the Tertiary families Belemnoseidae, Spirulirostridae, etc. These are now excluded from the Sepiida.

INTRODUCTION

The fossil record of many coleoid groups is very bad. Except for belemnites and aulacocerids, they are rare fossils, and the living fauna includes many families that are not known from the fossil record. Furthermore, there exist puzzling fossils which do not easily fit into any coherent story. Much of the descriptive literature is very old, and needs revision, and there have been few recent attempts to construct phylogenies.

The present review is no more than an outline of the problem. After a brief account of geological factors, there follows, first, a summary of extinct coleoid groups which possessed normal phragmocones, among which the ancestors of living forms must presumably be sought; and, second, a review of fossil evidence for cephalopods of modern dibranchiate type.

LOCALITIES FOR FOSSIL COLEOIDS

Fossil coleoids, apart from belemnites and aulacocerids, are rare objects in later Palaeozoic and Mesozoic rocks. Under most sedimentary conditions squid gladii were never fossilized, and phragmocones, although calcareous, appear to have stood little better chance of preservation. Pro-ostraca of belemnites are even rarer than gladii, and seem to have been of extremely delicate construction. Almost all the fossil material comes from a handful of localities, and, for this reason, from a small selection of geological horizons. Most famous of these is the neighbourhood of Solenhofen, in Bavaria, where the lithographic limestone of Upper Jurassic date (Kimmeridgian Stage) accumulated. It is thought to have accumulated as calcareous mud in lagoons protected from the open ocean of the Tethys by a barrier reef. Numerous remarkable fossils were entombed: *Archaeopteryx*, pterodactyls, various other reptiles, crustaceans, insects, and numerous fish have been recovered as well as fossil coleoids. Rare specimens among these show the outline of the soft parts and help to fill in our inadequate knowledge of the anatomy of these animals.

In the Lower Jurassic, the Lower Lias on the coast of Dorset, England, was worked by professional dealers and collectors in the nineteenth century. Notable here were ink sacs, gladii of more than one type, and arm-crowns showing double rows of hooks. These were used by dealers to concoct improbable animals which usually incorporated a belemnite guard at the other end (e.g. Huxley, 1864: pl. 1, figs 1, 2); their real nature is discussed below. The stratigraphical horizons of these fossils were seldom recorded, and as professional collecting has virtually died out the chief source of material is old museum collections. Many of the finds were made by breaking open nodules, and occasional ammonites in the matrix show that some of these came from the Black Ven Marls, of Turneri and Obtusum Zone age.

The Upper Lias (Toarcian Stage) provided several localities where fine-grained sediments were laid down and delicate fossils preserved. Holzmaden in Württemberg is the most famous, and has yielded a fauna second only to Solenhofen in its variety of rare or exceptionally-preserved forms (Hauff, 1953). A number of early gladii have been found here. In England, the "Fish Beds" of the Upper Lias of Dumbleton, Gloucestershire, had a more local fame, and yielded well preserved *Belemnoteuthis* and "*Geoteuthis*" now to be found in the British Museum and at Oxford (Crick, 1921). Much of this material is unstudied.

In the Upper Cretaceous the Fish Beds of the Lebanon (Couches à poissons du Liban) are fine-grained limestones somewhat like those of Solenhofen, in which delicate structures and outlines of soft parts are preserved. The localities are of two different ages belonging to the Cenomanian and Senonian Stages. Both have yielded cephalopods, but the greater variety, including the only known examples of *Palaeoctopus newboldi* (Woodward), comes from the Senonian locality.

Material from localities and stratigraphical horizons other than those mentioned above is restricted to a very few lucky finds, such as the unique specimen of *Jeletzkya* from the Upper Carboniferous of Illinois.

The Tertiary record is disappointingly sparse. There are sepiids (*Belosepia*) and some fossils commonly referred to the Sepiida but here separated from them (p. 34). There are a few late Tertiary argonauts but squid gladii are rare.

Apart from the fragility of many coleoid skeletons, there is another factor which limits their preservation in the geological record. The present ocean floors are of Mesozoic and later age. They have not yet been elevated above sea level for our inspection. Earlier ocean floors have largely been destroyed, their mangled remains being found in orogenic belts. Thus the prospect of finding fossil records of groups which may have always been oceanic in their distribution is poor.

EXTINCT COLEOID GROUPS

No attempt is made here to formally define the Coleoidea in such a way as to include the various fossil groups commonly included in them. The term is used for forms which have either a guard, or a reduced body-chamber, or both, indicating that soft tissues to some extent covered the outside of the shell. The latter was not necessarily internal in the sense that it is in living dibranchiate cephalopods. As thus defined Coleoidea may well be polyphyletic, different groups having different ancestry in later Palaeozoic orthocones.

The ranges in geological time of the extinct coleoid groups discussed in this section are shown in Fig. 4.

*Aulacocerida**

These are conical phragmocones with body-chambers, the apical end being enveloped by a guard or telum. Because of the latter, they have been generally lumped together with belemnites. Jeletzky (1966: 12–24) made it clear that there were two groups of animals with belemnite-like guards: the earlier Aulacocerida (Carboniferous–Jurassic) with complete body-chamber, and the later Belemnitida (Jurassic and Cretaceous) in which the body-chamber is reduced to a dorsal pro-ostracum.

Jeletzky regarded the Aulacocerida as "the most primitive Coleoidea known". He found fundamental morphological differences between them and Belemnitida which caused him to reject them as the parent stock of belemnitids. He supposed, rather, that Aulacocerida and Belemnitida were "convergent, but independent offshoots of some orthoconic ectocochlians, presumably bactritids" (Jeletzky, 1966: 23). He did not regard them as ancestral to any other group.

Belemnitida

The order was restricted by Jeletzky (1966: 107) by the exclusion of Aulacocerida, Phragmoteuthida, and other forms which had usually been placed in it. As restricted, the order consists of shells with the guard, phragmocone and pro-ostracum all well developed. The guard consists of thicker lamellae formed of primary calcite alternating with thinner layers of organic (?chitinous) material. The protoconch is roughly spherical and is completely sealed by a closing membrane, there being no caecum as is present in ammonites, *Groenlandibelus* and *Spirula* (Fig. 1). The siphuncle is ventral and the septa closely spaced.

The shape of the pro-ostracum is known chiefly from the course of growth lines on the conotheca (outer wall of the phragmocone), marking the positions of former shell apertures. These turn sharply forwards forming a hyperbolar zone. Because the pro-ostracum is nearly parallel-sided, it is difficult to follow growth lines across the hyperbolar zone and thus to estimate the length of the pro-ostracum. Rare pro-ostraca which have been preserved in fossils (Mantell, 1848; Crick, 1896a; Hölder, 1973) are between two and three times as long as their greatest breadth, which is near the

* All formal systematic names are used in the same sense as by Jeletzky (1966) except in the case of the Sepiida. To save repeating nomenclatural and bibliographic detail, the reader is referred to his paper for authorship, synonymy and diagnoses of units mentioned.

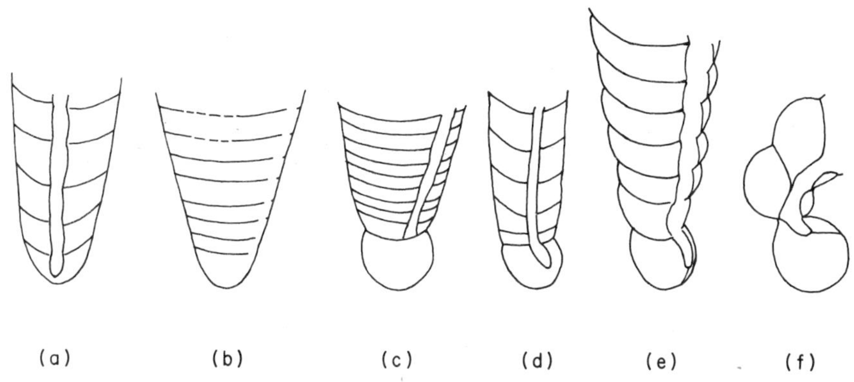

FIG 1. Diagrammatic drawings of protoconchs. All × 16. (a) *Pseudorthoceras* (Silurian) based on Ristedt, 1968: text-fig. 2; (b) *Belemnoteuthis* (Upper Jurassic) based on Makowski, 1952: text-fig. 11A; (c) a belemnite, *Megateuthis* (Middle Jurassic), based on Jeletzky, 1966: pl. 14, fig. 1A; (d) *Sphaerorthoceras* (Middle Silurian), based on Ristedt, 1968: pl. 3, fig. 7; (e) *Groenlandibelus* (Upper Cretaceous), based on Jeletzky, 1966: pl. 20, fig. 1A; (f) *Spirula* (Recent), based on Naef, 1922: text-fig. 27 (after Appellöf, 1893). The guards of (b) and (e) are not shown. The siphuncle of (b) is not preserved but a caecum was presumably present because the first septum is perforated.

anterior end. The pro-ostracum is always simple in contrast to that of Phragmocerida.

Flower (1945) reported a "belemnite" from the late Lower Carboniferous but although Flower & Gordon (1959) later confirmed the presence of abundant Aulacocerida from this horizon, the presence of true Belemnitida seems very uncertain, especially as they are not found again until the Jurassic, intervening fossils formerly classified as Belemnitida having been removed to Aulacocerida. Belemnitida first become abundant in rocks of the Sinemurian Stage of the Lower Jurassic in Europe, and remain common fossils until the very end of the Cretaceous. There is a single, rare genus of early Tertiary belemnites (*Bayanoteuthis*; Jeletzky, 1966: 106, 140).

Jeletzky (1966: 107) thought that the Belemnitida evolved from "phragmoteuthid-like coleoids" and became extinct without leaving any descendants.

Phragmoteuthida

This group was first distinguished from other extinct coleoids by Jeletzky (in Sweet, 1964). He pointed out that the group is characterized by a distinctive pro-ostracum which sprang from about three-quarters of the circumference of the phragmocone (that is, it was like a body-chamber with a strip of ventral wall missing) and which had a three-lobed anterior margin (Fig. 2a). The shape of the body-chamber is not capable of being proved in all the fossil material assigned to the group, but the shape of the anterior margin, which can sometimes be ascertained from growth lines, is sufficiently diagnostic.

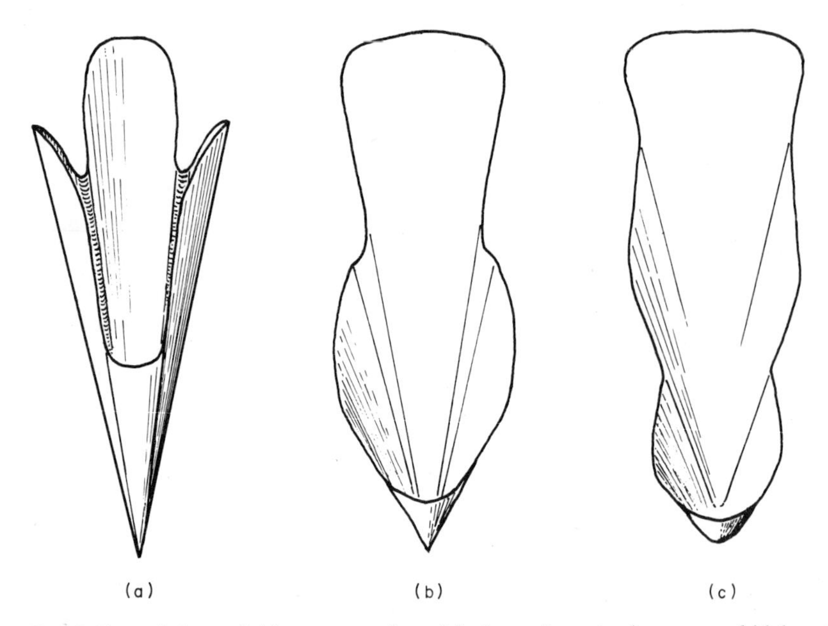

FIG. 2. Ventral views of: (a) reconstruction of the lower Jurassic phragmoteuthid, based on material in the British Museum (Natural History); (b) *Geoteuthis simplex* Voltz, Lower Jurassic (Toarcian) of Metzingen, Württemberg, Germany, based on Naef (1922: fig. 45); (c) the gladius of *Vampyroteuthis* drawn from a specimen sent by Dr Clyde Roper to Professor J. Z. Young. All reduced to the same length for comparison. The posterior part is a phragmocone in (a); a conus in (b) and (c).

The earliest member is *Permoteuthis* (Rosenkrantz, 1946: 161, fig. 6: Lectotype, designated by Jeletzky, 1966: 38) from the Upper Permian of East Greenland, known only by a pro-ostracum fragment. The next fossils are from the black shales of the Wengener Schiefer, Upper Trias, near Villach, Austria (authors cited by Naef, 1922: 261). Several of these fossils show phragmocone and pro-ostracum in association (e.g. Jeletzky, 1966: fig. 4A, 33). They also clearly show that the median lobe of the anterior margin was separated from the lateral lobes by notches, which leave a characteristic trace in the growth lines comparable with the selenizone of pleurotomariids, and were termed hyperbolar zones by Jeletzky, who reviewed the confusing terminology (1966: 32).

Both Naef (1922: 262) and Jeletzky (1966: 38) referred to a phragmocone from the Lower Jurassic (Lower Lias) of Dorset, England [British Museum Natural History, (BM 83963)], which had been figured by Huxley (1864: pl. 1, figs 4, 4A) who noted that it appeared to show a hyperbolar zone (Huxley, 1864: 14). L. Bairstow (pers. comm. to Jeletzky) had noted several other phragmocones of the same provenance with phragmoteuthid-like growth lines. Study of this material in the British Museum enables the detailed morphology to be worked out. There are a number of phragmocones, with apical angles around 27°, of which the apical ends are commonly missing, but there is no sign of a belemnite-type guard or of the "sheath" which enveloped the phragmocone of *Belemnoteuthis.* A few critical specimens (e.g. BM 39873) show association of phragmocone with pro-ostracum. The pro-ostracum is made of very thin, iridescent, presumably aragonitic material; the ventral margins are extremely thin and crinkly, and the writer has the impression that they may have been flexible. The anterior margin of the pro-ostracum is not preserved but the growth-lines on several specimens show that it was of the three-lobed phragmoteuthid type.

The collection includes a number of distinctive iridescent pro-ostraca, most of which are so badly damaged that no structure can be made out. However, BM C10811 shows a pro-ostracum of the same type as BM 39873 enclosing an ink sac and associated with hooked arms, and BM 48894 confirms some of these details. Specimens which show badly preserved parts of the iridescent pro-ostracum and/or ink sac in juxtaposition with hooked arms are not uncommon (Fig. 3). The arms only of six of these were figured by Crick (1907). The arm-crowns figured are associated with damaged pro-ostraca and/or ink sacs. The greatest number of arms which Crick could count in the British Museum specimens was six, but Jeletzky has recorded and figured (1966: 138, pl. 16, fig. 3) a specimen in the Sedgwick Museum, Cambridge, England, (no. J37,812) with eight arms, though he mistook it for a belemnitid.

The numerous fossils showing one or more of the arms agree in proving that each arm had a double row of hooks (Fig. 3). Crick (1907: 269) pointed out that the form of these hooks was different from those of the Upper Jurassic *Belemnoteuthis* (see p. 27).

I have dwelt at length on this material because it is undescribed and is essential for the understanding of the Phragmoteuthida, which in turn are important, as Jeletzky noted (1966: 37) for understanding coleoid evolution. I conclude that the material from

FIG 3. A phragmoteuthid showing ink sac and hooked arms, from the Lower Lias of the Dorset coast (BM 82895); ×0·5. Previously figd. in part by Crick (1907: pl. 23, fig. 3).

the Dorset Lower Lias, for which there seems to be no suitable generic name, is evidence for the existence of animals with the following characteristics.

There was a well developed phragmocone with ventral siphuncle. There is no evidence for a guard. The pro-ostracum was longer than the phragmocone and sprung from about three-quarters of its circumference. It must have been conical like the phragmocone, for as the animal grew in size the pro-ostracum was progressively incorporated in the phragmocone. The anterior margin was three-lobed with deep notches between lobes. An ink sac was present. There were (at least) eight arms, each bearing two rows of hooks.

Evidence for later Phragmoteuthida is scarce. An ink sac and associated arm-crown from the Upper Lias (early Toarcian Stage) of Holzmaden (Müller, 1965: 284, text-fig. 393C; wrongly stated to be *Acanthoteuthis*) is exactly like the Dorset Lower Liassic forms as far as its characters can be seen. The same kind and arrangement of hooks again recur in the Upper Cretaceous (Cenomanian Stage) of the Lebanon (Roger, 1944b, 1946: pl. 2). The body is preserved but critical details of morphology cannot be made out. It is probably unsafe to assume, on the basis of arm-hooks, that typical Phragmoteuthida survived into the Upper Cretaceous.

To consider for a moment the life-habits of these phragmoteuthids: first of all, they clearly possessed buoyancy. The apparent lack of a guard suggests that they would have had difficulty in maintaining a horizontal swimming position. This has three alternative implications: (i) there was in fact a guard but it has not been preserved in association with the other parts of the animal; (ii) the soft parts extended posteriorly to the phragmocone, which lay in the central region of the body; (iii) the animal habitually lived with the long axis of the body vertical, like *Spirula*. There seems to be no way at present of deciding between these alternatives.

We do not know whether the phragmoteuthid shell was internal or external, though further study may resolve this point. The incomplete ventral wall suggests that there was a muscular mantle cavity wall, such as is possessed by modern dibranchiates and is thought to have been present in belemnites. Phragmoteuthids may, therefore, have been quite good swimmers by jet propulsion, perhaps not as effective as the belemnites.

The deep notches in the anterior margin must have fulfilled a function but it is not known what this was. They would have corresponded in position roughly to the sinuses in the mantle margin shown by *Sepia* and some squids on either side of the mid-dorsal line, but it is not possible to assert that the phragmoteuthid notches accommodated a respiratory current. The arms were relatively short, but not more so than in some modern squids.

Groenlandibelidae

The genus *Groenlandibelus* was set up by Jeletzky (1966: 92) for *Belemnoteuthis rosenkrantzi* Birkelund (1956), from the Upper Cretaceous (Upper Maestrichtian) of west Greenland, a long narrow phragmocone with small apical angle (14 to 15°) and a thin guard enveloping the protoconch and phragmocone. Jeletzky showed that the species has the siphuncle beginning in the protoconch in the form of a caecum, while in all Belemnitida where the protoconch is known it is closed by an unperforated septum and the siphuncle does not penetrate into it (Fig. 1). He enumerated a number of other differences in microscopic structure.

The adoral part of *Groenlandibelus* has not been recovered and so it is not known whether or not there was a body-chamber. The phragmocone was presumably covered by the mantle which secreted the "guard". Growth lines show that the aperture was prolonged dorsally into a narrow projection, pro-ostracum, or rhachis, but it is impossible to determine whether this was long or short (Birkelund, 1956: pl. 1, figs 9a, 9g).

The guard in *Groenlandibelus* is very small and would have been quite inadequate to keep the animal horizontal in the water by weighting its tail. Its normal attitude of life is, therefore, uncertain.

The only other genus included in Groenlandibelidae by Jeletzky is *Naefia* (Wetzel, 1930) from the Upper Cretaceous (Upper Campanian) of Chile.

Jeletzky (1966: 103) placed *Groenlandibelus* in the order Sepiida because of a number of characters which he regarded as typically sepiid. These included: presence of caecum and prosiphon; relatively wide siphuncle; marginal position of the siphuncle in the earliest chambers; and conotheca and septa formed of a single shell layer. However, these characters can only be regarded as sepiid if *Spirula* is included in the order. This is questioned by the present author, the most obvious character that *Spirula* and *Sepia* have in common being the possession of an internal phragmocone. The phragmocones are so different that it seems more reasonable to separate them on this basis rather than to group them together.

If this view is adopted, as it is in this paper, then one must look again at the Tertiary "sepiids" of authors, e.g. the genera included in Sepioida by Naef (1922) or in Sepiida by Jeletzky (1966). In fact these are clearly divided into *Sepia*-like forms, with strongly oblique septa and complete suppression of the ventral wall of the siphuncle; and genera with septa of conventional form, making a large angle with the phragmocone, and tubular siphuncle. The former group includes *Belosepia* and possibly one or two other genera (Sepiidae of Jeletzky, 1966: 107). The latter include the genera usually cited in textbooks (e.g. Swinnerton, 1947: 173) as transitional between belemnites and *Spirula* (e.g. *Belemnosis*, *Spirulirostra*, *Spirulirostrella*, and a number of others: Naef, 1922: 48–75). Most of these dozen or so genera have been founded on specimens representing the apical parts only of shells which have been preserved because the apical part of the phragmocone is enveloped by a calcareous deposit (rostrum of Naef, sheath of Jeletzky). There appears to be no foundation for the reconstructions of Naef (1922: e.g. text-figs 12, 14, 21, 23) in which the whole phragmocone is enveloped by the sheath. The earliest part of the phragmocone in some genera is curved, but is already straightening out by the end of the short preserved part, and the writer believes that they were more or less straight shells. Possibly one of these genera gave rise to *Spirula* by loss of the sheath and increased curvature; more probably, they become extinct about the middle of the Tertiary, and *Spirula* is descended from some parallel lineage which never had a massive sheath.

The phylogenetic origin of Groenlandibelidae is obscure. Earlier groups which possess the same distinctive protoconch are: (i) the Ammonoidea, whose specialized and complex shells seem to preclude them as ancestors; and (ii) some Upper Palaeozoic orthocones placed in Orthocerida and/or Bactritida (Ristedt, 1968) (Fig. 4). The latest of these are Devonian in age, and there is an

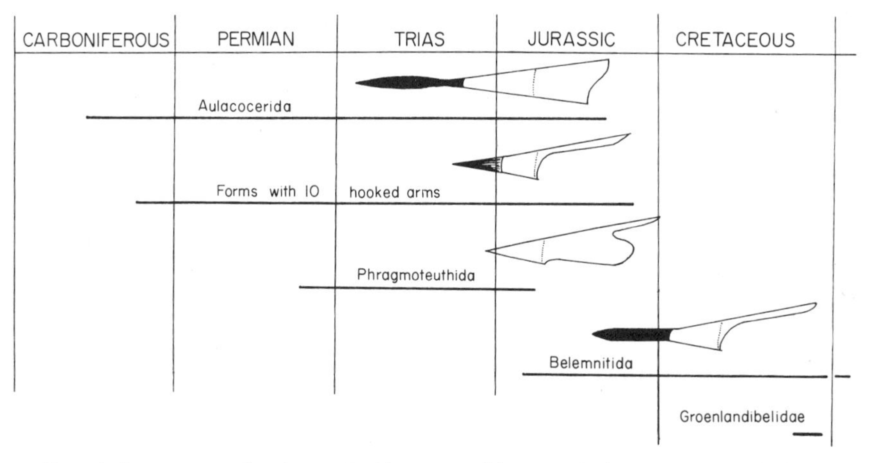

FIG. 4. Age ranges of extinct coleoid groups with normal phragmocones. Black area: guard or telum; dotted line: last septum.

enormous time gap between them and the Groenlandibelidae even by the standards of the sporadic fossil record of coleoid groups. It may be, of course, that intermediate forms exist but have not been recognized as such because their protoconchs and microscopic shell structure have not been investigated. In their absence, eventual origin in the late Palaeozoic Orthocerida must be presumed.

Forms with ten similar hooked arms

Ten-armed cephalopods are rare fossils in the late Palaeozoic and Mesozoic rocks. The oldest is *Jeletzkya* from the Middle Pennsylvanian of Illinois (Johnson & Richardson, 1968). The unique specimen comprises a well preserved arm-crown in which the arms, each about 10 or 11 mm long, are subequal in length and all of similar type. Each bears two rows of hooks (Fig. 5). Radiography of the

FIG. 5. (a) Part of an arm, ×7 and (b) individual hooks, ×15·5 of *Jeletzkya* from the Upper Carboniferous of Illinois, USA, based on Johnson & Richardson (1968).

rock enclosing the specimen revealed an elongated object which the authors interpreted as a gladius, although it seems difficult to exclude the possibility that it is part of an orthocone body-chamber, or a pro-ostracum.

The next well preserved arm-crown appears to be *Phragmoteuthis? ticinensis* Rieber (1970) from the Middle Trias of Canton Ticino, Switzerland. It has ten arms, about 8 mm long, each bearing two rows of hooks. There are between 30 and 36 pairs of hooks on each arm, and the largest hooks were in the middle part of each arm. At the bases of the arms are structures which appear to be pointed mandibles. An ink sac is preserved, in about the right position to belong to the same animal, though the body outline is not visible. The whole animal may have been about 50 mm long.

From the same geological formation *Phragmoteuthis bisinuata* (Bronn), refigured by Rieber (1970: pl. 4), was an animal with different shaped hooks, but it is not so well preserved and the number of arms cannot be counted. It is doubtful whether either of these forms belongs to the genus *Phragmoteuthis* as understood by Jeletzky (1966) (see p. 19 of this paper).

The remaining examples are from the Jurassic of Europe. Pinna (1972) described a find from the fossiliferous Lower Jurassic (Lower Sinemurian Stage) of Osteno, on the lake of Lugano, Lombardy, Italy. The locality, discovered by Pinna, resembles the more famous localities of Solenhofen and the Lebanon in yielding numerous fish and decapod crustaceans, as well as land plants and rare cephalopods. As with the Triassic forms, the only clearly preserved part of the fossil is the arms, of which nine can be counted. Each bears a double row of hooks. A maximum of about 23 pairs of hooks can be counted on any one arm, and they seem to have been largest in the middle part of each arm, as with the Triassic examples.

A series of remarkably-preserved fossils was discovered in finely laminated shales of the Oxford Clay (Upper Jurassic: Callovian Stage) at Christian Malford, Wiltshire, England, in the 1840s during the construction of the Great Western Railway from London to Bristol. From here came the material described as *Belemnoteuthis antiqua* Pearce (1847). As well as phragmocones, soft parts, apparently the mantle, head, and arms, have been mineralized and preserved (Fig. 6). A number of important specimens were described by Owen (1844) but the species has not been re-studied since.

All the English Oxford Clay specimens were crushed, but interpretation of the phragmocone is aided by Polish material (*Belemnoteuthis polonica* Makowski, 1952: 41). Makowski had ten phragmocones of which several were uncrushed and well preserved.

The protoconch was seen in two, sectioned specimens (Makowski, 1952: text-fig. 11A, B). It is hemispherical or dome-shaped (Fig. 1b), and there is no sign of the constricted aperture which characterizes all known protoconchs of Belemnitida (Jeletzky, 1966: *passim*).

The apical angle of the phragmocone is between 20 and 21°. The apical portion of the phragmocone is enveloped by a shell layer or "guard". At the apex it is less than 1 mm thick; it gets thinner forwards and about half-way along the phragmocone is barely detectable, and the septa can be seen through it. It bears two ridges, semicircular in cross-section, close together on either side of the mid-dorsal line. The "guard" is composed of layers of prismatic aragonite (Judith Milledge, pers. comm.) except for the apical region, where Makowski's sections show a region of granular crystalline material.

In most English specimens the anterior margin of the phragmocone is obscure, but Jeletzky (1966: pl. 16, fig. 2) figured one in the Sedgwick Museum (J24,841) with a pro-osctracum, a little longer than the phragmocone, with semicircular anterior margin. Three of Makowski's figured specimens (1952: text-figs 7A, 8B, 9) have growth lines which indicate the existence of a bluntly pointed pro-ostracum, flanked by hyperbolar zones whose dorsal boundaries diverge at an angle of about 10°.

In more complete English examples, the pro-ostracum is concealed by the transversely-striated mantle, already mentioned. Several specimens (e.g. BM 25966 and 32350) show that the mantle was not continuous across the mid-ventral line, though the ventral margins may overlap as a consequence of crushing during fossilization. Posteriorly these ventral mantle margins curve away rather like the front of a man's jacket (Fig. 6). The ink sac lies near the junction of phragmocone and mantle.

Some specimens suggest that the front margin of the mantle was more or less straight, and there does not appear to be any trace of the funnel. The head is recognizable but obscure in structure. Conspicuous are two semicircular structures, best preserved in BM C5020 (Owen, 1844: pl. 6, figs 1, 3), which Owen believed to be the eyes, but this seems improbable.

The arms were up to about 10 cm long and each was armed with a double row of hooks, of the same general type as in the earlier forms mentioned in this section. Each arm bore at least 25 pairs of hooks. Eight arms can be made out in BM 25966 (Fig. 6). In most specimens fewer arms are visible.

Phragmocones and "guards" identical with the Oxfordian *Belemnoteuthis* are known from the Lower Jurassic (Toarcian Stage) of Dumbleton, Gloucestershire, England, by unpublished material in Oxford University Museum (OUM J14,802 and others) and from the same horizon near Ilminster, Somerset (OUM J17,281–2, showing ink sac).

The latest forms in this group occur in the Lithographic Stone at Solenhofen, and their relationship to *Belemnoteuthis* requires further investigation. A number of remains of hooked arms have been found at Solenhofen, and are usually referred to as *Acanthoteuthis speciosa* Münster. Naef (1922: 252, text-fig. 91) illustrated one with ten arms each bearing a double row of hooks. There seem to have been at least 20 pairs of hooks on each arm. As in earlier forms of this group, the hooks were largest in the central part of each arm as noted by Crick (1897: 2) who also recorded that the largest hooks in a specimen in the British Museum (BM C5783) were 7·5 mm long. The arm-crown of *Acanthoteuthis speciosa* is almost identical with that of *Belemnoteuthis antiqua* and several authors have considered the possible identity of the two genera, usually without coming to a firm conclusion. Most specimens of *Acanthoteuthis* do not show a phragmocone, but it is rarely present and shows dorsal, apical ridges like those of *Belemnoteuthis* (BM 83734). Naef (1922: 253) considered this evidence and rejected it, because he thought the ridges might be artifacts or accidental. However, a specimen at the Technische Hochschule, Darmstadt, figured by Angermann (1902: pl. 6, fig. 1) (Fig. 7) clearly shows a *Belemnoteuthis* type of phragmocone and "guard". At most, there seems to be little difference between the forms from the Oxford Clay and the Lithographic Stone.

Systematic position

Pre-Jurassic members of this series either consist only of arms or arm-hooks, or at best the rest of the animal is obscure (*Jeletzkya*). We therefore have to study the group on the basis of its Jurassic members, i.e. *Acanthoteuthis* and *Belemnoteuthis*, assumed to be related to the earlier forms because of the identical number of arms and shape and arrangement of the arm-hooks.

The group is differentiated from Belemnitida (*sensu* Jeletzky, 1966) by (a) the shape of the protoconch, and (b) the composition of the guard, aragonitic in *Phragmoteuthis*, calcite in Belemnitida. Confusion has been rife because ever since the mid-nineteenth century, when classification was less advanced, the belemnoteuthids have been referred to as "belemnites", as indeed have the English Lower Liassic forms here referred to as Phragmoteuthida (p. 21).

As Jeletzky noted (1966: 145) there are several features in common with Belemnitida: apical angle, spacing of septa, structure of the siphuncle and the presence of cameral deposits. Because of these he placed Belemnoteuthidae as a family of Belemnitida. The

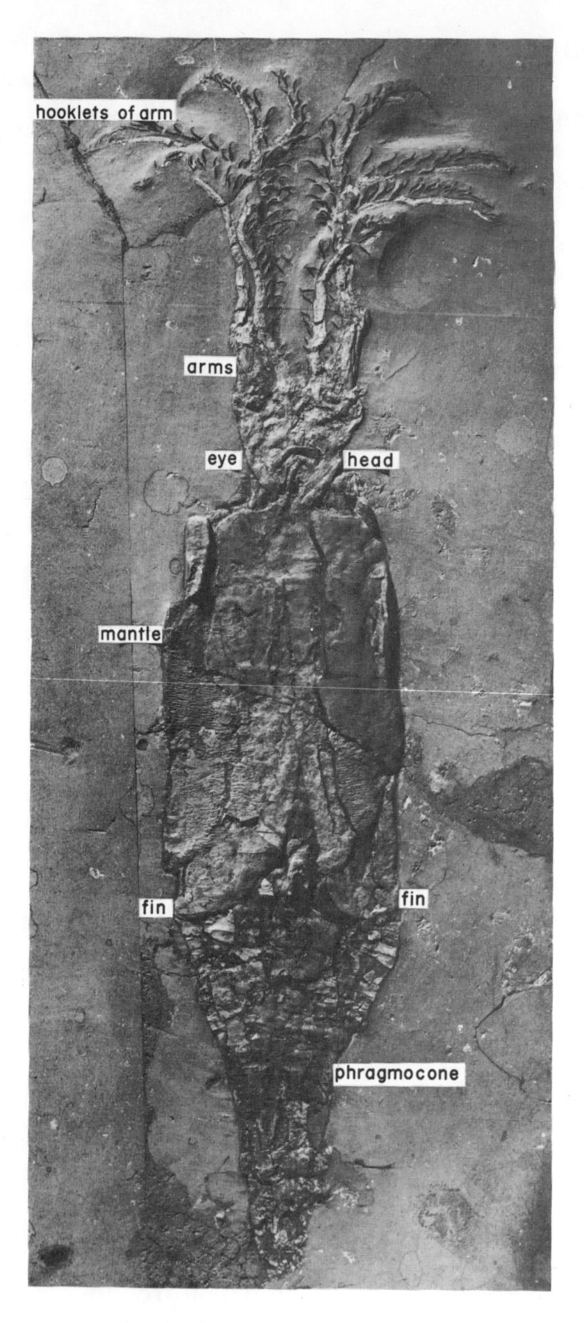

FIG. 6. *Belemnoteuthis antiqua* Pearce. Ventral view. Oxford Clay, Christian Malford, Wiltshire (BM 25966); ×0·49. The labels are old and should be ignored.

FIG. 7. ?*Belemnoteuthis* sp. ("*Acanthoteuthis*"). Dorsal view. Lithographic Stone (Upper Jurassic), Eichstädt, Bavaria. Previously figd. Angermann (1902: pl. 6, fig. 1) (Technische Hochschule, Darmstadt); ×0·25.

present writer regards the features mentioned in the previous paragraph as of primary importance, and the points of similarity as indications of the characters possessed by a common ancestor of belemnites and belemnoteuthids.

The number of arms possessed by belemnoteuthids and belemnites may be considered briefly. A group of cephalopods with ten similar, hooked, arms existed from Upper Carboniferous to Upper Jurassic or Lower Cretaceous, and at least the later members of this group had the characters of *Belemnoteuthis.* The number of arms in Belemnitida is unknown. Jeletzky (1966: 108) stated that "so far as is known, all . . . possess ten equal or subequal, arms . . . provided with two rows of arm hooks". However, the present writer does not know of any fossils in which arms or arm-hooks are preserved in association with belemnitid guard or phragmocone. Specimens which purport to show such an association (e.g. Huxley 1864: pl. 1, figs 1, 2; BM 74106; 39855) are spurious, having been manufactured by collectors or dealers from belemnite guards, ink sacs which may have belonged to loligosepiids or phragmoteuthids, and phragmoteuthid arm-crowns. Further confusion was introduced by Crick (1907) who figured six arm-crowns with hooked arms. The figured specimens are in the British Museum, and while several have much damaged remains of phragmocone and/or pro-ostracum, none has a belemnite guard. As noted above, they are Phragmoteuthida.

FOSSIL CUTTLEFISH AND SQUIDS

Cuttlefish

The earliest known cuttlefish is represented by two specimens, of which only one has been figured, from the Jagua Formation (Upper Jurassic; Upper Oxfordian Stage) of Cuba (Schevill, 1950). The dorsal view of this fossil is that of a typical cuttlebone, differing from some species of living *Sepia* only in the greater prominence of the postero-lateral "wings". A natural break shows part of the phragmocone in cross-section. This fossil was given the new generic name *Voltzia* by Schevill, who considered it to be distinct from the slightly later European form *Trachyteuthis,* referred to below. It is doubtful whether this distinction can be maintained.

European Upper Jurassic examples come from the succeeding Kimmeridgian Stage of England and Germany. The English ones come from the Kimmeridge Clay formation and one (BM 5018) was described by Owen (1855) and named *Coccoteuthis latipinnis*; the others (BM 39701, C46843, C46844) are unpublished but similar. The dorsal surfaces of these fossils are identical in structure with that of *Sepia*. There is no trace of phragmocone in any of them.

The German specimens all come from the Lithographic Stone of Solenhofen, and examples were figured by several nineteenth century German authors (references in Crick, 1896b: 440). Like the English examples, they are almost identical with the dorsal shield of *Sepia* (Figs 8 and 9). In most of the specimens, as authors have remarked, there is no trace of the phragmocone, but one or two (e.g. BM 83730) show structures which appear to be the remains of septa. The name *Trachyteuthis* was given to the Solenhofen objects by von Meyer (1846). The name was rejected by Crick (1896b: 440), as inadequately defined, in favour of *Coccoteuthis* Owen 1855, but most authors have accepted *Trachyteuthis* as valid.

Schevill, as already noted, rejected identity between these European fossils and his *Voltzia* because the former did not appear to possess phragmocones; however, this appears a perverse conclusion in view of the otherwise virtual identity of these objects. The general absence of the phragmocone is not surprising; in living *Sepia* it is made of very delicate aragonite plates, and would be readily destroyed by solution after burial; in fact it is generally absent from the Tertiary fossils which are universally accepted as sepiids.

One of the Solenhofen *Trachyteuthis* in the British Museum (BM C5775) exhibits the outlines of some of the soft parts, notably the arms, of which eight are clearly visible (Crick, 1896b). A similar specimen, showing at least six arms, was figured by Abel (1927: text-fig. 475). Neither has hooks on the arms.

Upper Cretaceous sepiids have been described from several countries. *Trachyteuthis libanotica* (Fraas) (Roger, 1946: 17; = *Libanoteuthis* Kretzoi, 1942) is reported from both the Cenomanian and Senonian localities of the Lebanon. From the United States comes *Actinosepia* Whiteaves (Campanian and Maestrichtian Stages of west-central North America), recently redescribed from new material by Waage (1965). It has a broad cuttlebone with ridges on the dorsal surface. Forms from Bohemia were named *Glyphiteuthis* by Reuss (1854) and also figured by Fritsch (1910: pl. 5, fig. 6).

In the Tertiary (Eocene) *Belosepia* Voltz 1830 is very similar to *Sepia*, and was placed in the family Sepiidae by Naef (1922) and by Jeletzky (1966). The part commonly preserved corresponds to only the most posterior part of the *Sepia* cuttlebone and may appear

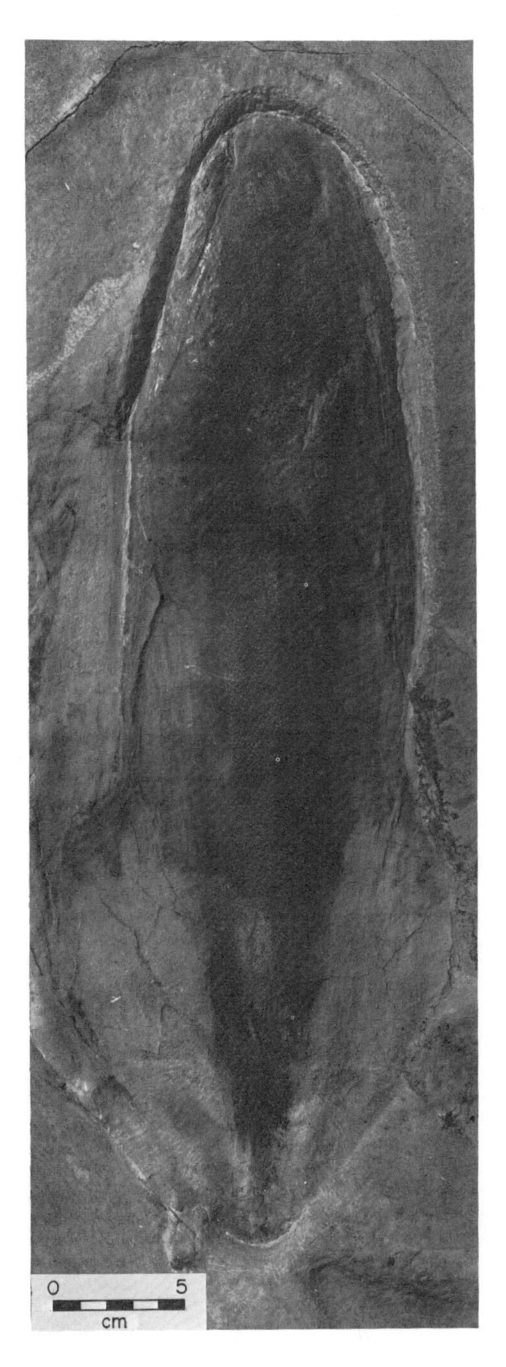

FIG. 9. *Trachyteuthis* sp. Dorsal view. Detail showing marginal notch and hyperbolar zone between lateral (left) and median fields. Lithographic Stone (Upper Jurassic), Solenhofen, Bavaria (BM C46858); ×0·67.

FIG. 8. *Trachyteuthis* sp. Dorsal view. Lithographic Stone (Upper Jurassic), Solenhofen Bavaria (BM C2291); ×0·36.

superficially unfamiliar, but in fact corresponds closely in detail. The remainder of the phragmocone and the dorsal shield were hardly ever fossilized, the survival of the posterior portion being probably due to its being of more massive construction than in most sepiids. Other Tertiary sepiids recorded by several authors were listed by Naef (1922: 92).

The other Tertiary genera commonly included in the Sepiida (*Beloptera, Belemnosis, Spirulirostra, Vasseuria* and others) are here excluded, for reasons given elsewhere (p. 24).

Loligosepiina

This group is characterized by a gladius with a broad median field, with a nearly straight anterior margin, bounded by asymptotes diverging at from 12 to 20°, and wings or lateral fields separated from the median field by "hyperbolar zones" which are the loci of marginal notches or inflections (Fig. 10). There was no phragmocone, but a conus was present in at least some species. The dorsal views of these gladii usually given by authors do not show the strong curvature of the posterior region.

Ink sacs are often preserved in association with these gladii, but no other traces of soft parts have been observed.

The earliest known specimens are from the Lower Lias of the coast of Dorset, England (Lower Jurassic: Lower Sinemurian Stage). Several were figured by Buckland (1836; 1858: pl. 44, fig. 6, 7: pl. 45; figs 1–3; pl. 46). They have not been described or figured since, although there is fairly abundant museum material. Many specimens are incomplete, being preserved in nodules that are smaller than the original gladius, though their shape can be reconstructed from their growth lines.

From the Upper Lias (early Toarcian Stage) fossils are known from several localities, chiefly Holzmaden, a small town near Metzingen, in Germany (Fig. 2b) and the "Fish Beds" of Dumbleton, Gloucestershire, England. A number from the former place were figured by Quenstedt (1849: pls 32–35, *passim*) and other nineteenth century German authors. The Dumbleton material has not been described except for an isolated example figured by Crick (1921: pl. A). There is greater variety of form at this stratigraphical level than in the Lower Lias, affecting the relative proportions of the median and lateral fields and some detailed points of morphology. This has resulted in a number of genera being set up, and indeed, two families were recognized by Jeletzky (1966: 42).

FIG. 10. *Loligosepia* sp. showing anterior part of gladius, with ink sac and duct. From Drift, presumed derived from the Upper Lias. (BM C46819); ×0·75.

Loligosepiina found in the Lithographic Stone were figured by Münster (copied by Naef, 1922: fig. 46a); narrower forms figured by Quenstedt (1849: pl. 35, figs 3, 4) may also belong. These forms have lost the marginal notches, but otherwise are like some of the Lower Jurassic forms in shape. The genus *Geoteuthinus* was proposed for them by Kretzoi (1942). Slightly earlier forms from the Oxford Clay of England (e.g. BM 67808) have not been described.

A gladius from the Upper Cretaceous of the Lebanon (Senonian Stage), as reconstructed by Roger (1946: 16, text-fig. 9) from its posterior half (all that is preserved), is similar to the Upper Jurassic *Geoteuthinus*, except that the median field is broader and the posterior end is apparently more pointed; the latter difference is of doubtful importance because the appearance of the posterior end probably depends on the degree of crushing or distortion which it suffers during fossilization.

From the Oligocene near Budapest, Hungary, Kretzoi (1942) described a new genus *Necroteuthis*. It has a convex or pointed anterior margin in contrast to the straight one of most earlier Loligosepiina.

Origin of the Loligosepiina

Jeletzky (1966: 37) proposed that Loligosepiina evolved from Phragmoteuthida by loss of the phragmocone. His reason was that both groups have median and lateral fields separated by hyperbolar zones. He pointed out that the only other change necessary to effect the transition was the reduction in relative size of the lateral fields. Geological relationships are all right, most phragmoteuthids being older than most loligosepiids (Fig. 11). Unless the similarity of morphology is supposed to be accidental, no other possible ancestors are known. Jeletzky then went on to reject a direct origin because of the presence of arm-hooks in Phragmoteuthida and their absence in Loligosepiina and thought that the two groups must have had a common origin back in the Permian. It seems

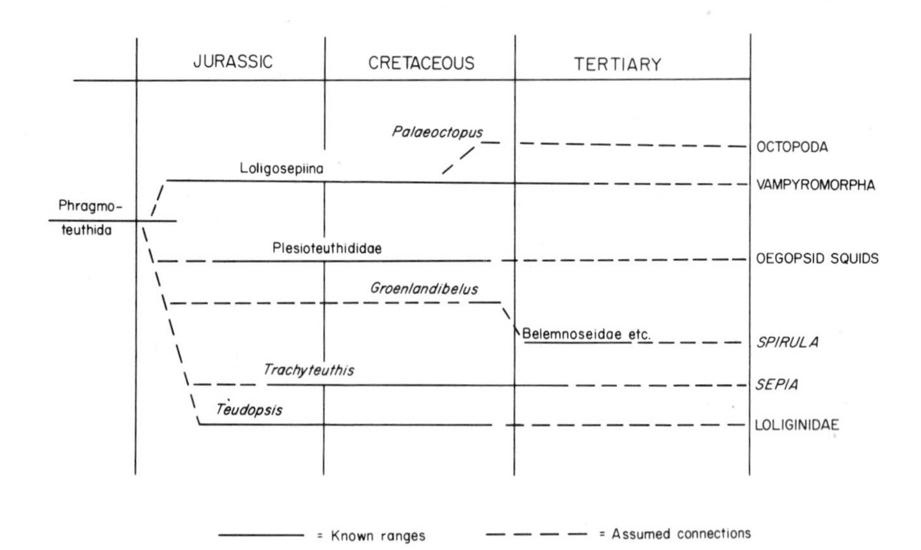

FIG. 11. Possible phylogeny of some coleoid groups.

equally possible, in view of the inadequate fossil record of both groups, that phragmoteuthid genera existed which did not have hooks, or that hooks were lost during the evolutionary transition.

Teudopseidae and Palaeololiginidae

Among the fossil cephalopods in the Upper Lias shales of Holzmaden and other German localities are a series of broad but pointed gladii. The posterior ends of these are like *Loligosepia*, with lateral fields or wings clearly marked off from the median field, but the latter is pointed instead of having a straight anterior margin as in *Loligosepia*. Along the midline of the median field is a strong rib which appears to correspond to that sometimes present, though never prominent, in Loligosepiina, the median keel (Jeletzky, 1966: 33, explanation to fig. 4). Naef (1922: 142) included these fossils in *Beloteuthis* Münster, 1843, probably a synonym of *Teudopsis* Deslongchamps 1835 (Roger, 1952: 740). There is a series of forms, all from the Upper Lias, in which the anterior part of the median field gets progressively narrower and the median keel stronger (Naef, 1922: text-figs 53a, b; 54a). One of the narrower forms has been figured from the Upper Lias of Gloucestershire, England, by Crick (1921: pl. B).

In the Upper Jurassic the trend is continued further in fossils from the Lithographic Stone of Solenhofen known as *Palaeololigo* (Naef, 1922: 148, text-fig. 55) in which the median field is relatively shorter and the median keel projects to form a rhachis about three-eighths of the total length of the gladius. A similar form is represented by a solitary gladius from the Purbeck Beds (Upper Jurassic: Volgian Stage of southern England) figured as *Teudopsis brodiei* by Carruthers (1871) (BM C5251).

In the Upper Cretaceous (Senonian Stage) Fish Beds of Sahel-Alma, Lebanon, cognate forms occur and were named *Beloteuthis libanotica* by Naef (1922: 146, text-fig. 154b). Better material was the basis of a reconstruction published by Roger (1946: 18, text-fig. 11) as *Palaeololigo libanotica*, and a similar fossil is in the British Museum (BM C46838) from the same beds.

The shape of the gladius of these forms, with lateral fields and degenerate hyperbolar zones, are points of similarity to the Loligosepiina, and origin in that group seems probable.

Plesioteuthididae

The earliest fossils in this group come from the Upper Lias (Toarcian Stage) of Germany. Naef (1921) gave the name *Paraplesioteuthis* to gladii with a broad median field, differing from the contemporary *Loligosepia* only in the absence of notches between the median and lateral fields, and in the relative smallness of the latter. In the type species, *P. sagitta* (Münster), the asymptotes diverge at about the same angle as in *Loligosepia* (about 15°). In *P. hastata* (Münster) the angle is smaller (10°) and the lateral fields also smaller.

Thus it is reasonably certain that the plesioteuthids became differentiated from Loligosepiina during the later part of the Lower Jurassic by relative elongation and narrowing of the median field, reduction of lateral fields and loss of notches.

The generic name *Plesioteuthis* was proposed by Wagner (1860) for gladii from the Lithographic Stone of Solenhofen. The two sides of the gladius are thickened and diverge at a small angle (between about 5 and 12°); the central rhachis is a stout rod which does not extend as far forward as the lateral thickenings. There is a leaf-shaped expansion at the posterior end and rare specimens with suitable preservation show that this was a conus (Fig. 15). The overall length of the gladius ranges up to about 30 cm.

Some individuals were fossilized in side view (e.g. BM C1046) and show that the gladius has appreciable curvature in sagittal section (Fig. 12).

A number of specimens show the outline of the body, and sometimes a transversely striated layer, like that of *Belemnoteuthis*, which is usually interpreted as the mantle. The apparent breadth of the body has presumably been increased by flattening, but the outline appears to have been cigar-shaped (Barthel, 1964: pl. 9, fig. 2). The example figured by Crick (1915) (BM C15118) shows a pair of long, pointed lateral fins close to the posterior end. Naef (1922: text-fig. 42a) showed fins which are a little shorter and blunter. The outline of an ink sac is often to be seen, about half-way along the body, with a duct about a quarter of the length of the body.

Some specimens show the outlines of several arms (Naef, 1922: text-fig. 42a; Crick, 1915: pl. 9; Klinghardt, 1943: text-fig. 4; BM 83733). In several puzzling examples an arm-crown with eight arms is seen spread out on the bedding-plane at varying distances from the rest of the animal (Barthel, 1964: pl. 9, fig. 2; Klinghardt, 1943: text-figs 9, 14) (Fig. 13). The explanation seems to be, as suggested by Klinghardt (1943: 13) that the dying animal touched down on the sediment head first and then drifted a short distance before coming to rest. The impression of the inner surfaces of the arms thus recorded show lateral projections which could indicate

FIG. 12. *Plesioteuthis prisca* (Rüppell). Side view. The gladius is on the left. The whitish area to its right represents the mantle, with the outline of the ink sac visible through it. Lithographic Stone (Upper Jurassic) Bavaria; (BM C1046); ×0·82.

FIG. 13. *Plesioteuthis prisca* (Rüppell). The gladius is visible within the body outline. Lithographic Stone (Upper Jurassic) Eichstätt, Bavaria. Institut für Paläontologie, Munich. Previously figd. Barthel (1964: pl. 9, fig. 8); ×0·5.

structures similar to the trabeculae which support the protective membranes along the side of the arms of some modern ommastrephid squids.

In the Upper Cretaceous, *Plesioteuthis*-like gladii occur in the Lebanon (BM C2919 figd. Woodward, 1896: 233; BM C5017, figd. Woodward, 1883: pl. 1; C28408). The form recorded as *Leptoteuthis syriaca* (Woodward) by Roger (1946: 14) may belong here, for his photographs (pl. 4, figs 5, 6; pl. 9, figs 1, 2) show no sign of the lateral fields shown in his diagram (text-fig. 6). It is possible that impressions of the soft parts have been misinterpreted.

Kelaenidae

The genus *Kelaeno* is known only from the Lithographic Stone of Bavaria. It differs from all other fossil gladii in consisting predominantly of a shallow, spoon-shaped conus, the rest of the gladius being much more reduced but of a standard type consisting of a narrow central rhachis flanked by lateral fields (Fig. 14). Examples

FIG. 14. *Kelaeno* sp. Seen from the left side. The dendritic deposits probably mark the outline of the soft parts. Lithographic Stone (Upper Jurassic), Bavaria (BM C46524); ×0·78.

which include the outline of the body (Naef, 1922: fig. 56b; van Regteren Altena, 1949: figs 1, 2) show it as a rather squat animal, two to three times as long as broad, with a blunt posterior end unlike that of most squids. It possessed an ink sac. The number of arms cannot be counted.

Celaenoteuthis (Naef, 1922: 153, fig. 57), from the same locality and horizon, may be a related genus. The gladius of this form is not very different from Loligosepiina (see p. 34) except that it possessed a shallow conus and a long, narrow, free rhachis like *Kelaeno.*

Niobrarateuthis (Miller, 1957) from the Upper Cretaccous of Kansas, USA, was also included in Kelaenidae by Jeletzky (1966: 45).

If *Celaenoteuthis* is really homologous with *Parabelopeltis,* then the family may have evolved from the Lower Jurassic Loligosepiina by the addition of a conus and by the development of the "median keel" into a rhachis.

ORIGIN OF LIVING GROUPS

Sepiida

The origin of the Sepiida (as here restricted, see p. 34) is perhaps the biggest single problem in trying to reconstruct coleoid phylogeny. The derivation from early Tertiary genera with normal phragmocones such as *Belemnosella* has been noted and rejected above (p. 24). It was in fact questioned by Wagner (1938), by Waage (1965: 26) and by Jeletzky (1969: 26) on stratigraphical grounds. In any case, the supposed sequence of genera does not even exhibit a good morphological transition, there being no genera showing phragmocones intermediate in type between the normal and the sepiid.

Such an origin is put out of court if one accepts the series of sepiid dorsal shields dating from the Upper Jurassic onwards. The author accepts this because both shape and detailed structure, especially the appearance and microscopic structure of the tuberculate area, link up all the forms and differentiate them from other cephalopods, although few of the earlier ones show any convincing trace of the septa.

Waage (1965: 26) adopted a different explanation. He supposed that the late Cretaceous *Actinosepia* really lacked septa, but was convinced, by the virtual identity of microscopic structure between the dorsal shields of *Actinosepia* and *Sepia,* that they were related. He therefore supposed that sepiids might be derived from trachyteuthids *via Actinosepia*, the earlier forms perhaps having a non-calcified buoyancy apparatus. This suggestion runs counter to all that is now known about the cephalopod buoyancy system, which is dependent on the presence of rigid camerae for the withdrawal of liquid and the creation of gas spaces. It seems unlikely that a non-calcified buoyancy apparatus could exist, and equally so that a calcified phragmocone could re-appear in evolution having once been lost.

We therefore come back to the fact that there were sepiid-like forms in the Upper Jurassic. The only other forms which are at all similar are some of the Jurassic loligosepiids, which like the Jurassic sepiids, have prominent posterior wings. Some examples of *Trachyteuthis* from the Lithographic Stone (e.g. BM C46858) (Fig. 9) show a notch, giving rise to a hyperbolar field, between the wing and the median area, just like the earlier loligosepiids. No one has supposed that the latter possessed phragmocones, but the notches in both cases seem to indicate derivation from the Phragmoteuthida.

The only phragmoteuthid phragmocones that are well known, from the Lower Jurassic of Dorset, England, are normal; furthermore the pro-ostracum is very thin, in contrast to the sepiid dorsal shield. The phragmoteuthid fossil record, however, is obviously very incomplete, and there may have been other genera that we do not know of. At the present time a phragmoteuthid origin during the Jurassic, unsupported by any transitional forms, is the best that can be proposed. Fossil material bearing on such an origin would be highly desirable.

Vampyromorpha

Pickford (1949: 27) noted the resemblance between loligosepiid gladii ("Belemnosepiidae") and that of *Vampyroteuthis*, which is the only extant cephalopod with a broad median field like loligosepiids (Fig. 2). The similarity extends to the strong posterior curvature in both cases. The rarity of loligosepiid gladii after the Jurassic could result from loss of calcification, leading to the thin, transparent gladius of *Vampyroteuthis*.

Jurassic Loligosepiina were large animals, some of them probably a metre or more in length, which lived in shallow water. *Vampyroteuthis* is small and shows adaptations to a deep-water existence. It may thus be an example of a once-successful animal surviving in the deep sea after it has been superseded elsewhere by more highly evolved forms.

Octopoda

The only Mesozoic octopod is *Palaeoctopus* from the Upper Cretaceous (Senonian Stage) of the Lebanon, known by one specimen in the British Museum (Woodward, 1896) and two in Paris (Roger, 1944a). All three are very similar, and show a body nearly circular in outline (after crushing) with a posterior pair of fins. The body narrows down to a small head and the eight arms appear to show suckers and cirri. There is a U-shaped "shell" or gladius. Authors have generally compared it with the living cirrate octopods, though Naef (1922: 286) saw resemblance to young *Octopus*. It does not help much except to show that primitive Octopoda were in existence by the late Cretaceous.

This is not the place to review the anatomical evidence. The fact that *Vampyroteuthis* has several octopod characters (not least, of course, the eight arms) might suggest that the Octopoda diverged from the Mesozoic vampyromorphs (i.e. the Loligosepiina). Such a hypothesis would be supported by the fact that the gladius of *Palaeoctopus* could be derived from the wings or lateral fields of the loligosepiine gladius (Fig. 2b), the median field having been lost.

A few fossil argonaut shells are known (Naef, 1922: 294) from the Upper Tertiary, and look very like the modern ones.

Squids

The variety of form of gladii among living squids is not matched among fossils. Most fossil gladii are Mesozoic and in view of the time gap between them and living species one must beware of attaching too much importance to similarities of form. The Tertiary must have seen much evolution of, and within, the numerous oegopsid families, and one may expect appearance of new forms, on the one hand, and parallel or convergent evolution on the other. The relatively simple structure of the gladius, and the limits on the form of an effective gladius, lead one to anticipate homeomorphy. It will therefore be unwise to make more than general statements, at least until more detailed comparative work on modern, as well as ancient, gladii is available.

The most convincing comparison is between the Jurassic *Plesioteuthis* and living Ommastrephidae, which have almost identical gladii (Fig. 15) which stand apart from most others. Thus, ommastrephid-type squids may have existed in the Upper Jurassic, and the family may represent the root-stock of the oegopsids, or of some of them. The living *Thysanoteuthis* has a similar gladius, apparently without a conus.

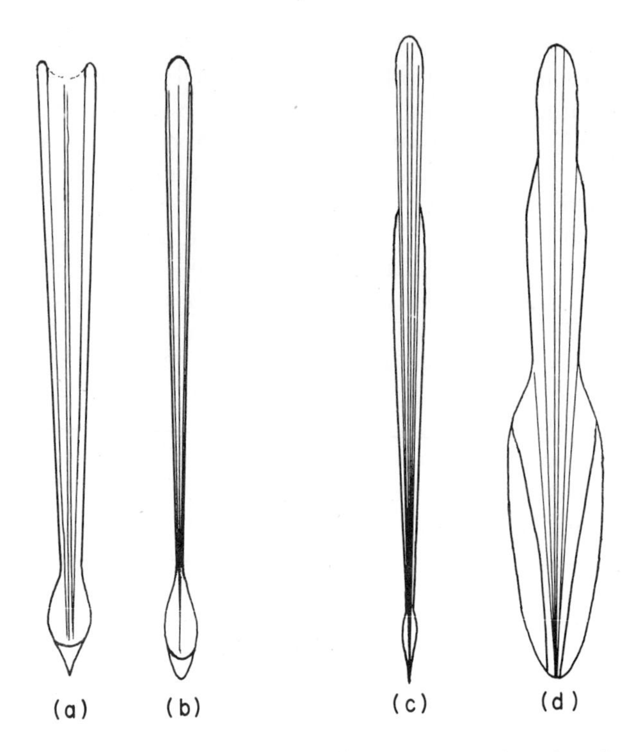

FIG. 15. Gladii of: (a) *Plesioteuthis*; (b) *Ommastrephes*; (c) *Onychoteuthis*; (d) *Rossia*. Not to scale.

The other families which may show affinities with Jurassic gladii are Gonatidae and Onychoteuthidae. Both have gladii with lateral and median fields (Fig. 15), like Loligosepiina although much narrower. These are two of the small minority of living families in which hooks are present on arms and/or tentacles. This is not the place to consider whether hooks are, or are not, primitive features; several fossil groups possessed them, but others are not known to have done so. Insofar as Phragmoteuthida (or some of them) had hooks, Gonatidae and Onychoteuthidae may be said to retain phragmoteuthid features in hooks and gladius.

The other relatively common type of gladius in the Mesozoic is that of the Teudopseidae and Palaeololiginidae, leaf-shaped with prominent rhachis which may project well forward. Many modern squids have this type of gladius, both myopsids and oegopsids, and it does not seem safe to link any of them closely with fossil forms.

Spirula

This genus is often placed in Sepiida but is very different from Sepiidae. The supposed palaeontological evidence (i.e. supposed common ancestors in early Tertiary "sepiids") has already been rejected (pp. 24–25). In many characters *Spirula* seems to resemble oegopsid squids (Huxley & Pelseneer, 1886). It should be removed from Sepiida, though no formal taxonomic proposal is made here.

There is a reasonable case for the traditional link between *Spirula* and the Eocene Spirulirostridae, and perhaps the late Cretaceous Groenlandibelidae. Earlier than this we cannot at present go.

CONCLUSION

We are obliged to assume that Oegopsida, together with Myopsida and Sepiida, had common ancestors in which the fourth pair of arms was already modified to form the tentacles, unless we can believe such modifications took place independently in different lines. This seems unlikely because of the large number of other features that these three groups have in common, despite their differences. It has been explained above that the origins of Sepiida and of some oegopsids are to be sought in Phragmoteuthida, which thus appear to be the ultimate ancestors. The Phragmoteuthida are therefore likely to have had the fourth pair of arms differentiated from the others, but this may pose a problem because they also appear to be ancestors of *Vampyroteuthis* and octopods, and we would then have to assume that the tentacles had been lost in these forms. If this is not acceptable, we may perhaps think that there were different phragmoteuthid groups, one with tentacles which led to the Decapoda, and another which gave rise to the octopods. Here, at the moment, we are in the realm of speculation.

The phylogeny of the living coleoids has to be compiled from hopelessly inadequate palaeontological evidence. It is not surprising that attempts to work it out have been made at long intervals and have been more or less unconvincing. The present author is under no illusion that his work is better than that of his predecessors. The purpose of this contribution will have been fulfilled if it stimulates neontologists to look critically at the anatomical evidence and try to harmonize it with that of the fossils, to produce a more acceptable solution.

Acknowledgements

I thank Dr M. K. Howarth and Mr D. Phillips for facilitating study of the collections in the British Museum (Natural History). I owe a debt of gratitude to the published works of Dr J. A. Jeletzky whose careful morphological work forms the basis of much that I have written here.

References

Abel, O. (1927). *Lebensbilder aus der Tierwelt der Vorzeit.* Jena: G. Fischer.

Angermann, E. (1902). Über das Genus Acanthoteuthis Münst. aus den lithographischen Schiefern in Bayern. *Neues Jb. Miner. Geol. Paläont.* Teil Bd. **15**: 205–230.

Appellöf, A. (1893). Die Schalen von *Sepia, Spirula* und *Nautilus.* Studien über den Bau und das Wachsthum. *K. svenska Vetensk.-Akad. Handl.* **25**(7): 1–105.

Barthel, W. (1964). Zur Entstehung der Solnhofener Plattenkalke (unteres Untertithon). *Mitt. bayer. St. Paläont. Hist. Geol.* **4**: 37–69.

Birkelund, T. (1956). Upper Cretaceous belemnites from West Greenland. *Meddr Grønland* **137**(9): 30.

Buckland, W. (1836). *Geology and mineralogy considered with reference to natural theology.* London: William Pickering. 2 vols.

Buckland, W. (1858). *Geology and mineralogy considered with reference to natural theology.* 3rd edition. London: George Routledge. 2 vols.

Carruthers, W. (1871). On some supposed vegetable fossils. *Q. Jl geol. Soc. Lond.* **27**: 443–449.

Crick, G. C. (1896a). On the proöstracum of a belemnite from the Upper Lias of Alderton, Glos. *Proc. malac. Soc. Lond.* **2**: 117–119.

Crick, G. C. (1896b). On a specimen of *Coccoteuthis hastiformis,* Rüppell, sp., from the Lithographic Stone, Solenhofen, Bavaria. *Geol. Mag.* (4) **3**: 439–443.

Crick, G. C. (1897). On an example of *Acanthoteuthis speciosa,* Münster, from the Lithographic Stone, Eichstädt, Bavaria. *Geol. Mag.* (4) **4**: 1–4.

Crick, G. C. (1907). On the arms of the belemnite. *Proc. malac. Soc. Lond.* **7**: 269–279.

Crick, G. C. (1915). A dibranchiate cephalopod (*Plesioteuthis*) from the Lithographic Stone (Lower Kimmeridgian) of Eichstädt, Bavaria. *Proc. malac. Soc. Lond.* **11**: 313–318.

Crick, G. C. (1921). On some dibranchiate Cephalopoda from the Upper Lias of Gloucestershire. *Proc. Cotteswold N.F.C.* **20**: 249–256.

Flower, R. H. (1945). A belemnite from a Mississippian boulder of the Caney Shale. *J. Paleont.* **19**: 490–503.

Flower, R. H. & Gordon, M. (1959). More Mississippian belemnites. *J. Paleont.* **33**: 809–842.

Fritsch, A. (1910). *Miscellanea palaeontologica. 2. Mesozoica.* Prague.

Hauff, B. (1953). *Das Holzmadenbuch.* Öhringen.

Hölder, H. (1973). Miscellanea cephalopodica. *Münster. Forsch. Geol. Paläont.* **29**: 39–76.

Huxley, T. H. (1864). On the structure of Belemnitidae; with a description of a more complete specimen of Belemnites than any hitherto known, and an account of a new genus of Belemnitidae: *Xiphoteuthis. Mem. geol. Surv. U.K.* **2**: 1–22.

Huxley, T. H. & Pelseneer, P. (1886). Report on the specimens of the genus *Spirula* collected by H.M.S. Challenger. *Rep. scient. Results Voy. Challenger* (Zoology) No. 83.

Jeletzky, J. A. (1966). Comparative morphology, phylogeny, and classification of fossil Coleoidea. *Paleont. Contr. Univ. Kans.* Article Mollusca No. 7: 1–162.

Jeletzky, J. A. (1969). New or poorly understood Tertiary sepiids from southeastern United States and Mexico. *Paleont. Contr. Univ. Kans.* Paper No. 41: 1–39.

Johnson, R. G. & Richardson, E. S. (1968). Ten-armed fossil cephalopod from the Pennsylvanian of Illinois. *Science, N.Y.* **159**: 526–528.

Klinghardt, F. (1943). Vergleichende Untersuchungen über Tintenfische und belemnitenähnliche Weichtiere. *Sber. Ges. naturf. Freunde Berl.* **1942**: 5–17.

Kretzoi, M. (1942). *Necroteuthis* n.g. (Ceph. Dibr., Necroteuthidae n.f.) aus dem Oligozän von Budapest und das System der Dibranchiata. *Földt. Közl.* **72**: 124–138.

Makowski, H. (1952). La faune callovienne de Łuków en Pologne. *Palaeont. pol.* **4**: 1–64.

Mantell, G. A. (1848). Observations on some belemnites and other fossil remains of Cephalopoda, discovered by Reginald Neville Mantell, C. E., in the Oxford Clay, near Trowbridge in Wiltshire. *Proc. R. Soc.* **5**: 746–748.

Meyer, H. von (1846). [Über *Trachyteuthis*] *Neues Jb. Miner. Geogn. Geol. Petrefakt.* **1846**: 598.

Miller, H. W. Jr. (1957). *Niobrarateuthis bonneri*, a new genus and species of squid from the Niobrara formation of Kansas. *J. Paleont.* **31**: 809–811.

Müller, A. H. (1965). *Lehrbuch der Paläozoologie.* Bd.2. Invertebraten. Teil 2. Mollusca 2, Arthropoda 1. Jena: G. Fischer.

Naef, A. (1921). Das System der dibranchiaten Cephalopoden und die mediterranean Arten derselben. *Mitt. zool. Stn. Neapel* **22**: 527–542.

Naef, A. (1922). *Die Fossilen Tintenfische.* Jena: G. Fischer.

Owen, R. (1844). A description of certain belemnites, preserved, with a great proportion of their soft parts, in the Oxford Clay, at Christian-Malford, Wilts. *Phil. Trans. R. Soc.* 1844: 65–85.

Owen, R. (1855). Notice of a new species of an extinct genus of dibranchiate cephalopod (*Coccoteuthis latipinnis*) from the Upper Oolitic shales at Kimmeridge. *Q. Jl geol. Soc. Lond.* **11**: 124–125.

Pearce, C. (1847). On the fossil Cephalopoda constituting the "Genus *Belemnoteuthis*". *Geol. J. Lond., Rec. of Discoveries, British & foreign, Palaeont.* **1**: 75–78.

Pickford, G. E. (1949). *Vampyroteuthis infernalis* Chun, an archaic dibranchiate cephalopod. II. External anatomy. *Dana-Report* No. 32: 1–132.

Pinna, G. (1972). Rinvenimento di un raro cefalopode coleoideo nel giacimento sinemuriano di Osteno in Lombardia. *Atti Soc. ital. Sci. nat.* **113**: 141–149.

Quenstedt, F. A. (1849). *Petrefactenkunde Deutschlands. Die Cephalopoden.* Tübingen.

Regteren Altena, C. O. van (1949). Teyler's Museum, systematic catalogue of the palaeontological collection; sixth supplement (Teuthoidea). *Archs Mus. Teyler* (3) **10**: 53–62.

Reuss, A. E. (1854). Loliginidenreste in der Kreideformation. *Abh. K. böhm. Ges. Wiss.* (5) **8**: 29–32.

Rieber, H. (1970). *Phragmoteuthis? ticinensis* n.sp. ein Coleoidea-Rest aus der Granzbitumenzone (Mittlere Trias) des Monte San Giorgio (Kt. Tessin, Schweiz). *Paläont. Z.* **44**: 32–40.

Ristedt, H. (1968). Zur Revision der Orthoceratidae. *Abh. math.-nat. Kl. Akad. Wiss. Mainz* J. 1968: 211–287.
Roger, J. (1944a). Phylogénie des Céphalopodes Octopodes: *Palaeoctopus newboldi* (Sowerby 1846) Woodward. *Bull. Soc. géol. Fr.* (5) **14**: 83–98.
Roger, J. (1944b). *Acanthoteuthis* (*Belemnoteuthis*) *syriaca* n.sp. Céphalopode Dibranche du Crétacé supérieur de Syrie. *Bull. Soc. géol. Fr.* (5) **14**: 3–10.
Roger, J. (1946). Les Invertébrés des couches à poissons du Crétacé supérieur du Liban. Etude paléobiologique des gisements. *Mém. Soc. géol. Fr.* (n.s.) 51.
Roger, J. (1952). Sous-classe des Dibranchiata Owen 1836, In *Traité de Paléontologie* **2**: 689–755. Piveteau, J. (ed.). Paris: Masson.
Rosenkrantz, A. (1946). Krogbaerende cephalopoder fra Østrønlands Perm. *Meddr. dansk. geol. Foren.* **11**: 160–161.
Schevill, W. E. (1950). An Upper Jurassic Sepioid from Cuba. *J. Paleont.* **24**: 99–101.
Sweet, W. C. (1964). Cephalopoda—General features. In *Treatise on invertebrate paleontology*, Part K, Mollusca 3, K4–K13. Moore, R. C. (ed.). Kansas: Geol. Soc. Am. & Univ. Kansas Press.
Swinnerton, H. H. (1947). *Outlines of palaeontology*. Third edition. London: Edward Arnold.
Voltz, M. (1830). Observations sur les belémnites. *Mém. Soc. Sci. nat. Strasb.* **1**: 70.
Waage, K. M. (1965). The Late Cretaceous coleoid cephalopod *Actinosepia canadensis* Whiteaves. *Postilla, Peabody Museum, Yale.* No. 94: 1–33.
Wagner, A. (1860). Die fossilen Ueberreste von nackten Dintenfischen aus dem Lithographischen Schiefer und dem Lias des suddeutschen Juragebirges. *Abh. bayer. Akad. Wiss.* (Math. phys. Kl.) **8**: 751–821.
Wagner, J. (1938). A Kiscelli Közép-Oligocén (Rupélien) rétegek kétkopoltyús cephalopodáiés új Sepia-félék a Magyar Eocenbol. *Ann. nat. Mus. hung.* **31**: 179–199.
Wetzel, W. (1930). Die Quiriqina-Schichten als Sediment und paläontologisches Archiv. *Palaeontographica* **73**: 49–104.
Woodward, H. (1883). On a new genus of fossil "Calamary" from the Cretaceous formation of Sahel-Alma, near Beirut, Lebanon, Syria. *Geol. Mag.* (N.S.) **10**: 1–5.
Woodward, H. (1896). On a fossil octopus (Calais Newboldi J. de C. Sby. MS.) from the Cretaceous of the Lebanon. *Q. Jl geol. Soc. Lond.* **52**: 229–234.

Symp. zool. Soc. Lond. (1977) No. 38, 49–60.

PRESENT STATUS AND NEW TRENDS IN CEPHALOPOD SYSTEMATICS

G. L. VOSS

Rosenstiel School of Marine and Atmospheric Science, University of Miami, Miami, Florida, USA

SYNOPSIS

Since the publication of d'Orbigny's monograph in 1848, 190 genera and 854 species of cephalopods have been described. During 126 years, new genera and species have been added at an average annual rate of 1·5 and 6·7 respectively. The total number of genera and species, including d'Orbigny's, are 206 genera and 983 species. Approximately 333 species have been synonymized at a rate of 2·6 species annually.

Eighty-five percent of the genera of living coleoids contain five or less species and nearly 50% are monotypic. Only *Sepia* and *Octopus* contain over 100 species each. The monotypic genera are nearly all represented by oceanic species while genera with many representatives are inhabitants of shallow or deep benthic waters.

Lack of detailed morphometric studies of large numbers of individuals on a population basis is indicated by lack of subspecific taxa; only about 10 subspecies, most of a dubious nature, have been recorded.

Critical revisions based upon adequate study material, analysis of character variations, and above all examination of type material, are sorely needed. Only nine revisions, some only partially fulfilling these criteria, have been published since 1900.

The present trend of clarifying and up-dating the classification through a literature review only causes further confusion as much of it is so uncritical, and identifications dubious. Future systematic work should continue the task of enumerating and describing the members of the class, but should be accompanied by critical studies of comparative anatomy and development, and by detailed analyses of morphometric characters, especially the limits of individual, specific and interspecific variations. Every effort should be made towards critical revisionary work that will lead eventually to a clarification of cephalopod phylogeny and classification.

When I was asked by the conveners of this symposium to review the present status of cephalopod systematics, I little realized what I had let myself in for, and, as I progressed, what a sad state of affairs exists in this field. In order to see the full picture it is necessary to start from some early vantage point, not so early that it would include those impossible nominal species of the early post-Linnean students, but early enough that trends can be seen. For this I have chosen the date 1848, the year in which Alcide d'Orbigny's famous *Histoire naturelle* was completed.

This work contains descriptions of 16 genera and 129 species of living cephalopods, discounting a number of taxa that even then d'Orbigny considered dubious. Thirty-one years later George Washington Tryon (1879) published a monograph of the cephalopods that listed 18 genera and 144 species. This was followed seven years later by the first of Hoyle's three-part Catalogue of Recent Cephalopoda (1886a, 1897, 1909), which surveyed new taxa at ten-year intervals. After considering the last of these, recourse was made to the *Zoological Record*, which was searched for new taxa in ten-year intervals through the last volume, and finally my own catalogue which is up-to-date through 1974.

The results of this tabulation are shown in the accompanying table (Table I). This gives the number of genera and species added during each interval as well as the number of genera and species described each year. The results are interesting. Over the span of 126 years, an average 1·5 genera and 6·7 species have been described annually.

TABLE I

Number of genera and species of cephalopods described from 1848 to 1974

	Years	Genera	Species	No./year	Workers
	1835/1848	16	129		d'Orbigny
	1849–1879	18	144	4·8	Steenstrup
	1879–1886	34	115	14·3	Hoyle
	1887–1896	7	79	7·9	Verrill
	1897–1906	29	69	6·9	
	1907–1916	41	128	12·8	Chun, Pfeffer
	1917–1926	20	96	9·6	Joubin, Berry
	1927–1936	22	79	7·9	Robson, Sasaki
	1937–1946	2	15	1·5	
	1947–1956	6	28	2·8	
	1957–1966	7	56	5·6	
	1967–1974	4	45	5·6	
Total	1848–1974	190	854	6·7	
Since	1758	206	983		

For the species, this figure is quite close to the number described annually over the last 20 years. If one considers that the average of 6·7 probably exceeds the latter figure because of the high number of species described during the years of the great research voyages in the late 1800s and early 1900s, one is forced to the belief that we have nowhere near exhausted the number of undescribed species either on our museum shelves or in the sea.

Further study of the table shows three periods of special interest—1879–1886, 1907–1926 and 1937–1956. The first two are periods of large increases in new species; the third represents a kind of systematic doldrums. The explanation for these fluctuations seems to be concerned with two factors: the number of systematists active during those periods, and the findings of the major biological expeditions.

The first surge in systematics was probably in the interval from 1879 to 1886. Those were the years when Japetus Steenstrup in Denmark, William Hoyle in England, and Addison Verrill in the United States were first drawing the scientific world's attention to the oceanic and deep-sea cephalopods, and Georg Pfeffer in Germany and Louis Joubin in France were just starting their careers. During this brief span of only eight years, 115 new species were described, an average of 14·3 species per year.

The next major period of activity occurred between 1907 and 1926. During the first decade the results of two great expeditions were published, the *Valdivia* cephalopods by Carl Chun (1910) and the magnificent monograph of the oegopsid squid by Georg Pfeffer (1912) based upon the collections of the German Plankton Expedition. The next two decades spanned the primacy of the so-called Leipzig school of cephalopod research; this was accompanied by the studies of Louis Joubin on the Monaco collections and the investigations of Adolf Naef in Naples. Robson was at work at the British Museum, Sasaki was investigating the cephalopods of Japan, and Berry was reporting on the fauna of western America and the Hawaiian Islands. It was the zenith of cephalopod systematics. In all, 83 new genera and 303 new species were described during the three decades, something of a record.

But why the slump for 1937–1956? Only 8 genera and 43 species were described in 20 years. Undoubtedly, World War II brought most work to a standstill, but even more important was the lack of systematists. Grimpe published his last paper in 1933, Hoyle in 1912, Joubin in 1933, Massy in 1928, Naef in 1928, Pfeffer in 1912, Robson in 1933, Sasaki in 1929 and the new crop of students were not yet very productive. The surge since 1957 represents an almost totally new interest and from all indications will remain at about its present level for a number of decades to come, barring some unforeseen international holocaust.

This table, however, is also somewhat misleading in the totals, which show 206 genera and 983 species. No absolutely accurate listing of the present numbers of probably valid genera and species is available, but based upon my own catalogue and revisionary work now in progress, it seems likely that the totals should be corrected to about 144 to 150 genera and about 650 species. This lower figure accounts for the genera and species synonymized during the same period. This reveals that 333 species have been synonymized at an average annual rate of 2·6 over the last 126 years.

Table II shows the number of genera and the number of species in each, again from my catalogue. This reveals some startling facts. Eighty-five percent of the total number of genera contain five or fewer species and nearly 50% of the genera are monotypic. Only two genera, *Sepia* and *Octopus*, contain over 100 species each, accounting for one-third of the described species. It would have been useful to give the number of subgenera and subspecies but, because of conflicting usage, this was difficult to determine. However, subspecies have rarely been used in cephalopods and probably less than ten are now recognized.

It might be well to pause here for comment on the deductions to be drawn from the data just presented.

First, it seems clear that we are indeed still in the descriptive or *alpha* stage of systematics and that we are far from knowing the total number of species of cephalopods. The number of new species to be added is not necessarily dependent upon new collections and cruises but may well be dependent upon the time available to systematists to work up old material. A quick count of undescribed species in my own laboratory tallied three new species in press, about 15 in manuscript, and at least another ten on my shelves awaiting study. How many more are in my unworked collections I have no idea. This situation may obtain to a greater or lesser degree in the laboratories of a number of my colleagues.

TABLE II

Number of species contained in the genera of recent cephalopods

No. of species/genus	No. of genera	Percent
1	69	48
2	30	21
3	9	6
4–5	14	10
6–10	12	8
11–20	6	4
21–30	2	1
100+	2	1

Second, from the disparity in the numbers of species in the various genera, it may be concluded that, either the cephalopods show a remarkable diversity of form resulting in the large number of monotypic genera, or the work of analysis of generic and specific characters is insufficient for proper classification. While I agree in part with the latter choice, a review of some of the families containing primarily monotypic genera reveals such great diversity, and the differences are of such magnitude, that the genera could not be united without destroying in large part our conceptions of species levels in other groups. It is probable that the generic diversity is natural. Support is given to this viewpoint by the fact that, with few exceptions, the monotypic genera consist of oceanic midwater species while polytypic genera are represented by shallow-water continental-shelf or deep-benthic forms, both areas apparently conducive to a high degree of speciation.

Third, we might question why there are so few subspecies. One of my colleagues bluntly stated that he did not believe in subspecies and that the use of subgenera and subspecies destroys the beauty of Linnean binomial nomenclature. On the other hand, the failure to use subgenera and subspecies makes it simpler to pigeonhole taxa without having to exercise our brains as biologists in the search for signs of relationships and pathways of evolution and speciation.

This raises another question concerning the status of systematics. Although my search has not been exhaustive, it is very clear that we are sadly deficient in monographs, revisions and reviews and that those that exist are, with few exceptions, far outdated. Many of the large cephalopod works on which we rely so heavily are either voyage or cruise reports (Chun, 1910; Hoyle, 1886b; Joubin, 1895, 1900, 1920, 1924) or faunal reports (Verrill, 1882; Jatta, 1896; Berry, 1912a,b; Naef, 1923; Sasaki, 1929; Grimpe, 1925). Only four major monographs have been produced (d'Orbigny, 1835–48; Tryon, 1879; Pfeffer, 1912; Robson, 1929, 1932) and even these are largely uncritical compilations whose greatest value is in bringing together the mass of genera and species, thus making the information more readily accessible. Only two of these monographs warrant special consideration.

Pfeffer's monograph is a landmark study of the oegopsid squid. It has, however, two great weaknesses; Pfeffer did not examine the type specimens of many of the species, and he felt compelled to give names to every specimen mentioned by previous authors despite the fact that the specimens were in such poor condition that they could not be identified or were larvae unconnected to any developmental series. As a result, his monograph is difficult to use and often misleading.

Robson attempted a monumental task, which was doomed to failure before it was started. The number of species was too large and contained too few critical reviews; too many of the species were known only from unique specimens, often females; and he was now suffering from the mental difficulties that shortly forced his retirement from the British Museum. To this latter cause may be laid the odd, often conflicting, and many times erroneous, statements in his monograph that have so often confused and trapped the unwitting reader. Nonetheless, used with caution, it is an invaluable work for the octopod specialist.

Mention must also be made of revisions, although I exclude those of a local nature that, while adding to our knowledge of the systematics of a region do not make the contributions necessary for updating or evaluating cephalopod classification as a whole. In this category must be considered Pickford's (1945) revision of the octopods of the tropical western Atlantic, Roeleveld's (1972) revision of the sepiids of South Africa and Adam's (1954) *Siboga* studies, to mention a few.

In 1962, I published the first of a series of revisions of the cephalopods of the Atlantic but I immediately found, not unexpectedly, that it was impossible to do critical revisionary work without including the world-wide distribution. This is necessary especially if the genera are to be evaluated. Using the criteria of both completeness and world-wide scope, a review of the literature reveals an amazing lack of modern revisions.

Based upon a survey of the literature cited in Clarke's (1966) review and my unpublished review of the systematics of the cuttlefishes and squids, I have been able to find only nine critical revisions of cephalopod genera or families published since 1900 (Joubin, 1902; Berry, 1932; Adam, 1939; G. Voss, 1962; Adam & Rees, 1966; Roper, 1969; N. A. Voss, 1970; Young & Roper, 1969a,b). These are also the main papers in which authors have studied the types of the species they have discussed, an absolutely essential part of any truly critical taxonomic study. It is little wonder that the non-specialist often searches in vain for descriptions and figures of the specimens before him, or that the student beginning his studies of the cephalopods is baffled by the lack of order and the difficulty of assessing the conflicting statements in the literature.

From the brief review of the present numbers of families, genera, and species of the cephalopods, it can be seen that the systematics of the class is still far from the stage when the broad evolutionary picture can be seen. This despite the fact that numerous distinguished specialists, Steenstrup, d'Orbigny, Joubin, Hoyle, Pfeffer, Chun, Sasaki, and more recently Adam and Berry, to mention only a few, devoted long periods of their lives to the study of the group. Specialists in other fields might well wonder why the state of the art is so poor when there are so few species. There are a number of reasons, a few of which I will enumerate.

First, with a few exceptions, cephalopods are either swift animals or, conversely, very secretive in their habits. I am not certain that Professor Young's studies of their intelligence and behaviour can be interpreted to say that they have enough intelligence to deliberately avoid nets and traps, but there are times when the ship-board collector may well believe so. Probably as a result of our poor collecting devices, many species are still rare in collections. Only about 20 or 30 years ago over one-third of the cephalopod species were known only from unique specimens. Newer, larger, and more swiftly towed nets have caught many more individuals in the last 20 years and the problem of scarcity of specimens is disappearing. Even the octopods are being collected in large numbers as the result of the development of fish poisons and better field techniques.

Second, despite all these advances in collecting, there are many species known only from larval or juvenile specimens and for which no adults have even been captured.

Third, these animals with only a few exceptions (such as the large-egged octopods and many of the sepioids) go through developmental stages that are so varied that many of the larvae and juveniles of both octopods and teuthoids defy identification with known adults. Some biologists deplore the use of the term "larva" in cephalopods but I use it in the sense in which it is used by ichthyologists among whom I was once counted: that stage in the development of the young before it possesses the characters of the adult. In other words, attempts to identify a larva using keys and descriptions based upon adult specimens will inevitably lead only to confusion.

Unfortunately, my predecessors were very prone to naming these strange larvae and juveniles. Georg Pfeffer abhorred specimens without names and created new genera and species wholesale, many based upon ontogenetic stages. The taxonomic and nomenclatural mazes created as a result of this propensity have given work to many later students but helped systematics not at all.

Fourth, teuthologists seem to have a predilection for publishing works drawing attention to the importance of various structures and characters such as sucker dentition, spermatophores, radulae, funnel organs, beaks, gill lamellae, etc. and then walking away from the subject leaving it to someone else to determine the actual value of the character by doing what is then termed pedestrian research, attempting, through the study of sufficient numbers of specimens, to determine the range of variation and the reliability of the characters for systematic use.

Fifth and last, because I could go on at length, cephalopod literature lacks almost entirely that wealth of comparative anatomical study that has been so invaluable to vertebrate systematics. Naef (1923) briefly touched upon it in his great work, but no later student has tackled it. But comparative anatomy is the base upon which the higher classifications and phylogenies are built and until this has been provided I see little hope of being able to build a better phylogenetic classification.

Let me give a few examples of the difficulties involved in the reliability of what have been considered strong taxonomic characters. Almost every teuthologist, myself included, feels required in descriptions of new species to figure the radula, the beaks, the spermatophores, and the hectocotylized arm.

The radula, or more usually part of it, is one of the most figured structures from molluscs. Typically, one is extracted with considerable difficulty, mounted, and usually crudely figured. Robson (1925) devised a formula for depicting whether the radulae of octopods had symmetric or asymmetric cusps and we have used this slavishly ever since. All but Adam, that is, who published a seldom-quoted paper (1941) on the diagnostic value of the radula in octopods that should give us all pause for thought. Just how valid a taxonomic character is the radula if no one knows what amount of individual variation it possesses? Recently this was brought home to me when studying specimens of a new genus of *Graneledone* from Antarctica. A set of teeth from one individual had only five nearly homodont teeth in each row with no marginal plates while another had seven heterodont teeth with only the slightest vestige of plates. The type of another new species from the Falkland Islands has a radula that does not look like the radula from any of the other *Graneledone* that have been figured, and might pass for a

Pareledone. About 30 specimens of *Benthoctopus* from the northeastern Pacific collected by William Pearcy, while showing some general features in common, are so variable that without other characters the radulae would be useless for identification. The illustration of only a single radula would possibly confuse generations of later workers.

Beaks have been illustrated by many people and these were reviewed by Clarke (1962) who himself drew attention to their use as taxonomic characters. His paper gave me great hope as an aid in certain puzzling groups but either I lack his perspicuity or, as I now suspect, after having extracted some hundreds of them from known specimens, there is a fairly wide range of variation within any one species while between other species there may be great similarity.

But surely spermatophores, being a result of sexual differentiation and thus, perhaps, clues to the degree of reproductive isolation, are more reliable. Pickford (1945) has presented us with a system of proportions and indices for comparing spermatophores of various species. For years I have measured the length and width, the sperm mass, and the horn, of selected spermatophores and listed these in my tables. But what do they mean? Perhaps nothing, according to one of my graduate students who is engaged in a detailed comparative study of spermatophores throughout the class. The earliest spermatophores formed are much shorter than the later ones and thus, the spermatophore length index is extremely variable. Far more important are the internal structural details that must be well figured to be of value.

And then of course there is the little, flimsy, semi-gelatinous funnel organ that everyone has ignored except in grossest detail. This, unexpectedly, is coming to the fore as one of the most reliable, conservative characters showing, in some groups, strong specific differences quite out of keeping, it would appear, to any inherent value it may have to the animal.

What is needed at this time and stage in cephalopod systematics, is a careful study of the variation of the characters employed, at the individual, population and specific levels including geographical variation. Such a study would provide means of evaluating the significance of generic characters as well.

One of the most confusing problems in specific determinations is the complicated life history of many of the species. It is possible for there to be as many as four generic names and as many or more specific names in a single specific developmental line. Especially in the cranchiids this has led to confusion in the nomenclature and has resulted in the situation that almost every specimen of poorly known forms is placed in a new species.

The time has long since passed when we can permit such things to occur. We now have enough specimens in the world collections to permit the working out of these life histories. This requires painstaking work and careful attention to details, but the result will be a clarification of the species and a great reduction in the number of species names to be dealt with.

So far in systematics we have been dealing mainly with species as separate entities. I believe that we are nearing the time when we should be looking at the possibility of evaluating and depicting speciation patterns. In the more advanced classes such as birds and fishes, subspecies are commonly used. Some cephalopod systematists refuse to use subspecies; if they can find the slightest degree of difference, they describe a new species. In my opinion, this destroys or confuses any possible picture of relationships that we might be able to construct.

Study of considerable numbers of specimens has convinced me that many of the genera, if treated in this fashion, would contain not a series of independent species but a series of subspecies or *Rassenkreis* far better exemplifying relationships than at present.

An example might be made of the situation with regard to the genus *Illex* in the Atlantic, where I believe at present relationships are confused. On the other hand, detailed studies of *Octopus vulgaris*, using careful statistical approaches, would, I believe, show that it is a series of closely related species or perhaps a world-wide series of subspecies. Certainly some of the species complexes in *Loligo* would be clarified using subspecies instead of a series of closely related allopatric species.

In conclusion I hope that what I have said today will encourage some new students in this field to use some of these approaches and lead us into a new era in cephalopod systematics.

Acknowledgements

I wish to thank the National Science Foundation for its generous support of my cephalopod research upon which this paper is based and particularly my present grant GB 24994. This paper is a Scientific Contribution from the Rosenstiel School of Marine and Atmospheric Science, University of Miami, Florida, USA.

References

Adam, W. (1939). Cephalopoda. Partie I. Le genre *Sepioteuthis* Blainville, 1824. *Siboga Exped.* No. 55a: 1–33.

Adam, W. (1941). Notes sur les Céphalopodes XV. Sur la valeur diagnostique de la Radule chez les Céphalopodes Octopodes. *Bull. Mus. r. Hist. nat. Belg.* **17**(38): 1–19.

Adam, W. (1954). Cephalopoda. 4. Céphalopodes à l'éxclusion des genres *Sepia, Sepiella* et *Septioteuthis. Siboga Exped.* No. 55c: 121–198.

Adam, W. & Rees, W. J. (1966). A review of the cephalopod family Sepiidae. *Scient. Rep. John Murray Exped. 1933–34.* **11**: 1–165.

Berry, S. S. (1912a). A review of the cephalopods of western North America. *Bull. Bur. Fish. Wash.* **30**: 263–336.

Berry, S. S. (1912b). The Cephalopoda of the Hawaiian Islands. *Bull. Bur. Fish. Wash.* **32**: 255–362.

Berry, S. S. (1932). Cephalopods of the genera *Sepioloidea, Sepiadarium,* and *Idiosepius. Philipp. J. Sci.* **47**: 39–53.

Chun, C. (1910). Die Cephalopoden. *Wiss. Ergebn. dt. Tiefsee-Exped. "Valdivia"* **18**: 1–552.

Clarke, M. R. (1962). The identification of cephalopod "beaks" and the relationship between beak size and total body weight. *Bull. Br. Mus. nat. Hist.* (Zool.). **8**: 421–480.

Clarke, M. R. (1966). A review of the systematics and ecology of oceanic squids. *Adv. mar. Biol.* **4**: 91–300.

Grimpe, G. (1925). Zur Kenntnis der Cephalopodenfauna der Nordsee. *Wiss. Meeresunters., Kiel* (N.F.) **16**(3): 1–124.

Hoyle, W. E. (1886a). A catalogue of recent Cephalopoda. *Proc. R. Phys. Soc. Edinb.* **9**: 205–268.

Hoyle, W. E. (1886b). Report on the Cephalopoda. *Rep. scient. Results Voy. "Challenger"* (Zoology) **16**(44): 1–245.

Hoyle, W. E. (1897). A catalogue of recent Cephalopoda. Supplement, 1887–96. *Proc. R. phys. Soc. Edinb.* **12**: 363–375.

Hoyle, W. E. (1909). A catalogue of recent Cephalopoda. 2nd Supplement, 1897–1906. *Proc. R. phys. Soc. Edinb.* **17**: 254–299.

Jatta, G. (1896). I cefalopodi viventi nel Golfo di Napoli (Sistematica). *Fauna Flora Golf Neapel* **23**: 1–268.

Joubin, L. (1895). Céphalopodes. *Résult. Camp. scient. Prince Albert I.* **9**: 3–63.

Joubin, L. (1900). Céphalopodes. *Résult. Camp. scient. Prince Albert I.* **17**: 1–135.

Joubin, L. (1902). Revision des sépiolides. *Mém. Soc. zool. Fr.* **15**: 80–145.

Joubin, L. (1920). Céphalopodes provenant des campagnes de la "Princesse Alice" (1898–1916) (3e Série). *Résult. Camp. scient. Prince Albert I.* **54**: 1–95.

Joubin, L. (1924). Contribution à l'étude des céphalopodes de l'Atlantique Nord. (4e Série). *Résult. Camp. scient. Prince Albert I.* **67**: 1–113.

Naef, A. (1923). Die Cephalopoden. *Fauna Flora Golf Neapel* **35**: 1–863. (1972, Translated into English by the Israel Program for Scientific Translation, Jerusalem.)

d'Orbigny, A. (1835–1848). In *Histoire naturelle générale et particulière des céphalopodes acétabulifères vivants et fossiles.* 2 vols. text and atlas. Ferussac, A. & d'Orbigny, A. (eds). Paris.

Pfeffer, G. (1912). Die Cephalopoden der Plankton-Expedition. Zugleich eine monographische Übersicht der oegopsiden Cephalopoden. *Ergebn. Plankton-Exped.* **2**: 1–815.

Pickford, G. E. (1945). Le Poulpe américain: a study of the littoral Octopoda of the Western Atlantic. *Trans. Conn. Acad. Arts Sci.* **36**: 701–811.

Robson, G. C. (1925). On seriation and asymmetry in the cephalopod radulae. *J. Linn. Soc.* (Zool.) **36**: 99–108.

Robson, G. C. (1929). *A monograph of the recent Cephalopoda.* Part I. Octopodinae. **1**. London: British Museum (Nat. Hist.).

Robson, G. C. (1932). *A monograph of the recent Cephalopoda.* Part II. The Octopoda, excluding the Octopodinae. **2**. London: British Museum (Nat. Hist.).

Roeleveld, M. A. (1972). A review of the Sepiidae (Cephalopoda) of southern Africa. *Ann. S. Afr. Mus.* **59**: 193–313.

Roper, C. F. E. (1969). Systematics and zoogeography of the world wide bathypelagic squid genus *Bathyteuthis* (Cephalopoda: Oegopsida). *Bull. U.S. nat. Mus.* **291**: 1–208.

Sasaki, M. (1929). A monograph of the dibranchiate cephalopods of the Japanese and adjacent waters. *J. Fac. Agric. Hokkaido imp. Univ.* **20** (Supplement) **10**: 1–357.

Tryon, G. W. (1879). *Manual of conchology* **1**. Cephalopoda. Philadelphia.

Verrill, A. E. (1882). Report on the cephalopods of the north-eastern coast of America. *Rep. U.S. Commnr Fish.* **1879**(7): 211–450.

Voss, G. L. (1962). A monograph of the Cephalopoda of the North Atlantic. I. The family Lycoteuthidae. *Bull. Mar. Sci. Gulf Carib.* **12**: 264–305.

Voss, N. A. (1970). A monograph of the Cephalopoda of the North Atlantic. The Family Histioteuthidae. *Bull. Mar. Sci.* **19**: 713–867.

Young, R. E. & Roper, C. F. (1969a). A monograph of the Cephalopoda of the North Atlantic: the family Cycloteuthidae. *Smithson. Contr. Zool.* **5**: 1–24.

Young, R. E. & Roper, C. F. E. (1969b). A monograph of the Cephalopoda of the North Atlantic: the family Joubiniteuthidae. *Smithson. Contr. Zool.* **15**: 1–10.

Symp. zool. Soc. Lond. (1977) No. 38, 61–87.

COMPARATIVE CAPTURES OF PELAGIC CEPHALOPODS BY MIDWATER TRAWLS

C. F. E. ROPER

Department of Invertebrate Zoology,
Smithsonian Institution,
Washington, DC, USA

SYNOPSIS

The captures of pelagic cephalopods by the 3 m Isaacs-Kidd midwater trawl (IKMT), the 8 m^2 rectangular midwater trawl (RMT 8), and the small (1400 mesh) Engel trawl (EMT) are compared. The sampling site was a one-degree square area in the North Atlantic Ocean east of Bermuda known as Ocean Acre. The IKMT and the RMT 8 were equipped with closing devices. Comparative samples were taken on the same cruise or at least during the same season of the year. The comparisons were made on net captures taken at 13 standardized depth increments from the surface to 1250 m for both day-time and night-time. Comparisons were developed for catch rate (standardized to number of specimens captured per hour of trawling), species composition, size distribution, and co-occurrence of species.

The comparison of IKMT and the RMT 8, nets with nearly equivalent mouth openings, indicates that the IKMT catches slightly larger specimens of the same species than the RMT 8. The RMT 8, however, catches more specimens per hour of a given species than the IKMT, and it tends to catch a greater diversity of species.

The Engel trawl, a net with a much larger area of mouth opening than the other nets, catches a significantly greater number of species, more specimens of each species, and very much larger specimens than either the IKMT or the RMT 8.

INTRODUCTION

The development of the Isaacs-Kidd midwater trawl (IKMT) in 1953 (Isaacs & Kidd, 1953) initiated a new era in sampling and analysis of midwater macroplanktonic and nektonic organisms. Since that time several modifications of the IKMT have appeared (e.g., Aron, Raxter, Noel & Andrews, 1964) and other midwater trawls, both high-speed and low-speed varieties, have been constructed (e.g., Schärfe, 1960; McNeely, 1963; Clarke, 1969a; Baker, Clarke & Harris, 1973). Summaries on the development and use of the broad spectrum of midwater sampling gear are presented in Harrison (1967) and in Gehringer & Aron (1968).

A number of studies comparing catch characteristics of various kinds of midwater trawls have been conducted with analyses concentrated primarily on fishes, crustaceans, and plankton (e.g., Aron, 1962; King & Iverson, 1962; Berry & Perkins, 1966; McGowan & Fraundorf, 1966; Badcock, 1970; Friedl, 1971).

No specific comparative study of cephalopods from different midwater trawls has been undertaken. Those reports in which comparisons have been attempted are concerned primarily with the vertical distribution of pelagic cephalopods. In an initial study, Clarke (1969b) reported catches of cephalopods taken during the SOND cruise in a 1 m ring net (N113H) and a 3 m IKMT, each equipped with a closing device in the form of a catch-dividing bucket (Foxton, 1963, 1969). Catches were recorded only as numbers of individuals captured in each net-tow; that is, they were not standardized for fishing effort (e.g., numbers per hour of trawling). The N113H is a plankton net, and therefore caught larvae primarily, while the 3 m IKMT, designed to sample macroplankton and nekton, caught juveniles primarily. Clarke (1969b: 976) concluded that a thorough sampling of the cephalopod fauna was "very inadequate" using either net. Because of inherent limitations of the catch-dividing bucket system on the 3 m IKMT, principally contamination (Foxton, 1969; Clarke, 1969a and pers. comm.), sampling with this device was terminated, and the rectangular midwater trawl (RMT) with a mouth-closing device was developed (Clarke, 1969a; Baker *et al.*, 1973).

Clarke & Lu (1974) described the vertical distribution of cephalopods at 30°N 23°W utilizing specimens captured with the RMT, the IKMT with catch-dividing bucket, open IKMT, and open British Columbia midwater trawl. The number of specimens captured for each species was recorded, but again captures were not standardized for fishing effort, although tows ranged from less than half an hour to over four hours in duration. The RMT tows were taken during a one-week period in March–April, 1972, while all remaining comparative tows were taken during August, September and October, 1961, and April and June of 1962. Although differences in catch between the similarly sized IKMT and RMT were thought to be due to net selection (Clarke & Lu, 1974: 983), the wide variability in years between samples, the different seasons, and the absence of standardization for fishing effort, make meaningful comparisons difficult.

The vertical distribution of pelagic cephalopods was reviewed by Roper & Young (1975) based primarily on midwater trawling programs using a non-closing 3 m IKMT, the Ocean Acre 3 m IKMT with closing device and a mouth-closing "Tucker trawl", similar to the RMT 8. Captures were analysed by catch rate, but no direct comparisons between trawls were attempted.

The studies upon which this paper is based were also designed to determine vertical distributions of nektonic forms. As several different types of gear were employed, however, an opportunity arose to determine fishing characteristics in a comparative manner with regard to cephalopod sampling. Since each design of midwater trawl samples the fauna differently, the ecological information derived is biased by net type. The relative sophistication of the IKMT and RMT systems and methods gave the opportunity to make valid comparisons and so attempt initial evaluation of the sampling effectiveness of the nets involved. Thus, a comparison of the cephalopods captured in a 3 m IKMT, and RMT 8, and a small (1400 mesh) Engel trawl (EMT) is presented here.

GEAR AND METHODS

The results of this study are particularly important because the IKMT and RMT *in situ* monitoring systems permitted precise depth placement and control of the nets and this resulted in refined sampling strategies that could be strictly adhered to, consequently allowing a reasonable degree of comparability.

Despite this sophistication of the gear and methods, it is recognized that certain limitations may be imposed by the complex environment that prohibit complete realization of the ideal program. For example, sampling with different nets could not be simultaneous, environmental conditions may not be stable during a given season from year to year, filtration rates may not be consistent because of differences in ships, weather conditions or current configurations. Nevertheless, these programs provide the most comprehensive and comparative data available on midwater cephalopod populations to date and should form a stimulus for further analyses.

The 3 m Isaacs-Kidd midwater trawl

The sampling program of which these studies formed a part, took place in a one-degree square area east of Bermuda, centered at 32°N 64°W, known as the Ocean Acre. The Ocean Acre program (1967–1972) utilized a 3 m IKMT equipped with the discrete-depth plankton sampler, a closing device at the cod end of the net (Aron *et al.*, 1964). Details of the methods and equipment of the Ocean Acre program are presented in Gibbs & Roper (1970) and in Gibbs, Roper, Brown & Goodyear (1971).

Briefly, the IKMT cod-end closing device consists of a 15 cm diameter, four chambered cylinder (Aron *et al.*, 1964). This trawling system allows the collection of three sequential samples from one depth; the fourth chamber contains the sample captured during oblique retrieval of the net from the fishing depth to the surface. The gates separating the chambers are closed by a solenoid-actuated triggering mechanism upon receipt of a frequency-coded signal via the conductor towing cable. During the later stages of the 14-cruise Ocean Acre program a system of simultaneous, *in situ* monitoring of depth of net, ambient temperature, light intensity, and flow of water through the net was employed.

The mesh size (bar measure) of the 3 m IKMT liner was 6·0 mm throughout with a 3 m-long 0·75 mm cod-end net. The area of the opening of the mouth during the fishing procedure was 7·44 m^2. Brooks, Brown & Scully-Power (1974) determined experimentally that the 3 m IKMT, apparently not equipped with a cod-end device, has a filtering efficiency of 92%. Possibly, the addition of the cod-end sampler would slightly reduce efficiency.

The major limitation of the IKMT with the cod-end closing device is that contamination of chambers can occur when specimens caught in the mesh of the net during sampling with one chamber are later washed into a subsequent chamber. The design of the sampling program requires that all three chambers fish sequentially for one hour at one depth, which greatly reduces the problem of contamination. The net essentially does not fish while being set with all chambers open (Aron *et al.*, 1964) and specimens that may enter chambers from a previous tow are identifiable by their poor condition and can be eliminated from analysis.

The rectangular midwater trawl 8

The National Institute of Oceanography (NIO), now the Institute of Oceanographic Sciences (IOS), in Great Britain had carried out studies of midwater organisms in the eastern Atlantic for several years using the mouth-closing RMT nets from RRS *Discovery* (Foxton, 1969; Baker *et al.*, 1973).

The sampling program conducted in the eastern North Atlantic by IOS (Currie, Boden & Kampa, 1969) has, since 1968, utilized primarily the rectangular midwater trawls described by Baker *et al.* (1973). The mesh size (bar measure) of the RMT 8 was 4·5 mm throughout its length with a 1·5 m section of 0·75 mm mesh ahead of the cod-end bucket. The area of the mouth opening in fishing configuration was 8·0 m^2. A 1·0 m^2 RMT was rigged on the frame above the RMT 8, but catches from this net are not analysed here. The RMT 8 closing is achieved at the mouth of the net, a design intended to eliminate contamination during set and retrieval. One sample per tow is taken at each depth, rather than three sequential samples. The monitoring system acoustically telemeters information on depth of net, opening-closing events, flow (relative velocity and distance travelled), and temperature.

The opportunity to compare the Ocean Acre program and the IOS sampling techniques and nets arose in 1973 when a sampling survey was conducted in the Ocean Acre area from RRS *Discovery*.

The Engel midwater trawl

The Engel trawl fished during the Ocean Acre program is a very large midwater trawl originally developed for the commercial herring fishery (Schärfe, 1960). It is fished entirely as an open net. The 1400 mesh model of the Engel trawl is somewhat smaller than the standard 1600 mesh EMT net, but its manner of operation is the same as described by Schärfe (1964). The mesh size (bar measure) of the EMT is 101 mm in the wings; it tapers to 38 mm in the cod end. The last 15·2 m of the cod end is lined with a 12 mm-mesh bag. A precise measure of the mouth opening of the EMT is difficult to obtain because the behaviour of these large midwater trawls under tow is not fully understood. Mouth opening can be altered by variations in ship speed, type of doors, diameter of warps and so on. Measurements of 11 m of vertical mouth opening were made on the 1400 mesh EMT during trials prior to the Ocean Acre cruise. It was not possible to take measurements of horizontal spread, but it was estimated to be about twice that of the vertical opening, or 22 m, giving a cross-sectional area of the mouth of 242 m^2 (K. Smith, pers. comm.). The fishing depth was determined by wire angle and recorded by time-depth recorder.

Table I lists the specifications of the three nets.

TABLE I

Mesh size (bar measure) and area of mouth of midwater trawls

Gear	Mesh size		Mouth
	Main body	Cod end	opening (m^2)
3 m IKMT	6·0 mm	3·0 m of 0·75 mm	7·44
RMT 8	4·5 mm	1·5 m of 0·75 mm	8·0
1400 mesh EMT	101 tapered to 38 mm	15·2 m of 12·0 mm	242·0

Sampling procedure

Initially, biologists of the IOS and the Smithsonian Institution planned to conduct a two-ship cruise during the spring of 1973 in order to take simultaneous tows with the IKMT and RMT at identical depths (strata). This type of comparative data would have eliminated seasonal or temporal variability. Unfortunately, the second ship was unavailable so the IKMT could not be used.

The RMT 8 data were accumulated during 13th to 26th March 1973 aboard *Discovery*. The IKMT data were taken during Ocean Acre Cruise 13, 23rd February to 3rd March 1972. So, while the dates of the two collections do not coincide, they do derive from the same season, early spring. Ocean Acre Cruise 6 also occurred in the spring, 25th to 30th April 1969, but fishing effort was only about one-third that of Ocean Acre 13. Combining data from Ocean Acre 6 and 13 will partially reduce the effects of annual fluctuations in populations.

The comparative data for the IKMT versus the EMT were accumulated during consecutive legs of Ocean Acre Cruise 12 from 20th August to 8th September 1971.

Each sampling program required a slightly different technique of trawling. The Ocean Acre sampling strategy called for closing-net samples to be taken at discrete depths (14 depths between 0 and 1500 m) and the range at each depth seldom varied more than 10 m (Gibbs & Roper, 1970; Gibbs *et al.*, 1971). The trawl was set to the desired depth in a non-catching mode, then three sequential one-hour samples in separate chambers were secured from that depth. Sampling speed was 2·5 to 3 knots.

The IOS sampling scheme required the closing-net sampling of 16 depth strata, of varying thickness depending on depth, between 0 and 2000 m. Sampling time was two hours in the 10–1000 m strata and four hours in the 1000–2000 m strata. The RMT was set in a closed mode into the desired depth stratum (e.g. 100 m thick between 100 and 1000 m), opened and fished partly horizontally, partly obliquely for two hours at 2 knots, then closed and retrieved (Baker *et al.*, 1973). During the *Discovery* cruise to the Ocean Acre area the regular IOS sampling strategy was applied, after which a series of tows was made with the RMT 8 fished at discrete depths following the Ocean Acre strategy. We had hoped to make a comparison of the different techniques of sampling, but, unfortunately, foul weather terminated the program early.

The EMT was fished at most of the depths established for the Ocean Acre program (25–1000 m), and fishing time at any given depth varied from half an hour at shallowest depths to two hours at greatest depths. Fishing speed was 1·5 to 2 knots.

Because of the variation in duration of sampling between the types of nets, and because the same nets were fished for varying lengths of time, all captures of cephalopods have been standardized on a catch per effort basis to numbers of specimens captured per hour of trawling. This, at least, makes possible a more meaningful comparison of catches of the same net and between different nets.

RESULTS

All individuals taken by each net (except the EMT) were analysed separately as night-time or day-time captures. Captures made during crepuscular periods, one hour before and one hour after sunrise and sunset, were excluded from the analysis. To facilitate inter-net sample comparison, the water column was divided into contiguous sampling strata similar to those of the IOS sampling procedures, and discrete samples taken by the IKMT and RMT from within any particular stratum were compared.

The following comparisons were made of captures by the 3 m IKMT versus the RMT 8 and the 3 m IKMT versus the EMT 1400: (1) total numbers of species, (2) composition of catch by species, (3) co-occurrence of species, (4) catch rate, (5) size range of specimens by species.

Isaacs-Kidd versus rectangular midwater trawl

Tables II and III present the number of species captured at each depth horizon for day and night, and the average number of specimens captured per hour of trawling at each depth.

Day (Table II). The RMT captured specimens in all 13 depth strata while the IKMT captured specimens in eight of the 12 strata in which it was fished during the day. In ten of the 12 strata co-sampled during the day, the RMT captured a greater number of species than the IKMT. Within the IKMT group the maximum numbers of species were caught in the 51–100 m, 301–400 m, and 401–500 m strata; below 500 m, numbers of species fell off sharply. The numbers of species caught by the RMT, however, remained relatively high throughout the water column, with a maximum of 14 species captured in the 101–200 m stratum. The total number of species captured by both nets shows a maximum number of species (13–14) in the day-time in the 51–200 m strata; below 200 m numbers are nearly constant (6-8 species) with reduced numbers at

TABLE II

IKMT and RMT captures by species and catch rate during the day

Depth (m)	IKMT					RMT					No. of species in both nets
	OA 6		OA 13		IKMT Total	A-A		A		RMT Total	
	No. spp.	Av. no./hr	No. spp.	Av. no./hr	no. spp.	No. spp.	Av. no./hr	No. spp.	Av. no./hr	no. spp.	
12–25	—	—	—	—	—	—	—	6	0·7	6	6
26–50	—	—	2	15·2	2	—	—	4	9·7	4	4
51–100	9	2·2	2	3·2	9	—	—	6	6·0	6	13
101–200	—	—	0	0	0	—	—	14	1·25	14	14
201–300	0	0	—	—	0	—	—	4	0·6	4	4
301–400	3	1·7	1	2·0	4	—	—	5	1·3	5	8
401–500	0	0	5	3·4	5	—	—	2	4·0	2	6
501–600	0	0	—	—	0	1	0·5	2	9·5	3	3
601–700	—	—	3	1·8	3	—	—	5	1·9	5	7
701–800	0	0	2	0·8	2	—	—	5	0·6	5	6
801–900	—	—	1	1·0	1	5	0·7	2	0·5	6	6
901–1000	0	0	—	—	0	5	0·7	2	1·0	6	6
1001–1250	—	—	2	1·0	2	4	0·3	5	0·5	7	7
(lumped) 0–100	9	2·2	3	9·1		—	—	10	4·6		13

Number of species, average number of specimens/hr of trawling, total number of species/net, grand total number of species for both nets.
— Indicates no tow made at that depth.
0 Indicates no catch in the tow at that depth.
OA 6, OA 13: Ocean Acre Cruise 6 and 13 respectively.
A-A: RMT tows made under Ocean Acre sampling regime.
A: RMT tows made under IOS sampling regime.

201–300 m and 501–600 m (Fig. 1). During the day the IKMT caught a total of 18 different species and the RMT caught 37 species, with seven species common to both nets.

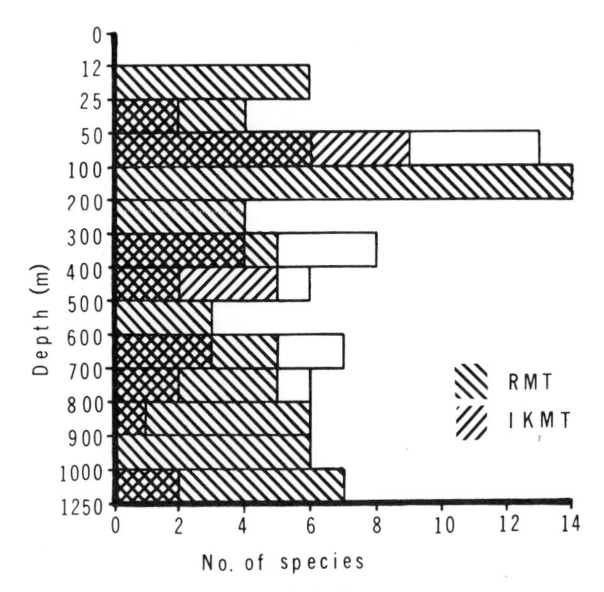

FIG. 1. Number of species captured by IKMT, RMT, and both trawls combined during the day. Total numbers of species represented by open area.

The catch rate of the IKMT, in terms of the average number of specimens per species per hour, exceeded that of the RMT in four of the eight strata in which both trawls caught specimens; catch rates were nearly equal in two horizons, and those of the RMT exceeded the IKMT at two horizons.

Night (Table III). The IKMT captured specimens in ten of the 11 strata in which it fished and the RMT caught specimens in all the 13 depth strata. The nets co-sampled in 11 strata; the RMT captured more species in seven of the strata, the IKMT caught more in two strata, and catches were equal in two strata. The IKMT data showed species diversity maxima, 12 and 16, in the 51–100 m and the 101–200 m horizons, respectively; between 201 and 600 m a reduced but fairly constant diversity of four to six species occurred. In the upper 300 m the RMT captured between eight and ten species, except at 101–200 m where a maximum catch of 16 species was recorded. Species decreased irregularly below 301 m. The total number of species captured by both nets was high (12–21) in

TABLE III

IKMT and RMT captures by species and catch rate during the night

	IKMT					RMT					
	OA 6		OA 13			A-A		A			
Depth (m)	No. spp.	Av. no./hr	No. spp.	Av. no./hr	IKMT Total no. spp.	No. spp.	Av. no./hr	No. spp.	Av. no./hr	RMT Total no. spp.	No. of species in both nets
10–25	—	—	7	16·8	7	—	—	8	11·0	8	12
26–50	4	1·4	4	9·1	8	8	2·6	5	13·3	10	15
51–100	—	—	12	2·13	12	5	6·2	7	12·1	8	13
101–200	11	1·6	12	1·0	16	11	1·7	12	2·8	16	21
201–300	3	1·3	5	2·4	6	—	—	9	1·2	9	12
301–400	—	—	5	2·3	5	—	—	7	1·1	7	10
401–500	—	—	4	1·0	4	4	1·2	3	0·7	6	9
501–600	—	—	5	1·1	5	—	—	4	0·7	4	8
601–700	—	—	—	—	—	—	—	2	2·0	2	2
701–800	—	—	2	1·3	2	0	0	5	0·7	5	5
801–900	—	—	—	—	—	—	—	1	0·5	1	1
901–1000	—	—	1	1·0	1	—	—	1	0·5	1	2
1000–1250	0	0	0	0	0	1	0·3	2	1·1	3	3
(lumped) 0–100	—	—	15	8·1		11	4·0	11	12·48		22

Number of species, average number of specimens/hr of trawling, total number of species/net, grand total number of species for both nets.
— Indicates no tow made at that depth.
0 Indicates no catch in the tow at that depth.
OA 6, OA 13: Ocean Acre Cruise 6 and 13 respectively.
A-A: RMT tows made under Ocean Acre sampling regime.
A: RMT tows made under IOS sampling regime.

the upper 300 m, peaking at 21 species in the 101–200 m stratum. Below 301 m total species diversity decreased (Fig. 2). At night the IKMT captured 30 different species while the RMT caught 28 species with 15 species captured in common.

The catch rate of the IKMT exceeded that of the RMT in seven of the 11 co-sampled strata at night; the RMT had greater catch rates at four strata.

Figures 1 and 2 also present a vivid demonstration of the diel vertical migrations which many species undertake.

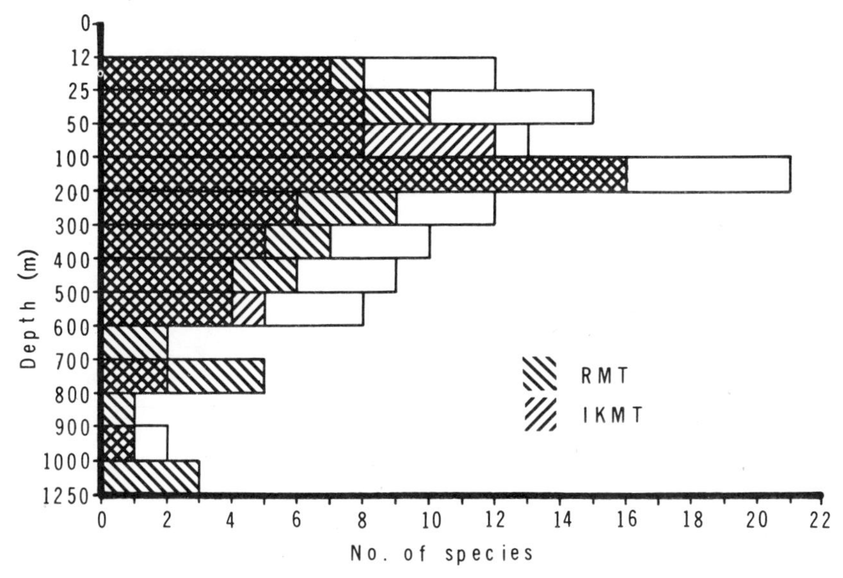

FIG. 2. Number of species captures by IKMT, RMT, and both trawls combined during the night. Total numbers of species represented by open area.

A comparison was made of the species composition found in the two nets. During the comparative cruises a total of 50 species was captured, and 22 species were common to both the IKMT and the RMT. Table IV lists the co-occurring species with number of specimens, size range and mean for each net, and the per cent difference of number of specimens and average size between the captures of the IKMT and the RMT. Size is recorded in mm as mantle length (ML). Number of specimens is listed to provide an indication of the sample size for the calculation of mean mantle lengths; it is not indicative of the catch rates of the nets because effort is not taken into account.

TABLE IV

Co-occurring species in IKMT and RMT

Species	IKMT		RMT		% Difference	
	No. of specimens	ML	No. of specimens	ML	No.	ML
1. *Selenoteuthis scintillans*	4	10-*13*-17	3	12-*15*-17	−25	+15
2. *Abraliopsis pfefferi*	5	6-*14*-22	21	4-*12*-34	+320	−14
3. *Pyroteuthis margaritifera*	66	4-*15*-60	47	2-*6*-15	−29	−60
4. *Pterygioteuthis giardi*	15	5-*13*-19	45	3-*11*-18	+200	−15
5. *Octopoteuthis danae*	1	*17*	1	*16*	0	−6
6. *Octopoteuthis* sp.	12	7-*10*-13	3	2-*5*-7	−75	−50
7. *Onychoteuthis banksi*	4	13-*16*-20	14	2-*9*-18	+250	−44
8. *Histioteuthis dofleini*	1	*23*	10	4-*19*-115	+900	−17
9. *Histioteuthis corona*	1	*12*	2	9-*11*-13	+100	−8
10. *Ctenopteryx sicula*	12	4-*10*-16	15	2-*7*-21	+25	−30
11. *Brachioteuthis riisei*	13	10-*15*-27	15	5-*10*-20	+15	−33
12. *Ommastrephes* sp.	7	4-*14*-27	2	4-*5*-6	−71	−64
13. *Mastigoteuthis magna*	1	*23*	6	10-*20*-35	+900	−13
14. *Mastigoteuthis* sp.	2	21-*21·5*-22	16	4-*15*-45	+700	−30
15. *Leachia cyclura*	609	5-*23*-50	503	8-*25*-55	−17	+9
16. *Bathothauma lyromma*	10	5-*13*-21	8	4-*6*-10	−20	−54
17. *Helicocranchia* sp.	9	6-*14*-60	212	2-*10*-55	+2256	−29
18. *Egea inermis*	3	8-*13*-17	1	*46*	−67	+254
19. *Eledonella pygmaea*	10	5-*11*-26	32	3-*12*-21	+220	+9
20. *Vitreledonella richardi*	3	10-*10·6*-11	1	*11*	−67	+4
21. *Argonauta argo*	1	*9*	2	7-*8*-9	+100	−11
22. *Alloposus mollis*	2	5-*6·5*-8	1	*3*	−50	−54

Number of specimens, minimum-*mean*-maximum size by mantle length (ML) in mm, per cent difference in numbers and mantle length.

In the per cent difference columns, a positive value indicates that the RMT caught larger average size or greater numbers of specimens and a negative value indicates that the IKMT caught more or larger specimens. Occasionally one net provided the largest average size of a species, but the other net will have caught the largest specimen, for example, *Abraliopsis pfefferi*. The tendency exists for the IKMT to catch specimens of a slightly larger mean mantle length than the RMT catches. Large differences in mean size often occurred in species in which very few specimens were caught, for example, *Alloposus mollis*, or where there was a significant difference between the numbers of specimens caught, as in *Octopoteuthis* sp. In the case of *Egea inermis*, only one specimen was caught in the RMT, but it was much larger than any of the three caught in the IKMT. The difference in catches of *Ommastrephes* sp. may indicate an ability of the IKMT to catch faster swimming forms, as it caught more and significantly larger specimens than did the RMT. Such differences could be a reflection of the small sample size rather than of major differences between the nets.

Where sample size is large enough for both nets, specimens of the co-occurring species that occur in reasonable numbers generally do not exhibit a wide disparity in mean or maximum mantle length. Specimens of the well represented *Pyroteuthis margaritifera* caught in the IKMT, however, are notably larger in average mantle length and maximum mantle length than the specimens from the RMT.

An examination of the minimum size of specimens indicates that the RMT caught smaller individuals in 17 of the 22 species. In many species the difference was of the same magnitude as occurs for differences in maximum size. The difference in mesh size between the two nets is probably sufficient to account for this phenomenon, as the RMT has a mesh size of 4·5 mm while that of the IKMT is 6·0 mm. The mesh size of the cod-end liners of both nets is 0·75 mm.

Several techniques for measuring co-occurrence exist, one of which is the Index of Similarity, calculated from the formula:

$$S = \frac{2C}{A + B}$$

where A is the number of species in the IKMT (32), B is the number of species in the RMT (40), and C is the number of species common to both trawls (22). In this case $S = 0{\cdot}61$. If the 0·85 level is considered the limit of similarity, an index of similarity of 0·61 indicates that, so far as species composition is concerned, the two nets are relatively dissimilar. In order for the nets to be regarded as similar, at least 31 co-occurring species would be required. Another comparison that can be applied to measure the similarity of the two nets in relation to species composition is Jaccard's Coefficient of Community. This measure is expressed by the formula

$$cc = \frac{c}{a+b-c} \times 100$$

where a and b are the total number of species in the IKMT and RMT, respectively, and c is the number of co-occurring species. Therefore,

$$cc = \frac{22}{32+40-22} \times 100 = 44\%,$$

on a scale where 100% represents identical species representation and 0 corresponds to no relationship. To achieve a level of 85%, a reasonable limit of similarity, 33 species would have to co-occur in the two nets. The coefficient of community of 44% also indicates that the IKMT and the RMT are relatively dissimilar in species composition.

Of the 50 species captured during the study period, 22 were co-captured in both nets. Ten species were captured solely by the IKMT, and 18 species were captured only by the RMT (Table V). Most of these species were represented by low catch numbers; only five of these 28 species consisted of more than five specimens (6, 6, 8, 10, 48).

The catch rates and size ranges of co-occurring species within each depth stratum have been compared. During the day (Table VI), for example, *Leachia cyclura* was co-captured in the 26–50 m stratum at a nearly equivalent catch rate of 28·9 specimens per hour for the IKMT and 32·5 specimens per hour for the RMT. The size ranges and means also were similar. At 51–100 m, however, a significant difference occurs in that the RMT caught many more specimens per hour and of somewhat larger maximum size than the IKMT, although the mean size was about the same as that of the Ocean Acre 6 sample. *Pyroteuthis margaritifera* co-occurred at the 301–400 m and 401–500 m strata, where the RMT had a

TABLE V

Non-co-occurring species in IKMT and RMT

Species captured by IKMT but not RMT	No. of speci-mens	ML	Species captured by RMT but not IKMT	No. of speci-mens	ML
1. *Heteroteuthis dispar*	1	*19*	1. *Octopoteuthis sicula*	8	3-*8*-15
2. *Lampadioteuthis megaleia*	6	5-*11*-19	2. *Onykia caribaea*	4	2-*4*-6
3. *Abralia redfieldi*	2	15-*17*-19	3. *Discoteuthis laciniosa*	1	*13*
4. *Taningia danae*	2	8-*8*-8	4. *Lepidoteuthis grimaldi*	4	16-*23*-27
5. *Tetronychoteuthis dussumieri*	1	*21*	5. *Histioteuthis meleagroteuthis*	2	7-*7·5*-8
6. *Brachioteuthis* sp.	5	8-*9*-10	6. *Histioteuthis* sp.	3	4-*4*-5
7. *Taonius pavo*	10	13-*23*-47	7. *Neoteuthis* sp.	1	*11*
8. *Megalocranchia megalops*	1	*8*	8. *Bathyteuthis abyssicola*	6	2-*14*-40
9. *Ocythoe tuberculata*	1	*6*	9. *Brachioteuthis beani?*	1	*4*
10. Octopod	1	*7*	10. *Chiroteuthis veranyi*	3	40-*48*-63
			11. *Mastigoteuthis hjorti*	1	*20*
			12. *Grimalditeuthis bomplandi*	1	*55*
			13. *Joubiniteuthis portieri*	1	*5*
			14. *Galiteuthis* sp.	48	5-*19*-43
			15. *Egea inermis*	5	7-*12*-16
			16. *Tremoctopus violaceus*	2	6-*7·5*-9
			17. *Scaeurgus unicirrhus*	1	*11*
			18. *Vampyroteuthis infernalis*	4	10-*19*-30

Number of specimens, minimum-*mean*-maximum size by mantle length (ML) in mm.

TABLE VI

Co-occurring species in IKMT and RMT during the day by depth stratum

		No./hr				Size range			
Depth, (m)	Species	OA 6 IKMT	OA 13 IKMT	A-A RMT	A RMT	OA 6 IKMT	OA 13 IKMT	A-A RMT	A RMT
26-50	*B. riisei*	—	1·4	—	2·0	—	12-*12·5*-13	—	14-*17*-20
	L. cyclura	—	28·9	—	32·5	—	10-*21*-35	—	10-*22*-37
51–100	*L. cyclura*	6·3	5·4	—	42·5	25-*38*-47	11-*22*-37	—	19-*34*-55
	Octopoteuthis sp.	3·0	0	—	1·5	8-*9*-10	0	—	7-*10*-15
	Ommastrephes sp.	1·8	1·0	—	0	*27*	*21*	—	0
301–400	*P. margaritifera*	2·0	0	—	3·5	10-*11*-13	0	—	4-*5*-6
401–500	*P. margaritifera*	0	5·2	—	7·5	0	6-*15*-24	—	2-*6*-11
601–700	*L. cyclura*	—	3·0	—	4·0	—	12-*26*-34	—	31-*36*-45
701–800	*E. pygmaea*	0	0·8	—	0·5	0	7-*8*-9	—	*4*
801–900	*Mastigoteuthis* sp.	—	1·0	0	0·5	—	*22*	0	*20*
1000–1250	*E. pygmaea*	—	0·9	0·3	1·0	—	18-*22*-26	13-*14*-15	15-*16*-19
	Mastigoteuthis sp.	—	1·0	0	0·2	—	*21*	0	*45*

Total number of specimens captured/hr of trawling, size range of mantle length (ML) in mm, as minimum-*mean*-maximum.
— Indicates no tow made at that depth.
0 Indicates no catch in the tow at that depth.
OA 6, OA 13: Ocean Acre Cruise 6 and 13 respectively.
A-A: RMT tows made under Ocean Acre sampling regime.
A: RMT tows made under IOS sampling regime.

superior rate of capture. The differences in sizes were marked, however, in that the mean and maximum sizes captured by the IKMT were twice those of the RMT, and the minimum sizes captured by the RMT were considerably smaller than those of the IKMT. In general the RMT had a greater catch rate than the IKMT, especially at depths shallower than 700 m, but the differences for the most part were not large. No firm pattern of difference in size ranges occurs between nets, other than for *P. margaritifera*.

At night a larger number of species co-occur (Table VII) than during the day, especially in the 51–100 m and the 101–200 m strata. Also, the catch rates of both nets are markedly higher. *Pyroteuthis margaritifera*, caught in four of the upper five strata, again demonstrates a consistently larger size of specimens in the IKMT than the RMT, and in three of the four captures the IKMT caught a notably higher number of specimens per hour. Save for the 10–25 m stratum, the highest catch rates in the upper 200 m were recorded by the RMT (14 of 19 co-occurring species). Below 201 m seven of the nine species were caught at a greater number per hour by the IKMT than the RMT. The depth range below 201 m also corresponds to markedly reduced capture rates by both nets.

The general trend, as in other comparisons, is for the IKMT to catch a larger mean size of specimens per species than the RMT, but this trend is reversed in *Leachia cyclura*. *L. cyclura* was present in eight of the nine horizons in which co-occurring species were recorded, and in nearly every case its minimum, mean, and maximum sizes in the RMT exceeded those of specimens in the IKMT.

Isaacs-Kidd versus Engel trawl

The comparative tows of the IKMT versus the EMT were made consecutively on Ocean Acre Cruise 12 in the same depth strata over a three-week period during August–September, 1971. Since the EMT was fished as an open net no discrete depth comparisons were possible, and neither were there any day/night comparisons.

During Ocean Acre 12 a total of 54 species was captured, 20 of which were co-occurring species in the IKMT and the EMT (Table VIII). The EMT fished only about half the number of hours (55%) of the total at depth trawling time of the IKMT. So, while Table VIII does not show effort as specimens per hour *per se*, it should be noted that the IKMT catches do represent about twice the effort of the EMT. The extreme difference in catches both in terms of

TABLE VII

Co-occurring species in IKMT and RMT during the night by depth stratum

Depth (m)	Species	No./hr				Size range			
		OA 6 IKMT	OA 13 IKMT	A-A RMT	A RMT	OA 6 IKMT	OA 13 IKMT	A-A RMT	A RMT
10–25	*L. cyclura*	—	106·0	—	89·0	—	7-*15*-29	—	12-*21*-37
	P. margaritifera	—	5·0	—	1·0	—	10-*13*-20	—	*4*
	B. riisei	—	2·5	—	1·0	—	14-*16*-18	—	*5*
25–50	*P. giardi*	2·0	0	1·5	4·5	5-*7*-11	0	8-*10*-12	6-*10*-15
	B. riisei	0	1·0	3·0	0·5	0	*12*	9-*9·5*-10	*20*
	L. cyclura	0	33·2	10·5	59·5	0	5-*18*-30	10-*13*-15	12-*24*-40
51–100	*P. giardi*	—	1·6	2·5	11·5	—	8-*13*-17	8-*13*-18	3-*9*-15
	P. margaritifera	—	1·3	0·5	5·0	—	4-*18*-28	*3*	2-*4*-6
	L. cyclura	—	11·6	25·0	11·0	—	9-*17*-36	8-*19*-22	16-*21*-30
	Helicocranchia sp.	—	1·6	0	54·0	—	6-*9*-11	0	3-*9*-55
	A. pfefferi	—	1·0	2·5	2·5	—	*6*	15-*16*-18	7-*22*-34
	O. banksi	—	1·0	0	0·5	—	*13*	0	*5*
	C. sicula	—	1·0	0	0·5	—	*16*	0	*2*
101–200	*P. margaritifera*	7·0	1·0	1·0	3·5	8-*12*-20	14-*18*-26	10-*11*-12	2-*8*-15
	C. sicula	1·5	1·0	0	2·5	4-*6*-10	*10*	0	4-*8*-21
	Mastigoteuthis sp.	1·0	0·4	0·5	2·0	6-*15*-25	*23*	*10*	4-*7*-12
	L. cyclura	1·4	1·4	2·5	5·0	24-*29*-45	21-*27*-39	22-*25*-30	14-*29*-44
	Helicocranchia sp.	1·0	1·2	9·0	14·0	13-*17*-21	13-*20*-30	4-*11*-20	4-*10*-20
	Octopoteuthis sp.	1·0	1·0	1·5	0·5	*7*	*10*	*3-4-5*	*5*
	O. banksi	1·0	0·8	2·0	0·5	*14*	*20*	11-*15*-18	*5*
	B. riisei	1·0	0	0·5	0	*27*	0	*7*	0
	B. lyromma	0	1·3	1·0	1·0	0	5-*12*-22	6-*6·5*-7	4-*4*-4

201–300	*L. cyclura*	1·5	2·0	—	0	33-*42*-48	26-*32*-36	—	0
	V. richardi	1·0	1·0	—	0	*10*	11-*11*-11	—	0
	P. margarilifera	0	3·0	—	0·5	0	22-*39*-60	—	*12*
	E. pygmaea	0	3·0	—	0·5	0	6-*7*-8	—	*5*
301–400	*B. lyromma*	0	3·0	—	0·5	0	8-*11*-15	—	*10*
	H. dofleini	—	3·0	—	1·0	—	*23*	—	4-*5*-6
401–500	*L. cyclura*	—	5·2	—	4·0	—	17-*22*-32	—	4-*24*-40
	L. cyclura	—	1·0	3·0	0	—	8-*18*-25	19-*26*-50	0
501–600	*H. dofleini*	—	0	1·0	1·0	—	0	4-*8*-12	6-*6·5*-7
701-800	*L. cyclura*	—	1·5	—	1·0	—	15-*19*-23	—	30-*30*-30
	L. cyclura	—	1·6	0	0·5	—	14-*22*-35	0	*45*
	E. pygmaea	—	1·0	0	1·5	—	*9*	0	4-*6*-8

Total number of specimens captured/hr of trawling, size range of mantle length (ML) in mm, as minimum-*mean*-maximum.
— Indicates no tow made at that depth.
0 Indicates no catch in the tow at that depth.
OA 6, OA 13: Ocean Acre Cruise 6 and 13 respectively.
A-A: RMT tows made under Ocean Acre sampling regime.
A: RMT tows made under IOS sampling regime.

TABLE VIII

Co-occurring species in IKMT and EMT, Ocean Acre 12

Species	IKMT		EMT		% Difference	
	No. of specimens	ML	No. of specimens	ML	No.	ML
1. *Selenoteuthis scintillans*	5	8-*11*-14	23	11-*29*-45	360	164
2. *Abraliopsis pfefferi*	11	5-*18*-39	239	12-*25*-38	2073	39
3. *Abralia redfieldi*	1	*8*	158	18-*24*-38	15 700	200
4. *Thelidioteuthis alessandrinii*	5	6-*8*-10	7	18-*22*-31	40	175
5. *Pyroteuthis margaritifera*	44	4-*12*-41	199	10-*24*-49	352	100
6. *Pterygioteuthis giardi*	12	9-*13*-17	17	14-*16*-20	42	23
7. *Octopoteuthis danae*	1	*27*	3	34-*115*-159	200	326
8. *Onychoteuthis banksi*	25	5-*9*-21	64	12-*26*-45	156	189
9. *Discoteuthis laciniosa*	1	*42*	3	52-*82*-134	200	95
10. *Histioteuthis dofleini*	7	8-*41*-176	40	8-*41*-125	471	0
11. *Brachioteuthis riisei*	2	13-*16*-18	4	40-*54*-68	100	237
12. *Mastigoteuthis magna*	6	20-*29*-50	14	27-*77*-151	133	166
13. *Taonius pavo*	2	39-*64*-88	33	55-*100*-231	1550	56
14. *Bathothauma lyromma*	5	6-*35*-95	29	12-*74*-130	480	111
15. *Helicocranchia pfefferi*	52	6-*13*-34	39	13-*27*-49	−25	108
16. *Egea inermis*	43	5-*24*-43	71	12-*33*-425	65	37
17. *Eledonella pygmaea*	4	13-*20*-32	102	13-*28*-43	2450	40
18. *Vitreledonella richardi*	2	6-*11*-15	1	*57*	−100	418
19. *Tremoctopus violaceus*	3	6-*7*-8	1	*12*	−200	71
20. *Argonauta argo*	1	*7*	1	*10*	—	43

Number of specimens, minimum-*mean*-maximum size by mantle length (ML) in mm, percentage difference in numbers and ML.

numbers of specimens captured per hour per unit area of mouth-opening and of size ranges is readily apparent. The difference in numbers captured, however, is not so great in the smaller species, such as *Pterygioteuthis giardi*, and with three small and/or "rare" species the IKMT caught more specimens than the EMT, for example, *Vitreledonella richardi*. The EMT caught considerably larger specimens in 19 of the 20 species, the only exception being *Histioteuthis dofleini*. Otherwise specimens ranged from slightly larger, for example, *Pterygioteuthis giardi*, to several times larger, as in *Octopoteuthis danae*, and up to nearly ten times larger in maximum size, as in *Egea inermis*.

An index of similarity of 0·54 and a Jaccard's coefficient of community value of 37% for co-occurring species between the IKMT and the EMT indicate a predictable dissimilarity.

The EMT captured 25 species that were not captured by the IKMT, while the IKMT caught nine species not taken by the EMT (Table IX). The species caught in the IKMT generally were represented by quite small specimens, and only two of the nine species were represented by more than two specimens. Most species taken by the EMT, on the other hand, were large and some of them, for example, *Ommastrephes bartrami* and *Todarodes sagittatus*, represent the largest specimens ever taken in the Ocean Acre program. Numbers of specimens range from one to nearly 100. *Hyaloteuthis pelagica* is a very rarely caught squid, represented by only a few records in the literature. The EMT captured 96 specimens which ranged from 9 to 93 mm in mantle length. These captures exceed the numbers (fewer than 20) and maximum size (71 mm mantle length) of all previously recorded specimens (Clarke, 1966).

DISCUSSION

A comparison of captures of cephalopods was conducted using three different midwater trawls, the 3 m IKMT, the RMT 8, both closing nets, and the 1400 mesh EMT, a non-closing net. Although the data are somewhat limited, they do indicate that the three nets sample cephalopods differently. As a result they depict different aspects of ecological communities, because of differences in design, techniques of fishing, and so on.

In the IKMT versus RMT comparison, 50 species were captured in all, 40 in the RMT and 32 in the IKMT; 22 species co-occurred. Analysis of the species composition of both nets based on co-occurrence of species indicates that the captures of the two nets are not highly similar in species content. The index of

TABLE IX

Non-co-occurring species in IKMT and EMT, Ocean Acre 12

Species captured by EMT but not by IKMT	No. of specimens	ML	Species captured by IKMT but not by EMT	No. of specimens	ML
1. *Enoploteuthis leptura*	2	26-*44*-63	1. *Spirula spirula*	1	*18*
2. *Enoploteuthis anapsis*	27	12-*48*-93	2. *Octopoteuthis* sp.	2	8-*8·5*-9
3. *Octopoteuthis sicula*	8	33-*115*-167	3. *Bathyteuthis abyssicola*	1	*12*
4. *Taningia danae*	1	*26*	4. *Ommastrephes* sp.	4	5-*6*-8
5. *Onykia caribaea*	1	*9*	5. *Grimalditeuthis bomplandi*	2	45-*53*-61
6. *Cycloteuthis sirventi*	1	*134*	6. *Liocranchia reinhardti*	5	7-*8*-12
7. *Tetronychoteuthis dussumieri*	2	44-*72*-100	7. *Leachia* sp.	1	*55*
8. *Histioteuthis meleagroteuthis*	29	27-*50*-109	8. *Megalocranchia megalops*	2	8-*9·5*-11
9. *Histioteuthis corona*	7	22-*42*-56	9. *Thysanoteuthis rhombus*	1	*4*
10. *Neoteuthis* sp.	2	20-*30*-40			
11. *Ctenopteryx sicula*	31	12-*26*-45			
12. *Ommastrephes bartrami*	3	310-*380*-518			
13. *Ornithoteuthis antillarum*	2	188-*192*-196			
14. *Hyaloteuthis pelagica*	96	9-*49*-93			
15. *Todarodes sagittatus*	1	*341*			
16. *Chiroteuthis veranyi*	3	75-*79*-81			
17. *Chiroteuthis* sp. A.	3	107-*116*-133			
18. *Chiroteuthis* sp. C.	5	39-*59*-83			
19. *Chiroteuthis* sp. B.	1	*84*			
20. *Mastigoteuthis hjorti*	12	27-*66*-181			
21. *Mastigoteuthis grimaldi*	2	?			
22. *Cranchia scabra*	1	*30*			
23. *Eledonella* sp. A.	5	33-*36*-44			
24. *Vampyroteuthis infernalis*	1	*28*			
25. Octopod sp. A.	7	13-*14*-15			

Number of specimens, minimum-*mean*-maximum size by mantle length (ML) in mm.

similarity = 0·61 and the coefficient of community = 44%. Caution should be applied in interpreting these results, however. The co-occurring species are the most commonly caught species, and conversely, the non-co-occurring species tend to be caught much more rarely. Insufficient sampling time may be a factor. Also, since the IKMT/RMT comparative tows were taken one year or more apart, even though they occurred during the same season, some differences may be expected due to annual fluctuations of occurrence and abundance of species. In the total Ocean Acre program the IKMT has caught all the species that are recorded only from the RMT in this study.

The IKMT captured 821 specimens in 112·6 hr of trawling for a catch per effort measure of 7·29 specimens/hr. The RMT captured 1057 specimens in 83·0 hr of trawling for a catch rate of 12·73 specimens/hr. In order to determine the relationships of the two catch rates, the catch per unit area of the mouth opening of each net has been calculated. The mouth area of the 3 m IKMT is 7·44 m^2, so the catch per area is:

$$\frac{7{\cdot}29 \text{ specimens/hr}}{7{\cdot}44 \text{ m}^2} = 0{\cdot}89 \text{ specimens/hr/m}^2.$$

The area of the mouth of the RMT 8 is 8 m^2 divided into the catch rate of 12·73 specimens/hr yields a value of 1·59 specimens/hr/m^2. In other words, the differences in the catch rates are not due to the differences in the area of the mouth opening of the nets. The RMT does catch more specimens per unit effort than the IKMT. Several explanations are possible. For example, the smaller mesh size of the RMT may be more efficient at catching larger numbers of smaller specimens, or the escape rate may be higher because of the bridle arrangement on the IKMT.

The IKMT and RMT differ somewhat in the size of specimens captured. The IKMT tends to catch animals of slightly larger mean mantle length, while the RMT catches markedly smaller minimum length specimens within species. These differences possibly are attributable to the smaller mesh size of the RMT net, 4·5 mm versus 6·0 mm.

The RMT caught a higher number of species during the day than the IKMT, but at night the IKMT caught slightly more than the RMT. No pattern of catch rates by day or night was evident between the two nets.

Finally, individual species differences seem to occur between nets. Specimens of *Pyroteuthis margaritifera*, for instance, were consistently larger and more frequently caught in the IKMT than in the RMT. *Leachia cyclura* was consistently larger in minimum, mean, and maximum size in the RMT than in the IKMT, a specific reversal of the general trend.

Studies of the IKMT versus the EMT had the advantage that the comparative tows were taken sequentially on the same cruise, but are limited because the EMT could be fished only as an open net. A total of 54 species was captured during the comparative tows, and 20 species co-occurred in both nets. The IKMT caught a total of 29 species, while the EMT captured 45 species. Little doubt exists as to the ability of the EMT to make superior captures in terms of numbers of species, numbers of specimens, and size of specimens.

The index of similarity of 0·54 and the coefficient of community of 37% indicate the relatively dissimilar catches in terms of species composition. The cautions mentioned in the discussion of the IKMT and RMT comparison may apply here, as well, although certainly to a lesser degree, because the EMT did catch some species that had not previously been recorded from the Ocean Acre program.

The EMT caught 1306 specimens in 68·83 hr of trawling for a catch rate of 18·9 specimens/hr, while the IKMT caught 250 specimens in 122·38 hr, or 2·0 specimens/hr. The value of catch rate by mouth area for the EMT is 0·78 specimens/hr/m^2 and that of the IKMT is 0·27 specimens/hr/m^2. Size range differences within species varied from specimens of about equal length, up to ten times larger in the EMT than in the IKMT.

As would be expected, a greater dissimilarity exists between the IKMT and the EMT than between the IKMT and the RMT in terms of species composition, numbers, and sizes of specimens and catch rates.

An interesting comparison is noted between the results of this study and the work of Clarke & Lu (1974) who reported on the vertical distribution of cephalopods in the eastern Atlantic at 30°N 23°W. Information extracted from their data indicates that 28 species were captured in the IKMT and 27 species were caught in the RMT for a total of 38 species, with 17 co-occurring species. These figures were tested for similarity with the following results:

Index of Similarity = 0·62
Coefficient of Community = 45%.

The measurements of similarity are nearly identical with those of the current study, i.e. 0·61 and 44%, indicating that a certain degree of predictability is justified when these two types of nets are fished in the same area.

The IKMT and the RMT used in the study reported by Clarke & Lu (1974) at 30°N 23°W caught a combined total of 618 specimens in 160 hr of trawling for an average catch rate of 3·86 specimens/hr. The same types of nets in the Ocean Acre area off Bermuda (32°N 64°W) captured a combined total of 1878 specimens in 195 hr of sampling for an average catch rate of 9·63 specimens/hr. Based on the two IKMT/RMT studies, a greater number of species occurs in the Bermuda area than in the eastern Atlantic at a similar latitude (50 to 38°), and a greater catch rate is recorded. These data indicate a greater species diversity and a greater relative abundance of cephalopods in the waters east of Bermuda.

Certain limitations occur that may affect the results of this comparative study of midwater trawls as many variables exist in the sampling of nektonic organisms. The attempts at reducing the variables in this study represent a reasonable beginning but fall short of the most desirable conditions. Ideally, we would like to have simultaneous tows of two types of gear from two vessels running parallel courses with nets fishing at identical depths and straining identical quantities of water. This is seldom, if ever, possible. Even then problems of the patchiness in the distributions of midwater organisms must be taken into account. A further limitation to thoroughly assessing the community structure of oceanic cephalopods is that the systematics and life histories of many families are inadequately known. Sufficiently thorough sampling often is a limitation. Both the IOS and the Ocean Acre studies were designed to overcome this problem. The sampling strategies of both the Ocean Acre program and the IOS require 13 tows each for day-time and night-time sampling at 0–1250 m, which should be adequate for the most abundantly captured species, at least. The comparison of samples from the same season or from sequential tows and the standardization of closing-net captures for sampling effort help to reduce the aforementioned limitations.

We know from the experience of this work and from previous work, including studies of sperm whale stomach contents (see Clarke, his chapter, this volume), that we are still not sampling the total pelagic cephalopod fauna. Any net used exclusively gives only a truncated view of the populations of cephalopods. Several different nets are required for a complete assessment.

The present comparative study of midwater trawl captures represents the first attempt at a quantitative analysis of the midwater cephalopod fauna based on standardized capture data. Such comparisons are necessary if we are to interpret the results of various midwater trawl studies, and ultimately, if we are to understand the communities of pelagic cephalopods.

Acknowledgements

I wish to express my sincere gratitude to the following individuals and institutions who have contributed to this study. Support of the Ocean Acre program has come largely from the US Navy Underwater Systems Center through C. L. Brown (Contract No. N00140-73-C-6304). R. H. Gibbs Jr., Smithsonian Institution, and all the Ocean Acre personnel aided in collecting specimens and data. The Institute of Oceanographic Sciences UK, through the kindness of P. M. David, invited me to participate in the RRS *Discovery* cruise to the Ocean Acre area in 1973. Mr David, the biologists and crew were most generous and helpful. Keith Smith, National Marine Fisheries Service, provided information concerning the specifications of the Engel trawl. G. M. Cailliet, Moss Landing Marine Laboratories, suggested the statistical tests for similarity. Discussions with C. C. Lu, Memorial University, concerning presentation of data and the text were most helpful. R. E. Young, University of Hawaii, made helpful comments. C. Lamb, National Museum of Natural History, aided measurably in the final preparation of the tables and editing; and I am especially grateful to M. J. Sweeney, of the same museum, for processing the large amount of data, preparing it for analysis and presentation, together with the figures and tables, and for many helpful discussions throughout the work.

The manuscript was read and valuable comments given by C. C. Lu, R. H. Gibbs Jr., J. Badcock and R. Aldred.

References

Aron, W. (1962). Some aspects of sampling the macroplankton. *Rapp. P.-v. Réun. Cons. perm. int. Explor. Mer* **153**: 29–38.

Aron, W., Raxter, N., Noel, R. & Andrews, W. (1964). A description of a discrete depth plankton sampler with some notes on the towing behavior of a 6-foot Isaacs-Kidd mid-water trawl and a one-meter ring net. *Limnol. Oceanogr.* **9**: 324–333.

Badcock, J. (1970). The vertical distribution of mesopelagic fishes collected on the SOND cruise. *J. mar. biol. Ass. U.K.* **50**: 1001–1044.

Baker, A. de C., Clarke, M. R. & Harris, M. J. (1973). The N.I.O. combination net (RMT 1+8) and further developments of rectangular mid-water trawls. *J. mar. biol. Ass. U.K.* **53**: 167–184.

Berry, F. H. & Perkins, H. C. (1966). Survey of pelagic fishes of the California Current area. *Fishery Bull. Fish Wildl. Serv. U.S.* **65**: 625–682.

Brooks, A. L., Brown, C. L. & Scully-Power, P. H. (1974). Net filtering efficiency of a 3-meter Isaacs-Kidd Mid-water Trawl. *Fishery Bull. Fish Wildl. Serv. U.S.* **72**: 618–621.

Clarke, M. R. (1966). A review of the systematics and ecology of oceanic squids. *Adv. mar. Biol.* **4**: 91–300.

Clarke, M. R. (1969a). A new midwater trawl for sampling discrete depth horizons. *J. mar. biol. Ass. U.K.* **49**: 945–960.

Clarke, M. R. (1969b). Cephalopoda collected on the SOND cruise. *J. mar. biol. Ass. U.K.* **49**: 961–976.

Clarke, M. R. & Lu, C. C. (1974). Vertical distribution of cephalopods at 30°N 23°W in the North Atlantic. *J. mar. biol. Ass. U.K.* **54**: 969–984.

Currie, R. I., Boden, B. P. & Kampa, E. M. (1969). An investigation on sonic-scattering layers: the R.S.S. *Discovery* SOND Cruise, 1965. *J. mar. biol. Ass. U.K.* **49**: 489–514.

Foxton, P. (1963). An automatic opening-closing device for large midwater plankton nets and midwater trawls. *J. mar. biol. Ass. U.K.* **45**: 295–308.

Foxton, P. (1969). SOND Cruise 1965: Biological sampling methods and procedures. *J. mar. biol. Ass. U.K.* **49**: 603–620.

Friedl, W. A. (1971). The relative sampling performance of 6- and 10-foot Isaacs-Kidd midwater trawls. *Fishery Bull. Fish Wildl. Serv. U.S.* **69**: 427–432.

Gehringer, J. W. & Aron, W. (1968). Field techniques. In *Zooplankton sampling. UNESCO-Monogr. Oceanogr. Methodol.* **2**: 87–104.

Gibbs, R. H. Jr. & Roper, C. F. E. (1970). Ocean Acre. Preliminary report on vertical distribution of fishes and cephalopods. In *Proceedings of an international symposium on biological sound scattering in the ocean.* Farquhar, G. B. (ed.). Washington, D.C.: U.S. Dept Navy.

Gibbs, R. H. Jr., Roper, C. F. E., Brown, D. W. & Goodyear, R. H. (1971). *Biological studies of the Bermuda Ocean Acre. I. Station data, methods and equipment for cruises* 1 *through* 11, *October* 1967–*January* 1971. Washington, D.C.: Smithsonian Institution.

Harrison, C. M. H. (1967). On methods for sampling mesopelagic fishes. *Symp. zool. Soc. Lond.* No. 19: 71–126.

Isaacs, J. D. & Kidd, L. W. (1953). Isaacs–Kidd Midwater Trawl. Final Report. *Scripps Inst. Oceanogr.*, Ref. 53–3: 1–21.

King, J. E. & Iverson, R. T. B. (1962). Midwater trawling for forage organisms in the central Pacific 1951–1956. *Fishery Bull. Fish. Wildl. Serv. U.S.* **62**(210): 271–321.

McGowan, J. A. & Fraundorf, V. J. (1966). The relationship between size of net used and estimates of zooplankton diversity. *Limnol. Oceanogr.* **11**: 456–469.

McNeely, R. L. (1963). Development of the *John N. Cobb* pelagic trawl—a progress report. *Comml Fish Rev.* **25**(7): 17–27.

Roper, C. F. E. & Young, R. E. (1975). The vertical distribution of pelagic cephalopods. *Smithson. Contr. Zool.* No. 209: 1–51.

Schärfe, J. (1960). A new method for "aimed" one-boat trawling in mid-water and on the bottom. *Stud. Rev. gen. Fish. Coun. Mediterr.* No. 13.

Schärfe, J. (1964). Discussion on fish detection. In *Modern fishing gear of the world.* **2**: 418. Kristjonsson, H. (ed.). London: Fishing News (Books).

Symp. zool. Soc. Lond. (1977) No. 38, 89–126.

BEAKS, NETS AND NUMBERS

M. R. CLARKE

The Laboratory, Plymouth, England

SYNOPSIS

Comparisons are made between collections of cephalopod beaks obtained from the stomachs of a variety of predators, including sperm whales, porpoises, seals, sharks, tuna and birds, and collections of cephalopods obtained with several kinds of nets.

Over 140 000 lower beaks from the stomachs of sperm whales from various regions, and more than 4000 specimens from nets form the main basis for comparisons, together with selected published work. Comparisons are made between the family composition of various samples from different regions, and the coverage is often sufficient to distinguish between regional effects and differences due to type of sampler.

In the North Atlantic, nets with small mouths take a large proportion of cranchiids while nets with large mouths catch a large proportion of enoploteuthids. Sperm whales, on the other hand, generally take a far bigger proportion of histioteuthids and octopoteuthids, families that are relatively rare in net samples; near surface lining in temperate and tropical regions in the North Atlantic yields very little but ommastrephids, which is also true of midwater photography.

The relative numerical importance of families in the diet of the sperm whale in different regions is assessed. Because sperm whales are so large and numerous they must play an important part in their trophic level. Cephalopods form a considerable part of the food of the sperm whale so that they too must be very important in the ecology of the oceans. The relationship, by weight, of cephalopod families in the sperm whale diet is assessed for various regions sampled by the author and other workers. A rough estimate of the total weight of cephalopods eaten by sperm whales each year is over 110 million tons, but is likely to have been twice as much in the years before 1946.

The possible significance of the increase in the number of species caught by opening-closing nets between 60 and 11°N in the North Atlantic is discussed.

INTRODUCTION

Inshore and neritic species of cephalopod such as *Octopus*, *Sepia* and *Loligo* species spend part of their day on or near the sea bed in moderate depths and this, together with their close proximity to land, makes them easy prey to the nets of fishermen and marine biologists. On the other hand, oceanic species, which show a much greater diversity of structure than those inshore, often live entirely in midwater, may be at considerable depths and are often remote from marine laboratories. These factors make capture far more difficult and more expensive, since powerful ships with special large capacity winches and fishing gear are required. Thus, while much remains to be investigated about inshore and neritic species, our knowledge of the deep-water species (with one or two exceptions) is even poorer.

Methods of obtaining information about oceanic cephalopods include observation of living animals at the surface or by cameras and submersibles, observation of animals captured by nets or by lines, and examination of specimens removed from the stomachs of predators such as birds, fish, seals and cetaceans. A broad comparison of the information obtained by each of these methods (except submersibles) is made here; for this purpose, only families are considered and it is hoped from this to obtain a better idea of which are really important in the deep-ocean ecosystem. As might be expected, the three methods give a different impression of the relative importance of the families of squid captured. The remarkable feature of these comparisons is, however, the magnitude of the differences displayed. This shows we are only just beginning to sample many families that make large contributions to the biomass.

MATERIALS AND METHODS

The surface observations and line catches considered here include the results of many hundred hours spent observing squid at the surface during a number of North Atlantic, and one Indian Ocean, cruise, and observations and captures made for the author by scientists and crew aboard weather ships. Sub-surface photographs were taken with the NIO (National Institute of Oceanography now the Institute of Oceanographic Sciences, IOS) deep-sea camera adapted for midwater (Baker, 1957a, b; Clarke, 1966).

The day and night series of net hauls, from 0 to 2000 m made at seven localities in the North Atlantic shown in Fig. 1, were carried out as part of the biological programme of the Institute of Oceanographic Sciences, Wormley, Surrey, to study the vertical distribution of nekton. Vertical distribution of the cephalopods from these series has been described elsewhere (Clarke, 1969a; Clarke & Lu, 1974, 1975; Lu & Clarke, 1975a,b). Samples came from 61 metre-ring net hauls, 83 Isaacs-Kidd midwater trawl hauls (IKMT) fitted with catch-dividing buckets and 427 acoustically opening-closing rectangular midwater trawl hauls (RMT) with effective fishing mouths of 1 m^2 (196 hauls), 8 m^2 (226 hauls) and 25 m^2 (5 hauls). Details of the nets and methods of using them have been described by Foxton (1963, 1969), Clarke (1969b) and Baker, Clarke & Harris (1973). In total, 571 samples yielded 4189 cephalopods. Other net hauls considered here were made by 66 open rectangular midwater trawls (RMT 1, RMT 8 and RMT 90, with effective fishing mouths of 1, 8 and 90 m^2 respectively), four

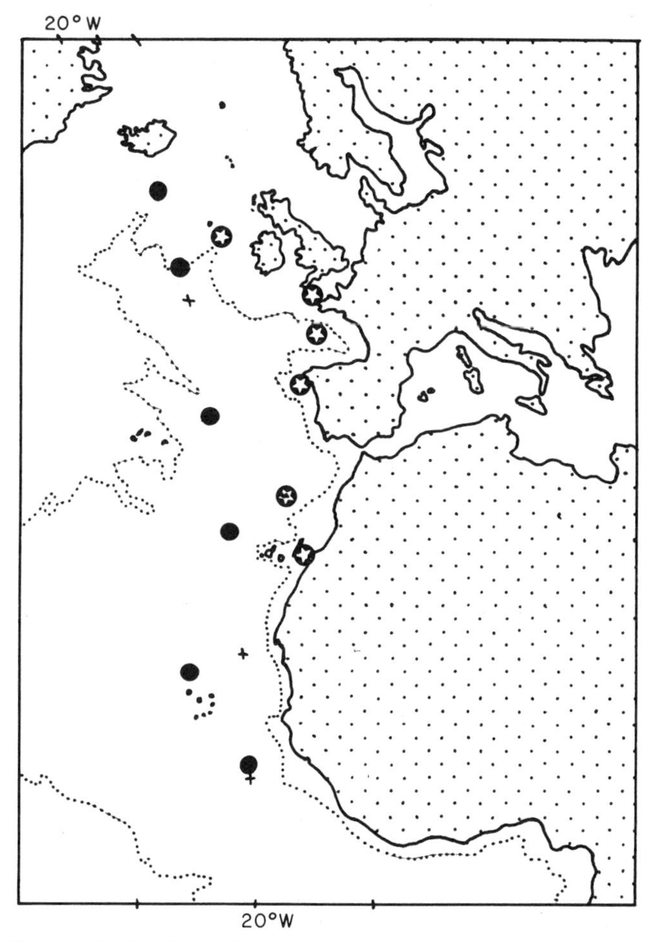

FIG. 1. Eastern North Atlantic showing the positions of a series of opening–closing nets sampling horizons of 10 m to 500 m (according to depth) down to 2000 m (filled circles); and positions at which predators were caught whose stomach contents were examined (stars).

British Columbia midwater trawls [BCMT, mouth measuring 50 ft × 50 ft (15·2 m × 15·2 m)] and 28 Engels midwater trawls [EMT, mouth measuring 116 ft × 65 ft (35·4 m × 19·8 m)]. Brief notes on the trawls are given in Appendix B.

Published reports on some collections made with nets in areas other than the North Atlantic have been utilized for comparative purposes. They include reports by R. E. Young (1972) and by Okutani & McGowan (1969) on material collected off California, by Roper (1969) on Antarctic material, and by Nesis on three collections from the west Arabian Sea, the eastern South Pacific and the western South Atlantic (1974b, 1973, 1974a respectively). These particular collections are included because they are from the same areas as those for which data from stomach contents were available; they are also total collections from which family percentage composition could be calculated at least roughly. [R. E. Young (1972) states that only 33 of 40 species are described, but as he treated over 5352 specimens and the programme yielded "over 5000 specimens", the missing seven species would probably not have too drastic an effect on the percentage calculations.] The published reports considered here are not all that are available from the broad areas considered, but they have not been selected for species composition and probably give a reasonably good idea of what would be obtained with the same methods by any expedition to the same place.

The soft bodies of cephalopods are rarely found in the stomachs of their predators. However, the chitinous beaks resist attack by the digestive juices and are often retained in the stomach for some time after the other tissues have been digested. Over 18 000 beaks have been found in the stomach of a single sperm whale by the author, and Akimushkin (1955) reported as many as 28 000 in another sperm whale. To obtain the maximum information from stomach contents, beaks were studied to determine criteria for their identification (Fig. 2) and to relate beak size to total wet weight, and consequently for each family (Clarke, 1962a,b,c and in press). The present comparisons (Appendix A) include analysis of stomach contents of sperm whales caught off Western Australia (Fig. 3), the east and west coasts of South Africa, South Georgia and elsewhere in the Antarctic (Clarke, in press), off western South America (Clarke, MacLeod & Paliza, 1976), Spain (Clarke & MacLeod, 1974), Madeira (Clarke, 1962a), in the Tasman Sea (Clarke & MacLeod, in prep. a) and off western Canada (Clarke & MacLeod, in prep. b). In total, these analyses have involved the identification of over 140 000 lower beaks (probably exceeding the total number of oceanic cephalopods in all the catalogued collections of the world) and the measurement of over 67 000 beaks.

Stomach contents of seals (Clarke & MacLeod, in prep. c), bottom-living sharks (Clarke & Merrett, 1972) and blue sharks (Clarke & Stevens, 1974) are also included in some comparisons.

While observations, catches and stomach contents are not strictly comparable since time, effort, place and depth differ, the number of samples taken over a broad seasonal, geographical and

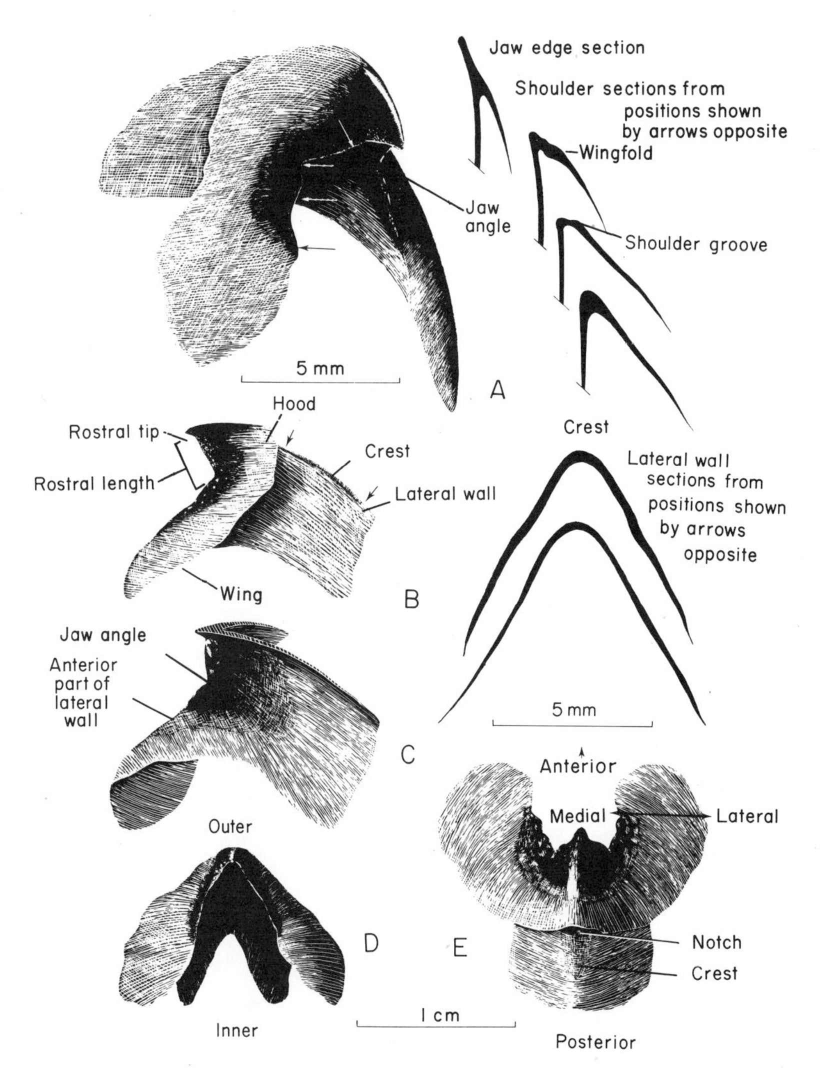

FIG. 2. Lower beaks of cephalopods with terms used to describe their most important features. A. Antero-lateral view of the lower beak drawn with its inner, morphologically dorsal, end pointing down. The arrows indicate the sites of the four sections drawn opposite. B. The lower beak in profile to show the rostral length, hood, crest and lateral wall. The arrows indicate the sites of the two sections of the lateral wall shown opposite. C. A medial cut seen from the inside. D. An anterior view. E. A view of the ventral surface.

depth range permit conclusions of a general nature. In a few instances, more precise comparisons can be made but in the main only general conclusions on a global scale are profitable.

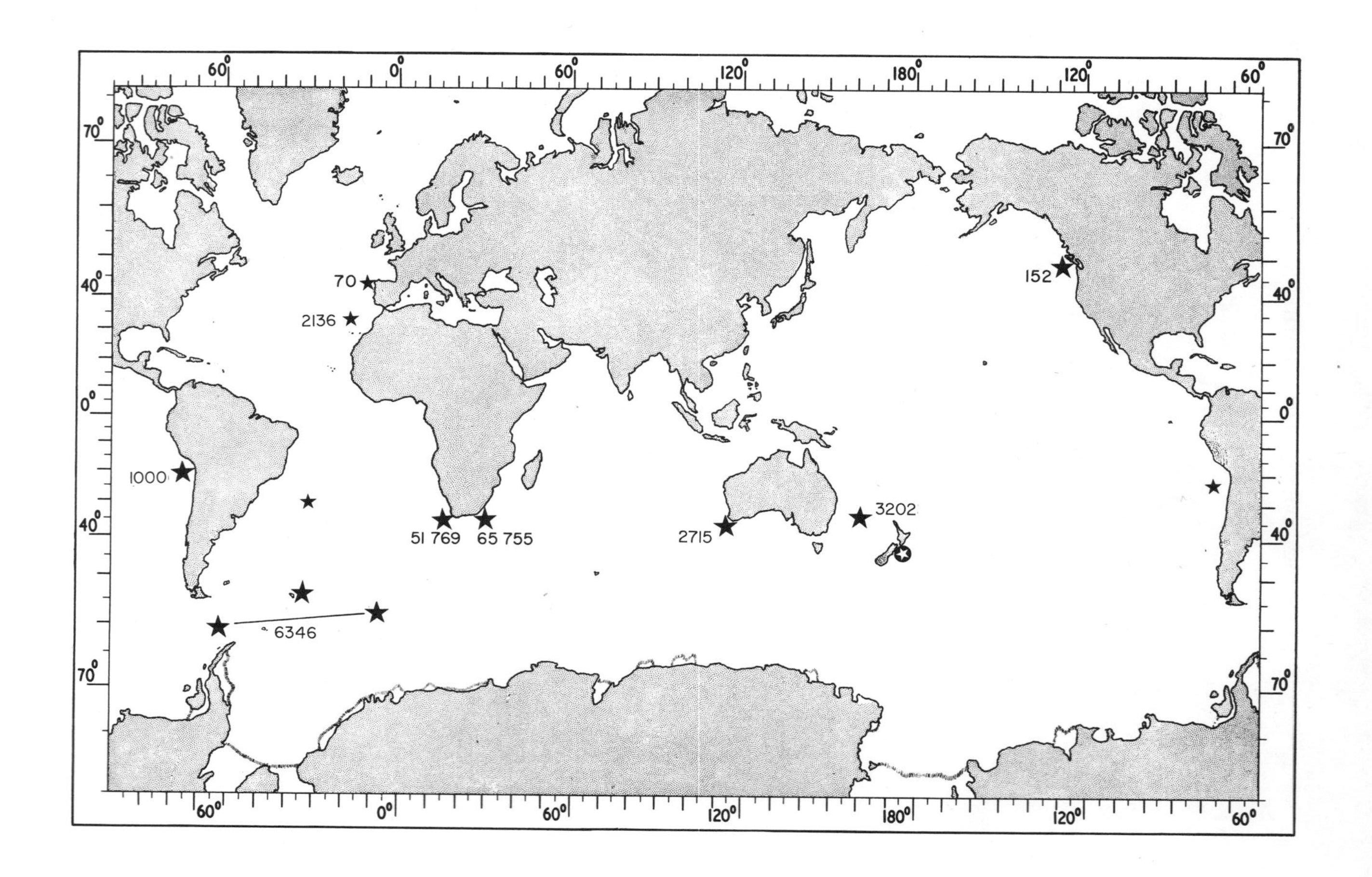

70
2136
152
1000
51 769
65 755
2715
3202
6346

Some published work on the stomach contents of sea birds (Imber, 1973; Harris, 1973), small Cetacea (Perrin, Warner, Fiscus & Holts, 1973), tuna (Williams, 1966; Perrin *et al.*, 1973) and sperm whales (Gaskin & Cawthorn, 1967; Akimushkin, 1955; Betesheva & Akimushkin, 1955; Tarasevich, 1968) have also been included where they relate to the present study.

One of the measurements employed is the mantle length (ML) of the intact animal; this is the distance between the anterior end of the body and the posterior tip of the body in most squids. However, in Chiroteuthidae, Mastigoteuthidae and Joubiniteuthidae the measurement is taken to the posterior end of the fins, and in the octopods it is taken from the centre of the lens of the eye to the posterior end of the body. The lower rostral length (LRL) (rostral length on Fig. 2) has been measured to express the size of the beak. The body weight of cephalopods has been estimated from the LRL using curves that relate these factors for families and are published elsewhere (Clarke, in press). These curves are sometimes derived from only a few specimens but they give better estimates than alternative methods, and also provide a means of finding the relative importance by weight of the species and/or families represented, and the average weight in the whales' diet.

While the consideration of families rather than species is unusual and obscures many extra differences between samples being compared, in a general description and comparison such as this, it has the merit of removing confusing details. In this way the apparently most important families in the ocean by number and weight can be shown and these should provide useful comparisons for future studies of cephalopod ecology.

Details of the samples are given in Appendix A and notes on nets used are given in Appendix B.

IMPORTANCE BY NUMBERS

Bay of Biscay region

Comparison may be made between 97 cephalopods from 32 open hauls with RMT 8 and a Boris otter trawl in the Bay of Biscay, 92 beaks taken from the stomachs of blue sharks (*Prionace glauca* L.) caught off Looe, Cornwall, England, and in the Bay of Biscay

---

FIG. 3. The regions in which sperm whales' stomach contents have been studied by the author and colleagues (filled stars). Areas studied by Gaskin & Cawthorn and used here (open stars). Numbers of lower beaks identified are shown.

(Clarke & Stevens, 1974), and 70 beaks from the stomach of a sperm whale (*Physeter catodon*) caught off Portugal at 41° 32′N, 9°48′W (Clarke & MacLeod, 1974), (Appendix A). About 75% of the beaks from the stomachs of the blue shark were from deep-water cephalopods which had probably been eaten in the Bay of Biscay, although many of the sharks were caught off Looe.

Figure 4 shows cumulative percentage curves of cephalopod families for these three sets of samples. The families are arranged in order of decreasing percentage for the net samples.

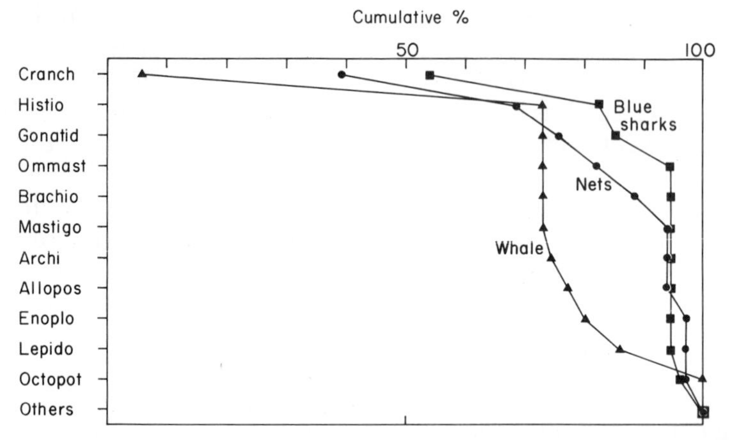

FIG. 4. Cumulative percentage curves for oceanic squids from the stomachs of blue sharks caught off south-west England, a whale caught off northern Spain and samples from open nets fished in the Bay of Biscay. The cumulative curves were derived by arranging the families in an arbitrary order down the left-hand side. Each point on a curve is derived by adding the percentage represented by the family in the samples to the sum of the percentages represented by families above it. For example, 6% of the sample from the whale were cranchiids while 67% were histioteuthids, and the point for the histioteuthids is plotted at 73%. Thus, in drawing comparisons between the samples, the slopes of the curves immediately above the points for each family are the important factor and not the values of the points themselves. (The abbreviations used in Figs. 4–19 are given on p. 126.)

The principal differences between these samples are that the whale had far fewer cranchiids but many more histioteuthids and octopoteuthids than were found in nets or sharks. The whale also took *Architeuthis*, *Alloposus* and *Lepidoteuthis*, which were not present in the nets or shark stomachs. Gonatids and ommastrephids were present in nets and shark stomachs but not in the stomachs of the whale. The only enoploteuthid present in the nets and whale stomach was *Ancistrocheirus lesueuri*.

In the eastern North Atlantic, 53°–55°N

Comparison may be made between 373 cephalopods from the series of 34 hauls of opening-closing RMT 1+8 to 2000 m at 53°N 20°W, surface observations and fishing from a weather ship at 53°N 20°W, and 84 cephalopods caught with open RMTs near 55°N 12°W (Appendix A).

Figure 5 shows that all the surface observations and collections from the weather ship only demonstrated the presence of ommastrephids. The series of nets taken at the same place revealed many more species in the water column but include no ommastrephids.

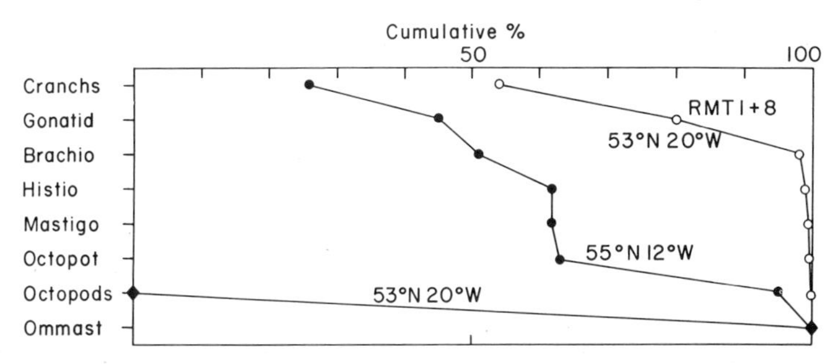

FIG. 5. Cumulative percentage curves for a series of 34 hauls with the opening-closing RMT 1+8 combination net at 53°N 20°W, a collection of 84 cephalopods caught near 55°N 12°W with an open RMT 8, and results of surface observations and surface fishing from a weather ship at 53°N 20°W. Families arranged in order of decreasing percentage for the RMT 1+8 series.

The open RMT 8 hauls had relatively fewer cranchiids, fewer brachioteuthids, more histioteuthids, more octopoteuthids, more ommastrephids and many more octopods. These differences are less likely to be due to the differences in operation (open and partly oblique as compared with horizontal hauls) than to the position relative to the continental slope, and this is discussed below (p. 117).

In the eastern North Atlantic, 37°–40°N, 20°–25°W

Comparisons may be made between the 555 cephalopods from ten Engels trawls fished in the Azores region near 37°N 25°W, cephalopod flesh collected from sperm whales caught near the Azores (Robert Clarke, 1956) and a series of RMT 1+8s fished to 2000 m at 40°N 20°W (Lu & Clarke, 1975a) (Appendix A). The series was unusual in collecting very little of all phyla and only 53 cephalopods.

Figure 6 shows that the contents of the nets differed tremendously; the Engels trawls were extraordinarily rich in enoploteuthids (almost 80%) which were barely represented (6%) in the RMT 1+8s. On the other hand the RMT 1+8s included

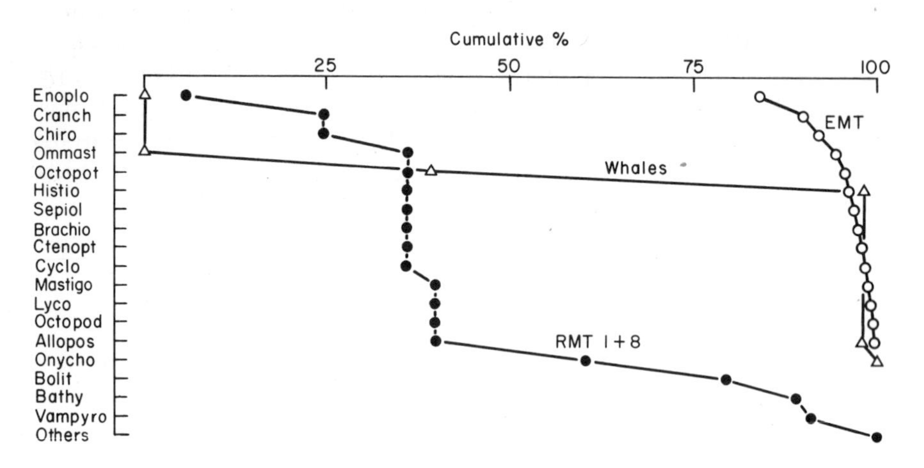

FIG. 6. Cumulative percentage curves for ten EMT nets fished in the Azores region, a series of RMT 1+8s fished to 2000 m and a collection of flesh from the stomachs of sperm whales caught near the Azores and reported by R. Clarke (1956). Families arranged in order of decreasing percentage for the EMT samples.

onychoteuthids, bathyteuthids, vampyroteuthids and octopods which were not present in the EMT nets. The samples from whales were dominated by octopoteuthids and histioteuthids, both of which were absent from the RMT samples and barely represented in the EMT samples.

The Madeira region and 30°N 20°W

Comparisons can be made between 255 cephalopods collected near 30°N 20°W with 32 opening-closing RMT 1+8s fished to 2000 m, 215 cephalopods from 45 IKMT nets with a catch-dividing bucket fished to 1400 m (both in Clarke & Lu, 1974) and 148 cephalopods from three British Columbia midwater trawls [50ft square mouth (15·2 m)] fished to a maximum of 100 m. Together with these we may include, from near Madeira at 32°–33°N, 16°–17°W, 122 cephalopods from nine Engels trawls fished to a maximum of 590 m, and 2136 lower beaks from the stomach contents of a sperm whale (Clarke, 1962a; Clarke & MacLeod, 1974) (Appendix A).

Figure 7 shows the cumulative percentage curves for each method of collection. The whale and the Engels midwater trawl samples have a far smaller percentage of cranchiids than the smaller nets (RMT 1+8, IKMT and BCMT). All the nets have a

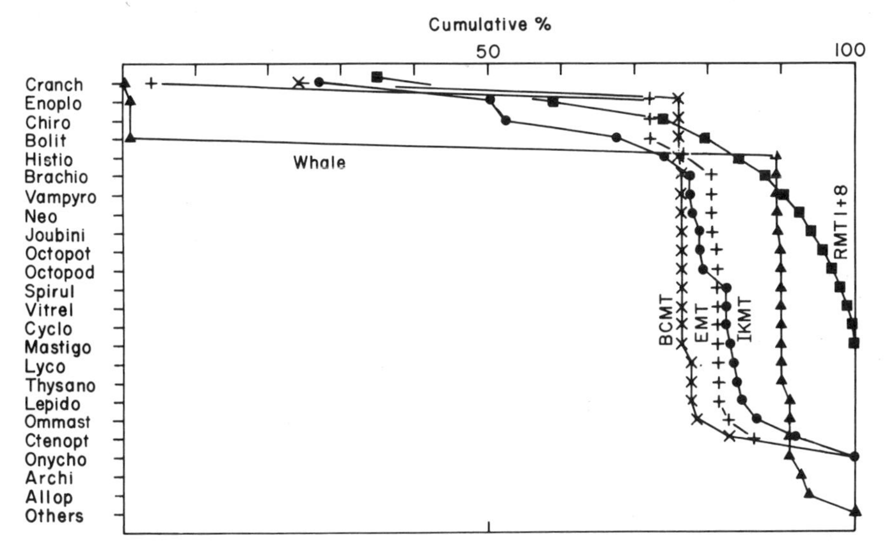

FIG. 7. Cumulative percentage curves for samples collected near 30°N 20°W with 32 opening-closing RMT 1+8s, 45 IKMT and three BCMT nets, and for nine EMT samples fished near Madeira and a sample of 2136 beaks from the stomach of a sperm whale caught off Madeira. Families arranged in order of decreasing percentage for the RMT 1+8 series.

large percentage of enoploteuthids (>20%) and a rather small percentage of histioteuthids (<7%) while there are very few enoploteuthid beaks in the whale stomach and almost 90% of the beaks are histioteuthid. The whale sample has no onychoteuthids, which are represented in three types of net, while alloposids are only present in the whale. Thus, the largest trawl (EMT) is similar to the whale and dissimilar to the other nets with respect to cranchiids, and as both the Engels and the whale were close to Madeira, and the other nets were not, this could possibly be due to the former's proximity to land and slope areas.

Canary Islands, east of Fuerteventura, 28°N 14°W

A comparison can be made between samples comprising 185 cephalopods from a series of 38 IKMT nets and 38 N113 nets (1m² ring nets) with catch-dividing buckets to 975 m (Clarke, 1969a), and nine samples comprising 833 specimens from EMT nets (Appendix A).

Figure 8 shows that the large nets (EMT) catch a larger proportion of enoploteuthids and histioteuthids but fewer or no spirulids, sepiolids, cranchiids and bolitaenids.

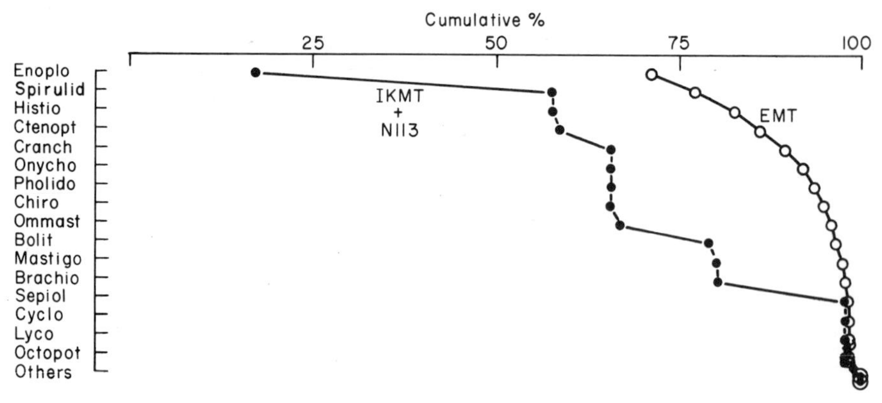

FIG. 8. Cumulative percentage curves for samples from a series of 38 IKMT nets and 1 m^2 ring nets (N113) and for nine EMT nets fished at 28°N 14°W to the east of Fuerteventura, Canary Islands. Families arranged in order of decreasing percentage for the EMT nets.

Eastern North Atlantic

Comparisons can be made between all the series of opening-closing trawls (three series of 1 m^2 ring nets, 1 RMT 8, 6 RMT 1+8 and 2 IKMT series comprising 593 samples) put together (Clarke & Lu, 1974, 1975; Lu & Clarke, 1975a,b), sperm whale stomach contents comprising 2206 lower beaks (Clarke, 1962a; Clarke & MacLeod, 1974), surface observations and fishing (Baker, 1960; Clarke, 1966) and stomach contents of bottom-living sharks representing at least 15 cephalopods (Clarke & Merrett, 1972) (Appendix A). While few cephalopods were found in bottom-living sharks (little food of any kind was found) the composition is of interest.

Figure 9 shows the cumulative percentage composition of samples, with families, arranged in decreasing order of magnitude for the opening-closing nets (RMT and IKMT). Surface observations and near-surface line and hand-net fishing yields over 95% ommastrephids. A very few onychoteuthids are caught and a cranchiid, two histioteuthids and a chiroteuthid have been found dead at the surface by the author, but these had probably been regurgitated by sperm whales. In contrast, almost 80% of beaks from the sperm whales are histioteuthids while the Cranchiidae is

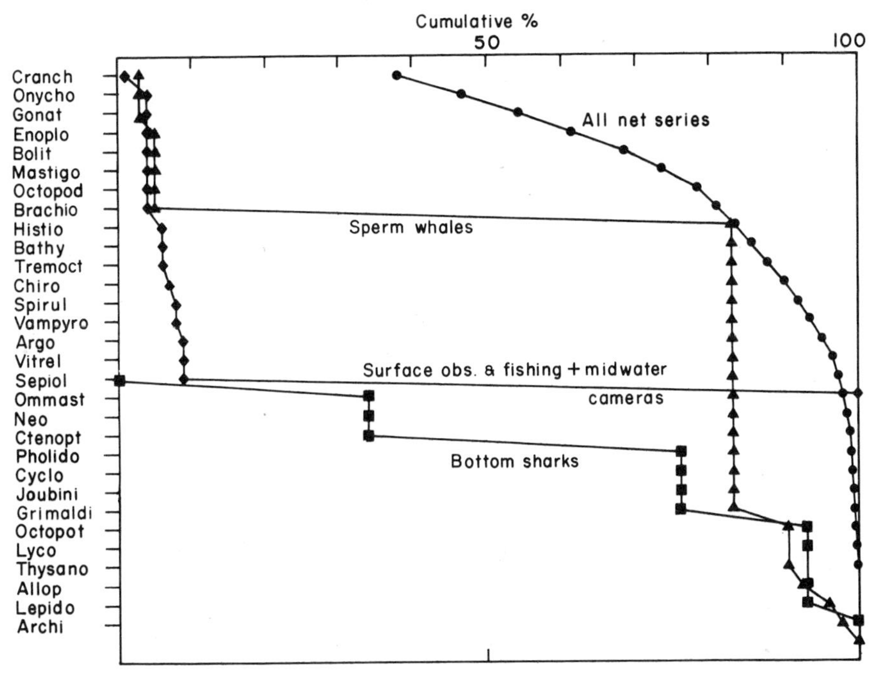

FIG. 9. Cumulative percentage curves for samples of all series of opening-closing trawls in the North Atlantic considered here, from sperm whale stomachs caught off Madeira and Spain, from the stomachs of bottom-living sharks and from surface fishing and observations. Families arranged in order of decreasing percentage for the opening-closing series.

the best represented family (>37%) in the opening-closing net samples. The sperm whale samples are notably different from the nets in having higher percentages of octopoteuthids, alloposids, lepidoteuthids and architeuthids, all very rare families in net hauls. The presence of octopoteuthids and architeuthids in the stomachs of bottom-living sharks shows an interesting similarity to the whale samples. This is emphasized by the ommastrephid and pholidoteuthid genera present in these sharks since these (*Todarodes* and *Pholidoteuthis*) are extremely rare in RMTs and IKMTs, but are important in samples from sperm whales in regions other than the North Atlantic. *Pholidoteuthis* (also known as *Tetronychoteuthis*) is found in sperm whales caught in the Azores (see above, p. 98). These similarities support other evidence that sperm whales probably catch some species on or close to the sea floor.

Figure 10 summarizes the change in family composition according to latitude from the data derived from the RMT and

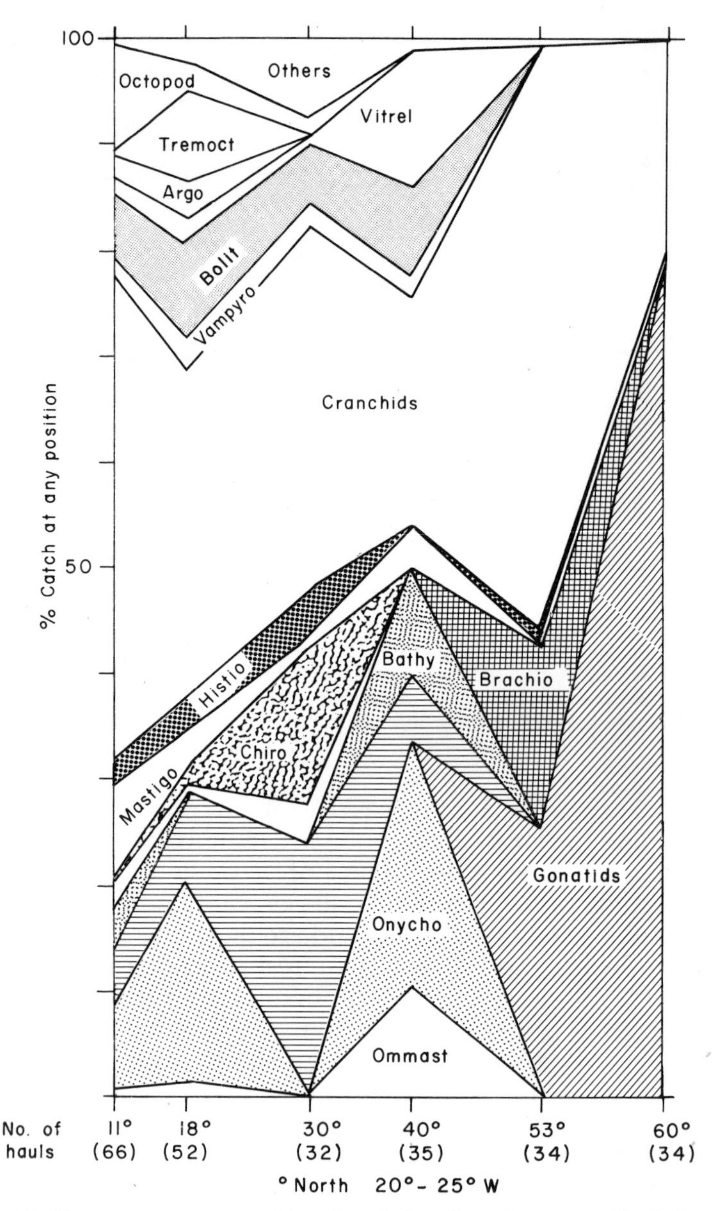

FIG. 10. The percentage composition of cephalopods in the eastern North Atlantic at 20°–25°W and at 60°, 53°, 40°, 30°, 18° and 11°N showing the change in numerical importance of families. Only opening-closing RMT nets were included. The white area below "Chiro" is Brachioteuthidae. The area with horizontal lines represents the Enoploteuthidae.

Isaacs-Kidd trawl series at 20°–25°W. (See Clarke & Lu, 1974, 1975; Lu & Clarke, 1975a,b.) This shows clearly the small proportion of histioteuthids and octopoteuthids and the very large proportions of cranchiids and enoploteuthids present in the samples.

South Atlantic

From the South Atlantic the author has little material for comparison with 49 402 lower beaks and 152 other specimens from sperm whales caught off Donkergat in South Africa and a small collection (11) of squid from sperm whales caught at 27°–31°S, 32°–33°W (Clarke, in press). A small collection of squids from birds at Ascension Island yielded only ommastrephids. Nesis (1974a) described 274 cephalopods from the western South Atlantic caught with a variety of nets and I have included percentages for the families derived from his paper (Appendix A).

Figure 11 shows cumulative percentage curves for these collections in order of decreasing magnitude for Donkergat. The most

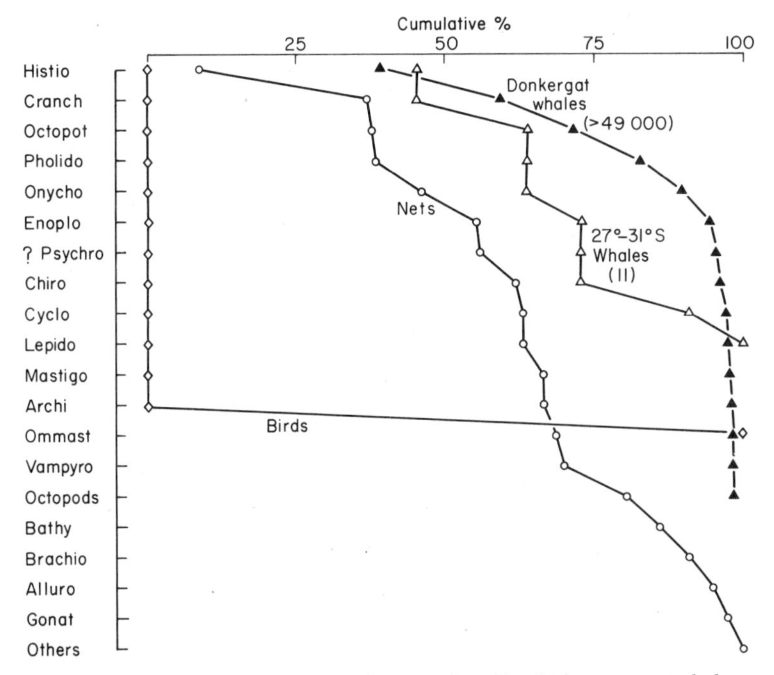

FIG. 11. Cumulative percentage curves for samples of beaks from sperm whales caught off South Africa (49 402 beaks), 11 squids from whales caught at 27°–31°S, 32°–33°W, a small collection of squids from birds at Ascension Island and samples from a variety of nets fished in the south-west Atlantic described by Nesis (1974a). Families arranged in order of decreasing percentage for the Donkergat, South Africa, sperm whale stomach content samples.

interesting feature of these curves is that although the mid-Atlantic collection is extremely small, it included histioteuthids, octopoteuthids, cycloteuthids and a lepidoteuthid, all of which are rare in net hauls. Thus these forms are not limited to the neighbourhood of the continental slope, as might be supposed from the South African material. However, 27°–31°S, 32°–33°W is near the Bromley Plateau rising to within 700 m of the surface, so that we cannot exclude the possibility that the squids are associated in some measure with the shallower plateau region.

The net hauls are richer in families than the stomach content samples and have fewer histioteuthids, octopoteuthids and pholidoteuthids (*Tetronychoteuthis* is included here for reasons described elsewhere, Clarke, in press), but more chiroteuthids, mastigoteuthids, octopods, bathyteuthids, brachioteuthids, alluroteuthids and gonatids. Some of these differences are probably due to the inclusion of a few net stations from the Antarctic, and Antarctic species were not included in the South African (Donkergat) percentages. However, this does not greatly influence the proportion of histioteuthids and octopoteuthids.

Indian Ocean

Comparisons can be made between 62 794 lower beaks and 1347 other specimens from sperm whales caught off Durban, South Africa, and 2477 lower beaks and 67 other specimens from Albany, Western Australia (Clarke, in press), a collection of 1122 squid from tuna stomachs collected off East Africa between 2°50′S and 8°40′S and identified by the author (Williams, 1966), surface observations and surface lining and hand-netting by the author during a three-month cruise on RRS *Discovery* in the Indian Ocean, and a collection of 127 young squid caught in the western Arabian Sea off the Somali coast and described by Nesis (1974b) (Appendix A).

Figure 12 shows cumulative curves for these collections with families arranged in order of decreasing percentage for the Durban collection. Beaks known to belong to species which are not from the Durban region are omitted (i.e. Antarctic species, see Clarke, 1972). Surface observations were almost entirely ommastrephids, the tuna from East Africa contained almost entirely enoploteuthids (65%) and ommastrephids, and the dominance of these two families is also clear in the larvae caught in nets from much the same region. The whale samples from the west and east Indian Ocean mainly differ in that cycloteuthids, ommastrephids and pholidoteuthids are all more important in the east but they

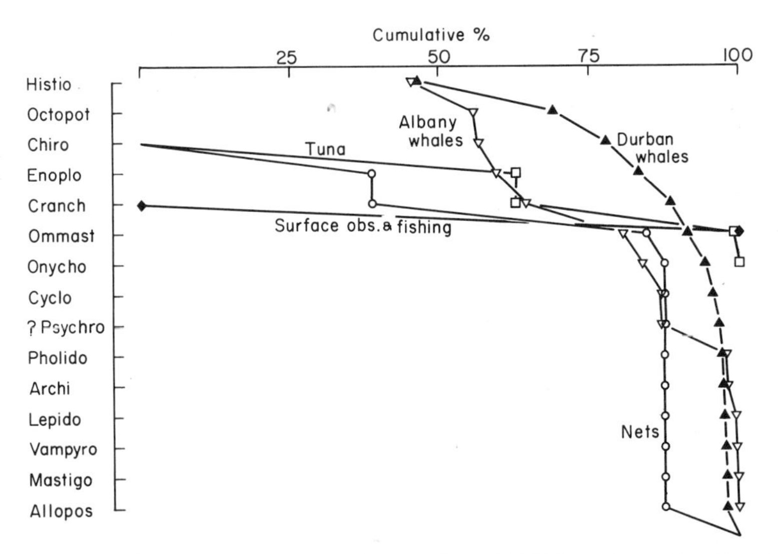

FIG. 12. Cumulative percentage curves for samples of 62 794 lower beaks identified from stomachs of sperm whales at Durban, 2477 lower beaks from Albany, 1122 cephalopods from tuna caught off East Africa, a small collection of 127 young squid from nets fished in the western Arabian Sea and the result of surface observation and fishing. Families arranged in order of decreasing percentage for the Durban collection.

both differ from samples derived by other methods in having a large percentage of histioteuthids, octopoteuthids and a larger percentage of cranchiids.

Western Pacific

Comparisons may be made between 3202 lower beaks and squids collected from sperm whales caught at various positions in the Tasman Sea (Clarke & MacLeod, in prep. a), and around New Zealand (2118 lower beaks partly re-identified from Gaskin & Cawthorn, 1967) and 907 lower beaks from the stomachs of New Zealand birds (Imber, 1973) (Appendix A).

Figure 13 shows that the 2118 beaks from whales caught near New Zealand have a far larger proportion of onychoteuthids and ommastrephids and a far smaller proportion (or none) of octopoteuthids, histioteuthids and cranchiids than the Tasman Sea samples. The large proportion of onychoteuthids is due to the Antarctic influence on the fauna near New Zealand (see Clarke, in press). The bird samples mainly differ from the New Zealand whale samples in having more histioteuthids, a large percentage of

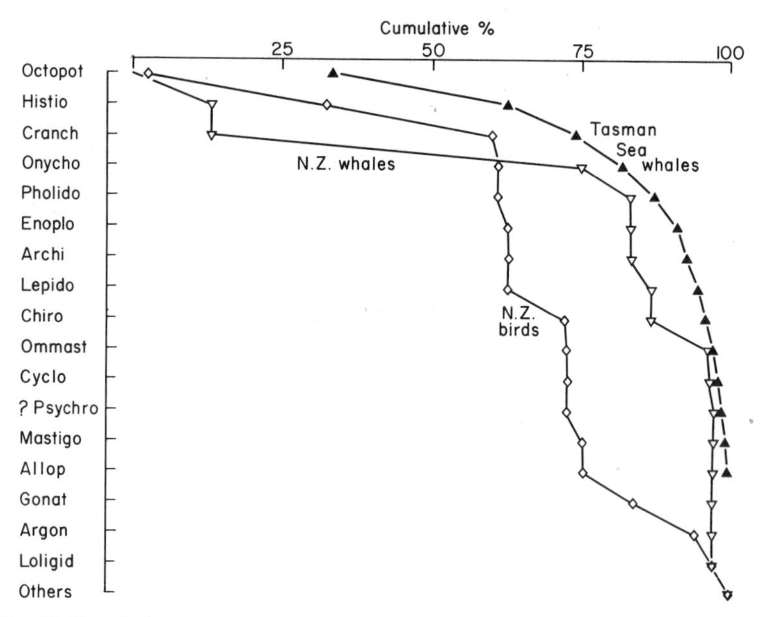

Fig. 13. Cumulative percentage curves for samples of beaks from sperm whales caught at several stations straddling the Tasman Sea, a collection of beaks from sperm whales from New Zealand waters described by Gaskin & Cawthorn (1967) and a collection of beaks from New Zealand birds described by Imber (1973). Families in order of decreasing percentage for the Tasman Sea collection.

cranchiids, more chiroteuthids, mastigoteuthids, gonatids and argonauts, and fewer onychoteuthids, pholidoteuthids and ommastrephids. The high proportions of histioteuthids and chiroteuthids in the birds are surprising since these squids do not usually live near the surface. Their occurrence may be a reflection of upwelling which takes place near New Zealand.

North-eastern Pacific

Comparisons can be made between 152 squid beaks from stomachs of sperm whales caught off western Canada (Clarke & MacLeod, in prep. b), over 400 lower beaks and cephalopods from stomachs of tuna (*Thunnus albacares*) and remains of 7882 cephalopods in two species of porpoise (*Stenella attenuata* and *S. longirostris*) caught to the west of Central America at 7°–12°N, 90°–93°W and described by Perrin *et al.* (1973), over 8400 squids taken from trawls during two large surveys off California (Okutani & McGowan, 1969; R. E. Young, 1972), and 299 lower beaks from stomachs of albatrosses examined in the Galapagos Islands (Harris, 1973).

Figure 14 shows that the net samples have far more enoploteuthids than the other samples and more gonatids and cranchiids than porpoises, birds or tuna. The whale and porpoise samples are similar to one another in having a large proportion of onychoteuthids, but they differ in that the porpoise has a large proportion of ommastrephids, which is the dominant family for the tuna. The birds differ from the other samples in having large proportions of octopoteuthids and histioteuthids. The presence of these is probably a consequence of local upwelling as suggested for the occurrence of the latter family in birds of New Zealand (p. 106).

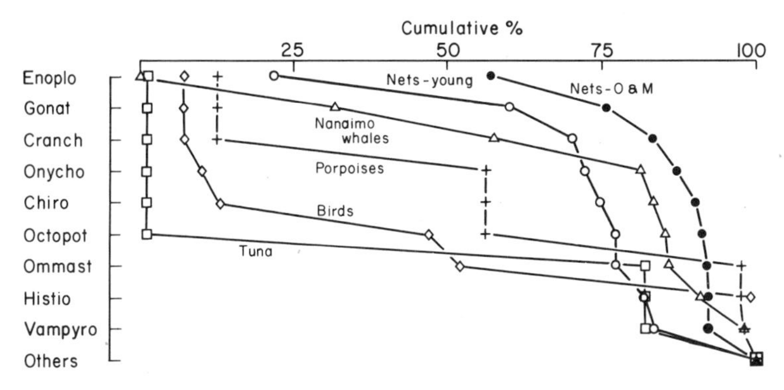

FIG. 14. Cumulative percentage curves for samples of beaks identified from tuna and porpoises west of Central America described by Perrin *et al.* (1973), beaks from birds of the Galapagos Islands described by Harris (1973), beaks from sperm whales caught off western Canada and two collections of squids from nets described by Okutani & McGowan (1969) and R. E. Young (1972). Families arranged in order of decreasing percentage for Okutani & McGowan's collection.

Absence of both these families from tuna and porpoises that were also feeding near the surface is difficult to explain unless it is due to the different locality.

South-eastern Pacific

Comparisons can be made between a collection of 1000 lower beaks from sperm whales caught off Peru and Chile (Clarke, MacLeod & Paliza, 1976) and a collection of 647 cephalopods from the same region made with a variety of nets and described by Nesis (1973). The net samples are very different from the whale samples in having a small percentage of histioteuthids, chiroteuthids and octopoteuthids and much larger percentages of enoploteuthids, cranchiids and octopods (Fig. 15).

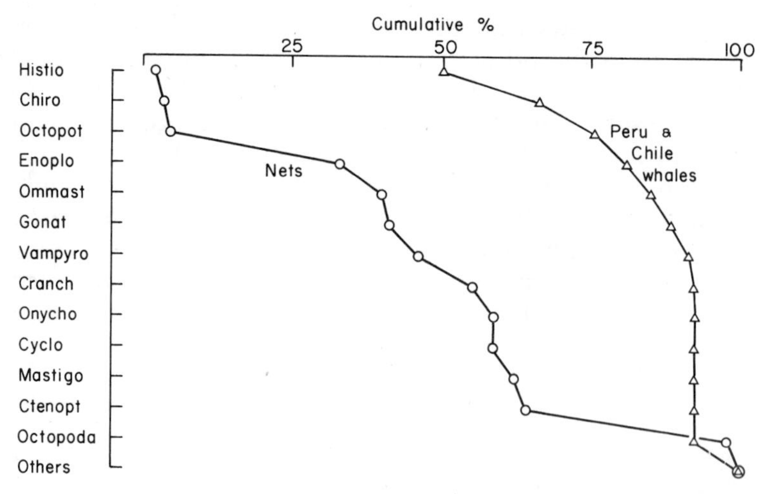

FIG. 15. Cumulative percentage curves for samples of beaks from sperm whales caught off Peru and Chile, and a collection of squid collected with a variety of nets reported by Nesis (1973). Families in order of decreasing percentage for the sperm whale stomach contents.

The Antarctic

Figure 16 shows a comparison between 6346 lower beaks and squids from sperm whales (Clarke, in press) and 366 lower beaks from Weddell seals (Clarke & MacLeod, in prep. c) in the Antarctic. The main contrast is the presence of Octopodinae in the

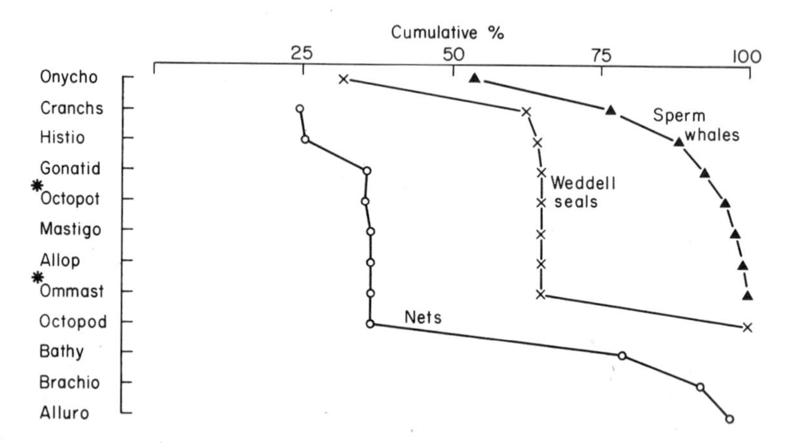

FIG 16. Cumulative percentage curves for samples of beaks from Weddell seals and sperm whales in the Antarctic, and for squids caught by IKMT nets on a cruise of *Eltanin* (Roper, 1969). Stars indicate two families which occur at South Georgia in whale stomachs but are probably near the southernmost limit of their distribution.

stomachs of seals which reflects their close proximity to land. Otherwise the four most numerous families in the whales' stomachs are also the four most numerous in those of the seals, and are even in the same order of importance. Data from cephalopods from samples taken with Isaacs-Kidd trawls by the USS *Eltanin* have not yet been fully published but Roper (1969) gave the proportions of the main species and they are included in Fig. 16. The most obvious differences between the nets and the predators are the large percentages of bathyteuthids and brachioteuthids in nets and the absence of onychoteuthids.

Numerical importance of families in whales' stomachs

Figure 17 summarizes the relative importance of the families occurring in sperm whale stomachs for each region. The New Zealand and North Pacific curves are derived from data given by Gaskin & Cawthorn (1967) and the Russian workers (Betesheva & Akimushkin, 1955; Tarasevich, 1968) respectively. The important

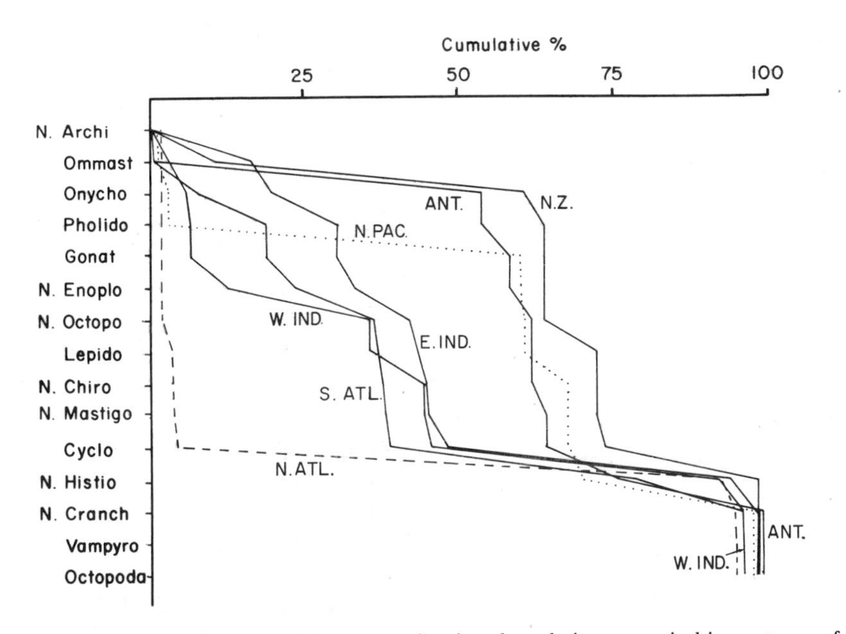

FIG. 17. Cumulative percentage curves showing the relative numerical importance of families identified from beaks. Each line represents a region. N.Z. = New Zealand; E. IND. = East Indian Ocean; W. IND. = West Indian Ocean; S. ATL. = South Atlantic; N. ATL. = North Atlantic (1 sample); ANT. = Antarctic; N. PAC. = North Pacific. See text for sources. "Ammoniacal" squid are indicated by N.

families for the Antarctic are the Onychoteuthidae and the Cranchiidae; for New Zealand, the Onychoteuthidae and the Histioteuthidae; for the North and South Atlantic and Indian Ocean, the Histioteuthidae; the South Atlantic has more cranchiids and the North Atlantic fewer octopoteuthids than the other regions. The North Pacific is characterized by having few Histioteuthidae but many Gonatidae.

SIZE OF CEPHALOPODS SAMPLED

Nets can only be pulled slowly through the water; they have a fairly constant and therefore predictable direction, and sampling with them is blind. Predators may move much faster, can change direction at will and can see and pursue their prey. Thus, it is not surprising that predators are often more efficient than nets in capturing the larger fast-swimming cephalopods. A comparison of nets shows that larger, faster nets are likely to catch more than smaller, slower ones. In practice, towing nets fast usually damages the catch and a speed of less than three knots is often preferred. The standard nets are usually about 8 m^2 or less in mouth area (RMT 8 and IKMT), a size which can be opened and closed for studies of vertical distribution. Much larger midwater nets have only been used in oceanic waters in the last 15 years and the cephalopods sampled have not yet been extensively reported upon.

Length

Figure 18 shows the range of the mantle lengths of squids by family from the stomachs of sperm whales caught off South Africa, Western Australia and in the Antarctic. The lower limits of these ranges have been joined by a heavy line. For comparison, the largest specimens of each family caught in the vertical series to 2000 m in the North Atlantic of both RMT and IKMT nets is also shown, and these are also joined by a heavy line. Clearly, except for the families Histioteuthidae and Cranchiidae, there is no overlap between the samples from the nets and the predator. If the beaks are examined, it is clear from larger beaks that the flesh remains do not span the full range taken by the whales and that the upper size limit should be extended for all families.

The range of size of ommastrephids caught by hand lines and dip nets at the surface of the North Atlantic is shown by a pecked horizontal line.

A pecked line in Fig. 18 joins the maximum sizes of squid in the families caught with a 10 ft (3·05 m) IKMT off California and

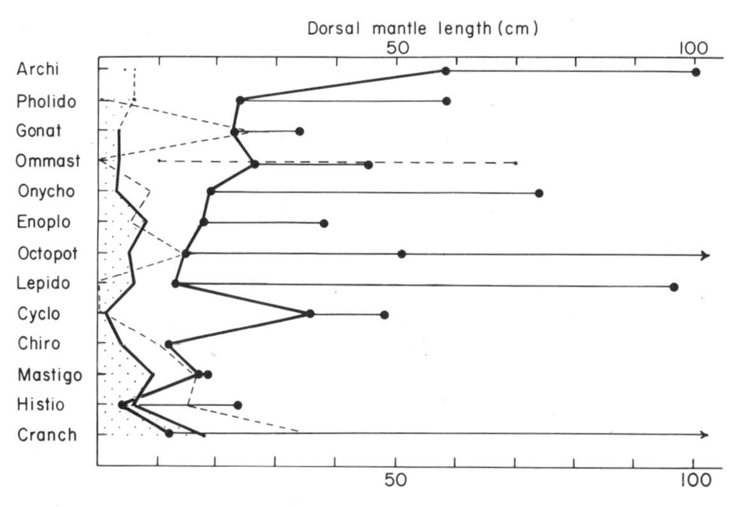

FIG. 18. Mantle lengths of squids caught in nets and by sperm whales. The range of specimens found in sperm whale stomachs considered here is shown by a horizontal line for each family; the lower ends of the ranges are connected by a heavy line. A second heavy line joins the maximum mantle length for each family for the North Atlantic opening-closing series of 593 nets. A pecked line indicates the maximum MLs for the families in R. E. Young's collection (1972). The range for ommastrephids caught at the surface of the North Atlantic is indicated by a horizontal pecked line. Also shown is the largest larval architeuthid described from nets in the literature.

reported on by R. E. Young (1972). In most cases where animals from the same family were caught by nets, the largest animal was greater in size than the largest one caught in the North Atlantic series of RMT and IKMT nets. Overlap with the whale samples occurs in the Gonatidae as well as in the Histioteuthidae and Cranchiidae, and it is greater in the last two families than is the case with the North Atlantic series.

The overall length of some lower beaks of *Mesonychoteuthis* (Cranchiidae), taken from sperm whale stomachs, is greater than the maximum mantle length of animals in any of the families caught in the North Atlantic series, except cranchiids.

It must be remembered that the species composition of the families is different in nets and sperm whale samples, and many of those taken in nets are small species and not young specimens of the large species sampled by whales.

Weight

By removing lower beaks from specimens obtained from both nets and sperm whale stomachs, it has been possible to relate total wet weight of the cephalopods to the lower rostral length of the beaks for the majority of families (Clarke, 1962b and in press). This provides a means of estimating the body weight from beaks and hence the mean weight and total weight for each family in the collections of stomach contents. While more specimens would undoubtedly improve the curves upon which the estimates are based, and in some families extrapolation of the curves is unavoidable, the estimates from beaks are likely to be much better than taking a mean or total weight of the very few intact specimens which are available in some families, particularly as the flesh of the different families is digested at different rates (Clarke, in press). Figure 19 shows the relative importance by weight of the families of cephalopods in stomachs of sperm whales killed in the seven regions studied (Clarke, 1962a, in press; Clarke & MacLeod, 1974, in prep. a, b; Clarke, MacLeod & Paliza, 1976).

These cumulative curves show that different families differ in their importance by weight in the diet of the whales from region to region. Octopoteuthids are important in all regions except the

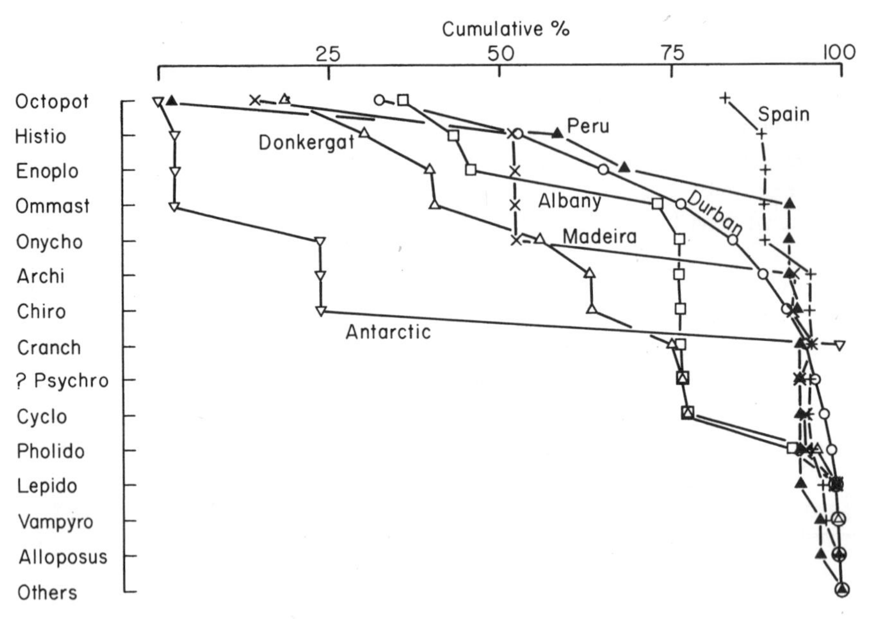

FIG. 19. Cumulative percentage curves to show the importance by weight of local cephalopods in the diet of sperm whales caught in different regions. Weights were estimated from the rostral lengths of lower beaks in the stomachs by the use of LRL: body weight relationships published elsewhere (Clarke, in press).

Antarctic and off Peru and Chile; these include the large *Taningia danae*. Histioteuthids, including several species, are important everywhere except in the Antarctic. Enoploteuthids are most important off South Africa and off Peru and Chile; these are all *Ancistrocheirus*. Ommastrephids are important off Durban, Albany and Peru; *Todarodes* in the Indian Ocean and *Dosidicus gigas* off Peru and Chile. Onychoteuthids are very important in the Antarctic and rather less so elsewhere; mainly *Kondakovia* and *Moroteuthis knipovitchi* in the Antartic and *Moroteuthis robsoni* off South Africa. Architeuthids are important off Madeira, Spain and South Africa. Cranchiids are of great importance in the Antarctic and of some importance off Donkergat, South Africa; the very large *Mesonychoteuthis hamiltoni* is important in the Antarctic, and *Phasmatopsis* and *Taonius* species in South African waters. Pholidoteuthids (these are confused with *Tetronychoteuthis* and it is not clear at present whether *Pholidoteuthis* is a correct name to use or not; Clarke, in press) are important off Donkergat, west South Africa, and Albany, Western Australia. Lepidoteuthids are of minor importance everywhere except the Antarctic and the Peru and Chile regions. Beaks of cycloteuthids were of minor importance off Donkergat, South Africa.

Because the whales are such large predators and consume considerable quantities of flesh (see below), it is likely that the cephalopods in their diet play a very important role in the ecology of the ocean.

Weight eaten by whales and nutrient dispersal

By combining recent estimates of the total number of sperm whales in the sea with estimates of the weight of cephalopods consumed by sperm whales of particular weights (Clarke, in press), it is possible to obtain some indication of the total weight of squid consumed by whales annually.

I have taken the estimate of the total number of sperm whales from a 1973 report of the International Whaling Scientific Committee (IWC/24/4R), in which the world's stock for 1972 was assessed. From a knowledge of natural mortality rates in the exploited population, Dr Gambell (pers. comm.) considers that a good estimate of the total number of sperm whales may be derived as follows (Table I). The total number of males can be obtained by multiplying the estimate of exploitable males of the southern hemisphere and the North Pacific by 2·5 and adding the result to the estimate for the North Atlantic (which I have taken as half the

TABLE I

Estimated body weights of whales and the weight of food they eat

	Males	Females	Total	
1. Estimated no. of exploitable whales in S. Hemisphere	128×10^3	295×10^3		
Estimated no. of exploitable whales in S. Hemisphere × 2·5 and 1·66	320×10^3	491×10^3	811×10^3	
Estimated no. of exploitable whales in N. Pacific	68×10^3	102×10^3		
Estimated no. of exploitable whales in N. Pacific × 2·5	170×10^3	255×10^3	425×10^3	
Estimated no. of exploitable whales in N. Atlantic	11×10^3	11×10^3	22×10^3	
	501×10^3	757×10^3	1258×10^3	
2. Weight—mean	15	5		tons
—total	$7{\cdot}5\times10^6$	$3{\cdot}8\times10^6$	$1{\cdot}1\times10^7$	tons
3. Weight consumed/whale/day				
(2% of weight)	0·3			tons
(3% of weight)		0·15		tons
Weight consumed/whale/yr	109·5	54·8		tons
Weight consumed by species/yr	$5{\cdot}5\times10^7$	$4{\cdot}1\times10^7$	$9{\cdot}6\times10^8$	tons
$\frac{\text{Weight of whale}}{\text{Weight of food}}$			0·103	

See text for explanation.

total estimate of 22 000). The total number of females can be obtained by multiplying the estimated number of exploitable mature females for the southern hemisphere by 1·66 and adding the result to 2·5 times the estimate of females for the North Pacific (taken as three-fifths of the total population since the ratio of males to females is 1 : 1·5 elsewhere) and half the total estimate for the North Atlantic.

Sergeant (1969) plotted daily consumption of captive cetaceans and concluded that large rorquals eat 3.5% of their body weight per day. It seems likely that, given food of the same calorific value daily consumption of sperm whales would lie between 2 and 4% of their body weight per day. I have taken 2% for male whales and 3% for females, as conservative estimates. Commercially killed sperm whales are mainly between 8 and 50 tons in weight, that is nearly all of those between 9·14 m and 15·24 m (30 and 50 ft) in length (according to Gambell, 1972, based on Omura, 1950). If we take a mean weight of 15 tons and 5 tons for males and females respectively to include both the large exploited whales and the smaller individuals we shall probably have a rather low estimate of weight. By multiplying the estimate of the total number of whales of each sex by these estimates of mean weight (Table I), and adding them together, we arrive at a total weight of sperm whales in the sea of 11 million tons.

These conservative estimates show that at least 100 million tons of squid were eaten by sperm whales in 1972. Calculations of consumption of squid are all based upon feeding or calorific values of muscular cephalopods, while in the ocean over 50% of squids eaten are ammoniacal cephalopods and from their low protein content and structure (Denton, 1974) clearly have less than half the calorific value. As a maximum figure, if one accepts a value of 15 tons for the average whale and a consumption of 3·5% of body weight per day as well as an average calorific value of 75% of that for muscular squids then the estimate of cephalopods eaten would be 320 million tons.

The present stock of sperm whales has declined since 1946 when it was virtually unexploited. An estimate of the 1946 exploitable stock is $9{\cdot}9 \times 10^5$ whales (IWC/24/4R) which, if treated in the same way as the 1972 stock estimate, suggests a total population of about 2 million sperm whales in 1946. Assuming an average weight of 15 tons for males and 5 tons for females, their total biomass would be 26 million tons; these whales probably consumed 260 million tons of cephalopods. These figures are very large when compared with estimates of world fish resources (Gulland, 1971). The total world catch of fish from the oceans is now between 60 and 70 million tons per year.

We know that the cephalopods eaten by sperm whales are mainly bathypelagic or bathybenthic and are carnivorous, so that we can be sure these squids are in the fourth or higher trophic level. Sperm whales must therefore be in the fifth or higher trophic level of Ryther (1969). While we cannot determine the proportion of the fifth trophic level represented by sperm whales, it is worth recalling some of the other squid-eaters which are very probably in this level. Other whales believed to be deep divers, such as the pigmy sperm whale (*Kogia breviceps*) and the bottlenosed whale (*Hyperoodon*), also eat squid, probably from the mesopelagic and bathypelagic depths; animals in which cephalopods are a very important constituent of the diet include several beaked whales, pilot whales, many other small cetaceans (Delphinidae), many seals (e.g. *Mirounga, Otaria*), and many deep-living fish, such as the scabbard fish *Aphanopus*, and *Alepisaurus ferox* (see Rancurel, 1970), several tuna species (e.g. Williams, 1966), and oceanic birds such as albatrosses, penguins (Murphy, 1936) and many other sea birds (Ashmole & Ashmole, 1967). The few estimates of population of these squid-eaters are not too reliable but brief reference to the seals gives some idea of the biomass of other members of the same trophic level as sperm whales. By far the most important squid-eating seals are the southern elephant seal weighing 1–4 tons and numbering $600–700 \times 10^3$ (derived from Laws, 1960; King, 1964), the fur seals weighing 0·01–0·4 tons and numbering $0{\cdot}7 \times 10^6$, and the sea lions weighing 0·15–1·0 tons and numbering $1{\cdot}2–1{\cdot}3 \times 10^6$ (King, 1964). By taking middle range estimates of their weight and numbers they still only comprise $2{\cdot}5 \times 10^6$ tons or 22% of the sperm whale weight, and eat perhaps something in the region of $10–20 \times 10^6$ tons of squid per year. Thus, taking this into consideration we might expect the sperm whale to comprise a large proportion by weight of its trophic level.

In conclusion, the weight of cephalopods consumed by whales is extremely large by any standards, and when one considers that much of the squid population must escape predation by sperm whales it is clear that cephalopods form a very important part of oceanic life. Any vertical or horizontal movement of the most numerous cephalopods must redistribute the nutrient in the deeper layers of the ocean. Where such species annually ascend the slope into shallow shelf regions, as does *Todarodes sagittatus*, they may introduce nutrient at high trophic levels into shallow waters.

Sperm whales probably carry nutrient upwards by eating deep-sea squid and defaecating near the surface. If only 5% of fat, carbohydrate and protein in the food remained in the faeces a considerable quantity would be involved, though this would be but a small fraction of the nutrient in the euphotic zone.

DISCUSSION

While samples from the various nets and predators considered here are not strictly comparable because of differences in time and space, the data from stomachs of sperm whales are so extensive, both geographically and numerically, that broad comparisons are valid. Clearly, there are many more cephalopods in the sea than net sample analysis suggests. Also, many species are large and they are much more important in the ecology of the sea than has been suspected from purely net samples. Only in the last decade or so have really large nets of commercial size and construction been used extensively in the deep sea for research into the larger fish and squid species in midwater and on the continental slopes. Samples from the largest of these nets give reason to hope that before too long man will be able to sample cephalopods as effectively as the whale. However, the largest Engels trawl used by the author is the 1600 mesh net (mouth 116 ft × 65 ft or 35·4 m × 19·8 m), and this is certainly not taking the same species or sampling the families in the same ratio as the sperm whale. This may not just be a deficiency of size and speed, since most of the whales examined are caught over the continental slope or in other areas where the bottom topography discourages deep-water fishing activities; the whales are not so restricted. Certainly in the author's experience when large nets have been accidentally fished close to the bottom in slope or island areas, some large species taken by sperm whales have been sampled; specimens of *Taningia danae* exceeding 20 cm mantle length were caught off the Azores in this way, and intentional bottom trawling on the slope has provided an extremely large *Octopoteuthis* measuring 51 cm ML and numerous large *Histioteuthis miranda*. In addition, while the RMT nets and IKMT nets in the series described above caught very small cephalopods, these nets occasionally catch large specimens (e.g. of *Todarodes sagittatus* west of Scotland and *Ommastrephes pteropus* near Madeira) and the definite impression is that small nets used near the slope stand a better chance of catching large specimens than the same nets over the abyssal plain. As pointed out by Cushing (1971) the catches of sperm whales by old American whaleships, reported by Townsend (1935), are a very good indicator of upwelling regions. As most of these are near land and slope regions, it is tempting to conclude that the squid are restricted to such regions and the sperm whales are not feeding as they cross the intervening ocean. So far, no material has come into the author's possession which answers this question completely; samples that straddle the Tasman Sea certainly show the cephalopod species to be present throughout the region, but much of it is less than 3000 m deep and must have topography rather similar to slope areas.

The whales certainly eat several midwater species but they also eat species in spawning condition, some of which they almost certainly take on the bottom (Clarke, in press). It is certain that sperm whales not infrequently dive to depths greater than 1000 m and there are indications that they go deeper (Clarke, 1976).

Many families of cephalopods have large amounts of ammonia in their tissues or coelom (Denton & Gilpin-Brown, 1973) and these are indicated by an N in Fig. 17. Ammoniacal squids comprise 53–78% of the number of cephalopods consumed by sperm whales; the percentage varies from region to region.

Samples from nets and predators are complementary in content and both are needed to obtain anything approaching a realistic picture of the cephalopods of a region; for this, a variety of nets and predators are required. Until now such integration of data in a quantitative manner has not been possible for any area.

For example, in the sperm whale diet, the Histioteuthidae and Octopoteuthidae are numerically the most important families in all areas except the Antarctic, New Zealand and the North Pacific where the Onychoteuthidae are most numerous. The Cranchiidae form a very large part of their diet in the Antarctic and the Tasman Sea. In the large Engels trawl used in the North Atlantic the Enoploteuthidae predominate, whereas in the smaller RMT and IKMT nets used for the series in the North Atlantic, the Cranchiidae are very important from 60°N to 11°N while the Octopoda, the Sepiolidae and the Spirulidae assume considerable importance near land. Histioteuthidae are usually of little importance in nets. Off California the Cranchiidae, Enoploteuthidae and Gonatidae are most important in the IKMT catches. In high latitudes the Brachioteuthidae and, in the Antarctic, the Bathyteuthidae become very important in both IKMT and RMT nets. In the diet of the other predators considered here, the Onychoteuthidae and Ommastrephidae are most important in that of porpoises off Central America, Enoploteuthidae and Ommastrephidae in tuna off East Africa and Central America, Cranchiidae in blue sharks off southern England, Onychoteuthidae, Cranchiidae and Octopodidae in Weddell seals of the Antarctic, and Ommastrephidae, Histioteuthidae, Octopoteuthidae, Cranchiidae in several bird species. Surface observations, hand-lines and hand-nets nearly always sample ommastrephids throughout the temperate and tropical deep oceans, and in the North Atlantic this is also true of cameras used in midwater (Baker, 1957a).

Significant advances in cephalopod ecology can only be made by utilizing information from a range of nets and predators. The nets used by most research workers at present must be augmented rather than replaced by large commercial-type trawls and predators should be actively sorted and studied as samplers, instead of the stomach contents being merely a part of a study of the predators' ecology; the emphasis often affects the detail.

The factors in a net's performance that make it able to catch a certain species with particular swimming, behavioural and distributional characters are still not known for any sub-surface oceanic species. We know that to increase size or speed of a net increases the size of specimens and the number of species caught. (See chapter by Roper, this volume.) There is also some unpublished evidence that the bigger trawls not only catch larger specimens of the big species, but also catch larger specimens of small species, such as *Spirula*, where size would not be expected to vary when the net mouth changes from 8 m^2 to 600 m^2. A sampling question is posed by the catches of the vertical series of the North Atlantic of RMT nets. There is a striking fall off in the number of species from the tropics to 60°N (Fig. 20) such that there is a reduction by half for each 10° of latitude. Now as we know that the total number of species present at depths trawled exceed by a few species those represented in the RMT series, it would appear that the net is taking a consistent slice of the squid community in spite of the change in the species over the latitudinal range (Clarke & Lu, 1974, 1975; Lu & Clarke, 1975a,b). Thus, the nets are possibly sampling squids with particular adaptations in the community, and

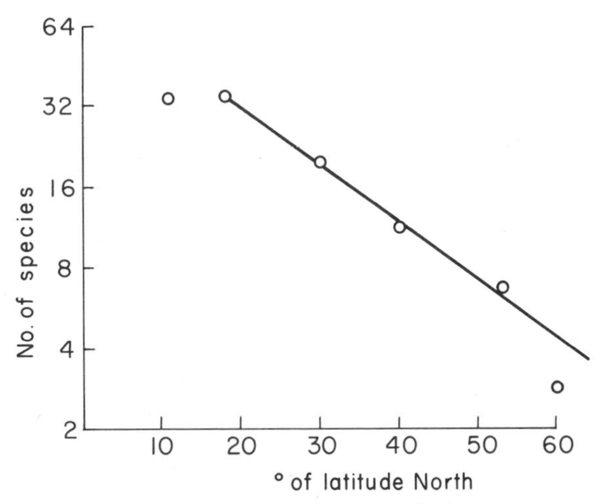

FIG. 20. The number of species caught in the RMT series in the eastern North Atlantic at 20°–23°W and the latitudes indicated.

the adaptations, irrespective of species, are the reasons for the nets' selection. Perhaps other, larger gear will move the species line for RMT nets upwards without changing its angle to the baseline. If this were found to be true, it would be possible to predict the number of species at one latitude from an extensive trawling programme at other latitudes so that it might prove possible to reduce time spent at sea in certain studies.

Finally, it should be pointed out that not all beaks in sperm whales can be identified even to a family, and some beaks of these are large, suggesting squids of perhaps 2 m in mantle length.

Much remains to be discovered about the cephalopods of the deep sea, but study of stomach contents of their predators provides some measure of the inefficiency of our own sampling methods.

Acknowledgements

The large collections described here were only made and analysed by the active co-operation of a large number of people to whom I owe my warmest thanks. In particular Mr Neil MacLeod has given tremendous help with collecting, identifying, counting and measuring beaks and I am extremely grateful for his help, enthusiasm and helpful criticism. I am also greatly indebted to Mr A. de C. Baker, Dr J. Bannister, Mr G. Battin, Dr P. Best, Mr S. G. Brown, Miss K. Chidgey, Dr R. Clarke, Dr R. Gambell, Dr C. C. Lu, Dr I. MacAskie, Mr N. Merrett, Dr Ohsumi, Dr G. C. Pike, Mr J. Stevens, as well as the staffs of the Union Whaling Co. Ltd, Durban, the Institute of Oceanographic Sciences, England, RRS *Discovery*, RRS *Challenger*, RV *Sarsia*, and the ocean weather ships. To Mr P. M. David and the Director of the Institute of Oceanographic Sciences, I am most grateful for encouragement and permission to work on the material described here.

References

Ahlstrom, E. H. (1948). A record of pilchard eggs and larvae collected during surveys made in 1939 to 1941. *Spec. Scient. Rep. U.S. Fish Wildl. Serv. Fish.* No. 54: 1–74.

Akimushkin, I. I. (1955). [Nature of the food of the cachalot.] *Dokl. Akad. Nauk SSSR* **101**: 1139–1140. [In Russian.]

Ashmole, N. P. & Ashmole, M. J. (1967). Comparative feeding ecology of sea birds of a tropical oceanic island. *Bull. Peabody Mus. nat. Hist.* **24**: 1–131.

Baker, A. de C. (1957a). Underwater photographs in the study of oceanic squid. *Deep-Sea Res.* **4**: 126–129.

Baker, A. de C. (1957b). Some observations on large oceanic squids. *Rep. Challenger Soc.* **3** (9): 34.

Baker, A. de C. (1960). Observations of squid at the surface in the N.E. Atlantic. *Deep-Sea Res.* **6**: 206–210.

Baker, A. de C., Clarke, M. R. & Harris, M. J. (1973). The N.I.O. combination net (RMT 1+8) and further developments of rectangular midwater trawls. *J. mar. biol. Ass. U.K.* **53**: 167–184.
Betesheva, E. I. & Akimushkin, I. I. (1955). [Food of the sperm whale (*Physeter catodon*) in the Kurile Islands region.] *Trudȳ Inst. Okeanol.* **18**:86–94. [In Russian.]
Clarke, M. R. (1962a). Stomach contents of a sperm whale caught off Madeira in 1959. *Norsk Hvalfangsttid.* **51**: 173–191.
Clarke, M. R. (1962b). The identification of cephalopod "beaks" and the relationship between beak size and total body weight. *Bull. Br. Mus. nat. Hist.* **8**: 419–480.
Clarke, M. R. (1962c). Significance of cephalopod beaks. *Nature, Lond.* **193**: 560–561.
Clarke, M. R. (1966). A review of the systematics and ecology of oceanic squids. *Adv. mar. Biol.* **4**: 91–300.
Clarke, M. R. (1969a). Cephalopoda collected on the SOND Cruise. *J. mar. biol. Ass. U.K.* **49**: 961–976.
Clarke, M. R. (1969b). A new midwater trawl for sampling discrete depth horizons. *J. mar. biol. Ass. U.K.* **49**: 945–960.
Clarke, M. R. (1972). New technique for the study of sperm whale migration. *Nature, Lond.* **238**: 405–406.
Clarke, M. R. (1976). Observation on sperm whale diving. *J. mar. biol. Ass. U.K.* **56**: 809–810.
Clarke, M. R. (in press). Cephalopoda in the diet of sperm whales of the Southern Hemisphere and their bearing on sperm whale biology. *Discovery Rep.*
Clarke, M. R. & Lu, C. C. (1974). Vertical distribution of cephalopods at 30°N 23°W in the North Atlantic. *J. mar. biol. Ass. U.K.* **54**: 969–984.
Clarke, M. R. & Lu, C. C. (1975). Vertical distribution of cephalopods at 18°N 25°W in the North Atlantic. *J. mar. biol. Ass. U.K.* **55**: 165–182.
Clarke, M. R. & MacLeod, N. (1974). Cephalopod remains from a sperm whale caught off Vigo, Spain. *J. mar. biol. Ass. U.K.* **54**: 959–968.
Clarke, M. R. & MacLeod, N. (in prep. a). *Cephalopod remains from the stomachs of sperm whales caught in the Tasman Sea.*
Clarke, M. R. & MacLeod, N. (in prep. b). *Cephalopod remains from the stomachs of sperm whales caught off western Canada.*
Clarke, M. R. & MacLeod, N. (in prep. c). *Stomach contents of twelve Weddell Seals with particular reference to the cephalopods.*
Clarke, M. R., MacLeod, N. & Paliza, O. (1976). Cephalopod remains from the stomachs of sperm whales caught off Peru and Chile. *J. Zool., Lond.* **180**: 477–493.
Clarke, M. R. & Merrett, N. (1972). The significance of squid, whale and other remains from the stomachs of bottom-living deep-sea fish. *J. mar. biol. Ass. U.K.* **52**: 599–603.
Clarke, M. R. & Stevens, J. D. (1974). Cephalopods, blue sharks and migration. *J. mar. biol. Ass. U.K.* **54**: 949–957.
Clarke, R. (1956). Sperm whales of the Azores. *Discovery Rep.* **28**: 237–298.
Cushing, D. H. (1971). Upwelling and the production of fish. *Adv. mar. Biol.* **9**: 255–334.
Denton, E. J. (1974). On buoyancy and the lives of modern and fossil cephalopods. *Proc. R. Soc.* (B) **185**: 273–299.
Denton, E. J. & Gilpin-Brown, J. B. (1973). Flotation mechanisms in modern and fossil cephalopods. *Adv. mar. Biol.* **11**: 197–268.

Foxton, P. (1963). An automatic opening-closing device for large midwater plankton nets and midwater trawls. *J. mar. biol. Ass. U.K.* **45**: 295–308.
Foxton, P. (1969). SOND Cruise 1965: Biological sampling methods and procedures. *J. mar. biol. Ass. U.K.* **49**: 603–620.
Gambell, R. (1972). Sperm whales off Durban. *Discovery Rep.* **35**: 199–358.
Gaskin, D. E. & Cawthorn, M. W. (1967). Diet and feeding habits of the sperm whale (*Physeter catodon* L.) in the Cook Strait region of New Zealand. *N.Z. Jl mar. Freshwat. Res.* **1**: 156–179.
Gulland, J. A. (1971). *The fish resources of the ocean.* London: Fishing News (Books) Ltd.
Harris, M. P. (1973). The biology of the waved albatross *Diomedea irrorata* of Hood Island, Galapagos. *Ibis* **115**: 483–510.
Imber, M. J. (1973). The food of grey-faced petrels (*Pterodroma macroptera gouldi* Hutton), with special reference to diurnal vertical migration of their prey. *J. Anim. Ecol.* **42**: 645–662.
King, J. E. (1964). *Seals of the world.* London: British Museum (Natural History).
Laws, R. M. (1960). The southern elephant seal (*Mirounga leonina* Linn.) at South Georgia. *Norsk Hvalfangsttid.* **49**: 466–476, 520–542.
Lu, C. C. & Clarke, M. R. (1975a). Vertical distribution of cephalopods at 40°N, 53°N and 60°N at 20°W in the North Atlantic. *J. mar. biol. Ass. U.S.* **55**: 143–163.
Lu, C. C. & Clarke, M. R. (1975b). Vertical distribution of cephalopods at 11°N 20°W in the North Atlantic. *J. mar. biol. Ass. U.K.* **55**: 369–389.
Murphy, R. C. (1936). *Oceanic birds of South America.* **1** & **2**. New York: American Museum of Natural History.
Nesis, K. N. (1973). [Cephalopods of the eastern equatorial and southeastern Pacific.] *Trudȳ Inst. Okeanol.* **94**: 188–240. [In Russian.]
Nesis, K. N. (1974a). [Oceanic cephalopods of the southwestern Atlantic Ocean.] *Trudȳ Inst. Okeanol.* **14**: 533–537. [In Russian.]
Nesis, K. N. (1974b). [Cephalopod larvae in the Western Arabian Sea.] *Okeanologiya* **14**: 537. [In Russian.]
Okutani, T. & McGowan, J. A. (1969). Systematics, distribution, and abundance of the epiplanktonic squid (Cephalopoda, Decapoda) larvae of the California Current April, 1954–March, 1957. *Bull. Scripps Instn Oceanogr.* **14**: 1–90.
Omura, H. (1950). On the body weight of the sperm and sei whales located in the adjacent waters of Japan. *Scient. Rep. Whales Res. Inst., Tokyo* **4**: 1–13.
Perrin, W. F., Warner, R. R., Fiscus, C. H. & Holts, D. B. (1973). Stomach contents of porpose *Stenella* spp., and yellowfin tuna, *Thunnus albacares*, in mixed species aggregations. *Fishery Bull. natn. oceanic atmos. Adm. U.S.* **71**: 1077–1092.
Rancurel, P. (1970). Les contenus stomacaux *d'Alepisaurus ferox* dans le sud-ouest Pacifique (Céphalopodes). *Cah. ORSTOM* (Oceanogr.) **8**: 4–87.
Roper, C. F. E. (1969). Systematics and zoogeography of the worldwide bathypelagic squid *Bathyteuthis* (Cephalopoda: Oegopsida). *Bull. U.S. natn. Mus.* **291**: 1–210.
Ryther, J. H. (1969). Photosynthesis and fish production in the sea. *Science, N.Y.* **166**: 72–76.
Sergeant, D. E. (1969). Feeding rates of Cetacea. *Fiskdir. Skr.* (ser. Havanders.) **15**: 246–258.
Stonehouse, B. (1962). Ascension Island and the British Ornithologists' Union Centenary Expedition 1957–59. *Ibis* **103b**: 107–123.

Tarasevich, M. N. (1968). Dependence of distribution of the sperm whale males upon the character of feeding. *Zool. Zh.* **47**: 1683–1688.
Townsend, C. H. (1935). The distribution of certain whales as shown by the logbook of American whaleships. *Zoologica, N.Y.* **19**: 1–50.
Williams, F. (1966). Food of longline-caught yellowfin tuna from East African waters. *East Afr. agric. For. J.* **31**: 375–382.
Young, R. E. (1972). The systematics and areal distribution of pelagic cephalopods from the seas off Southern California. *Smithson. Contrib. Zool.* No. 97: 1–159.

Appendix A

List of samples compared

A. *Bay of Biscay Region*
1. 92 beaks from 26 *Prionace glauca* L. June–September (Clarke & Stevens, 1974).
2. 70 beaks from one sperm whale. 41°32′N, 9°48′W. 22.6.66 (Clarke & MacLeod, 1974).
3. 97 cephalopods from 32 open RMT 8, RMT 1 and Boris trawls. June, September and November.

B. *Eastern North Atlantic, 53°–55°N*
1. 375 cephalopods caught with 24 opening-closing RMT 1 + 8 combination net at 53°N 20°W. Maximum depth 2000 m. 16–28.5.71 (Lu & Clarke, 1975a).
2. 84 cephalopods caught with an open RMT 8. July 1973. Various lengths of wire out up to 2000 m.
3. Over 200 cephalopods caught by hand-nets and hand-lines at 53°N 20°W from weather ship *Juliett.*

C. *Eastern North Atlantic, 37°–40°N, 20°–25° W*
1. 53 cephalopods caught by opening–closing RMT 1 + 8 combination net. 35 hauls to a maximum of 2000 m. 1–3.10.70 at 40°N 20°W (Lu & Clarke, 1975a).
2. 110 cephalopods removed from stomachs of sperm whales at the Azores (R. Clarke, 1956).
3. 555 cephalopods from 11 EMT nets: two at 38°54′N, 21°55′W, and nine at 37°N 25°W. 12–19.10.66. 305–980 m wire out.

D. *The Madeira Region and 30°N 20° W*
1. 255 cephalopods from 32 opening-closing RMT 1 + 8 combination net to maximum depth of 2020 m. 31.3.72–6.4.72 (Clarke & Lu, 1974).
2. 215 cephalopods from 45 hauls with an IKMT, fitted with a catch-dividing bucket, to a maximum depth of 1400 m near 30°N 23°W (Clarke & Lu, 1974).
3. 148 cephalopods from three hauls with a BCMT fished to a maximum depth of 100 m. 12.6.62–25.9.62 (Clarke & Lu, 1974).
4. 122 cephalopods from nine hauls with an EMT fished to a maximum depth of 590 m off Madeira. 387–800 m wire out.
5. 2136 lower beaks from the stomach of a sperm whale caught off Madeira (Clarke, 1962a; Clarke & MacLeod, 1974).

E. *Canary Islands, East of Fuerteventura, 28°N 14°W*
1. 185 cephalopods from 38 IKMT nets and 38 N113 nets (1 m^2 ring nets), with catch-dividing buckets, to a maximum of 975 m (Clarke, 1969a).
2. 833 cephalopods from nine hauls with an EMT. 5–11.11.66 28°N 14°W. 380–800 m wire out.

F. *Eastern North Atlantic*

1. >4000 cephalopods from all the 571 opening-closing net hauls taken at the stations selected for vertical series, i.e. 60°N 20°W, 53°N 20°W, 40°N 20°W, 30°N 20°W, 18°N 25°W, 11°N 20°W and 28°N 14°W (Clarke, 1969a; Clarke & Lu, 1974, 1975; Lu & Clarke, 1975a & b). *Note.* In table 4 of Clarke & Lu, 1975, and in tables 1, 4, 5, 6, 7, 8 of Lu & Clarke, 1975b, the volume of water is expressed in 10 000 m^3 and not in 1 km^3 as indicated.
2. 2136 lower beaks from one whale examined at Madeira, August 1959, and 70 beaks from another examined at Vigo, Spain and caught at 41°32′N, 9°48′W on 22.6.66 (Clarke, 1962a; Clarke & MacLeod, 1974).
3. Observations and specimens while lining and hand-netting during cruises of *Discovery II* and 17 cruises by the author (Baker, 1960; Clarke, 1966).
4. Remains of 15 cephalopods from stomachs of bottom sharks caught on longlines between 377 and 2866 m in several positions and months (Clarke & Merrett, 1972).

G. *South Atlantic*

1. 51 769 lower beaks and 152 other specimens from stomachs of sperm whales caught off Donkergat, South Africa (49 402 of these lower beaks were from local species) (Clarke, in press).
2. 11 cephalopods from stomachs of sperm whales caught at 27°–31°S, 32°–33°W (Clarke, in press).
3. Remains of squids in 13 samples, of either the stomach contents or regurgitated food, of birds examined at Ascension Island. December 1957 to May 1958, collected by Dr B. Stonehouse (1962).
4. 274 cephalopods from various nets in south-western Atlantic (Nesis, 1974a). Eleventh Biological cruise of R.V. *Akademik Kurchatov*.

H. *Indian Ocean*

1. 65 755 lower beaks and 1347 other specimens from stomachs of sperm whales caught off Durban, South Africa; 62 794 beaks were from local species (Clarke, in press).
2. 2715 lower beaks and 67 other specimens from stomachs of sperm whales caught off Albany, Western Australia; 2477 beaks were local (Clarke, in press).
3. 1122 cephalopods from yellowfin tuna (*Thunnus albacares*) stomachs collected off East Africa between 2°50′S and 8°40′S and identified by the author (Williams, 1966).
4. Numerous cephalopods caught by hand-nets and lines during a three-month cruise, June–August 1962.
5. 127 young cephalopods caught in an IKS-80 ichthyoplankton net between the Gulf of Aden and the Gulf of Oman at 0–100 m, July–September 1969 (Nesis, 1974b).

I. *Western Pacific*

1. 3202 lower beaks and specimens collected from stomachs of sperm whales caught at various positions in the Tasman Sea (Clarke & MacLeod, in prep. a).
2. 2118 lower beaks partly re-identified from New Zealand (Gaskin & Cawthorn, 1967).
3. 907 lower beaks from stomachs of grey-faced petrels [*Pterodroma macroptera gouldi* (Hutton)] from New Zealand (Imber, 1973).

J. *North-Eastern Pacific*

1. 152 lower beaks and specimens from stomachs of sperm whales caught off western Canada (Clarke & MacLeod, in prep. b).
2. Over 400 lower beaks and cephalopods from stomachs of tuna (*Thunnus albacares*) caught to the west of Central America at 7°–12°N, 90°–93°W (Perrin *et al.*, 1973).
3. 7882 lower beaks and cephalopod remains from stomachs of the porpoises *Stenella attenuata* and *S. longirostris* caught to the west of Central America at 7°–12°N, 90°–93°W (Perrin *et al.*, 1973).
4. Over 3400 cephalopods from 3895 tows with a 1 m-diameter net described by Ahlstrom (1948) from a region extending 20°–42°N and 107°–138°W (1240 hauls contained cephalopods). Maximum depth was about 140 m for nearly all tows (Okutani & McGowan, 1969).
5. Over 5000 cephalopods from 445 midwater tows of a 10 ft (3 m) wide IKMT from about 28°–34°N, 118°–122°W off California (Young, 1972).
6. 299 lower beaks from stomachs of waved albatrosses (*Diomedea irrorata*) at Hood Island, Galapagos Islands (Harris, 1973).

K. *South-Eastern Pacific*

1. 1000 lower beaks from stomachs of sperm whales caught of Peru and Chile (Clarke, MacLeod & Paliza, in 1976).
2. 647 cephalopods from a variety of nets used by the vessels *Akademik Kurchatov* and *Baikal*. 30.10.67–23.11.67 and 26.8.68–14.11.68 from about 8°N–30°S and 70°–100°W (Nesis, 1973).

L. *Antarctic*

1. 6346 lower beaks and 195 identifiable squid remains from stomachs of sperm whales caught at various positions in the Antarctic (Clarke, in press). Only Antarctic species are included.
2. 366 lower beaks from stomachs of Weddell seals (*Leptonycotes weddelli*) killed at Deception Island (10 seals) and Halley Bay (2 seals) Antarctica.

Appendix B

Notes on gear

RMT

Rectangular midwater trawls. These have a mouth cut at 45° to the horizontal with top and bottom bars to hold the net open and a large weight on the lower bar. They are fished at about two knots and the mouth is then at 45° to the water flow. Various sizes have been used and the size is indicated after "RMT" by the number of square meters facing the front when the mouth is at 45°, i.e. the effective fishing area. Open versions of RMT 1, RMT 8, RMT 25 and RMT 90 nets have been used. RMT 1, RMT 8 and RMT 25 nets, which open and close by means of an acoustic signal, have been used many times effectively, and the RMT 1 + 8 combination net has been adopted as the standard net of the Institute of Oceanographic Sciences, England, and has been successfully used many hundreds of times.

IKMT
Isaacs-Kidd midwater trawls. These have a large V-shaped depressor plate to hold the mouth open and to sink the net in the water. The depressor is usually either 5 ft (1·5 m) or 10 ft (3·05 m) across.

EMT
Engels midwater trawls. These are large commercial trawls used from two warps. The mouth is held open by a hundred or so aluminium alloy floats on the headline, heavy chain and weights on the footrope and heavy metal Sübergrub otter boards on bridles to open the net horizontally. These may be of various sizes, but nets that caught samples considered here were 1600 mesh and had mouths 35·4 m wide and 19·8 m deep. Latest models, which are not yet used extensively, have mouths 100 m wide and 30 m deep. A 2400 mesh version is used commercially and by German scientists.

Boris Trawl
This is a modified shrimp trawl with three bridles at each side and a headrope 92 ft (28 m) long.

BCMT
British Columbia midwater trawl. This was used a number of times from RRS *Discovery II* in the North Atlantic. It had a square mouth having sides of about 50 ft (15·2 m) and plywood otter boards on pennants from the junction of the bridles with the two warps.

Abbreviations used in Figures

Allop(os): Alloposidae
Alluro: Alluroteuthidae
Archi: Architeuthidae
Argo(n): Argonautidae
Bathy: Bathyteuthidae
Bolit: Bolitaenidae
Brachio: Brachioteuthidae
Chiro: Chiroteuthidae
Cranch(id)s: Cranchiidae
Ctenopt: Ctenopterygidae
Cyclo: Cycloteuthidae
Enoplo: Enoploteuthidae
Gonat(id)s: Gonatidae
Grimaldi: Grimalditeuthidae
Histio: Histioteuthidae
Joubini: Joubiniteuthidae
Lepido: Lepidoteuthidae
Loligid: Loliginidae
Lyco: Lycoteuthidae
Mastigo: Mastigoteuthidae
Neo: Neoteuthidae
Octopod(s): Octopoda
Octopo(t): Octopoteuthidae
Ommast: Ommastrephidae
Pholido: Pholidoteuthidae
Psychro: Psychroteuthidae
Sepiol: Sepiolidae
Spirul(id): Spirulidae
Thysano: Thysanoteuthidae
Tremoct: Tremoctopodidae
Vampyro: Vampyroteuthidae
Vitrel: Vitreledonellidae

Symp. zool. Soc. Lond. (1977) No. 38, 127–159

LUMINESCENCE IN CEPHALOPODS AND FISH

PETER J. HERRING

Institute of Oceanographic Sciences, Godalming, Surrey, England

SYNOPSIS

Bioluminescence in cephalopods is probably restricted to the Teuthoidea, the Sepioidea and Vampyromorpha. The majority of the genera within these orders are known to include luminous species. Luminescence may be produced either by intrinsic chemical systems or by symbiotic luminous bacteria in special organs. Light organs are found in a variety of anatomical positions and range from simple patches of photogenic tissue to highly elaborate organs with many accessory optical structures. The organization and disposition of cephalopod light organs is described and compared with that of the light organs of fishes. Though there are many similarities between the two groups there are also significant differences, particularly in the reflector systems. The chemistry of cephalopod luminescence, though poorly known, appears to be closely allied to that of many other marine organisms, and, with a few possible exceptions, luminescence is under nervous control. Light organs of different structural types within a single species may operate either independently or in concert. Possible functions considered include ventral camouflage, distraction, disruptive illumination and communication.

INTRODUCTION

Luminescence is a common phenomenon among many groups of deep-sea animals. It reaches its highest degree of diversity and complexity among the cephalopods and fishes of the meso- and bathypelagic realms. Notwithstanding the taxonomic gulf between the two groups, the selection pressures of these environments have in both taxa induced remarkably similar bioluminescent capabilities, whose functional convergence is reflected in the similar distribution and general morphology of the effector systems. Many of these similarities have been noted by Packard (1972) in a general review of the evolutionary convergence of fish and cephalopods, but there are also significant differences between the two groups, determined by dissimilar physiological and structural organizations and behavioural responses. This account is an attempt to delineate both the similarities and the differences in the bioluminescence of cephalopods and fishes, and, where practicable, to interpret them in functional terms. Analogies are drawn where possible, even though they may be superficial, because even this degree of recognition may help to expose further parallels between the two groups.

TAXONOMIC DISTRIBUTION OF LUMINESCENCE

Luminescence is probably confined to the Sepioidea, Teuthoidea and Vampyromorpha (see Table I, which amplifies and amends the previous list given by Harvey, 1952). Chun (1910) found structures in the cirrate octopod *Cirrothauma murrayi* which bore some resemblance to light organs, and Akimushkin (1965) described a new species, *Tremoctopus lucifer*, with orange light organs on the first pair of arms of the female only. The observations of luminescence of this species quoted by Akimushkin were made in the ship's lights and may perhaps have been reflectance rather than luminescence. More certain observations are therefore needed before this exceptional sexual dimorphism can be accepted as a true example of luminescence. Generally all species within a single genus are either luminescent or non-luminescent, but in a few, for example *Loligo*, *Gonatus* and *Grimalditeuthis*, both luminous and non-luminous species occur. Most records of luminescent species are based on the identification of light organs, not on direct observation of luminescence. Even many of the apparent records of luminescence are open to alternative interpretations (Gaskin, 1967; Akimushkin, 1965) and any list of luminous species must therefore be treated with circumspection.

Bacterial luminescence

The use of bacterial symbionts as a source of light is a practice restricted to cephalopods and fishes. The sepiolid genera *Sepiola*, *Rondeletiola*, *Rossia*, *Euprymna*, and the loliginids *Uroteuthis*, certain species of *Loligo* and probably *Doryteuthis* all harbour symbiotic luminous bacteria in paired organs lying against the ink sac near the anus. The light is usually visible as a secretion of the bacterial contents into the mantle cavity and the surrounding sea-water, but in some species at least it can also be observed directly through the walls of the organ. The source of luminescence in the sepiolids *Heteroteuthis* and *Sepiolina* is still unresolved. Although the light organs are apparently almost identical to those of other sepiolids it has not been possible to isolate self-luminous bacterial cultures from them; dried organs of *Heteroteuthis* and *Sepiolina* (Heteroteuthinae) luminesce on addition of fresh water, entirely counter to the concept of a bacterial source (Skowron, 1926; Haneda, 1956). Preliminary investigations involving electron microscopy have also failed to demonstrate any bacteria within the organ (Herring & Dilly, unpublished). The only other cephalopod

Table I

Distribution of luminescence in cephalopods

		Genera containing luminous species	Position of organs: Eyeball	Anal/Ink sac	Other viscera	Arm or tentacle	Arm or tentacle tip	Other integumentary organs		Genera not containing luminous species
TEUTHOIDEA										
OEGOPSIDA										
Ommastrephidae:	5	*Ommastrephes, Symplectoteuthis, Ornithoteuthis, Dosidicus, Hyaloteuthis*	+	+	+	+	–	+	4	*Todarodes, Notodarus, Illex, Todaropsis*
Bathyteuthidae:	1	*Bathyteuthis*	–	–	–	–	–	+	0	
Lycoteuthidae:	5	*Lycoteuthis, Selenoteuthis, Oregoniateuthis, Nematolampas, Lampadioteuthis*	+	+	+	+	+	+	0	
Enoploteuthidae:	9	*Enoploteuthis, Enigmoteuthis, Abraliopsis, Abralia, Watasenia, Pterygioteuthis, Pyroteuthis, Thelidioteuthis, Ancistrocheirus*	+	+	+	+	+	+	0	
Histioteuthidae:	1	*Histioteuthis*	–	–	–	+	+	+	0	
Cycloteuthidae:	2	*Cycloteuthis, Discoteuthis*	+	+	–	–	–	+	0	
Onychoteuthidae:	2	*Onychoteuthis, Chaunoteuthis*	+	+	+	–	–	–	4?	*Onykia, Ancistroteuthis, Kondakovia, Moroteuthis*?
Gonatidae:	1	*Gonatus*	+	–	–	–	–	–	2	*Gonatopsis, Berryteuthis*
Ctenopterygidae:	1	*Ctenopteryx*	+	+	+	–	–	–	0	

continued

TABLE I—*continued*

		Genera containing luminous species	Position of organs							Genera not containing luminous species
			Eyeball	Anal/Ink sac	Other viscera	Arm or tentacle	Arm or tentacle tip	Other integumentary organs		
Octopoteuthidae:	2	*Octopoteuthis, Taningia*	−	+	−	+	+	+	0	
Chiroteuthidae:	2	*Chiroteuthis, Chiropsis*	+	+	−	+	+	−	1	*Valbyteuthis*
Mastigoteuthidae:	2	*Mastigoteuthis, Echinoteuthis*	+	−	−	+	−	+	0	
Cranchiidae:	24	*Cranchia, Liocranchia, Leachia, Drechselia, Ascocranchia, Egea, Crystalloteuthis, Sandalops, Corynomma, Bathothauma, Taonidium, Taonius, Teuthowenia, Fusocranchia, Toxeuma, Megalocranchia, Helicocranchia, Phasmatopsis, Galiteuthis, Verrilliteuthis, Mesonychoteuthis, Hensenioteuthis, Anomalocranchia, Uranoteuthis*	+	+	−	−	+	−	0	
Grimalditeuthidae:	1	*Grimalditeuthis*	−	−	−	−	+	−	0	
Psychroteuthidae:	0		−	−	−	−	−	−	1	*Psychroteuthis*
Architeuthidae:	0		−	−	−	−	−	−	1	*Architeuthis*
Neoteuthidae:	0		−	−	−	−	−	−	2	*Neoteuthis, Alluroteuthis*
Thysanoteuthidae:	0		−	−	−	−	−	−	2	*Thysanoteuthis, Cirrobrachium*

Brachioteuthidae:	0		–	–	–	–	–	–	1	*Brachioteuthis*
Batoteuthidae:	1	*Batoteuthis*	–	–	–	–	+	–	0	
Lepidoteuthidae:	0		–	–	–	–	–	–	3	*Lepidoteuthis, Pholidoteuthis, Tetronychoteuthis*
Promachoteuthidae:	0		–	–	–	–	–	–	1	*Promachoteuthis*
Joubiniteuthidae:	0		–	–	–	–	–	–	1	*Joubiniteuthis*
MYOPSIDA										
Loliginidae:	3	*Loligo, Uroteuthis, Doryteuthis*	–	+	–	–	–	–	5	*Lolliguncula, Lolidopsis, Loliolus, Alloteuthis, Sepioteuthis*
Pickfordiateuthidae:	0		–	–	–	–	–	–	1	*Pickfordiateuthis*
SEPIOIDEA										
Sepiolidae:	7	*Sepiola, Sepiolina, Rossia, Euprymna, Inioteuthis, Heteroteuthis, Rondeletiola*	–	+	–	–	–	–	1	*Sepietta*
Sepiidae:	0		–	–	–	–	–	–	3	*Sepia, Sepiella, Hemisepius*
Sepiadariidae:	0		–	–	–	–	–	–	1	*Sepiadarium*
Idiosepiidae:	0		–	–	–	–	–	–	1	*Idiosepius*
Spirulidae:	1	*Spirula*	–	–	–	–	–	+	0	
VAMPYROMORPHA										
Vampyroteuthidae:	1	*Vampyroteuthis*	–	–	–	+	–	+	0	

As far as is known no member of the OCTOPODA is luminescent; possible exceptions, in the Philonexidae and the Cirroteuthidae, are discussed in the text. Taxonomy of Teuthoidea based on Roper, Young & Voss (1969) with additions. Cranchiid genera from Clarke (1966).

that may use luminous bacteria is *Spirula*, which has a single light organ between the tail fins. Self-luminous bacterial cultures have been obtained from the region of this organ, but the evidence for bacterial symbiosis is not yet unequivocal.

Bacterial luminescence in fishes is a more widespread phenomenon (Tett & Kelly, 1973) and there is considerable variety in the mode of operation of the organs. The expulsion of luminous bacteria into the water by the sepiolids and loliginids is paralleled by the action of the caruncles on the angler-fishes *Cryptopsaras* and *Ceratias*, and the anal or mid-ventral organs of some macrurids. The bacteria from angler-fishes have not yet been obtained as self-luminous cultures, whereas those of macrurids and the Opisthoproctinae, like those of most sepiolids, can readily be cultured in this form. In cephalopods no more than a pair of bacterial organs is ever found, and there are never any additional intrinsic organs. This generalization is also applicable to nearly all fishes. The only exceptions are certain angler-fishes, in which up to four bacterial organs may be found, as in *Cryptopsaras*, or a bacterial organ in the lure (or esca) may be allied with intrinsic organs in the hyoid barbel, as in *Linophryne* (Hansen & Herring, 1977).

Intrinsic luminescence

Most luminescent systems in cephalopods and fishes are not bacterial but intrinsic. The organs vary very greatly in their organization and in the complexity of the associated accessory optical structures. Simple light organs, groups of photocytes with no more than a simple pigment cup or an undifferentiated reflector, are only rarely present in cephalopods. The scattered body photophores of *Vampyroteuthis* provide one example, but the most widespread type is that found in the ommastrephids (Roper, 1963; Clarke, 1965) in which small patches of luminous tissue are found embedded in the general body musculature, usually on the ventral side. These simple units may be aggregated to form lines of tissue ventrally, as in *Symplectoteuthis*, or dorsal patches as in *Ommastrephes pteropus* (Fig. 5 below). No other light organs are known in the ommastrephids, though in *Symplectoteuthis* and *Ornithoteuthis* additional accessory structures have been developed (Okada, 1968). Simple organs are otherwise almost totally lacking in cephalopods, though an undescribed simple light organ is found posteriorly in the males of *Ctenopteryx*. In fishes, however, simple light organs are very common, particularly in the stomiatoids, many of which have the general body surface and fins covered with a great many such organs, either with or without a pigment cup.

Compound light organs, with accessory lenses, reflectors and pigment or chromatophore "filters", have attained a very high degree of development in cephalopods and in fishes, in both of which groups considerable morphological variation is to be found. Some species have only a single type of compound organ, for example *Thelidioteuthis, Mastigoteuthis* and *Bathyteuthis* among the cephalopods and the Sternoptychidae among the fishes. Others may bear several different types of compound organs, as in the Lycoteuthidae, *Abraliopsis, Chiroteuthis* and *Phasmatopsis* among cephalopods and many of the stomiatoids among fishes. Even some of the smallest light organs such as the mantle organs in the Enoploteuthinae, and the epidermal organs of sharks, are complex in organization with lenses and other associated optical structures (Chun, 1910; Iwai, 1960).

ANATOMICAL DISTRIBUTION OF LIGHT ORGANS

The position of light organs on an animal's body is determined by the purpose of the light, within the obvious confines of the organization of the species concerned. The most abundant organs in both groups of animals are those which lie in the superficial tissues of the body. These may occur in very large numbers both in stomiatoid fishes, several thousand in some species, and in the ommastrephids and enoploteuthids. In *Enoploteuthis, Abralia, Abraliopsis, Watasenia* and *Enigmoteuthis* the ventral surfaces of the head, mantle and some of the arms are covered with minute light organs, which may be arranged in longitudinal rows. Numerous superficial organs also characterize the histioteuthids, where they are larger, but similarly distributed, though often more regular in arrangement than in the enoploteuthids. The Mastigoteuthidae also bear a smaller number of superficial organs, while in *Bathyteuthis* they are found only at the bases of the arms (Roper, 1969). Most enoploteuthids differ from other cephalopods with superficial organs in that they also have sub-ocular organs of quite different appearance; this is not so in *Thelidioteuthis* or *Ancistrocheirus.* Superficial organs are generally ventral in distribution, but they may be dorsal in some ommastrephids. A few of the organs of enoploteuthids, mastigoteuthids and histioteuthids are also placed dorsally. The organs of *Bathyteuthis* are largely dorsal, and the superficial organs of *Vampyroteuthis* extend right round the body, though they are most numerous ventrally.

The small superficial organs in fishes are rarely confined to the ventral aspect of the body, usually occurring all over the surface as

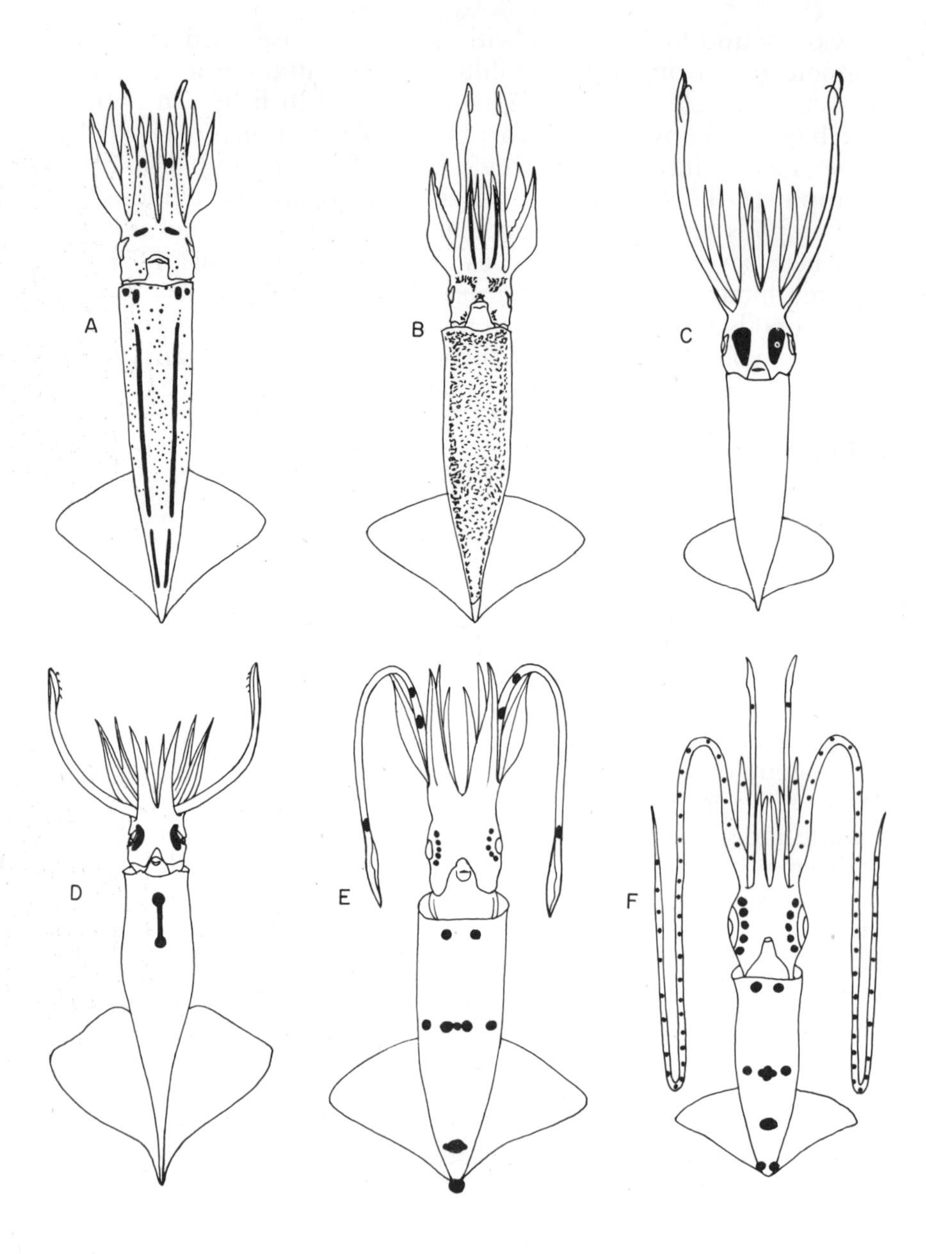

FIG. 1. Positions of the light organs of cephalopods. A. *Symplectoteuthis*; B. *Ommastrephes*; C. *Gonatus*; D. *Onychoteuthis*; E. *Selenoteuthis*; F. *Nematolampas*; (after Voss, 1962; Young, 1972) (not to scale).

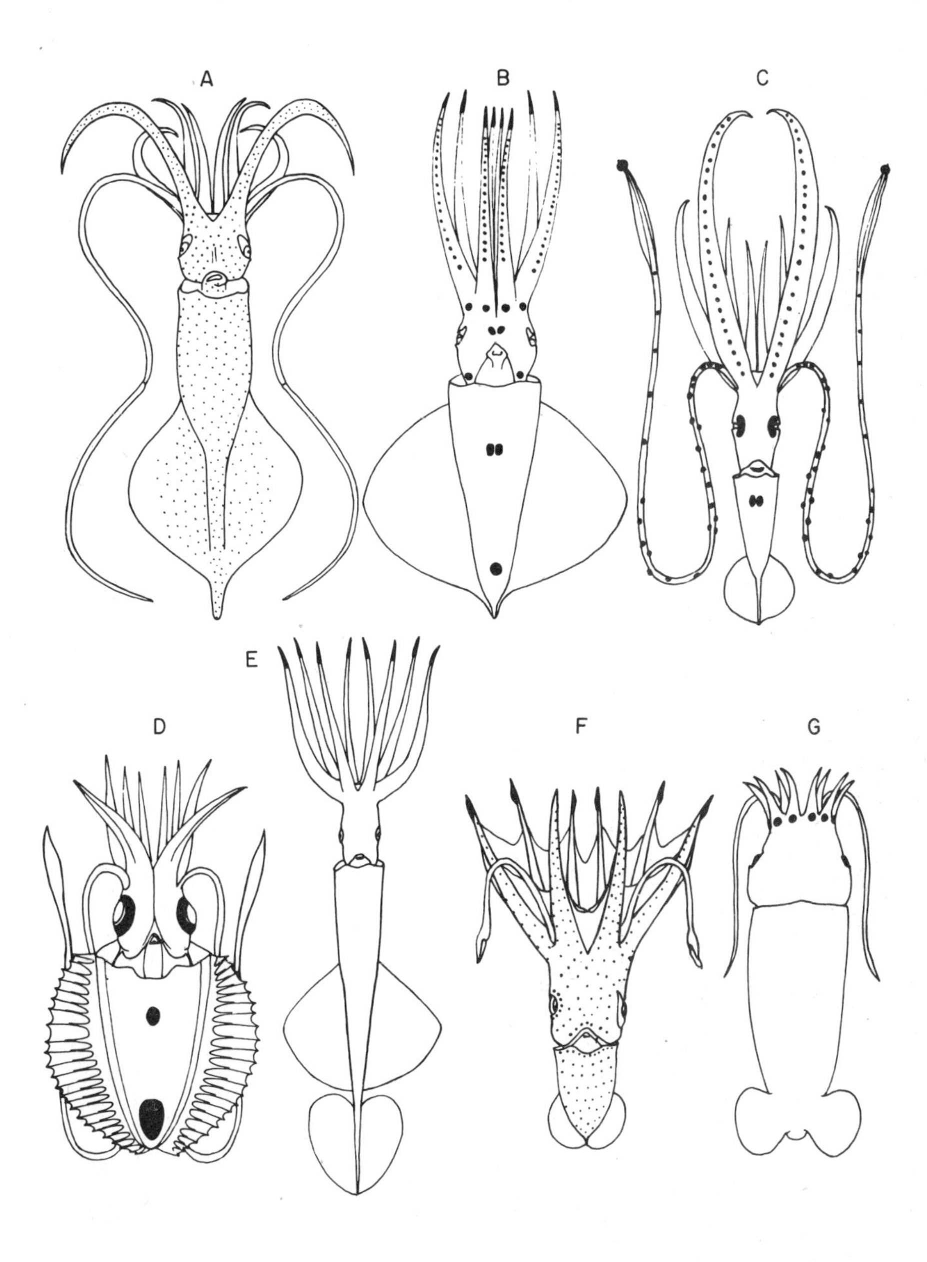

FIG. 2. Light organs of cephalopods. A. *Mastigoteuthis*; B. *Octopoteuthis*; C. *Chiroteuthis*; D. male *Ctenopteryx*; E. *Grimalditeuthis*; F. *Histioteuthis*; G. *Bathyteuthis*; (dorsal) (after Roper, 1969; Voss, 1969; Young, 1972).

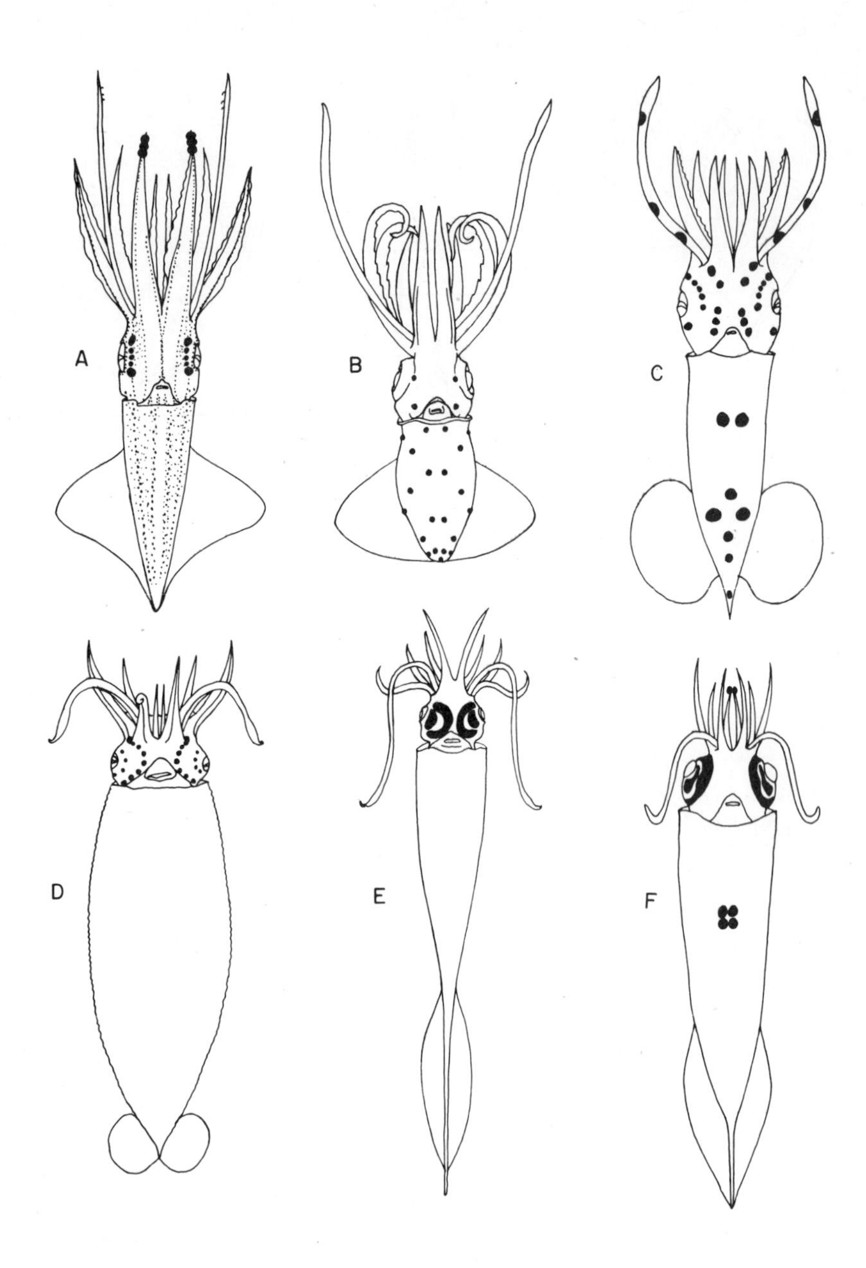

FIG. 3. Light organs of cephalopods. A. *Abraliopsis*; B. *Thelidioteuthis*; C. *Pterygioteuthis* (Enoploteuthidae); D. *Cranchia*; E. *Galiteuthis*; F. *Phasmatopsis* (Cranchiidae); (after Young, 1972; Voss, 1960; Clarke, 1962, 1966).

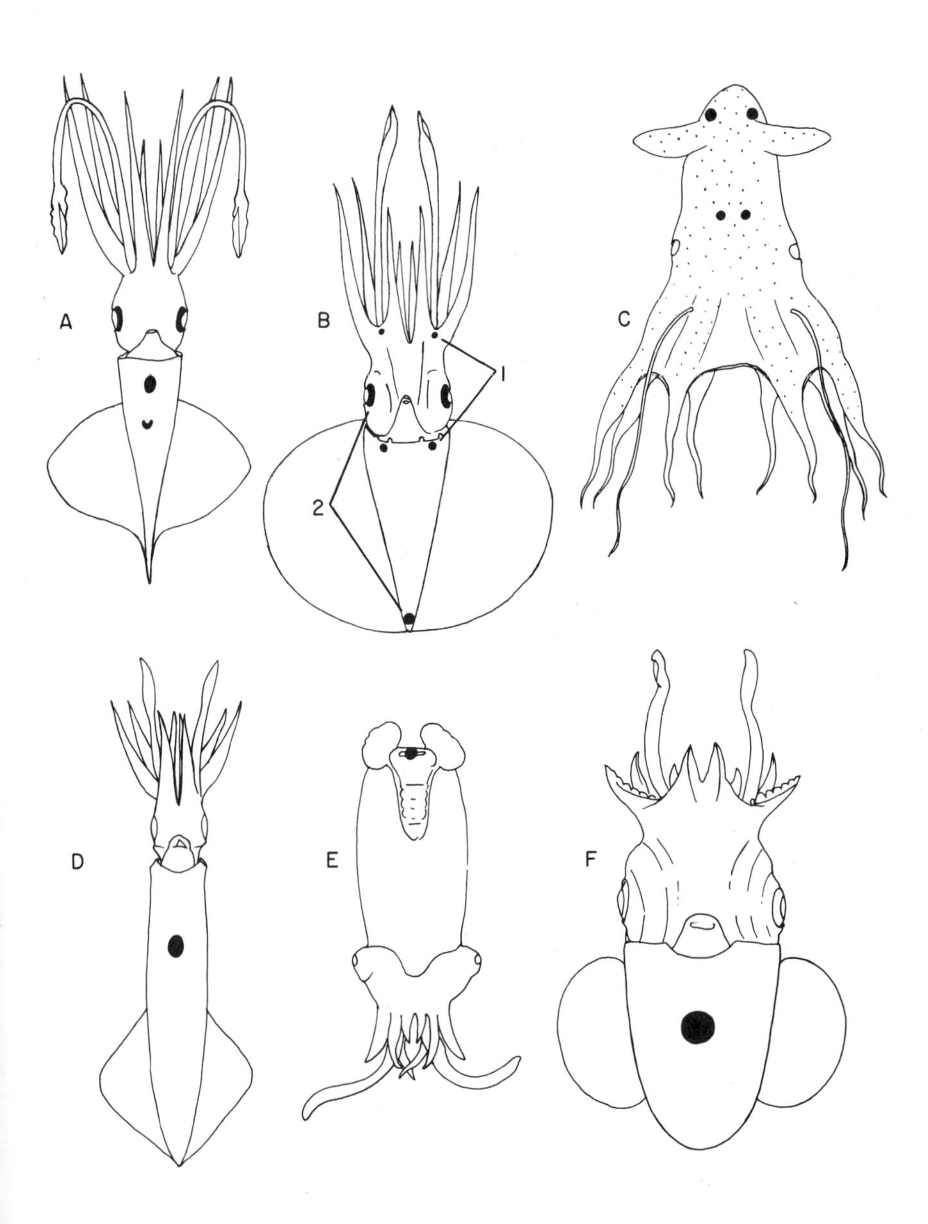

FIG. 4. Light organs of cephalopods. A. *Cycloteuthis*; B. *Discoteuthis*, light organ positions in two different species 1 and 2 (Cycloteuthidae); C. *Vampyroteuthis*; D. *Loligo*; E. *Spirula*; F. *Sepiolina*; (after Clarke, 1966; Young & Roper, 1969; Young, 1972).

in the Stomiatidae and *Porichthys*, often including the fins. Like those of the ommastrephids they may be aggregated to form larger patches or streaks, for example in *Astronesthes* and *Echiostoma*. The arrangement of the simple photogenic tissue in *Symplectoteuthis luminosa* is very reminiscent of the longitudinal organization of the organs in the paralepidids such as *Lestidium*. There is, of course, no serial arrangement of photophores in cephalopods similar to that in fishes, an inevitable consequence of the segmented organization of the latter, but it is interesting that secondary arrangement has provided a comparable result in some enoploteuthids. The condition of general scattering of superficial light organs on the ventral side in other cephalopods is perhaps more akin to that found in the sharks, in which they are also generally ventral in distribution.

Sub-ocular light organs are probably the most characteristic of all the cephalopod organs. They are to be found in many families, though the external appearance, and internal organization, varies very considerably. Light organs on the eyeball itself may be present as a row or rows of bead-like organs. In the Enoploteuthinae these are whitish, and very different from the superficial organs, whereas in the Pyroteuthinae they are similar in appearance to the visceral organs. Lines of sub-ocular organs are also found in the Lycoteuthidae, Cycloteuthidae, Ctenopterygidae, Cranchiidae (Cranchiinae) and Chiroteuthidae. The situation in certain species of the latter family, in which some of the sub-ocular organs have apparently coalesced to form two longitudinal streaks, provides a parallel to the situation in the Onychoteuthidae and the Cranchiidae (Taoniinae), in each of which the sub-ocular organ or organs are extended into crescent-shaped structures, sometimes covering the whole ventral surface of the eyeball. In the single luminous species in the Gonatidae, *Gonatus pyros*, a similar large luminous patch covers the ventral region of the eyeball (Young, 1972). Organs surrounding the eye, or associated with the eyelid, are to be found in *Mastigoteuthis* and *Histioteuthis*; in the latter they are greatly reduced round the enlarged left eye, but form a characteristic circlet around the right eye (Voss, 1969). Sub-ocular organs do not occur in fishes, but many species do have sub-orbital and/or post-orbital organs. These organs may be either bacterial, as in *Anomalops* and *Photoblepharon* (Haneda & Tsuji, 1971a), or self-luminous. The bacterial sub-orbital organs have no obvious parallels in the cephalopods, but like the sub-ocular organs of some cranchiids and *Gonatus* and the pre-ocular organs of *Mastigoteuthis glaukopis* they are the only light organs present in the animal. Most stomiatoids have substantial sub-orbital organs, as does the myctophid *Diaphus*; they are often the largest light organs in the animal, again like those of many cephalopods. Fish sub-orbital organs, however, as a consequence of the lateral compression of the body in most species, are laterally and not ventrally directed. Even where large sub- or post-orbital organs do occur there may be additional ventral ones on the lower jaw and isthmus. In many stomiatoids, and in some myctophids, there is a light organ which shines directly into the eye, but no equivalent organ has been described in the cephalopods, though in species with organs around the eyelid one or more might well cast at least some of their light into the eye.

Visceral light organs are common in several diverse groups of cephalopods. They are almost invariably complex in structure and are often quite large. They have been classified by Berry (1920b) as anal, branchial, gastric or axial depending on their situation. Other authors have grouped the gastric and axial organs together as abdominal organs. The anal organs are usually paired, though they may be fused into a single organ, and lie either side of the rectum on the ink sac. The bacterial organs of the sepiolids and loliginids come into this category, while anal organs also occur as the only visceral organs in species of *Chiroteuthis, Phasmatopsis, Taningia* and *Cycloteuthis*. Anal organs are also found in *Ctenopteryx, Onychoteuthis* and *Chaunoteuthis*, in each genus in conjunction with a more posterior visceral organ. Paired visceral light organs at the base of the gills (the branchial organs) are found only in the Lycoteuthidae and Pyroteuthinae, both of which also bear anal and other abdominal organs. A line of visceral organs in the Pyroteuthinae (the axial organs) extends to the tip of the tail, decreasing in size posteriorly, and similarly located organs are found in some of the Lycoteuthidae, of which *Selenoteuthis* has a particularly large one at the extreme tip of the tail. The (bacterial?) organ of *Spirula* is similarly placed, and ventrally directed visceral organs near the posterior tip of the body also occur in *Discoteuthis* and *Octopoteuthis*. Bacterial light organs associated with the rectum occur in the pelagic fishes *Opisthoproctus, Winteria* and *Rhynchohyalus*, but in these genera, unlike the cephalopods, the bacterial organ opens into the rectal lumen. The development of a line of organs down the length of the body, albeit within the mantle, in *Pterygioteuthis* and *Pyroteuthis* in particular, may perhaps be compared with the serial ventral photophores of many pelagic fishes. The enlarged posterior organ in *Selenoteuthis* and *Discoteuthis* has an anatomical counterpart in the enlarged caudal organs of *Gonostoma*, many myctophids, and perhaps the supposed luminous tissue in *Cyclothone* and *Eurypharynx*. Similarly the arrangement of mantle organs in *Onychoteuthis* and *Chaunoteuthis* may be likened to that of the ventral series, of a few large organs, in the scopelarchid *Benthalbella* (Merrett, Badcock & Herring, 1973). Cephalopods have light organs within the mantle cavity, shining through the mantle wall, but this type of disposition is unusual in fishes, in which internal organs, where they do occur, are usually associated with the gut and are very often bacterial in nature.

The anatomical arrangement of arms and tentacles around the mouth in cephalopods has given scope for the inclusion of light organs in or on these appendages. The arms bear a line of light organs in species of *Chiroteuthis*, *Octopoteuthis* and some lycoteuthids, while superficial organs have already been noted on the ventral arms of some ommastrephids, enoploteuthids, mastigoteuthids and histioteuthids. A characteristic of many unrelated genera is the presence of one or more specialized organs at the tips of some or all of the arms. These occur in species of *Abraliopsis*, *Watasenia*, *Histioteuthis*, *Octopoteuthis*, *Taningia*, *Phasmatopsis*, *Grimalditeuthis*, in the males only of *Selenoteuthis* and in the females only of *Leachia*. Light organs in the tentacles are found in *Pyroteuthis*, *Pterygioteuthis*, *Thelidioteuthis*, *Ancistrocheirus*, all the Lycoteuthidae and *Chiroteuthis*, while specialized organs at the tentacle tips are also found in the latter genus. There are no structures in fishes that can be directly compared with the arms and tentacles of cephalopods. It is nevertheless tempting to consider the provision of light organs in such appendages as the hyoid barbels of *Linophryne* and many stomiatoids, the escas of many angler fishes and the elongate fin rays of *Chirostomias* and *Chauliodus*, as perhaps indicative of some function similar to those of the arm and tentacle organs in cephalopods (see below).

There are some noteworthy differences in the distribution of the light organs in cephalopods and fishes. One feature of many fishes is the sexual dimorphism of the light organs. The anglerfishes in which the males completely lack light organs provide the most extreme example, but there are sexual differences in the size of the sub-orbital organs of many stomiatids, and in the position or presence of the large caudal light organs in some myctophids. Apart from Akimushkin's (1965) account of *Tremoctopus lucifer*, sexual dimorphism in the distribution of light organs in cephalopods is known only in *Selenoteuthis*, in which the male has organs on some arm tips, in *Leachia*, in which the female has a light organ on arm III (Voss, 1962; Young, 1972, 1975) and in *Ctenopteryx* (M. R. Clarke, pers. comm.). Sasaki (1914) counted over 100 more organs on the mantle of female *Watasenia* than on the male, but this may reflect normal variation or size effects and has not yet been confirmed as a true sexual dimorphism. Developmental changes occur in the light organ patterns of both fish and cephalopods, as might be expected, additional photophores being added as the animal matures. In *Bathyteuthis* there is a possibility that the light organs may regress as the animal grows, since they are much more prominent in smaller specimens (Roper, 1969). Regression of light organs in fishes is known in *Ceratias*, in which the dorsal caruncles are prominent in smaller specimens and are reduced in the adult, but is otherwise an equally rare phenomenon.

STRUCTURE OF CEPHALOPOD LIGHT ORGANS

Any organ capable of emitting light in an efficient manner is likely to be composed of certain optical elements. The structure of light organs in general is therefore constrained by the limitation of possible solutions to the physical problems involved. A typical general organization is exemplified by the bacterial organs of sepiolids and loliginids. The photogenic core of bacteria is surrounded by an elaborate reflector, and a structure described as a lens fills the aperture. The reflector is composed of a large number of overlapping plates, arranged in layers, and the structure of these plates varies in different parts of the organ. The ink sac lies immediately behind the organ and acts as a pigment screen. This basic structure of pigment screen, reflector, photogenic core and lens is common to most complex light organs and is therefore not surprisingly one of the more obvious parallels in development between fish and cephalopods.

Within the constraints of the overall organization there are, however, considerable differences. The bacterial light organs of fishes are often located in gut diverticula deep within the body, whose external surface is usually heavily pigmented by the slow aggregation or dispersion of melanophores. Fish light organs of this type may in consequence have very complex light guides and/or diffusing systems, reaching peaks of elaboration in *Opisthoproctus* and *Leiognathus* (Bertelsen & Munk, 1964; Hastings, 1971). The bacterial organs of macrurids and angler-fishes on the other hand are more akin to those of the sepiolids in general structure, but lack both the thick reflector and the lens. A feature common to many cephalopod organs is that there is much greater dependence upon an efficient reflector and less provision of a posterior pigment screen than is the case in fishes. Most cephalopods do not have large deposits of dark pigment within the tissues, but depend for colouration upon their chromatophores (see Packard & Hochberg, their chapter, this volume). Dark chromatophores are found most characteristically beneath the superficial light organs of enoploteuthids, whereas pigmentary layers, not necessarily organized as chromatophores, are found for example in the ocular organs of *Pterygioteuthis megalops* (Arnold & Young, 1974), in the arm light organs of chiroteuthids, cranchiids and octopoteuthids, and in the superficial organs of histioteuthids and *Vampyroteuthis*. Indeed in many of these a photogenic function has been assumed largely on the basis of the heavy pigment deposits in the walls. Most so-called anal organs lie on the ink sac itself, and this may be regarded as forming the pigment screen, as in the sepiolids.

Probably the most fundamental structural difference between the organs of fish and cephalopods is the reflector. In fish it is almost always made up of minute guanine crystals arranged in the form of platelets, themselves organized in parallel layers to provide a very efficient specular reflectance (Denton, 1970). In cephalopods, on the other hand, the major reflecting material is probably chitin (Denton & Land, 1971). Chitin is non-crystalline and therefore much more flexible in organization, often occurring as lozenge-shaped units, discs, ribbons or sheets. The refractive characteristics of chitin/cytoplasm layers in interference reflecting systems are such that more layers are required to form a surface of given reflectance than would be necessary for a guanine/cytoplasm system (Land, 1972). Guanine has not been identified in cephalopod light organs though Packard & Sanders (1971) found a "creamy guanine-like substance" in *Octopus* leucophores. The increased relative thickness of the reflectors of cephalopod light organs, compared with that in fishes is probably at least partly a function of this refractive difference. Chitin/cytoplasm multilayers have narrower spectral reflectance than do guanine/cytoplasm multilayers, but whether this is of any significance in the operation or organization of the reflector systems is not known.

Virtually nothing is known about the optical properties of cephalopod light organs, but histological work has shown structures variously interpreted as lenses or reflectors in the aperture of several of the more complex organs, notably in the Pyroteuthinae and Lyctoteuthidae (Chun, 1910; Mortara, 1921; Okada, 1970; Arnold, Young & King, 1974; R. E. Young, this symposium pp. 161–190). The beautiful iridescent colours of the light organs of these animals observed by reflected light in fresh specimens is often largely due to the presence of a thin superficial layer of reflecting platelets. Similar layers may also occur deeper in the light organ, but still between the photogenic tissue and the exterior, and their function is largely unknown, though Arnold *et al.* (1974) have speculated on their use in both intensity regulation and collimation. The pearly white lustre of many light organs, for example the sub-ocular organs of the Enoploteuthinae and some visceral organs of the Lycoteuthidae, is the result of a thick layer of light guides radiating from the core of the organ and thus providing a hemispherical cushion of diffusing surface. The contrast between these organs and the much more clearly collimated emission in those organs with enclosing reflectors is very marked, and though the reflective type of organ is usual in fishes such diffusing surfaces are not known. Fish light organs with accessory diffusing systems are common in species with bacterial organs, and are generally described as the "indirect" emission type (Haneda, 1950). In such cases the light is led to the exterior from the photogenic region deep within the body, in *Leiognathus* by modified muscle fibres (Hastings, 1971) and in *Opisthoproctus* by a wide hyaline body (Bertelsen & Munk, 1964). Light guide systems occur in many cephalopod light organs, in addition to those noted above, but not in bacterial organs. They have been described in the ommastrephids *Hyaloteuthis* and *Symplectoteuthis* (Okada, 1968) and reach a very great degree of elaboration in the sub-ocular organs of *Bathothauma* (Dilly & Herring, 1974). Similar systems are found in most other large sub-ocular organs such as in the Chiroteuthidae, Ctenopterygidae, and Onychoteuthidae. This structural manoeuvre for spreading light from a small photogenic core over a wide area inevitably results in diffuse rather than focused emission. It is very likely that many other of the more complex cephalopod photophores also contain light guide structures, as yet not recognized or wrongly identified as lenses or reflectors. There are no obvious cephalopod parallels to the simple organs of many fishes such as stomiatoids, in which a photogenic mass occurs within a pigment cup but without reflector, and the emission is defined simply by the aperture of the pigment screen. On the other hand a close parallel does exist between the superficial organs of ommastrephids and those in the gelatinous corium of many stomiatoids (O'Day, 1973). Neither have any associated optical structures whatever and both are simple spherules of photogenic tissue.

The photogenic tissues within the organs of different cephalopods have no one common feature, but within groups similarities do occur. Thus within the cranchiids probably all the Taoniinae have paracrystalline material in the sub-ocular organs similar to that described in *Bathothauma* (Dilly & Herring, 1974), in contrast to the Cranchiinae, which do not. It has long been noted how several organs of different appearance may occur in one species, and up to 15 or 16 different types of light organ have been described in some species of the Lycoteuthidae. Notwithstanding the apparent diversity it does appear that, in many cases at least, the photogenic material is similarly organized regardless of the variation in accessory structures. Thus the visceral organ(s) of such genera as *Onychoteuthis*, *Chiroteuthis* and *Ctenopteryx* have a very different appearance to the sub-ocular organs. Nevertheless the same photogenic material is found in both organs (Herring, unpublished). It appears probable that this is also true for most organs of the lycoteuthids and enoploteuthids and indeed Okada (1966) has described paracrystalline material in both the brachial and cutaneous organs of *Watasenia*. However, there is insufficient information to be certain yet. Differences nevertheless do occur; in *Phasmatopsis* the structure of the photogenic material is not paracrystalline in either the visceral organs of *P. oceanica* or the arm tip organs of *P. lucifer* though it is in the sub-ocular organs of both species. Organs with very different appearance in fishes also often possess a common photogenic material, at least at the histological level, as in the caudal and serial organs of myctophids. In other cases there are quite substantial differences, however, as in the stomiatoids in which the simple fin organs have a structure quite different from that of the serial organs. There is no analogy amongst the cephalopods to the multiple organs of the Maurolicidae, Gonostomatidae and Sternoptychidae in which a common photogenic mass illuminates several adjacent serial organs. Neither do the fishes have any parallel to the "double organ" described in the Lycoteuthidae (Chun, 1910; Okada, 1970).

MODE OF OPERATION

The mode of operation of cephalopod light organs is hardly known at all, and almost all conclusions about their function are based upon anatomical evidence. This is hardly surprising in view of the extreme difficulty of maintaining luminous cephalopods alive for any length of time. Berry (1920a,b) summarized the early observations of cephalopod luminescence and there have been relatively few recent accounts (e.g. Schmidt, 1922; Berry, 1926; Skowron, 1926; Haneda, 1956; Haneda & Tsuji, 1971b; Roper, 1963; Garcia-Tello, 1964; Clarke, 1965; Gaskin, 1967; Voss, 1967a,b). The most accessible luminous species has been the enoploteuthid *Watasenia scintillans*, and such information on control as is available relates primarily to this species (Harvey, 1952; Shoji, 1919). Light emission in cephalopods, as in other animals, may appear as flashes or glows, depending on the time-course of the luminescent response. Flashes generally imply a nervous control, whether direct or indirect, but glows are less easy to interpret. The third type of emission is that of a secretion into the surrounding medium, such as occurs in *Heteroteuthis* and other sepiolids, and as implied in Gaskin's (1967) account of *Moroteuthis*. This secretory mode is very common in other invertebrates but otherwise rare in cephalopods and most fishes, except the Searsiidae. The light of *Heteroteuthis* lasts for between several minutes and a few hours when discharged into the sea, often as large particulate clumps, but little is known about the kinetics of the luminescence. The luminescence of searsiid fishes is very similar, in that it is ejected as a mass of luminous points rather than as a uniform cloud of secretion, but, unlike that of *Heteroteuthis*, the luminescence is not visible from within the gland itself.

Although most reports of luminescence in cephalopods are of brief flashes, such as those from the brachial organs of *Watasenia* and the dorsal organ of *Ommastrephes pteropus* (Clarke, 1965; Voss, 1967a,b), a steady glow characterizes the luminescence of *Spirula*. As Schmidt (1922) noted, the light of *Spirula* remains at even intensity for periods of up to several hours, with only minor fluctuations imposed on the general level. The observed light intensity is generally unaffected by movements of the animal or changes in orientation but the intensity can be considerably increased by stimulation of the animal (Herring, unpublished). This is what might be expected from a bacterial source subject to limited control. The integumentary organs of *Histioteuthis* and enoploteuthids can also be observed to glow. These glows, though they may remain steady for a minute or two, can apparently be instantly extinguished. Whether glows observed in captive specimens give a true indication of normal light production is not known. The observed emission may be a manifestation of the "terminal glow" syndrome observed in dying specimens of other luminous animals. In healthy specimens of *Onychoteuthis banksi* shortly after capture all the light organs glowed brightly with superimposed intensity changes visible either as a gradual increase, or decrease, or as a rapid flickering of the mantle organs. Electrical stimulation of the body wall, when the organs were dark, initiated a brief glow from the ocular organs, and injection of acetylcholine stimulated a brief blue-green light from the anterior mantle organ. In dying specimens of the same species, however, only a steady glow could be observed, which rapidly declined in intensity and was extinguished in a few minutes.

Electrical stimulation has induced luminescence in both whole animals and isolated groups of brachial and body organs in *Watasenia* (Shoji, 1919), and Hasama (1941) has described action potentials associated with luminescence in this species. It is possible to obtain a feeble glow in *Histioteuthis* by electrical stimulation. Although the evidence is very fragmentary, it is probable that most cephalopod luminescence, with the possible exception of that of *Spirula*, is under nervous control, and Arnold & Young (1974) have described synapses on photocytes in *Pterygioteuthis*. This is a similar situation to that obtaining in fishes, though some bacterial systems may perhaps be under arterial control (Bertelsen, 1951). The luminous organs of *Ommastrephes pteropus*, and probably other ommastrephids, are composed of a very fine and elaborate capillary network to whose walls the photogenic material is applied. It is therefore anatomically possible that these organs might also be under arterial rather than direct nervous control. Most examples of serial ventral light organs in fishes glow steadily, for example *Argyropelecus*, whereas single organs, like the sub-orbital organs, and the dorsal organs, such as the glandular patches of astronesthids and myctophids, flash rapidly. The ventral integumentary organs of cephalopods such as *Histioteuthis* and enoploteuthids also apparently glow steadily, whereas those dorsal organs that have been observed in ommastrephids respond, like those of fishes, by flashing. The pharmacology of the organs is probably different in that many fish organs are stimulated by adrenaline or its analogues, whereas the very few observations on cephalopods indicate a cholinergic system. This is, of course, consistent with the more general neuromuscular transmission characteristics in the two groups.

Virtually nothing is known of the normal synchrony or otherwise in the responses of different types of organ within a single species. Ishikawa (1913) notes that in *Watasenia* the three types of organ, brachial, sub-ocular and integumentary, may shine together or quite independently. If different functions are to be inferred from different structures then it is unlikely that all organs will generally operate synchronously, and independent action of different groups of organs is to be expected. Control of light organs by the indirect action of superficial chromatophores has been observed in several cephalopods (Berry, 1920a; see Packard & Hochberg, their chapter, this volume and R. E. Young, this volume). Their rapid response makes these effectors much more suitable instruments for short-term control than are fish chromatophores. Thus each light organ in *Histioteuthis* has an associated chromatophore over the aperture which might easily control the output, and chromatophores are prominent, for example, on the surface of the fin organs of *Vampyroteuthis* and the tail organ of *Selenoteuthis*. It is not necessary to postulate either chromatophore or direct nervous control, as both could be effective in a single light organ depending upon the required time-course of the response. Chromatophore control of fish luminescence has been suggested in several species (Haneda, 1950) but is likely to be of a much slower nature than that in cephalopods. Haneda (1964) has made the interesting suggestion that the intensity of the organ of *Uroteuthis* is controlled by movements of ink in the overlying ink sac. There is no cephalopod equivalent to the rotation of light organs observed in stomiatoids (Nicol, 1960) or to the use of opaque shutters as in *Photoblepharon*. The organs at the arm tips of *Phasmatopsis* are so constructed, however, that they are probably shut off by the two pigmented lateral walls folding over the photogenic core, and the fin organs of *Vampyroteuthis* may be able to be retracted into a deep pocket of heavily pigmented tissue. Such mechanisms may be less a means of extinguishing the light than of preventing reflectance from the organ, thus betraying the animal's position, as Nicol (1960) has suggested for the rotation of some stomiatoid sub-orbital organs.

When its intensity is sufficiently high for colours to be readily distinguished, cephalopod luminescence is almost invariably described as blue or blue-green. Older observations described red, yellow and purple hues, but in many instances it is impossible to determine from the original account whether iridescence or luminescence was being observed. Many subsequent authors have intimated that Chun (1910) observed different luminescent colours from the different organs of *Lycoteuthis diadema,* and in particular that the anal organs gave out a red light. Chun himself made no such claims, describing the appearance of the organs only. His sole reference to emitted light describes a bluish one from an unidentified light organ "vermöchte ich von einem Organ ausstrahlendes bläuliches Licht wahrzunehmen". There is no evidence that this, or any other cephalopod, is capable of emitting red light, or indeed any colour other than blue or blue-green. Observations on the histology of *Lycoteuthis* (Okada, 1970) and on fresh specimens of the closely related *Selenoteuthis,* the colour of whose organs is very similar to that of *Lycoteuthis,* provide no support for the emission of red light. By contrast there are certain fishes which undoubtedly do emit red light, as well as blue, from separate organs, namely *Pachystomias* (Denton, Gilpin-Brown & Wright, 1970), *Aristostomias* (O'Day & Fernandez, 1974) and *Malacosteus.* Otherwise fish, like cephalopods, have the blue or blue-green emission so characteristic of almost all marine organisms.

It has been suggested that some cephalopods might be able to vary the colour of their light emission by the use of different chromatophores as colour filters. While this is theoretically possible, to be effective it would require an original broad spectrum emission from the light organ and most bioluminescent systems are restricted to a relatively narrow spectral band. On the other hand, the spectral output of organs of *Histioteuthis* must be somewhat modified by the presence of a purple "filter" pigment in the aperture of the organ. Similarly coloured material is found in the light organs of many stomiatoids.

CHEMISTRY OF LUMINESCENCE

The chemical characteristics of cephalopod luminescent systems are hardly known. Most cephalopods do not luminesce readily, even when captured in apparently good condition, and only limited information can be gained from non-luminescent corpses. In many cephalopod light organs luminescence can be elicited merely on exposing the organ to air or sea-water by cutting it open. Under these conditions luminescence is steady, not that of a rapid reaction. This can be observed in the organs of cranchiids, ommastrephids, *Ctenopteryx,* in some of the visceral organs of enoploteuthids, and in *Vampyroteuthis*. This luminescence is presumably a simple oxidation, an assumption strengthened by the fact that the addition of an oxidizing agent such as hydrogen peroxide invariably induces luminescence, whether or not the organs have been cut open. The luminescent systems of *Watasenia*, *Ommastrephes pteropus* and *Heteroteuthis* are oxygen sensitive (Shoji, 1919; Skowron, 1926; Girsch, Herring & McCapra, 1976) as are those of the known bacterial systems. All those luminescent systems whose characteristics are known in any detail have reaction mechanisms in which the formation of intermediate endoperoxides is an important feature. The response of the organs to oxidizing agents such as hydrogen peroxide is therefore not surprising; similar responses occur in many other animals, including fishes (Anctil & Gruchy, 1970; Merrett, Badcock & Herring, 1973). Only the systems of *Watasenia* and *Ommastrephes pteropus* have been investigated in any detail. The latter species has a very persistent luminescence after death and retains its activity when deep-frozen, as Haneda & Tsuji (1971b) noted in the related *Symplectoteuthis*. The system is extremely difficult to solubilize, and may be membrane-bound. Shimomura & Johnson (1972) reported the extraction of a luciferin/luciferase type of system from *Watasenia*, and Goto, Iio, Inoue & Kakoi (1974) have succeeded in extracting the oxidation products from the brachial organs of some 10 000 specimens. Certain of these products are similar to those of *Cypridina* luciferin and *Aequorea* photoprotein (see Baker, this volume). It therefore appears that the luminescent system of *Watasenia* is very closely related to that of the coelenterates and similar to that of *Cypridina* (Inoue, Kakoi & Goto, 1976). The chemistry is little known in fishes, but systems like that of *Cypridina* have been found in myctophids, *Porichthys* and others (e.g. Tsuji & Haneda, 1971; Cormier, Crane & Nakano, 1967). Fluorescence is a common feature of many light organs of both fish and cephalopods. The organs of *Mastigoteuthis*, *Selenoteuthis* and *Pterygioteuthis* have a bright greenish–blue fluorescence, rather like that of the fish *Maurolicus*, whereas the organs of *Ommastrephes pteropus* have a yellowish fluorescence (Fig. 5). An unexpected parallel occurs between the fluorescence of the "filter" pigments in the light organs of *Histioteuthis*, and in those of stomiatoids, both of which fluoresce bright red. The pigments themselves are not identical, that of *Histioteuthis* being probably a porphyrin, that of stomiatoids a soluble protein.

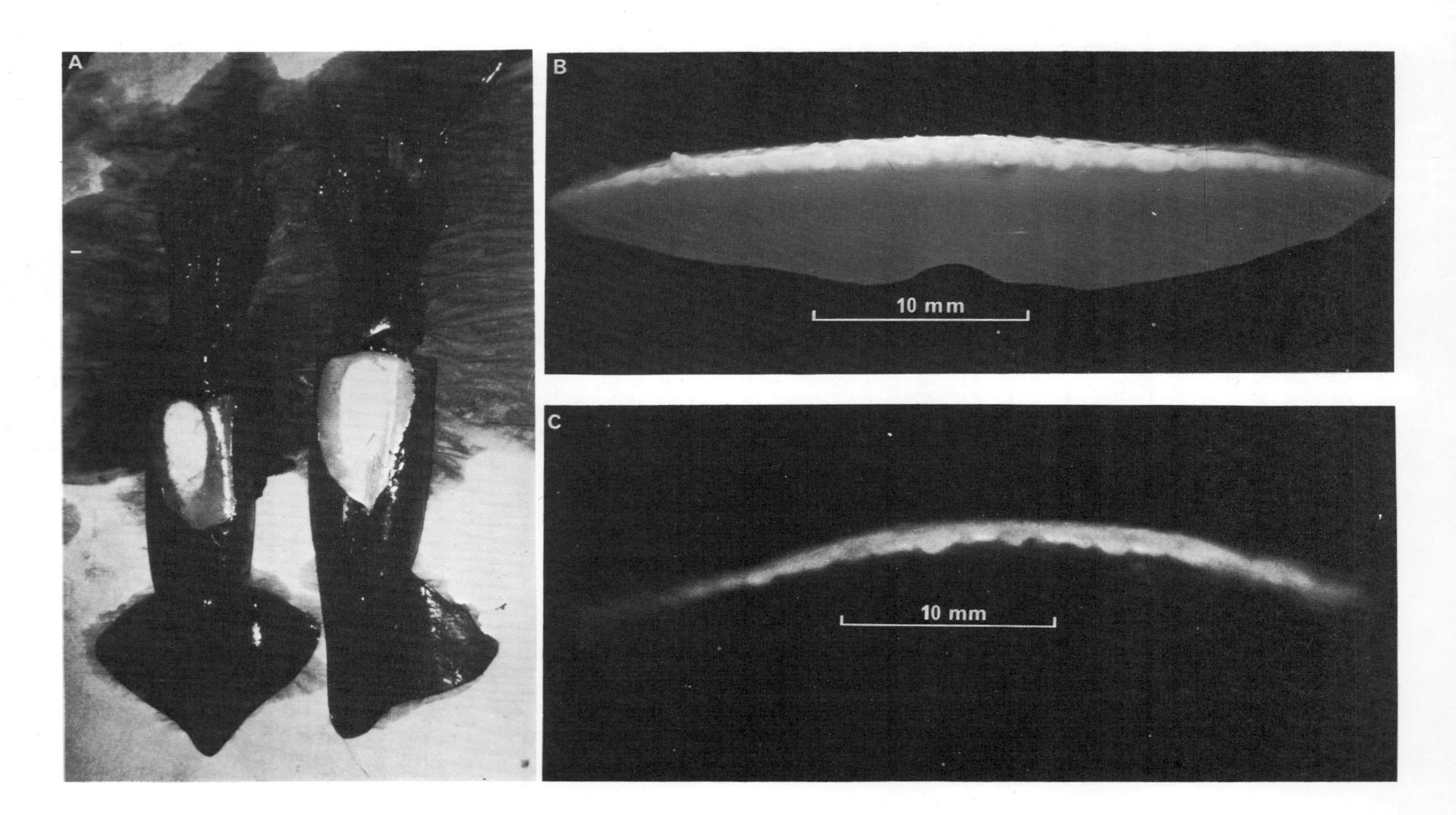
A
B
10 mm
C
10 mm

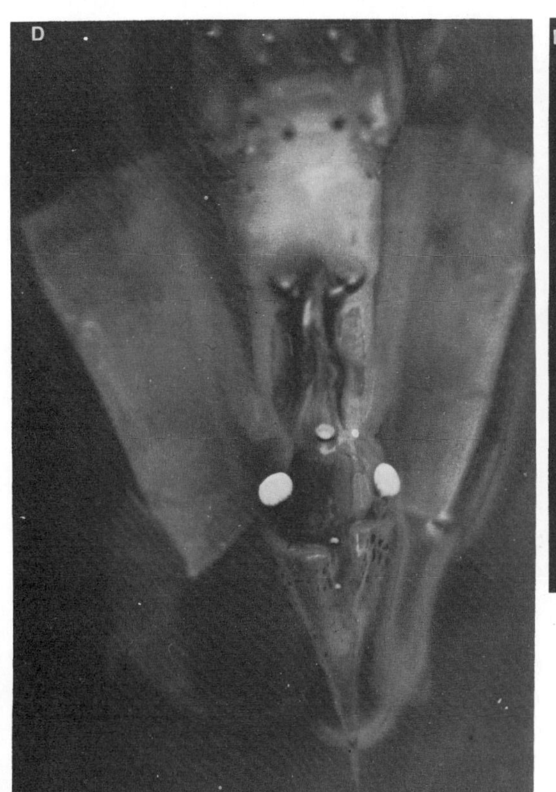

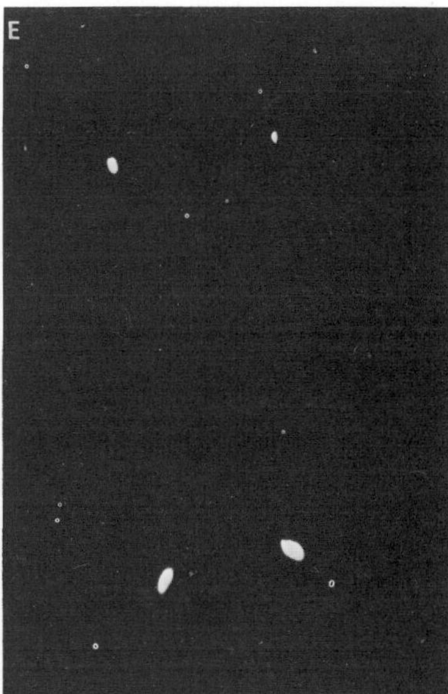

FIG. 5. A. Two fresh specimens of *Ommastrephes pteropus*; the epidermis has been removed above the large dorsal light organ. B. Section through the region of the light organ, photographed in ultra-violet light and showing the fluorescence of the thin layer of photogenic tissue. C. Similar section photographed by its own luminescence. D. Specimen of *Pterygioteuthis giardi* in ultra-violet light; note the strong fluorescence of the large pair of branchial light organs. E. The same species photographed by its peroxide-induced luminescence. Only the branchial organs (bottom) and one particular pair of ocular organs (top) respond readily to this treatment.

VERTICAL DISTRIBUTION

Only relatively recently have sampling systems been developed that can provide adequate detailed information about the depth distribution of cephalopods, and other pelagic animals (see Roper and Clarke, their separate chapters, this volume). Nevertheless Voss (1967a) formulated certain general principles concerning the types of luminescence associated with cephalopods in different depth horizons, and more recent data acquired with opening-closing systems have generally substantiated his conclusions (Clarke, 1969; Roper, 1972; Clarke & Lu, 1974, 1975; Lu & Clarke, 1975a,b). The effect of diurnal vertical migrations, and the probability of net avoidance, particularly by such groups as the Ommastrephidae and Onychoteuthidae, makes it difficult to assign precise depth ranges to many species. It may, however, be assumed that members of the above two families generally inhabit the upper mesopelagic or epipelagic water. They either have no light organs as in surface-living *Onykia*, or relatively undifferentiated organs as in the ommastrephids, or the visceral and sub-ocular organs of *Onychoteuthis*. Whatever their purpose, it is interesting that the same pattern of visceral organs also occurs in some specimens of *Symplectoteuthis* (Haneda & Tsuji, 1971b). Where organs are present they are primarily ventral in distribution, although some ommastrephids have large dorsal organs. The enoploteuthids and lycoteuthids, found mainly in the upper mesopelagic layers during the day, have numerous elaborate ventrally directed organs. They may be compared with fishes of the Maurolicidae, Sternoptychidae and Gonostomatidae, whose distribution is rather similar and which are generally also provided only with elaborate ventrally directed organs. At shallower day depths are to be found the smaller cranchiids, whose only light organs are sub-ocular. In the lower mesopelagic zone such genera as *Chiroteuthis*, *Octopoteuthis*, *Histioteuthis* and *Mastigoteuthis* are more frequent, most of which have a considerable number of elaborate light organs, and whose distribution may extend into the bathypelagic realm where *Vampyroteuthis* and *Bathyteuthis* are more usually taken. These latter genera have smaller light organs, not so restricted to the ventral surface as in most shallower cephalopods. Larger specimens of species showing an ontogenetic migration, such as *Bathothauma* (Aldred, 1974), also occur in the deeper layers.

In general, Voss' (1967b) thesis holds: light organs are absent in the shallowest species, they are restricted to the sub-ocular organs in the shallow cranchiids, they are numerous and often elaborate in the upper and lower mesopelagic species and they are very variable in the bathypelagic realm. The distribution of light organs in fishes shows somewhat similar trends. Perhaps the closest analogy to be drawn with fishes is that between the ommastrephids and myctophids, both of which undergo extensive migrations to the surface at night. Both have primarily ventrally directed light organs of varying number and complexity, but some also have large dorsal organs. No major differences in day depth have been found between the Pyroteuthinae and Enoploteuthinae, with visceral and superficial light organs respectively. The presence of a large number of tiny superficial organs, as in the Enoploteuthinae, is rare in fishes inhabiting the upper mesopelagic layers, in which they usually have the large complex organs typified by the Maurolicidae, Gonostomatidae and Sternoptychidae. The arrangement of light organs in the deeper cephalopods, such as the Histioteuthidae, is more analogous to that in lower mesopelagic fishes such as the stomiatoids. Internal bacterial organs are commonest among such coastal fishes as the leiognathids, but are also known in the mesopelagic argentinoids *Opisthoproctus*, *Winteria* and *Rhynchohyalus*. Bacterial organs are similarly distributed in cephalopods, being commonest among the coastal species of loliginids and sepiolids. The sepiolids are often associated with the bottom, and bacterial light organs are also found in the epibenthic macrurids, but these fishes extend into very much deeper regions than those from which the luminous sepiolids are known. In the bathypelagic realm, and below, the array of light organs in both cephalopods and fishes becomes very variable in distribution and complexity.

FUNCTION OF LUMINESCENCE

The diversity of luminous apparatus among the cephalopods has presumably been developed for many different purposes. Possible functions that will be considered here are ventral camouflage, distraction, disruptive illumination and communication.

Ventral luminescence has been postulated as a means of camouflage in many animals including cephalopods (Young, 1973 and his chapter, this volume; Young & Roper, 1976). The concept of ventral camouflage in cephalopods was first promulgated by Dahlgren (1916) who noted that "any animal situated below such a squid and looking upward at it from below would see a bluish light that would blend with the sunlight". The general ventral distribution of light organs in histioteuthids, ommastrephids and enoploteuthids is well suited for such a purpose, and the observations, in several of these animals, of steady glows such as occur in hatchet fish is just what might be expected. Young (1973) has offered an attractive solution to the feedback problems of such a system and he has further developed the concept of ventral camouflage in this volume. It is probably significant that the major development of serial, and secondarily longitudinally arranged, ventral light organs in both fishes and cephalopods occurs in similar depth horizons in the mesopelagic realm. The evidence for sub-ocular organs as ventral camouflage systems is less compelling, but seems likely for the cranchiids in which the eyes, and ink sac, are often the only pigmented parts of the body. The proportionally large eyes of the Taoniinae probably account for the development of the elaborate diffusing systems which seem a likely adaptation for ventral camouflage, but the multiple appearance of the sub-ocular organs of *Lycoteuthis* surely implies some diversity of function. A similar argument may be used to explain the position of anal or ink sac organs, but is less tenable for other visceral organs. Most cephalopods present a relatively greater ventral surface area than most fishes. Only the midwater sharks and rays have particularly large ventral surfaces, and it is interesting that the luminous species of these, such as *Isistius* and *Euprotomicrus*, have a multitude of tiny ventral light organs whose organization is somewhat similar to those seen in the Enoploteuthinae. Sub-orbital organs in fishes are not generally countershading devices, the array on the lower jaw and isthmus serving this purpose. In this respect at least the sub-orbital organs of fishes are functionally quite distinct from the sub-ocular ones of most cephalopods, though not perhaps those of the lycoteuthids and enoploteuthids with a more collimated output. Stalk-eyed fish larvae such as *Idiacanthus*, have no sub-ocular organs, unlike the stalk-eyed cephalopod juveniles of *Bathothauma* and *Leachia*.

The visceral organs, while perhaps acting as a ventral camouflage, may alternatively present a disruptive illuminated pattern, with no attempt at direct concealment. Certainly single or a very few visceral organs are unlikely to camouflage the whole underside effectively. The large posterior organs of *Selenoteuthis* and *Discoteuthis* may also perhaps be disruptive in function. Only four of the light organs of *Pterygioteuthis* respond readily to hydrogen peroxide treatment, the two branchial ones and a particular pair of ocular organs (Fig. 5). This selective response, if extrapolated to normal luminescence, also implies different functions for the various sets of light organs.

The sepiolid *Heteroteuthis* provides an example of luminescence acting as a distraction. When irritated the animal ejects a cloud of luminous particles from the light organ, paralleling the response in searsiid fishes. Nevertheless the provision of elaborate reflecting systems in the light organ itself, and the observations of luminescence from within the gland (e.g. Meyer, 1906), indicate that other uses are made of the light. Specialized arm or tentacle tip organs such as are found in *Chiroteuthis* and *Phasmatopsis* may act as lures for prey, like those of angler-fishes or stomiatoid barbels (Voss, 1967a). They may also perhaps be used as potentially sacrificial diversions to an attacker.

The virtual absence of luminous sexual dimorphism does not automatically rule out the use of light organs for sexual display (Hamabe & Shimizu, 1957; Young, 1975). The fish *Porichthys* does not have sexual differences in its light organ distribution, but has nevertheless been seen to use them in courtship displays (Crane, 1965). The large dorsal patches in some ommastrephids are certainly used by the animal when hooked, or attacked, and presumably function as a warning system. Reports that they may flash just prior to capture of flying-fish by the squid (Voss, 1967b) are less easy to interpret. The massive aggregation of small units to form these large organs allows the animal to put out a very bright, and therefore probably blinding, flash. There are no obvious functional parallels in cephalopods to the large sub-orbital organs of, for example, *Anomalops* and many stomiatoids, whose purpose seems to be that of local illumination, or of signalling. Rhythmic flashing such as that produced by these organs has not been observed in cephalopods, though *Watasenia* certainly can emit a series of flashes (Harvey, 1952). The specific patterns of light organs in related cephalopods are often as diverse, and as intricate, as those in, say, myctophids, and the possibility that they are used for specific recognition is equally plausible. An analogous communicative function in shoaling species, such as the ommastrephids and *Watasenia*, also cannot be ruled out. Packard (1972) suggested that some organs associated with the eye might actually illuminate it in a warning or dymantic display, but intra-ocularly directed organs are not a feature of most cephalopods, and in fishes a feedback function seems much more likely. Finally the possibility should not be overlooked that some cephalopod luminescence may be mimetic in function, though this would be extremely difficult to demonstrate.

So much of the information on luminescence in both cephalopods and fish is still purely anatomical, and so bewildering in its variety, that functional interpretations almost certainly err on the side of conservatism, an inevitable consequence of our restricted concepts of the normal lives of these remarkable animals. Any purpose that is fulfilled by colour or pattern in the illuminated terrestrial or coastal environment can also be achieved by luminescence in the dark of the deep sea. The cephalopod inhabitants of this environment almost certainly make far more extensive and varied use of their impressive luminescent abilities than we can at present envisage.

Acknowledgements

I am most grateful to Dr M. R. Clarke for reading the typescript and for his help in checking the taxonomic accuracy of Table I. Dr R. E. Young very kindly provided additional information, some of it unpublished, and checked Table I. Any remaining errors are entirely my responsibility.

References

Akimushkin, I. I. (1965). *Cephalopods of the seas of the U.S.S.R.* Israel Program for Scientific Translations Ltd: Jerusalem.

Aldred, R. G. (1974). Structure, growth and distribution of the squid *Bathothauma lyromma* Chun. *J. mar. biol. Ass. U.K.* **54**: 995–1006.

Anctil, M. & Gruchy, C. G. (1970). Stimulation and photography of bioluminescence in lantern-fishes (Myctophidae). *J. Fish. Res. Bd Can.* **27**: 826–829.

Arnold, J. M. & Young, R. E. (1974). Ultrastructure of a cephalopod photophore I. Structure of the photogenic tissue. *Biol. Bull. mar. biol. Lab., Woods Hole* **147**: 507–521.

Arnold, J. M., Young, R. E. & King, M. V. (1974). Ultrastructure of a cephalopod photophore. II. Iridophores as reflectors and transmitters. *Biol. Bull. mar. biol. Lab., Woods Hole* **147**: 522–534.

Berry, S. S. (1920a). Light production in cephalopods. I. An introductory survey. *Biol. Bull. mar. biol. Lab., Woods Hole* **38**: 141–169.

Berry, S. S. (1920b). Light production in cephalopods. II. An introductory survey. *Biol. Bull. mar. biol. Lab., Woods Hole* **38**: 172–195.

Berry, S. S. (1926). A note on the occurrence and habits of a luminous squid (*Abralia veranyi*) at Madeira. *Biol. Bull. mar. biol. Lab., Woods Hole* **51**: 257–268.

Bertelsen, E. (1951). The ceratioid fishes. *Dana Rep.* **39**: 1–276.

Bertelsen, E. & Munk, O. (1964). Rectal light organs of the argentinoid fishes *Opisthoproctus* and *Winteria. Dana Rep.* **62**: 1–17.

Chun, C. (1910). Die Cephalopoden. *Wiss. Ergebn. dt. Tiefsee-Exped. "Valdivia"* **18**: 1–552.
Clarke, M. R. (1962). A large member of the squid family Cranchiidae *Phasmatopsis cymoctypus* de Rochebrune 1884. *Proc. malac. Soc. Lond.* **35**: 27-42.
Clarke, M. R. (1965). Large light organs on the dorsal surfaces of the squids *Ommastrephes pteropus*, "*Symplectoteuthis oulaniensis*" and "*Dosidicus gigas*" *Proc. malac. Soc. Lond.* **36**: 319–321.
Clarke, M. R. (1966). A review of the systematics and ecology of oceanic squids. *Adv. mar. Biol.* **4**: 91–300.
Clarke, M. R. (1969). Cephalopoda collected on the SOND Cruise. *J. mar. biol. Ass. U.K.* **49**: 961–976.
Clarke, M. R. & Lu, C. C. (1974). Vertical distribution of cephalopods at 30°N 23°W in the North Atlantic. *J. mar. biol. Ass. U.K.* **54**: 969–984.
Clarke, M. R. & Lu, C. C. (1975). Vertical distribution of cephalopods at 18°N 25°W in the North Atlantic. *J. mar. biol. Ass. U.K.* **55**: 165–182.
Cormier, M. J., Crane, J. M. & Nakano, Y. (1967). Evidence for the identity of the luminescent systems of *Porichthys porosissimus* (fish) and *Cypridina hilgendorfii* (crustacean). *Biochem. biophys. Res. Commn* **29**: 747–752.
Crane, J. M. (1965). Bioluminescent courtship display in the teleost *Porichthys notatus*. *Copeia* **1965**: 239–241.
Dahlgren, U. (1916). The production of light by animals. Light production in cephalopods. *J. Franklin Inst.* **81**: 525–556.
Denton, E. J. (1970). On the organization of reflecting surfaces in some marine animals. *Phil. Trans. R. Soc.* (B) **258**: 285–313.
Denton, E. J., Gilpin-Brown, J. B. & Wright, P. G. (1970). On the "filters" in the photophores of mesopelagic fish and on a fish emitting red light and especially sensitive to red light. *J. Physiol., Lond.* **208**: 72–73P.
Denton, E. J. & Land, M. F. (1971). Mechanism of reflexion in silvery layers of fish and cephalopods. *Proc. R. Soc.* (B) **178**: 43–61.
Dilly, P. N. & Herring, P. J. (1974). The ocular light organ of *Bathothauma lyromma* (Mollusca: Cephalopoda). *J. Zool., Lond.* **172**: 81–100.
Garcia-Tello, P. (1964). Nota preliminar sobra una observacion de bioluminiscencia en *Dosidicus gigas* (D'Orb.) Cephalopoda. *Boln Univ. Chile* No. 46: 27–28.
Gaskin, D. E. (1967). Luminescence in a squid *Moroteuthis* sp. (probably *ingens* Smith) and a possible feeding mechanism in the sperm whale *Physeter catodon* L. *Tuatara* **15**: 86–88.
Girsch, S., Herring, P. J. & McCapra, F. (1976). Structure and preliminary biochemical characterization of the bioluminescent system of *Ommastrephes pteropus* (Steenstrup) (Mollusca: Cephalopoda). *J. mar. biol. Ass. U.K.* **56**: 707–722.
Goto, T., Iio, H., Inoue, S. & Kakoi, H. (1974). Squid bioluminescence. 1. Structure of *Watasenia* oxyluciferin, a possible light-emitter in the bioluminescence of *Watasenia scintillans*. *Tetrahedron Lett.* **26**: 2321–2324.
Hamabe, M. & Shimizu, T. (1957). The copulation behaviour of Yari-oka *Loligo bleekeri* K. *Rep. Japan Sea reg. Fish. Res. Lab.* No. 3: 131–136.
Haneda, Y. (1950). Luminous organs of fish which emit light indirectly. *Pacif. Sci.* **4**: 214–227.
Haneda, Y. (1956). Squid producing an abundant luminous secretion found in Suruga Bay, Japan. *Sci. Rep. Yokosuka Cy Mus.* No. 1: 27–32.
Haneda, Y. (1964). Observations on the luminescence of the shallow-water squid, *Uroteuthis bartschi*. *Sci. Rep. Yokosuka Cy Mus.* No. 8: 10–16.

Haneda, Y. & Tsuji, F. I. (1971a). The source of light in the luminous fishes, *Anomalops* and *Photoblepharon*, from the Banda Islands. *Sci. Rep. Yokosuka Cy Mus.* No. 18: 18–27.
Haneda, Y. & Tsuji, F. I. (1971b). Descriptions of some luminous squids from the water of northern New Guinea collected by the R/V Tagula. *Sci. Rep. Yokosuka Cy Mus.* No. 18: 29–33.
Hansen, K. & Herring, P. J. (1977). Dual bioluminescent systems in the Angler fish genus, Linophryne (Pisces: Ceratioidei). *J. Zool., Lond.* **182**.
Harvey, E. N. (1952). *Bioluminescence.* New York and London: Academic Press.
Hasama, B. (1941). Uber die Bioluminescenz bei *Watasenia scintillans* im bioelectrische sowie in histologischen Bild. *Z. wiss. Zool.* **155**: 109–128.
Hastings, J. W. (1971). Light to hide by: ventral luminescence to camouflage the silhouette. *Science, N.Y.* **173**: 1016–1017.
Inoue, S., Kakoi, H. & Goto, T. (1976). Squid bioluminescence III. Isolation and Structure of *Watasenia luciferia. Tetrahedron Lett.* **34**: 2971–2974.
Ishikawa, C. (1913). Einige Bermerkungen über den leuchtenden Tintenfisch, *Watasenia* nov. gen. (*Abraliopsis* der Autoren) *scintillans*, Berry, aus Japan. *Zool. Anz.* **43**: 162–172.
Iwai, T. (1960). Luminous organs of the deep sea squaloid shark *Centroscyllium ritteri* Jordan & Fowler. *Pacif. Sci.* **14**: 51–54.
Land, M. F. (1972). The physics and biology of animal reflectors. *Prog. Biophys. mol. Biol.* **24**: 75–106.
Lu, C. C. & Clarke, M. R. (1975a). Vertical distribution of cephalopods at 40°N 53°N and 60°N in the North Atlantic. *J. mar. biol. Ass. U.K.* **55**: 143–163.
Lu, C. C. & Clarke, M. R. (1975b). Vertical distribution of cephalopods at 11°N, 20°W in the North Atlantic. *J. mar. biol. Ass. U.K.* **55**: 369–389.
Merrett, N. R., Badcock, J. & Herring, P. J. (1973). The status of *Benthalbella infans* (Pisces: Myctophoidei), its development, bioluminescence, general biology and distribution in the eastern North Atlantic. *J. Zool., Lond.* **170**: 1–48.
Meyer, W. T. (1906). Über das Leuchtorgan der Sepiolini II. Das Leuchtorgan von *Heteroteuthis. Zool. Anz.* **32**: 505–508.
Mortara, S. (1921). Gli organi luminosi di *Pyroteuthis margaritifera* e lo loro complicazioni morfologische. *Memorie R. Com. talassogr. ital.* **82**: 1–30.
Nicol, J. A. C. (1960). Studies on luminescence. On the subocular light-organs of stomiatoid fishes. *J. mar. biol. Ass. U.K.* **39**: 529–548.
O'Day, W. R. (1973). Luminescent silhouetting in stomiatoid fishes. *Contr. Sci.* No. 246: 1–8.
O'Day, W. T. & Fernandez, H. R. (1974). *Aristostomias scintillans* (Malacosteidae): a deep-sea fish with visual pigments apparently adapted to its own bioluminescence. *Vision Res.* **14**: 545–550.
Okada, Y. K. (1966). Observations on rod-like contents in the photogenic tissue of *Watasenia scintillans* through the electron microscope. In *Bioluminescence in progress*: 611–625. Johnson, F. H. & Haneda, Y. (eds). Princeton: University Press.
Okada, Y. K. (1968). Studies of luminous Cephalopoda I. Luminous squids belonging to the Ommastrephidae. *Sci. Rep. Yokosuka Cy Mus.* No. 14: 88–94.
Okada, Y. K. (1970). Study of luminous Cephalopoda III. Lycoteuthidae. *Sci. Rep. Yokosuka Cy Mus.* No. 16: 58–66.
Packard, A. (1972). Cephalopods and fish: the limits of convergence. *Biol. Rev.* **47**: 241–307.

Packard, A. & Sanders, G. D. (1971). Body patterns of *Octopus vulgaris* and maturation of the response to disturbance. *Anim. Behav.* **19**: 780–790.

Roper, C. F. E. (1963). Observations on bioluminescence in *Ommastrephes pteropus* (Steenstrup, 1855), with notes on its occurrence in the family Ommastrephidae (Mollusca: Cephalopoda). *Bull. mar. Sci.* **13**: 343–353.

Roper, C. F. E. (1969). Systematics and zoogeography of the worldwide bathypelagic squid *Bathyteuthis* (Cephalopoda: Oegopsida). *Bull. U.S. natn Mus.* **291**: 1–210.

Roper, C. F. E. (1972). Ecology and vertical distribution of Mediterranean pelagic cephalopods. *Final Rep. Medit. biol. Stud.* **1**: 282–346. (Smithsonian Institution, Washington, D.C.)

Roper, C. F. E., Young, R. E. & Voss, G. L. (1969). An illustrated key to the families of the order Teuthoidea (Cephalopoda). *Smithson. Contr. Zool.* No. 13: 1–32.

Sasaki, M. (1914). Observations on Hotaru-ika *Watasenia scintillans*. *J. Coll. agric. Hokkaido imp. Univ.* **6**: 75–105.

Schmidt, J. (1922). Live specimens of *Spirula*. *Nature, Lond.* **110**: 788–790.

Shimomura, O. & Johnson, F. H. (1972). Enzymatic and non-enzymatic bioluminescence. In *Photobiology* **7**: 275–334. Giese, A. C. (ed.). London and New York: Academic Press.

Shoji, R. (1919). A physiological study on the luminescence of *Watasenia scintillans* (Berry). *Am. J. Physiol.* **47**: 534–557.

Skowron, S. (1926). On the luminescence of some cephalopods (*Sepiola* and *Heteroteuthis*). *Riv. Biol.* **8**: 236–240.

Tett, P. B. & Kelly, M. G. (1973). Marine bioluminescence. *Oceanogr. mar. Biol.* **11**: 89–173.

Tsuji, F. I. & Haneda, Y. (1971). Studies on the luminescence reaction of a myctophid fish *Diaphus elucens* Brauer. *Sci. Rep. Yokosuka Cy Mus.* No. 18: 104–109.

Voss, G. L. (1960). Bermudan cephalopods. *Fieldiana, Zool.* **39**: 419–446.

Voss, G. L. (1962). A monograph of the Cephalopoda of the North Atlantic. 1. The family Lycoteuthidae. *Bull. mar. Sci.* **12**: 264–305.

Voss, G. L. (1967a). Squids, jet-powered torpedos of the deep. *Natn. geogr. Mag.* **131**: 386–411.

Voss, G. L. (1967b). The biology and bathymetric distribution of deep-sea cephalopods. *Stud. trop. Oceanogr. Miami* **5**: 511–535.

Voss, N. A. (1969). Biological investigations of the deep sea. 47. A monograph of the Cephalopods of the North Atlantic. The family Histioteuthidae. *Bull. mar. Sci.* **19**: 713–867.

Young, R. E. (1972). The systematics and areal distribution of pelagic cephalopods from the seas off Southern California. *Smithson. Contr. Zool.* No. 97: 1–159.

Young, R. E. (1973). Information feedback from photophores and ventral countershading in midwater squid. *Pacif. Sci.* **27**: 1–7.

Young, R. E. (1975). *Leachia pacifica* (Cephalopoda: Teuthidae) spawning habitat and function of brachial photophores. *Pacif. Sci.* **29**: 19–25.

Young, R. E. & Roper, C. F. E. (1969). A monograph of the Cephalopoda of the North Atlantic: the family Cycloteuthidae. *Smithson. Contr. Zool.* No. 5: 1–24.

Young, R. E. & Roper, C. F. E. (1976). Bioluminescent countershading in midwater animals: evidence from living squid. *Science, Wash.* **191**: 1046–1048.

Symp. zool. Soc. Lond. (1977) No. 38, 161–190.

VENTRAL BIOLUMINESCENT COUNTERSHADING IN MIDWATER CEPHALOPODS

R. E. YOUNG

Department of Oceanography, University of Hawaii, Honolulu, Hawaii

SYNOPSIS

The requirements for effective bioluminescent countershading are examined in six species of midwater cephalopods. Bioluminescent countershading may occur in species that occupy depths where light levels are compatible with countershading during the day, night or twilight. *Pterygioteuthis microlampas* occupies such depths during both the day and night, *Sandalops melancholicus* (juveniles) during the day only, *Heteroteuthis hawaiiensis* (juveniles) during twilight and moonlit nights, *Enoploteuthis* sp. and *Histioteuthis dofleini* during the day and possibly at night. These species all seem to possess most of the basic mechanisms necessary for ventral countershading. They apparently have means of shielding opaque structures with bioluminescence, of producing blue, highly directional light, of detecting downwelling light and their own bioluminescent light, and of regulating the intensity of their bioluminescence. However, one species examined, *Thelidioteuthis alessandrinii*, probably does not countershade. A comparison of presumed countershading and non-countershading photophores in these species reveals only a single distinctive structural difference: countershading photophores possess "skylight" filters.

INTRODUCTION

Since W. D. Clarke (1963) first presented convincing arguments on the role of bioluminescent light in ventral countershading in midwater animals, considerable supporting evidence has accumulated. Badcock (1970) and Foxton (1970) found that countershading characteristics correlated with the vertical distribution of midwater fishes and decapod crustaceans (respectively), and they suggested that ventral countershading operates above 650 to 700 m off the Canary Islands. Denton, Gilpin-Brown & Roberts (1969) found that the complex structure of the photophores of the midwater fish *Argyropelecus* was compatible with a ventral countershading function. Arnold, Young & King (1974) came to similar conclusions concerning the structure of the ocular photophores in the midwater squid *Pterygioteuthis*. Denton, Gilpin-Brown & Wright (1972) demonstrated that the radiance field and the light intensity produced by the photophores of *Argyropelecus* and *Chauliodus* were comparable to that of their day habitats. Young (1973) found that the location of extra-ocular photoreceptors, the photosensitive vesicles, and the location of certain photophores in the midwater squids *Abraliopsis* spp. indicated that these animals could directly examine the luminous output of their photophores and the intensity of downwelling light. He suggested that they regulate their luminescence to match the downwelling light, and he suggested that photophores directed into the eyes of stomiatoid fish might be part of an analogous system. Lawry (1974) demonstrated this latter mechanism in myctophid fish and concluded that the eyes are used to measure downwelling light levels. He also noted that in an aquarium these fish luminesced under low-level laboratory illumination.

For ventral bioluminescent countershading to be most effective a number of conditions must be satisfied.

(1) The animals must live at depths where a bioluminescent countershading mechanism is feasible. If downwelling light is too low, countershading will be unnecessary, and if the downwelling light is too bright bioluminescent light will be inadequate for the job. Off Hawaii the changeover from predominately reflective to non-reflective fishes, and from half-red to all-red sergestids occurs at depths of 650 to 700 m (Amesbury, 1975; Walters, 1975), or about the same depth as found by Badcock (1970) and Foxton (1970) off the Canaries. Walters (1975), on the basis of the vertical distribution of sergestid shrimps, however, suggests that the lower limit of ventral countershading may be as deep as 775 m during the day off Hawaii. Therefore, we can conservatively place the lower limit of countershading during the day somewhere between 650 and 775 m off Hawaii. Presumably, whenever an animal encounters directional light with levels compatible with ventral countershading, whether during the day, night or twilight, bioluminescent countershading will be advantageous. Light intensity at depths of 650 to 775 m in clear water corresponds to levels of about 250 to 370 m at night under a full moon, or depths of about 65 to 190 m on a clear moonless night (G. L. Clarke & Denton, 1962).

The upper limit of ventral countershading is even less certain. Off Hawaii the upper limit of most of the micronekton is about 400 m (Maynard, Riggs & Walters, 1975) although a few species undoubtedly occur in shallower water (e.g., T. A. Clarke, 1973). Amesbury (1975) concluded that this upper limit of the midwater fish fauna is correlated with light intensity. This depth may represent the upper limit of ventral countershading for most shallow groups in the midwater micronekton. In the clearest ocean water, light intensity at 400 m is about 10^{-1} $\mu W/cm^2$ (G. L. Clarke & Denton, 1962). G. L. Clarke & Breslau (1960) have recorded bioluminescent light in excess of 10^{-2} $\mu W/cm^2$ from dense populations of dinoflagellates. Nicol (1967) suggests that bioluminescent light from fishes is possibly less than 5×10^{-4} $\mu W/cm^2$. Although these comparisons have limited value, they suggest 400 m is near (if not above) the upper limit for effective use of ventrally directed bioluminescent light.

(2) Bioluminescence must shield opaque structures in the animals when viewed from below. Photophores must either lie directly below opaque structures or some mechanism must exist for transferring light to such areas.

(3) Bioluminescent light must match the characteristics of the downwelling light. In midwaters the radiance field is symmetrical around the vertical and strongly dominated by light passing vertically downward (Jerlov, 1970). In clear oceanic waters light intensity peaks strongly in the blue region, being maximal about 478 nm, and its intensity decreases exponentially with depth (Kampa, 1971). For effective ventral countershading, therefore, the luminescence must be blue, comparable in intensity to the downwelling light and strongly directional. Blue luminescence can be achieved if the photocytes produce only these wave lengths, or if filters are used to restrict a broader band of light. Directionality is presumably a function of photophore location and the reflective and dioptric modifications of the photophores and/or associated structures.

To maintain a match between bioluminescent and downwelling light an animal may move up and down with an isolume as light intensity varies, or it may regulate the luminous output of its photophores. Evidence suggests that the latter alternative is the one generally adopted. Members of the micronekton commonly have vertical ranges during the day of 100m or more (Foxton, 1970; Badcock, 1970; T. A. Clarke, 1973; Roper & Young, 1975). In such populations members are exposed to more than a tenfold change in light levels (G. L. Clarke & Denton, 1962). Such distribution patterns suggest direct regulation of the photophores. In addition, echo-sounding records presumably of midwater fishes indicate that minor variations in light intensity from passing clouds or between sunny and overcast days do not affect the vertical position of these animals (Walters, in prep.). Walters (in prep.) also demonstrated with echo-sounding records that midwater animals can be passively carried up and down distances of at least 40 m by internal waves. The echo-sounding data suggests that these midwater animals do not respond to minor changes in light intensity by moving vertically.

(4) Bioluminescent light must be produced for extended periods of time. Although ventral countershading may be a response to detection of a predator and not a continuous process, even such intermittent countershading, presumably, requires a continuous glow for some minutes.

(5) Although not absolutely necessary, the ability to adjust ventral countershading mechanisms to correspond with a variety of body attitudes may be advantageous for some animals.

In this paper I will briefly survey six species of midwater cephalopods to determine if they meet these requirements for ventral bioluminescent countershading.

MATERIALS AND METHODS

Specimens were captured off the island of Oahu in the Hawaiian Archipelago at approximately 158°18′W, 21°23′N, over bottom depths of 1500 to 4000 m. Two types of trawls were used: a modified 3 m Tucker trawl and a 3 m Isaacs-Kidd midwater trawl (IKMT). The Tucker trawl opens and closes at the fishing depth; hence capture of specimens during setting and retrieval of the trawl (i.e. contamination) cannot occur. The opening-closing mechanism utilizes a mechanical release that is activated by weighted messengers sent down the towing cable.

The IKMT is always open, and occasionally specimens are captured during the raising and lowering of the trawl. To minimize this contamination, the trawl was dropped as rapidly as possible and retrieved with the ship moving slowly ahead. The net is pulled horizontally at 3 to 4 knots, and the modal depth of the tow is assumed to be the depth of capture. Depth records for both trawls were obtained with a benthos time-depth recorder. Details of the trawling programs can be found in T. A. Clarke (1973) and Walters (1975).

Material prepared for histological work was fixed in 3% gluteraldehyde in sea-water buffered with *S*-collidine for 1 to 24 hr, post-fixed in 2% osmium tetroxide in sea-water buffered with *S*-collidine for 1 to 4 hr, then dehydrated and embedded in Epon 812. Sections for light microscopy were cut between 1·5 and 10 μm.

A description of the ultrastructure of the photophores surveyed in this paper is being prepared by John Arnold and R. E. Young.

RESULTS AND DISCUSSION

Pterygioteuthis microlampas

The vertical distribution of this species based on Tucker trawl captures is presented in Fig. 1. During the day all captures were probably made between 450 and 575 m with most coming from depths between 450 and 500 m. The night captures indicate a range of 25 to 180 m with most captures coming from depths between 50 and 105 m. IKMT data show similar distribution patterns.

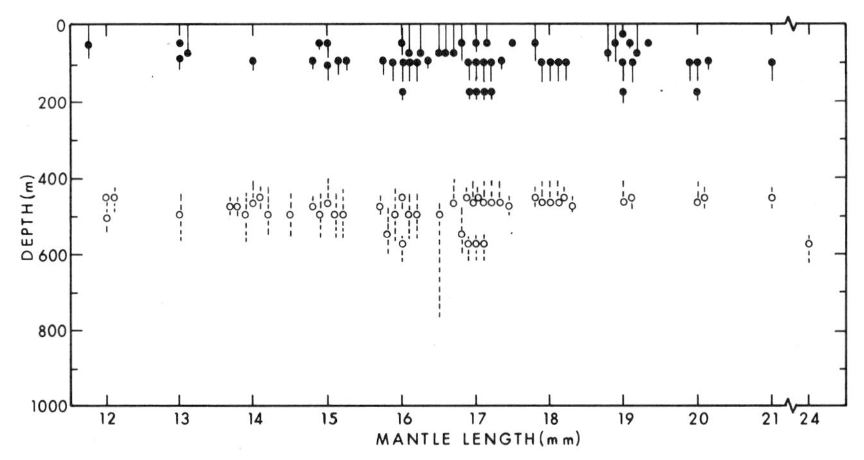

FIG. 1. Vertical distribution of *Pterygioteuthis microlampas*. Vertical bars: vertical range of opening-closing tow. Dots: most likely depth of capture, usually the modal depth of net. Dashed bars and open dots: day capture. Solid bars and solid dots: night capture.

The effects of moonlight on the night distribution were examined during new moon and full moon with the IKMT, and on separate occasions with the Tucker trawl. Data from the IKMT, which caught more specimens, are presented in Fig. 2. On neither occasion was a moonlight effect detectable. Peculiarly, the IKMT captured unusually large numbers of animals below 150 m during both the new moon and the full moon. The data clearly indicate that this species inhabits depths at night, as well as during the day, where bioluminescent countershading would be effective.

P. microlampas has a series of complex photophores. Fourteen photophores are present on each eye. Nine of these have a similar structure, and their arrangement is constant in all members of the

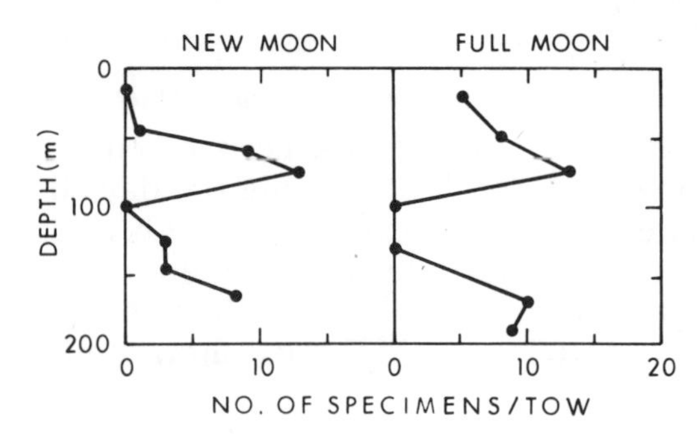

FIG. 2. Vertical distribution of *Pterygioteuthis microlampas* at night during new and full moon.

subfamily. These photophores are probably involved in countershading. The other five exhibit specific or generic variation among members of the subfamily and probably are not involved in countershading. Two large photophores, the anal photophores, whose structure is similar to the nine ocular photophores, lie within the mantle cavity to either side of the intestine. Further posteriorly on the midline within the body a series of four photophores is found. These organs have a distinctly different structure and their role in countershading is uncertain. Two large specialized photophores lie at the base of the gills; these and the photophores embedded in the tentacles probably play no role in countershading.

The structure of the presumed countershading photophores on the eyes has been examined by Arnold & Young (1974) and Arnold, Young & King (1974). They suggested that a variety of iridophores are utilized to regulate the direction, color and intensity of bioluminescent light and concluded that these photophores are probably involved in countershading.

P. microlampas is covered with an extensive array of iridophores, only part of which seems involved in ventral countershading. Much of the ventral surface of the head and adjacent funnel groove is highly silvered. The antero-ventral region of the head and the basal region of the ventral arms contains a pair of complex chambers, the brachial chambers, lined with iridophores and filled with a transparent, extracellular material (Fig. 3A). The concave dorsal wall of the brachial chamber is lined with a thick layer of iridophores. The ventral wall of the chamber is flatter and contains a dense array of unusually-oriented iridophores. The platelets, while irregular in spacing and not precisely aligned, generally lie almost perpendicular to the surface of the chamber

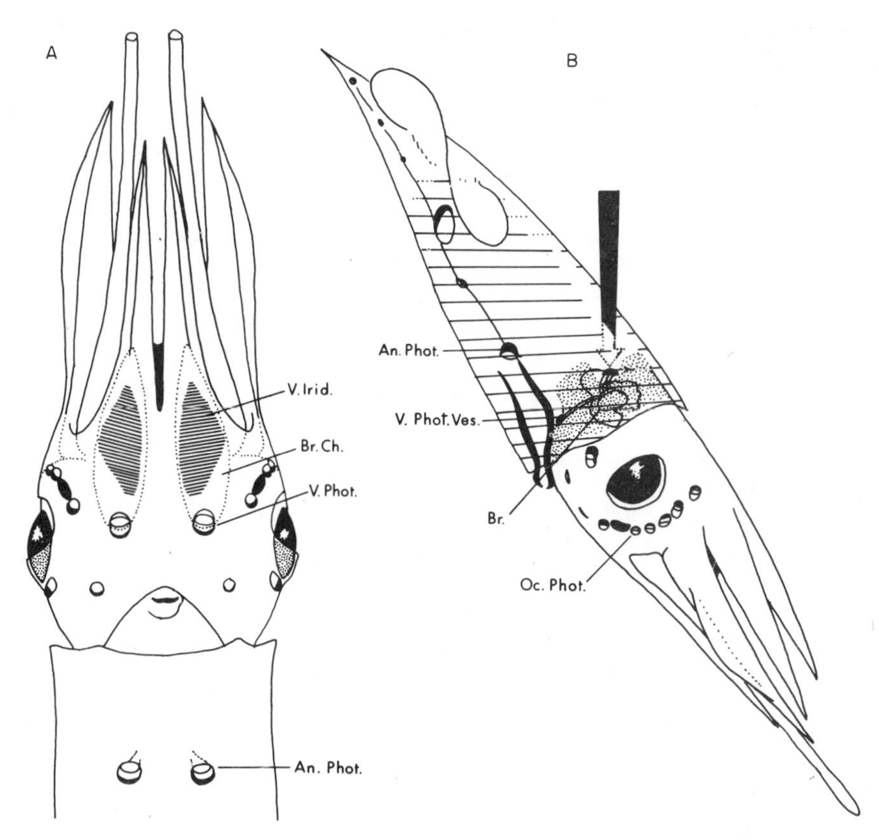

FIG. 3. *Pterygioteuthis microlampas*. A. Ventral view of head. B. Lateral view. Arrow points to dorsal photosensitive vesicles and represents the pathway of downwelling light. Cross-hatching on the mantle represents presence of iridophores. Symbols: An. Phot., anal photophores; Br., brain; Br. Ch., brachial chamber; Oc. Phot., ocular photophore; V. Irid., ventral iridophores of brachial chamber; V. Phot., ventral ocular photophore; V. Phot. Ves., ventral photosensitive vesicles. Dotted area in B indicates chromatophore cover on neck.

with their long axes oriented medio-laterally (Fig. 4A). A large ventral ocular photophore at the posterior edge of the chamber has its lens directed toward the chamber. Iridophores extend beyond the brachial chamber about half the length of the ventral arms. The silvered dorsal surface of the chamber is augmented by the silvered bases of the third arms and tentacles.

P. microlampas has two sets of photosensitive vesicles (Fig. 3B). The more dorsal set, the dorsal organs, are embedded in the postero-dorsal wall of the cephalic cartilage just posterior to the optic lobes. Each dorsal organ is compact and approximately

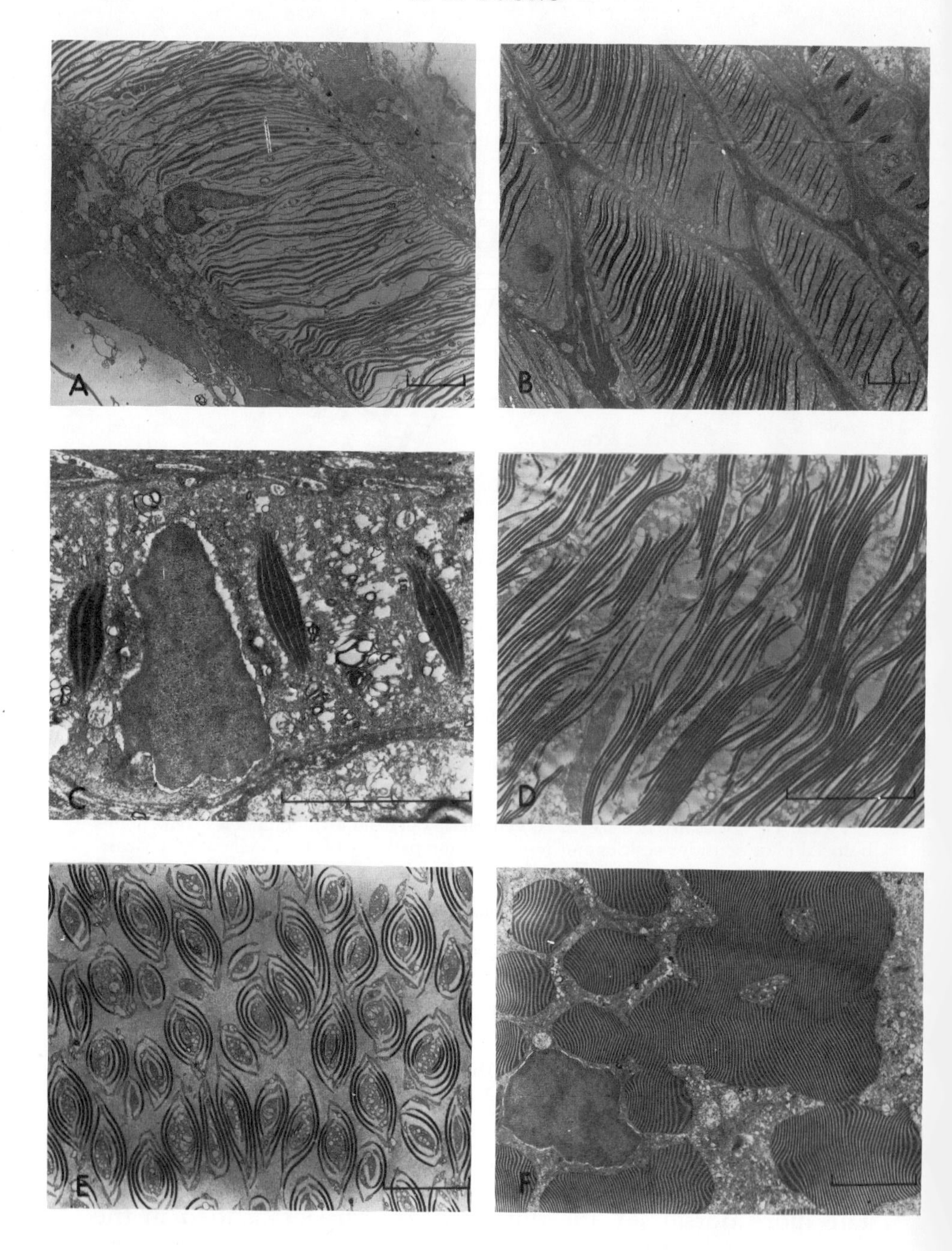
A
B
C
D
E
F

disc-shaped with a convex distal surface and a flat proximal surface. The disc is tilted some 30° or more from the longitudinal body axis. The ventral organs are deeply embedded in the postero-ventral surfaces of the cephalic cartilage. Each ventral organ is thick, compact and approximately square in outline with a broad, thin extension passing medially and dorsally for a short distance along the inner surface of the cephalic cartilage.

Each dorsal organ consists of numerous elongate vesicles that have an irregular shape and occasionally interconnect but generally form a single layer. Sensory cell-bodies are distributed on all sides of each vesicle, and sensory processes are oriented primarily at right angles to the cartilage, but the arrangement is not precise. Vesicles in the ventral organs are more irregular in shape and orientation, and the sensory processes tend to lie parallel to the cephalic cartilage.

Much of the neck region is covered with chromatophores. However, chromatophores are absent from a semicircular area that lies opposite each dorsal organ and forms a well-defined "window". Much of the mantle is covered by iridophores. However, iridophores are weak or absent along two dorso-lateral strips above the dorsal organs. These strips form a "window" in the mantle.

On several occasions *P. microlampas* has been observed alive in a shipboard aquarium. The countershading photophores produced a blue light and glowed continuously during the longest timed observation period (15 min). Light leaving the photophore was very directional but no attempt was made to quantify this. The brachial chamber did not seem to reflect light, presumably due to the very low intensity of the light. The animal exhibited a variety of attitudes from nearly vertical to nearly horizontal. The effect of varying attitudes on the direction of the luminous output could not be determined. The animal was forcibly tilted in the aquarium in an attempt to observe compensatory eye movements; however the eye did not appear to rotate. Under the microscope mechanical agitation of the ocular photophores indicated some degree of

---

FIG. 4. A. Iridophores near ventral surface of brachial chamber in *Pterygioteuthis microlampas*. Ventral surface of arm is in upper right-hand corner. B. Shield iridophores of *Heteroteuthis hawaiiensis*. Ventral surface of shield is just beyond upper right-hand corner. C. Iridophore with small iridosomes from ventral shield of *H. hawaiiensis*. D. Cross-section of iridophores of the fibrous reflector in *Sandalops melancholicus*. E. Slightly oblique cross-section of light guides in *S. melancholicus*. F. Iridophores from dorsal reflector in *S. melancholicus*. Scale is 5 μm.

independent movement over the surface of the eye. The posterior muscles of the anal photophores were pinched with forceps. This action resulted in contraction of the muscles and a ventral rotation of the photophores about 45° from an orientation nearly parallel to the body axis. In a fresh specimen a fibre optic light probe was directed into the ventral ocular photophore. The result was a strong ventrally-directed reflection from the ventral iridophores of the brachial chamber.

Only the basic outline of the ventral countershading mechanisms in this squid are apparent. The colour and intensity of the emitted light seem to be regulated within the photophores. Obvious windows allow downwelling light to reach the dorsal photosensitive vesicles. The ventral vesicles are clearly exposed to the anal photophores when the latter are directed anteriorly. The two anal photophores are well positioned to conceal much of the viscera, and some light from these photophores is probably reflected off the postero-ventral surface of the head.

On the head the large eyes and the heavily pigmented buccal membrane are completely opaque. Except perhaps medially, the eyes seem to have an adequate covering of photophores. Presumably the brachial channel and the other silvery areas on the ventral arms, tentacles and third arms are responsible for concealing the medial parts of the eyes and buccal membrane. Although it has not been possible to work out the exact geometry of the light paths in the brachial chamber, the ventral surface of the chamber apparently intercepts progressively more and more of the light emitted by the ventral ocular photophore and directs it ventrally. In *Pyroteuthis addolux*, a longer-armed relative of *Pterygioteuthis microlampas*, the brachial chamber extends almost the full length of the ventral arms.

Heteroteuthis hawaiiensis

The vertical distribution is presented in Fig. 5. During the day nearly all specimens of 17 mm mantle length (ML) or less were captured between 250 and 350 m. Specimens larger than 17 mm ML were taken between 375 and 650 m. At night over 90% of the specimens less than 17 mm ML were captured within a narrow zone between 150 and 200 m. This distribution is not affected by moonlight. Specimens 17 mm ML or larger were taken at depths of 110 to 550 m. At about 16 to 17 mm ML individuals become sexually mature, and their vertical distribution broadens perhaps as part of an ill-defined ontogenetic descent.

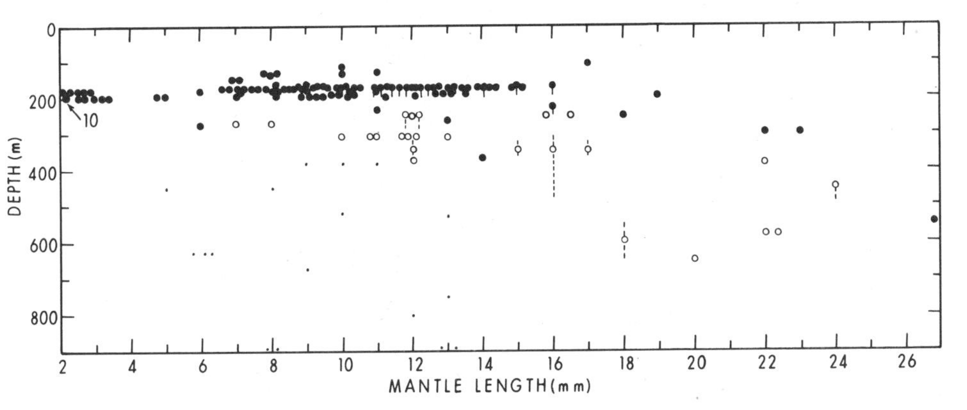

FIG. 5. Vertical distribution of *Heteroteuthis hawaiiensis*. Dots without bars represent captures from open nets at the modal depth of the net. Small dots represent probable contaminants, otherwise symbols as in Fig. 1.

H. hawaiiensis has a single large photophore located on the ventral surface of the viscera within the mantle cavity (Fig. 6). While this photophore in a related species was studied by a number of early workers, its structure remains poorly known. The organ consists essentially of a large cup-shaped reflector, the posterior cup, surrounding a central core containing branching glandular sacs embedded in a matrix of clear packing tissue, and an anterior cap containing numerous iridophores (Fig. 7D). The posterior cup lies embedded in the ink sac. The glandular sacs exit via four ducts in two papillae. A thick layer of muscle covers the anterior cap, and thick muscle bands pass to the photophore from posterior, lateral and anterior positions. Chromatophores are not present on the surface of the photophore.

This animal has a striking external complement of iridophores, many of which are involved in lateral and dorsal countershading and thus are beyond the scope of this paper. Some, however, function in ventral countershading. The antero-ventral surface of the mantle is highly modified and is commonly called the ventral shield. The ventral shield extends anteriorly beneath the funnel and part of the head. On the surface of the shield are thick circular patches of iridophores, each patch consisting of numerous small, thin disc-shaped iridophores. In some places there are five or six layers of discs. The platelets vary greatly in size and orientation (Fig. 4B,C). Those in the inner discs are relatively long, irregularly spaced and may be tilted as much as 45° to the mantle surface. The

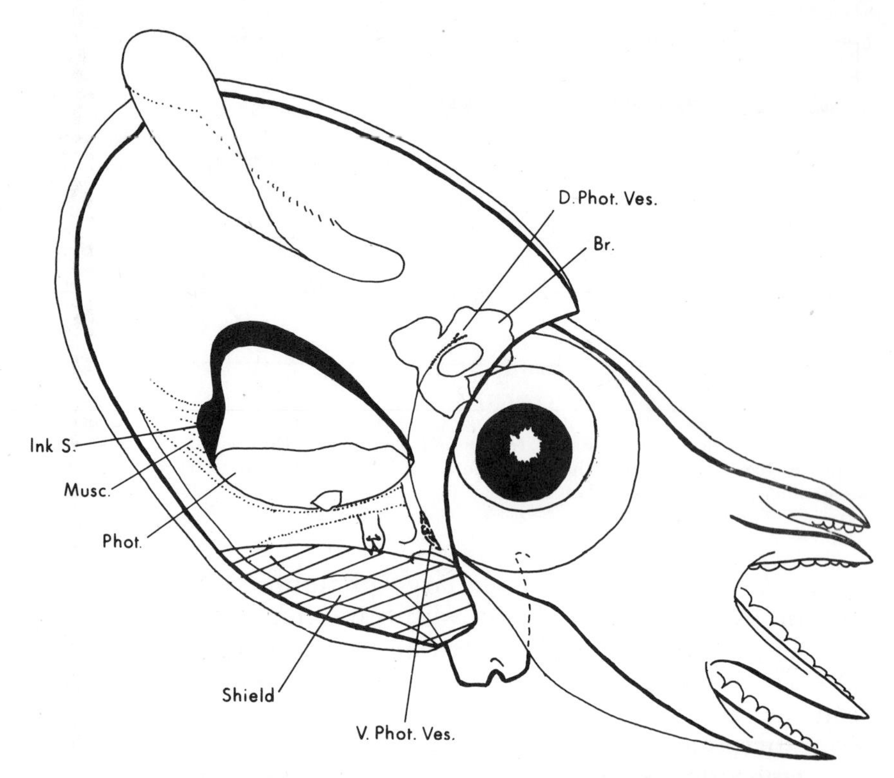

FIG. 6. *Heteroteuthis hawaiiensis* lateral view. Symbols: Br., brain; D.Phot.Ves., dorsal photosensitive vesicles; Ink S., ink sac; Musc., muscle; Phot., photophore; V.Phot.Ves., ventral photosensitive vesicles.

outer discs contain short platelets closely spaced and arranged in iridosomes, which may be oriented at 90° to the mantle surface. Around the periphery of the shield the patches of iridophores tend to merge together. The central portion of the shield between patches has only a few scattered iridophores. A single large

---

FIG. 7. A. Presumed countershading photophore of *Enoploteuthis* sp. B. Presumed non-countershading photophore of *Enoploteuthis* sp. C. Large photophore of *Thelidioteuthis alessandrinii*. D. Photophore of *Heteroteuthis hawaiiensis*. E. Photophore of *Histioteuthis* sp. F. Photophore of *Sandalops melancholicus*. Scale is 100 μm. Symbols: A.C., anterior cap; A.I., axial iridophores; A.R., anterior reflector; C, chromatophore; D, ducts; D.R., dorsal reflector; F.R., fibrous reflector; G.S., glandular sacs; I.S., ink sac; L, "lens"; L.G., light guides; M, muscle; P, pigment; P.B., photogenic body; P.C., photogenic crystalloids; P.Cu., posterior cup; P.R., posterior reflector; P.T., photogenic tissue; P.Ti., packing tissue; P.V., photosensitive vesicles (drawn in approximate position); R, reflector formed of light guides; R.D., ring-shaped depression.

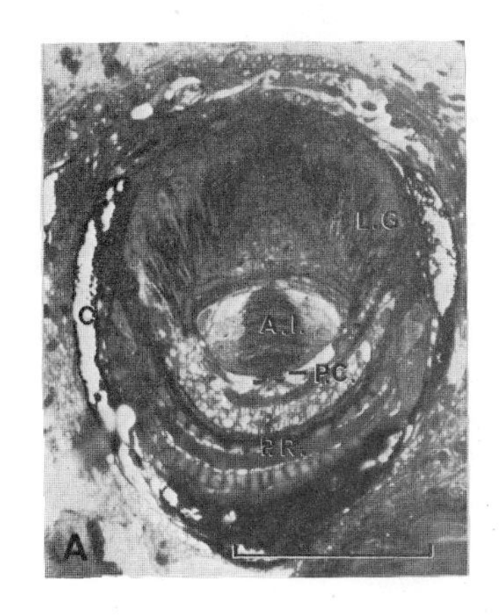
L.G.
C
A.I.
P.C.
P.R.
A

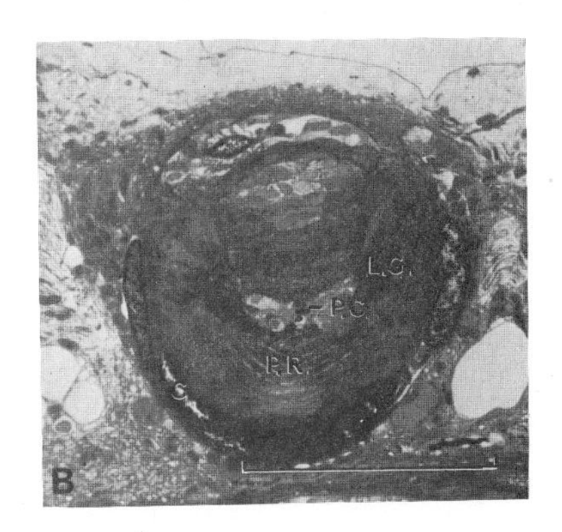
L.C.
P.C.
P.R.
C
B

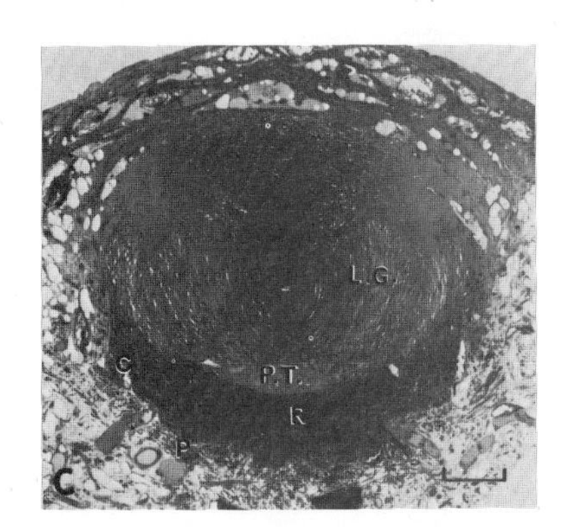
L.G.
C
P.T.
R
P
C

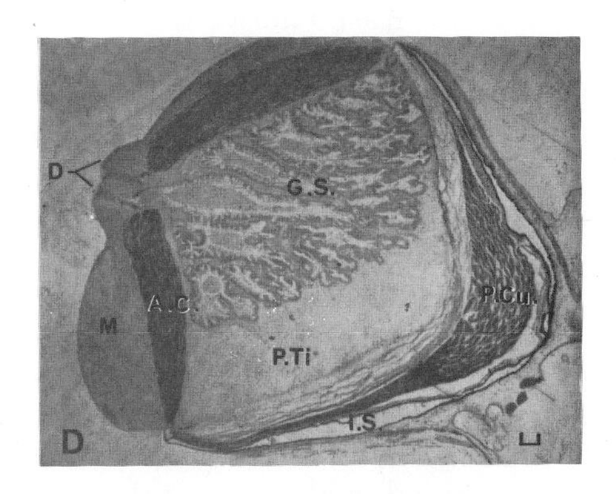
D
G.S.
A.C.
M
P.Cu.
P.Ti
I.S
D

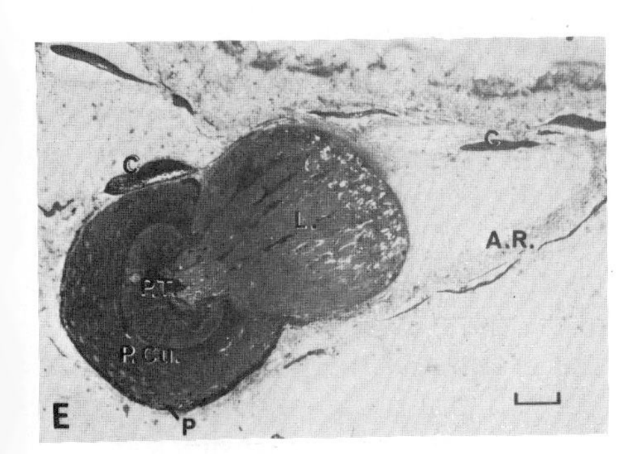
C
C
L.
A.R.
P.T.
P. Cu.
E
P

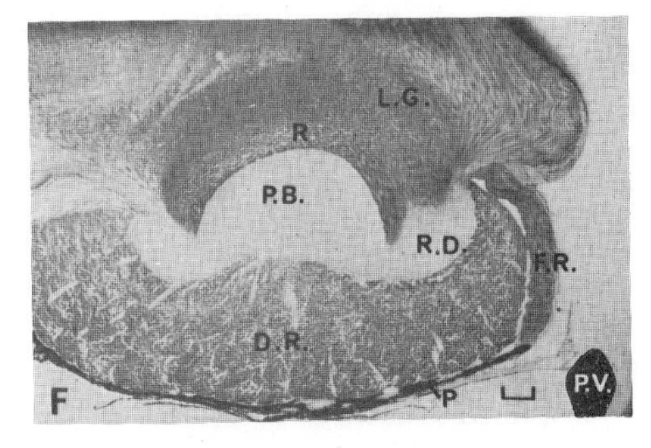
L.G.
R
P.B.
R.D.
F.R.
D.R.
F
P
P.V.

chromatophore is centered above each patch of iridophores. Other chromatophores are scattered over the areas largely devoid of iridophores although the distinction between these chromatophore patterns becomes slightly less apparent where the patches merge. These two groups of chromatophores have been observed in fresh specimens to contract or expand as separate units. A transition zone exists between the ventral shield and the lateral iridophores on the mantle. In this zone the lateral layer of iridophores tapers in thickness and reflects interference colours, with blue light being reflected in the thinner portions at oblique angles. Beneath this thin portion lies the thin and not entirely continuous layer of the disc-shaped iridophores that occupy the ventral shield.

Frequently in living as well as preserved specimens the ventral shield is raised above the surrounding mantle.

The antero-ventral surfaces of the head, ventral surfaces of the arms (beneath a thick layer of clear gelatinous tissue), and tentacle sheath carry a thick layer of iridophores. The posterior limit of iridophores on the ventral surface of the head coincides with the anterior margin of the ventral shield. Iridophores are also found near the ventro-medial surfaces of the eyes and on most of the oral surfaces of the dorsal pairs of arms and adjoining webs.

Two sets of photosensitive vesicles are present (Fig. 6). The more dorsal set lies on the posterior margin of the peduncle complex and consists of a short and narrow string of tiny vesicles. Still smaller vesicles are scattered along a nerve that runs ventrally past the dorsal organ and around the eye to a relatively large and extremely flat, loosely associated group of vesicles. Each of these ventral organs lies beneath the eye and just medial to the muscle that connects the lateral edge of the funnel to the head. The wall of each vesicle is narrow and seems to have a single layer of sensory cell-bodies. The sensory processes fill the lumen of each vesicle.

Much of the neck region, including the area lateral to the dorsal organs, is covered with chromatophores. Chromatophores are absent from the dorsal region of the neck. In an obliquely positioned animal, downwelling light entering the dorsal organs would pass through the dorsal surface of the mantle, pass lateral to the projecting tip of the liver and pass through the pigment-free region of the neck to enter the vesicles. Since the ventral vesicles are located beneath the eyes, they are shielded from downwelling light.

The photophore in the related *Heteroteuthis dispar* has long been known both to glow and to produce a luminous secretion

(Meyer, 1906; Harvey, 1952). I have observed *Heteroteuthis hawaiiensis* glow in a shipboard aquarium for long periods of time. The longest timed observation was 30 min, but intermittent observations indicate that the animal may luminesce for many hours continuously. The light is directed downward but is quite broad. No attempts have been made to measure the angle included by the beam. Blue light can be seen reflecting off the ventral arms, and some light can be seen passing through the base of the ventral arms to emerge anteriorly. When the animal was handled it strongly varied the light intensity nearly extinguishing it one moment and turning it back on the next. This variation almost certainly arose through regulation within the photophore and not through external chromatophore control. The iridophores of the shield glowed brightly. The animal continuously varied its orientation in the aquarium; however it was not possible to determine whether any compensatory orientation of the light beam occurred. The anterior cap iridophores reflect yellow light most strongly when illuminated with a microscope lamp at angles normal to their outer surfaces, and they clearly transmit blue bioluminescent light. The posterior cup iridophores reflect blue to blue-green light when illuminated with this same light.

Ventral countershading presumably operates in *H. hawaiiensis* in the following manner. The large photophore emits only blue light; the posterior cup and anterior cap iridophores probably function in much the same way as similarly-placed iridophores in the ocular photophores of *Pterygioteuthis microlampas*, the former by reflecting only blue light and the latter by transmitting only blue light. Much of the light leaving the photophore passes through the ventral shield. The optical properties of the complex arrangement of shield iridophores are poorly understood: they apparently intercept much of the light from the photophore and redirect it ventrally. The total effect is to spread the light over a broader surface, the shield being about twice the width of the photophore. This effect is somewhat comparable to placing a frosted piece of glass in front of a light bulb to produce a broadly illuminated surface. At least some directionality of the beam is maintained by the direct passage of some light through the shield without encountering iridophores. Some light from the photophore passes forward through the funnel, where it is detected by the ventral photosensitive vesicles (the more dorsal set detecting downwelling light), then the light is reflected downward off the ventral surfaces of the head and arms. Some light passes through the base of the arms and emerges in the buccal region, where it is reflected downward by iridophores on the oral surfaces of the more dorsal arms and webs. Thus light is spread over the antero-ventral surfaces of the animal to conceal an obliquely oriented animal. Although chromatophores on the surface of the ventral shield may contribute to the regulation of light intensity, most of the regulation probably takes place within the photophore itself; the actual mechanism, however, remains obscure.

Enoploteuthis sp.

Only five specimens of this undescribed species have been captured. One specimen (42 mm ML) was taken during the day at 515 m. The others (20–57 mm ML) were taken at night at depths between 60 and 125 m.

Enoploteuthis sp. like all members of its subfamily has many small photophores distributed primarily over the ventral surfaces of the arms, head, funnel and mantle, and with a few exceptions they are directed ventrally. The photophores in *Enoploteuthis* sp. consist of at least two major types. One type (Fig 7A), presumably the countershading photophore, has a large posterior reflector behind the presumed photocytes, which contain large crystalloids. Immediately distal to this photogenic area in the axis of the photophore is a small stack of iridophores, and distal laterally to the photogenic area are rod-like structures presumed to be light guides. The function of the axial stack of iridophores is unknown, but the irregular spacing of the platelets suggests a broad-band reflector. Along the axis distal to the axial iridophores lies a peculiar tissue whose function is unknown. Chromatophores occur around the photophore, and in preserved specimens an iris-like function for these chromatophores is indicated by their various states of contraction near the surface of the photophore. The second type of photophore has basically the same type of structure but lacks the axial iridophores (Fig. 7B).

Enoploteuthis sp. has four sets of photosensitive vesicles which are arranged similarly to those described in *Abraliopsis* sp. (Young, 1973). One set is located dorsally, one posteriorly and two ventrally. The dorsal organs are situated in concavities of the cephalic cartilage at the postero-dorsal edges of the head. Each organ is compact, dorso-ventrally flattened and circular to triangular in outline. The posterior organs are located on the medio-posterior surfaces of the optic lobes. Each of these organs is approximately elliptical and very flat. The ventral vesicles on each side consist of two narrowly joined, flattened lobes, one of which, the antero-ventral lobe, is located somewhat anterior and medial to the other, the midventral organ. The latter has a more irregular shape and is less compact than the former. Narrow strands of vesicles extend from the posterior organs to the dorsal organs and to the midventral organs.

Circular "windows" on the head, similar to those described in *Abraliopsis* sp. (Young, 1973) and characterized by a reduced number of chromatophores, are present above each dorsal organ. Large ventral windows, totally lacking chromatophores, lie on the ventral surface of the head above the funnel and below the ventral vesicles. Photophores are present on the dorsal surface of the funnel, directed at the ventral photosensitive vesicles.

Enoploteuthis sp. has not been observed to luminesce. The nearest relative for which observational data are available is *Abralia trigonura*, a member of the same subfamily with a similar vertical distribution and similar arrangement of photophores (although somewhat different in structure). This animal was seen to glow continuously in a shipboard aquarium during a 60-min observation period. Intermittent observations indicate that the animal continued to glow for at least another hour. The luminous output was faint and the colour of the light could not be reliably determined with the unaided eye. When the animal was held by hand, the light intensity could be seen to fluctuate. This apparent variation in luminous intensity may possibly have resulted from photophore rotation. The light beam was highly directional: viewed laterally, bioluminescence could barely be detected, and light intensity dropped very rapidly as the animal was rotated away from a ventral view.

In preserved material the posterior reflector from both types of photophores in *Enoploteuthis* sp. reflected blue light when illuminated with "white" light from a microscope lamp.

In *Enoploteuthis* sp. clear mechanisms exist for detecting the intensity of downwelling and bioluminescent light. The broad distribution of photophores seems capable of concealing all opaque structures. Although poorly understood, a possible mechanism for filtering emitted light exists in those photophores with both a posterior reflector and a stack of axial iridophores. Light leaving the photocytes in an anterior direction would be reflected back into the posterior reflector by the axial iridophores. As a result, all light produced in the photophore would encounter the posterior reflector, which would reflect only blue light. Chromatophores provide a potential means of regulating luminous output.

Histioteuthis dofleini

The vertical distribution of this species presented in Fig. 8 is from Young (1975a). During the day the vertical range is 375 to 850 m although most captures came from depths of 500 to 700 m. At night the range is 100 to 500 m with most captures coming from 150 to 300 m. During both day and night the larger individuals tend to be found at the greater depths.

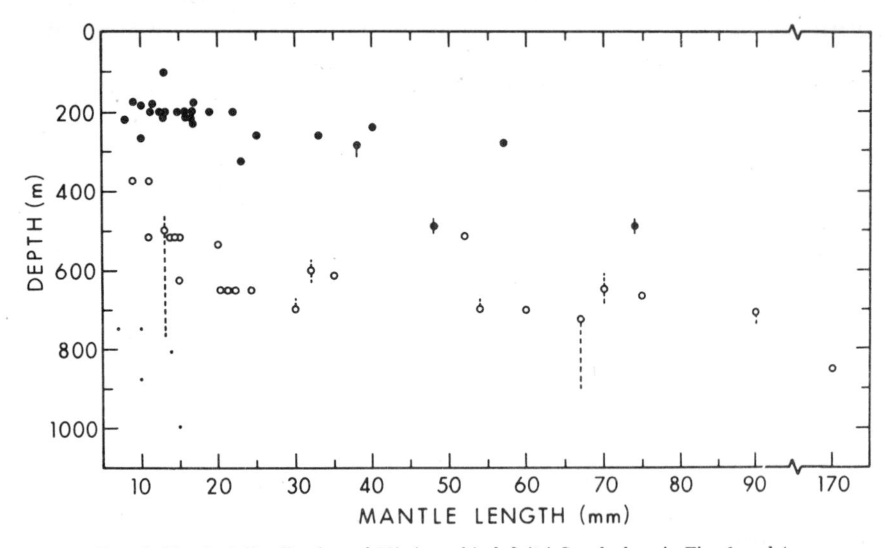

FIG. 8. Vertical distribution of *Histioteuthis dofleini*. Symbols as in Figs 1 and 4.

Young (1975a) has discussed the arrangement of photophores, chromatophores and iridophores in this animal. Briefly, large photophores are concentrated on the antero-ventral surface of the mantle, the ventral surface of the head and the aboral surfaces of the arms (Fig. 9). Photophores are directed anteriorly. Much of the head, arms and body is covered with iridophores. Overlying these iridophores is a layer of chromatophores.

Two rather rare species of *Histioteuthis*, *H. celetaria pacifica* and *H.* sp. nov., occur off Hawaii in approximately the same depth zone as *Histioteuthis dofleini* and have nearly identical photophores to those of *H. dofleini*. Although the following description and discussion is drawn from all three species only the rare ones have been observed live and sectioned histologically.

The photophore consists of a posterior portion containing a reflector, the posterior cup, which surrounds the photogenetic

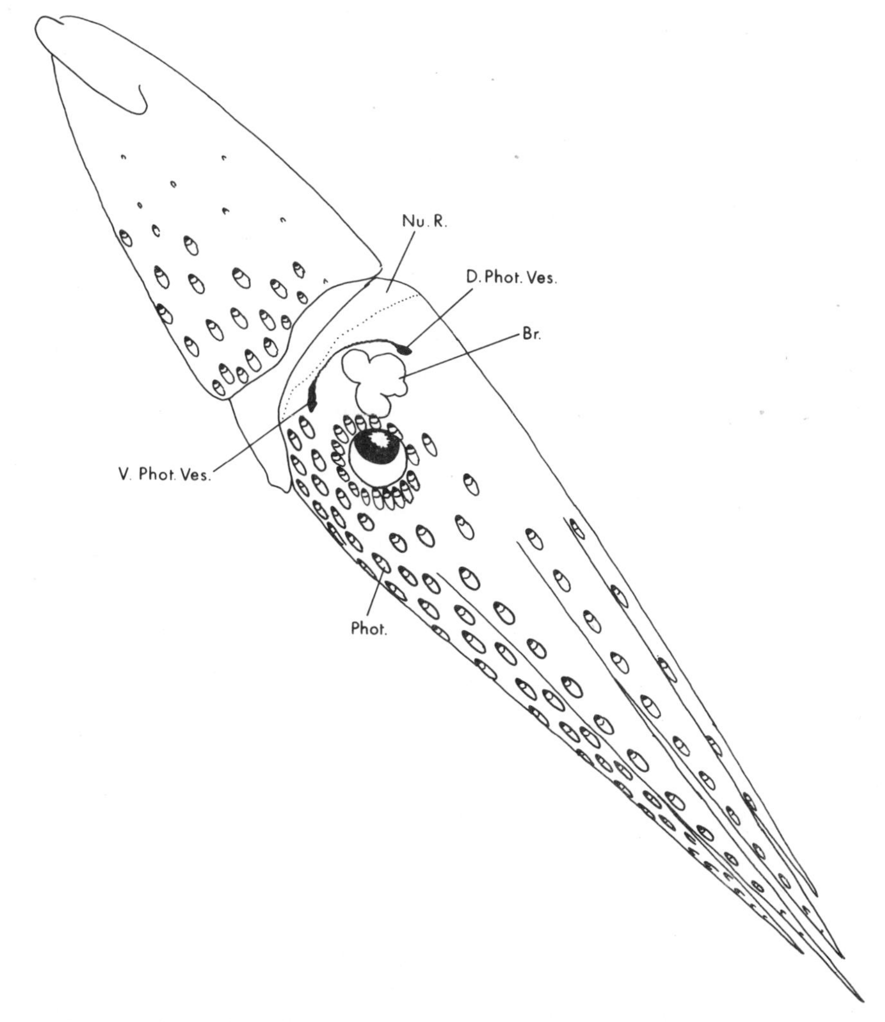

FIG. 9. *Histioteuthis dofleini*, lateral view. Symbols: Br., brain; D.Phot.Ves., dorsal photosensitive vesicles; Nu.R., nuchal region; Phot., photophore; V.Phot.Ves., ventral photosensitive vesicle.

tissue, a central portion consisting of a large "lens", and an anterior reflector (Fig. 7E). The posterior cup has much of its external surface covered with a dense layer of pigment. This pigment is not contained in chromatophores but in small membrane-bound packets containing spherical pigment granules and dispersed pigment. This pigment layer does not cover a circular patch on the side of the posterior cup adjacent to the integument. In the centre of this patch lies a large chromatophore which, when expanded, covers this "lateral window". The presumed photocytes are elongate cells forming a solid layer that fills the posterior cup and extends to the "lens". The lens consists of tightly packed cells whose entire cytosome contains a dense, coiled mass of membranes. The anterior reflector is scoop-shaped: it curves around the "lens", flares anteriorly and terminates in a rounded margin. The inner surface is lined with a thick layer of iridophores. The "scoop" is filled with a transparent packing tissue and covered by scattered irregular iridosomal platelets.

Histioteuthis dofleini has two sets of photosensitive vesicles, a dorsal and a ventral set, joined by a narrow strand of vesicles (Fig. 9). The dorsal and ventral organs are approximately the same size, consisting of a number of independent vesicles of various sizes and shapes formed into a flat and rather compact organ. The dorsal organ lies dorsal to the optic lobe and adjacent to the cephalic cartilage. Each ventral organ lies in the postero-ventral corner of the head adjacent to the cephalic cartilage and partially above the lateral base of the funnel. The walls of the vesicles have a single layer of sensory cell bodies, and the lumen is generally nearly or completely filled with tightly packed but irregularly arranged sensory processes.

The dorsal part of the nuchal crest, which may be a window to the dorsal organ, lacks chromatophores, and the surface of the head posterior to the nuchal crest carries only scattered chromatophores and no iridophores. Directly opposite the ventral organ in the ventro-posterior portion of the head, is an area devoid of chromatophores that is clearly a window to the ventral organ.

In fresh material the chromatophores surrounding the anterior reflector have been observed to expand and completely cover this structure. In dissection, the posterior reflector strongly reflects blue light when illuminated with a microscope light. The "lens" transmits bright red light when illuminated from the side. The "lens" does not seem to transmit blue light well laterally; however, when white light passes through the lens into the posterior reflector along the optical axis of the photophore, blue-violet light is reflected out through the lens. The slight violet hue probably results from scattered red light that did not enter the posterior reflector. The anterior reflector appears silvery in fresh specimens.

The ventral photosensitive vesicles are exposed to the photophores of the anterior margin of the mantle only when the mantle is strongly contracted or pursed inward. The latter behaviour might seem unlikely, yet I have observed a living specimen of *Histioteuthis heteropsis* off California in a shipboard aquarium repeatedly doing just this. I have not observed any species of *Histioteuthis* luminescing.

Apparently ventral countershading in *Histioteuthis* spp. operates in the following manner. The broad distribution of photophores will adequately conceal an obliquely oriented animal. The "lens" is probably a filter which passes only blue and red light (red light, presumably, is not produced by the photocytes). The filtering mechanism, however, is obscure. Light leaving the photocytes in other directions encounters the posterior cup which reflects blue light; other frequencies pass through and are absorbed by the surrounding pigment except in the region of the "lateral window" where apparently green or yellow light may emerge when the chromatophore screen is withdrawn. This latter luminescence would not be involved in countershading; its function is unknown. The anterior reflector presumably reflects some of the light in a more ventral direction.

The dorsal photosensitive vesicles are exposed to downwelling light and the ventral vesicles apparently can be exposed to the photophores on the anterior margin of the mantle. Chromatophores surrounding the anterior reflector provide a potential means for regulating the luminous output.

Thelidioteuthis alessandrinii

Only five specimens have been captured off Hawaii. The single-day capture (25 mm ML) was taken in an opening-closing tow between 720 and 780 m. At night four captures all from open tows were made from 50 to 500 m.

Photophores of two sizes cover much of the ventral and lateral surfaces of the mantle and head. Only the smaller photophores are present on the aboral surfaces of the arms and the surfaces of the fins. The larger photophores are directed away from (i.e. normal to) the body surface, while the smaller photophores are directed ventro-laterally on both sides of the animal. A few small photophores are present on the dorsal surface of the funnel and are directed dorso-laterally toward the ventral photosensitive vesicles.

The photophores have a relatively simple structure. The large organs have a centrally located mass of photocytes, which are backed by a thick reflector and are surrounded laterally by a concentrically arranged system of iridosomal platelets that probably act as light guides (Fig. 7C). The reflector is surrounded by a pigment layer consisting both of chromatophores and dispersed pigment. Numerous chromatophores cover the surface of the photophore. The smaller photophores consist essentially of a reflector backed by pigment and a mass of photocytes with an outer covering of chromatophores.

Three sets of photosensitive vesicles are present: a dorsal set, a posterior set, and a ventral set. The organs consist primarily of a loose association of variously-shaped independent vesicles or sometimes narrowly joined vesicles. The dorsal organs are broad flat structures, lying above the dorsal surface of the optic lobes. The vesicles are concentrated along the lateral margins of this organ. The posterior organs, although remaining within the confines of the broad cephalic cartilage, lie entirely lateral to the optic lobes and have vesicles concentrated near their lateral margins. A discontinuous strand of vesicles extends from each posterior organ to each dorsal organ, and ventrally a similar strand extends to each ventral organ. The ventral organs are broad, oval structures with vesicles concentrated along their anterior and lateral margins. The posterior surface of the head and the ventral surface above the base of the funnel lack iridophores and have only a sparse cover of chromatophores.

The photophores of living specimens when illuminated with white light appear silvery, although a slight tint of blue light is apparent in the larger photophores.

This species probably does not countershade ventrally.

Sandalops melancholicus

The vertical distribution of this species (Fig. 10) is essentially the same as given by Young (1975b), although a few additional captures have been added. Larvae were taken in the upper 400 m. Juveniles, characterized by having tubular eyes, were captured between 450 and 675 m. Only two adults were taken. These came from depths of about 800 and 1075 m. Vertical migration does not occur. Apparently only juveniles live in a zone where ventral bioluminescent countershading is possible.

S. melancholicus has two photophores, one large and one small, on each eye. In the tubular, upward directed eye of the juvenile, the large photophore covers most of the ventral surface of the eye. The smaller photophore is circular and lies on the anterior face of the eye with its ventral edge within an indentation in the larger photophore.

The large photophore has long, rod-like iridophores, composed of long concentric ribbons of iridosomal platelets, that

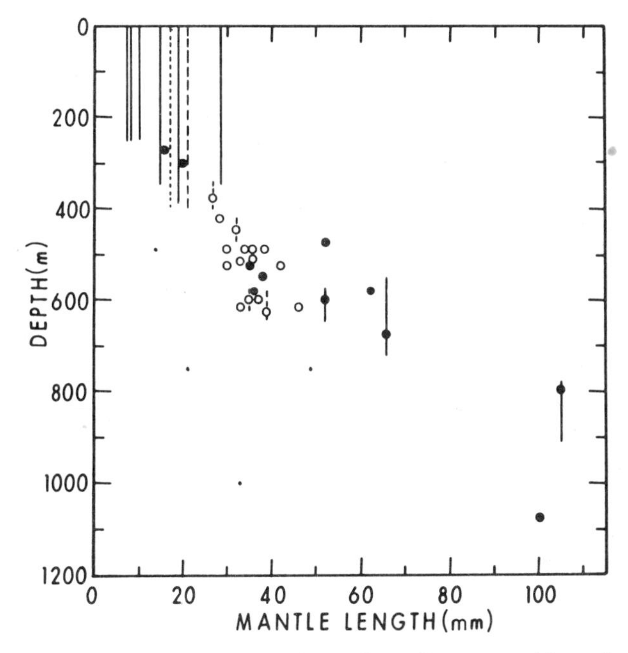

FIG. 10. Vertical distribution of *Sandalops melancholicus*. Bars without dots represent oblique tows, otherwise symbols as in previous figures.

occupy much of the volume of the photophore (Fig. 4E). Dilly & Herring (1974) assumed similar structures in *Bathothauma lyromma* to be light guides. At the medio-posterior end of the photophore a dorsal bulge marks the location of most of the other major structures within the photophore (Fig. 7F). A hemispherical tissue, the photogenic body, lies in the centre of the bulge. The flat dorsal surface of this tissue abuts against a thick reflector, the dorsal reflector (Fig. 4F). The platelets in this reflector are about 75 nm thick or about the same as in the blue-reflecting posterior cup iridophores of *Pterygioteuthis microlampas*. The ventral surface of the reflector has a broad ring-shaped depression that overlaps the peripheral area of the light-producing body (Fig. 7F). This depression is occupied by a homogenous substance which contains scattered cells. The light guides originate along the ventral margin of this circular zone between the light-producing body on the inside and the outer portions of the dorsal reflector on the outside. Those light guides that pass ventral to the photocytes are closely packed and form a concave reflector over the ventral surface of the light-producing body. Dissection of the photophore in preserved material confirms that this dense mat of light guides functions as an effective broad-band reflector. All of the light guides end on the broad ventral surface of the photophore.

The dorsal reflector is partially encircled along its outer ventral margin by a strip of iridophores (Fig. 4D). This strip, the fibrous reflector, consists of numerous slender iridophores which have a dorso-ventral orientation (Fig. 4F). The fibrous reflector covers the entire side of the dorsal reflector only in a short postero-medial zone. Except for this zone, a heavy pigment layer covers the outer surfaces of the dorsal and fibrous reflectors (Fig. 11). The fibrous reflector is composed of ribbon-like platelets of variable widths and lengths that are arranged in groups of variable size. The ventral margin of the fibrous reflector is adjacent to the light guides at their origin (Fig. 7F).

S. melancholicus has a single set of photosensitive vesicles. Each organ consists of a single bilobed vesicle and is located on the ventral surface of the peduncle complex (Fig. 11). The wall has three to five irregular layers of sensory cell-bodies. The sensory processes fill the lumen of the vesicle.

S. melancholicus has not been observed to luminesce. This animal is rather uncommon, and only dead and moribund specimens have been observed. In a fresh specimen, "white" light from a microscope lamp directed into the photophore will produce a blue glow in the dorsal half of the fibrous reflector. The fibrous reflector appears silvery when illuminated by "white" light incident normal to its outer surface.

S. melancholicus is very transparent; only the eyes and liver with its associated ink sac are opaque. The liver-ink sac is spindle-shaped and oriented vertically, thus reducing the ventral counter-shading problem (Young, 1975b). The eyes are shielded ventrally by the photophores. Presumably the large photophores operate in the following way. Light leaving the photocytes in a ventral direction is reflected dorsally by the closely-packed light guides into the dorsal reflector. Light leaving the photocytes dorsally passes directly into the dorsal reflector. Only blue light is reflected by this structure; other frequencies pass through and are absorbed by the surrounding pigment layer. Blue light leaving the dorsal reflector enters the ends of the light guides and is distributed over the ventral surface of the photophore to produce a broad, highly directional downward beam. Blue light also passes into the ventral ends of the fibrous reflector. Presumably the entire reflector transmits light; the platelets may provide light paths via internal reflection while the spaces between groups of platelets provide light paths with the help of reflection off the platelets. Considerable scattered light must be produced from the termination of platelets at various points along the reflector and possibly from some undulations in the platelets. The result is a glowing reflector which is detected by the nearby photosensitive vesicles (Figs 7F and 11).

This animal uses a single set of vesicles to detect both downwelling and bioluminescent light. Functional chromatophores scattered over the ventral surface of the photophore and the surface of the fibrous reflector provide a possible mechanism for regulating light intensity.

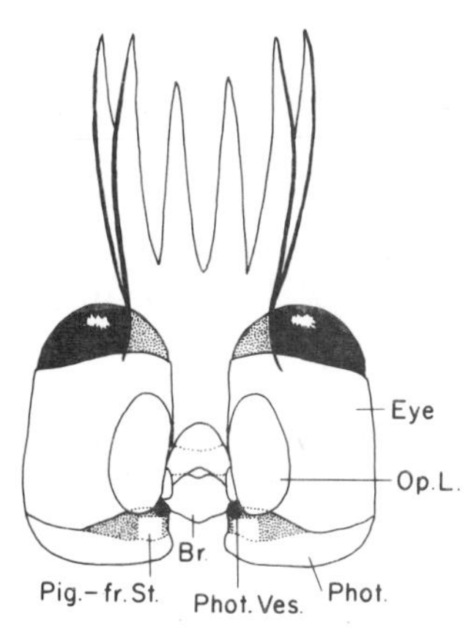

FIG. 11. Posterior view of the head of *Sandalops melancholicus*. Symbols: Br., brain; Op.L., optic lobe; Phot., photophore; Phot.Ves., photosensitive vesicle; Pig.-fr.St., pigment-free strip on fibrous reflector.

CONCLUSIONS

Only two of the species discussed, *Heteroteuthis hawaiiensis* and *Thelidioteuthis alessandrinii*, have day distributions in depths where bioluminescent countershading is questionable. While mature individuals of *Heteroteuthis hawaiiensis* may be encountered at ventral countershading depths during the day, immature animals are found primarily between 250 and 350 m, which may well be too shallow for bioluminescent countershading. At night their depth range indicates that bioluminescent countershading may occur on moonlit nights; however, the twilight period may be an especially critical time for ventral countershading. Much of the midwater fauna moves upward at twilight in approximate concert with changing isolumes to depths above 200 m (Boden & Kampa, 1967; T. A. Clarke, 1973). During this period *Heteroteuthis* is exposed to potential predators rising from below at a time when light levels fall to within countershading ranges. Thus ventral countershading may be highly advantageous during twilight while midwater animals are moving upward past *Heteroteuthis hawaiiensis*.

The little available depth data on *Thelidioteuthis alessandrinii* indicates that it lives near or below the lower limit of ventral countershading during the day. The possibility exists of ventral countershading at night; however, evidence discussed below indicates that the animal does not countershade ventrally.

Pterygioteuthis microlampas is clearly exposed to countershading light levels both during the day and night. One suspects that this animal indeed countershades during both of these periods although countershading at night might be somewhat more difficult since the radiance field produced by the moon will not always be symmetrical around the vertical in such shallow depths. *Sandalops melancholicus*, on the other hand, does not migrate, and countershading at night is not possible although it clearly resides within ventral countershading depths during the day. The remaining two species, *Enoploteuthis* sp. and *Histioteuthis dofleini*, clearly occupy countershading depths during the day and possibly at night for portions of the populations.

All species examined except *Thelidioteuthis alessandrinii* seem to have the necessary mechanisms for ventral countershading. These species all seem to have some type of "skylight" filter to restrict emitted light to the blue wavelengths. The filters invariably have two components, one located behind the photogenic tissue, a filter-reflector, and one in front of the photogenic tissue. The front component may reflect a broad band of light back to the rear component, which then becomes the sole filter, or it may duplicate the filtering function of the rear component. In all these species, opaque structures either have underlying photophores, or else have underlying light channels, diffusing structures or other reflectors lit by photophores.

All the species considered have photosensitive vesicles that are exposed to downwelling light as well as light from their own photophores. This arrangement provides them with the necessary information to match the intensity of downwelling light. *Sandalops melancholicus* provides an excellent example of the need for bioluminescent feedback in ventral countershading. If any animal countershades ventrally then juvenile *Sandalops* must certainly be one. Indeed, Young (1975b) has suggested that the tubular shape of the eyes in this squid is an adaptation for more effective countershading. In all squids the photosensitive vesicles lie within the cephalic cartilage. In *Sandalops*, however, the photophores are in such a position that vesicles within the cephalic cartilage cannot directly receive light from the photophores. In order to get direct feedback from the photophores, a rather exotic mechanism is employed, utilizing a reflector as a light guide to capture filtered light from the downward-directed photophore and transmit it to the more dorsally-located photosensitive vesicles.

Mechanisms for regulating light intensity are poorly understood, but most species have potentially suitable mechanisms involving either iridophores under muscular control or chromatophores, indeed Packard (see his chapter, this volume) also draws attention to the value of the chromatophores as a screening system. In those species that have been observed, bioluminescence is continuous and lengthy, in some cases extending over periods of hours, and the light is strongly directional. Mechanisms for adjusting the direction of the light beam to various body attitudes have not been demonstrated, although the muscular attachments of the photophores in some species suggests that such mechanisms exist.

Thelidioteuthis alessandrinii is the only species examined that apparently does not countershade ventrally. Its photophores have no obvious system for restricting the wavelengths of emitted light. Even the posterior reflector is a broad-band reflector. In addition the orientation of the photophores indicates that they produce a rather broad beam that extends laterally as well as ventrally. Yet even though the photophores of this species probably have other functions, they exhibit certain mechanisms common to some countershading photophores. Functional chromatophores are present on the surface of the photophores, providing a potential means of regulating light intensity (of course, they may merely function in concealing the photophores). In addition the photosensitive vesicles are apparently exposed to downwelling light and to the bioluminescent light of the smaller photophores. This arrangement suggests that bioluminescent feedback may be important for some functions in addition to ventral countershading. Indeed, in *Enoploteuthis* sp. two types of photophores are directed at the ventral photosensitive vesicles. (It should be mentioned that many species with photosensitive vesicles and non-countershading photophores—e.g. *Bathyteuthis* spp., *Vampyroteuthis infernalis*—have no bioluminescent feedback mechanism via the vesicles.) In *Thelidioteuthis alessandrinii* the structure of the photophores, especially the larger ones, suggests that light from each photophore is strongly directional, although the overall light field may be broad.

In *Enoploteuthis* the presumed non-countershading photophores are also highly directional and are ventrally directed.

Among the species examined, therefore, the only distinctive structural characteristic of presumed countershading photophores is the presence of skylight filters. However, skylight filters may not be an infallible indicator of a countershading function. Dilly & Herring (1974) have described the ocular photophores of the squid *Bathothauma lyromma.* Juveniles of this species live in ventral countershading depths. This distribution combined with the position of the photophores indicates, as the authors have suggested, that these photophores have a countershading function. While the photogenic tissue is backed by a large reflector, which, on the basis of its structure, is predicted to reflect blue light most strongly, no filter or reflector is present in front of the photocytes.

All photophores with skylight filters probably have a countershading function, but countershading need not be their only function. For example, the presumed countershading photophores of *Histioteuthis* and *Heteroteuthis* clearly have additional functions. The identification of presumed countershading photophores should be possible for those species that utilize skylight filters. Recognition of the patterns of these photophores may reveal patterns in the remaining photophores and thus provide clues to the bioluminescent functions of the latter.

While the evidence presented here suggests that potential countershading mechanisms exist in most of the squids examined, details of the mechanisms and their demonstrable involvement in ventral countershading for the most part remain for future investigation.

Since this manuscript was submitted for publication considerable information concerning ventral countershading has accumulated. Young & Roper (1976) have demonstrated that specimens of the squid *Abraliopsis* sp. responded to on–off sequences of overhead illumination in a shipboard aquarium by turning their photophores on and off. In addition, the glowing squid was invisible when beneath light of the proper intensity. Young & Roper (in press) have extended these observations to a number of squid (incl. *Pterygioteuthis microlampas*, *Heteroteuthis hawaiiensis* and *Enoploteuthis* sp.), one fish and one shrimp. They demonstrate that these animals can adjust their luminescence to match the overhead illumination. They further demonstrate that simple photophores without skylight filters may function in countershading at low light levels and that at the upper limit of an animal's ability to countershade, it may turn on nearly every downward directed photophore it possesses.

These results support the conclusions drawn in this paper except that evidence against ventral countershading in *Thelidioteuthis* is now not as convincing, and that the presumed non-countershading photophores in *Enoploteuthis* (i.e. those lacking skylight filters) possibly function in countershading either at very low light levels or near its upper depth limit for countershading.

Acknowledgements

I thank the following people for their assistance. John Walters, University of Hawaii, read and commented on the manuscript. Wilton Hardy, University of Hawaii, provided valuable information on the physical properties of light. John Arnold, University of Hawaii, helped with the electron microscopy and the interpretation of photophore structure, and Lois Williams-Arnold provided technical assistance in the preparation of histological material. Captain Bristol and the crew of the *Alpha Helix* were extremely helpful during a cruise when some of the material used in this paper was collected. Sherwood Maynard, Richard Spencer, John Walters, Vicky Ridge, Frederick Mencher, and Vernon Hu, University of Hawaii, worked hard and long during this cruise to collect material. All squid captured by the Isaacs-Kidd midwater trawl were provided by T. Clarke, University of Hawaii. This paper was supported by grant DES 72-01456 A02 from the National Science Foundation. This paper was Contribution No. 790, Hawaii Institute of Geophysics.

References

Amesbury, S. S. (1975). *The vertical structure of the micronektonic fish community off leeward Oahu.* Ph.D. Dissertation: Univ. of Hawaii.

Arnold, J. M. & Young, R. E. (1974). Ultrastructure of a cephalopod photophore. I. Structure of the photogenic tissue. *Biol. Bull. mar. biol. Lab. Woods Hole* **147**: 507–521.

Arnold, J. M., Young, R. E. & King, M. V. (1974). Ultrastructure of a cephalopod photophore. II. Iridophores as reflectors and transmitters. *Biol. Bull. mar. biol. Lab. Woods Hole* **147**: 522–534.

Badcock, J. (1970). The vertical distribution of mesopelagic fishes collected on the Sond cruise. *J. mar. biol. Ass. U.K.* **50**: 1001–1044.

Boden, B. P. & Kampa, E. M. (1967). The influence of natural light on the vertical migrations of an animal community in the sea. *Symp. zool. Soc. Lond.* No. 19: 15–26.

Clarke, G. L. & Breslau, L. R. (1960). Studies of luminescent flashing in Phosphorescent Bay, Puerto Rico, and in the Gulf of Naples using a portable bathyphotometer. *Bull. Inst. océanogr. Monaco* No. 1171: 1–32.
Clarke, G. L. & Denton, E. J. (1962). Light and animal life. In *The sea* **1**: 456–468. Hill, M. N. (ed.). New York: Interscience.
Clarke, T. A. (1973). Some aspects of the ecology of lanternfishes (Myctophidae) in the Pacific Ocean near Hawaii. *Fishery Bull. Fish Wildl. Serv. U.S.* **71**: 401–434.
Clarke, W. D. (1963). Function of bioluminescence in mesopelagic organisms. *Nature, Lond.* **198**: 1244–1246.
Denton, E. J., Gilpin-Brown, J. B. & Roberts, B. L. (1969). On the organization and function of the photophores of *Argyropelecus*. *J. Physiol., Lond.* **204**: 38P–39P.
Denton, E. J., Gilpin-Brown, J. B. & Wright, P. G. (1972). The angular distribution of the light produced by some mesopelagic fish in relation to their camouflage. *Proc. R. Soc. Lond.* (B) **182**: 145–158.
Dilly, P. & Herring, P. (1974). The ocular light organ of *Bathothauma lyromma* (Mollusca: Cephalopoda). *J. Zool., Lond.* **172**: 81–100.
Foxton, P. (1970). The vertical distribution of pelagic decapods (Crustacea: Natantia) collected on the SOND cruise 1965. *J. mar. biol. Ass. U.K.* **50**: 939–960.
Harvey, E. N. (1952). *Bioluminescence*. New York and London: Academic Press.
Jerlov, N. G. (1970). Light. General introduction. In *Marine ecology*: 95–102. Kinne, O. (ed.). New York: Interscience.
Kampa, E. M. (1971). Photoenvironment and sonic scattering. In *Proceedings of an internation symposium on biological sound scattering the ocean*: 51–59. Farquhar, G. B. (ed.). Washington: Dept. of the Navy.
Lawry, J. V., Jr. (1974). Lantern fish compare downwelling light and bioluminescence. *Nature, Lond.* **247**: 155–157.
Maynard, S. D., Riggs, F. V. & Walters, J. F. (1975). Mesopelagic micronekton in Hawaiian waters: faunal composition, standing stock, and diel vertical migration. *Fishery Bull. Fish Wildl. Serv. U.S.* **73**(4): 726–736.
Meyer, W. T. (1906). Ueber das Leuchtorgan der Sepiolini. *Zool. Anz.* **30**: 388–392.
Nicol, J. A. C. (1967). The luminescence of fishes. *Symp. zool. Soc. Lond.* No. 19: 27–55.
Roper, C. F. E. & Young, R. E. (1975). Vertical distribution of pelagic cephalopods. *Smithson. Contr. Zool.* No. 209: 1–51.
Walters, J. (1975). *Ecology of Hawaiian sergestid shrimp* (*Penaeidea*; *Sergestidae*). Ph.D. Dissertation: Univ. of Hawaii.
Walters, J. (in prep.). *An analysis of Hawaiian scattering layers.*
Young, R. E. (1973). Information feedback from photophores and ventral countershading in midwater squid. *Pacif. Sci.* **27**: 1–7.
Young, R. E. (1975a). Function of the dimorphic eyes in the midwater squid *Histioteuthis dofleini*. *Pacif. Sci.* **29**: 211–218.
Young, R. E. (1975b). Transitory eye shapes and the vertical distribution of two midwater squids. *Pacif. Sci.* **29**: 243–255.
Young, R. E. & Roper, C. F. E. (1976). Bioluminescent countershading in midwater animals: evidence from living squid. *Science, Wash.* **191**: 1046–1048.
Young, R. E. & Roper, C. F. E. (in press). Intensity regulation of bioluminescence during countershading in midwater animals. *Fish. Bull., U.S.* **74**.

Symp. zool. Soc. Lond. (1977) No. 38, 191–231.

SKIN PATTERNING IN *OCTOPUS* AND OTHER GENERA

A. PACKARD and F. G. HOCHBERG

University Medical School, Edinburgh, Scotland and *Santa Barbara Museum of Natural History, Santa Barbara, California, USA*

SYNOPSIS

Patterns of the skin on the dorsal surface of *Octopus* are built up hierarchically from morphological and physiological *elements, units, components* and *patterns.*

The *elements* comprise chromatophores, leucophores, iridocytes and melanophores. These elements are grouped into *units,* the skin patches. The patches, together with their surrounding grooves, are used by the nervous system to form dark and light *components,* whose details reflect their nervous connections. Components combined in series with like components and in parallel with unlike ones comprise whole *patterns.* A given pattern which may vary in intensity reflects a specific behaviour and can be evoked under given circumstances, suggesting unitary central programming.

Differences at the level of elements, units, components and patterns in various species of *Octopus* are discussed. Differentiation of mantle white spots is greater in *O. vulgaris* and *O. rubescens* than in *O. dofleini* and *O. bimaculoides.* Like patterns are considered homologous one with another, but do not necessarily involve the same structures, thus the ocellus of *O. bimaculoides* corresponds with the eye-ring of other species.

The principles of patterning observable in the genus *Octopus* can be generalized to other cephalopods. In coleoids, skin and body patterns and their associated behaviours appear to have evolved as single package deals, with natural selection acting at the level of the whole pattern.

The quantity of information contained in the photographs is many times that in the text; they have been selected from several hundred for the stories they tell.

INTRODUCTION

We are all of us, at one remove or another, concerned with pattern: either unravelling self-evident patterns or constructing them intellectually so as to bring order into our material. In this respect cephalopods are of special interest: their skin patterns, displayed two-dimensionally, are so *visible*—indeed they are created for seeing—and they can change before our eyes. Until recently, however, and despite the extensive work done on the physiology and structure of chromatophores over the last 150 years, there had been few systematic descriptions of the patterns. Cowdry (1911) had attempted a classification of the ones seen in *Octopus vulgaris* of the Caribbean. His account included coloured drawings (one of them of an animal in an "octopus car") but it was not until the 1940s that the first well-documented description appeared, illustrated by photographs and drawings, of the patterns to be seen in a decapod:

the cuttlefish (*Sepia officinalis*) (Holmes, 1940, 1955). William Holmes had been encouraged to take up the subject by J. Z. Young who realized that as cephalopod chromatophores are controlled by nerves with their cell bodies in the central nervous system, the patterns to which they give rise provide a skin-deep view of the brain. B. B. Boycott, in the same tradition, tackled the system at the other end in his series of anatomical and electrical stimulation studies of the brain (Boycott, 1953, 1954, 1961). Later, on the prompting of G. D. Sanders who was working at Naples as an assistant of Professor Young, Packard & Sanders (1969, 1971) published their accounts of patterning. These were favoured by current advances in colour and flash photography, and it is not surprising to find that, at the present time, the best illustrations of patterns in octopuses are not in the scientific journals but in magazines of underwater photographs and in semi-popular accounts (Lane, 1957; Voss & Sisson, 1971; Cousteau & Diolé, 1973).

In this paper we have relied exclusively on photographs of the skin. They were taken at magnifications ranging from 1/15th (Figs 6 and 27) to ×30 (Fig. 2), both in the field and in the laboratory where animals were placed in shallow water with material taken from the collecting sites. Film used included Kodak Plus X, Tri X; Kodacolor, Kodachrome and Ektachrome.

Of the species referred to, *Octopus vulgaris* (Cuvier), the common octopus of the Mediterranean, is too well known to require description. It was studied in Naples; it ranges in colour from reddish black through light brown to yellowish grey. The other octopuses were Eastern Pacific littoral and coastal forms from Puget Sound (*O. dofleini*) and the Santa Barbara channel. *O. bimaculoides* is the common two-spotted octopus of California, which breeds in the intertidal zone and hatches from a large egg (Pickford & McConnaughey, 1949). Its upper weight limit is about 1 kg. Its colour in the field is grey, light brown or green. *O. dofleini* is the giant Pacific octopus (Pickford, 1964), with a more northerly range, which reaches over 20 kg; its colour ranges from dark chocolate to brick-red. *O. rubescens* (Berry, 1953), the smallest species studied on this coast (it matures at less than 100 g), is also the most vivid and varied in its displays. It is characteristically orange-red (whence the name *rubescens*) but can go from black to light ochre. *O. californicus* (Berry, 1911, 1912) is a deep-water species not easy to keep and study alive. It is basically deep orange to brick-red and probably has few displays.

We have chosen to present the account in a synthetic manner, building up from the *elements* that form the basis of patterning through *units* and *components* to the body *patterns* themselves, though, in practice, our observations usually proceeded in the analytical way from the patterns downwards.

THE MAKE-UP OF THE SKIN

Elements

Colour elements and melanophores

Girod (1883) described the cephalopod skin as consisting of basically five layers: a transparent epidermis, and the dermis containing the well-known chromatophores and various kinds of iridocytes sandwiched between two fibrous layers. The general features of the *chromatophores* have been described many times and their structure is now understood in considerable detail (Cloney & Florey, 1968; Froesch, 1973b) (Table I). They contain pigments within a cyto-elastic sac that can expand and contract through the operation of radial muscles under nervous control. Our knowledge of the neuromuscular details of this control has been considerably extended by Florey & Kriebel (1969) and is reviewed by Florey (1969). The different colours to which the chromatophores give rise are a combination of the effects of the particular pigments

TABLE I

Pattern-giving elements *of the cephalopod skin and their main effects when illuminated*

Elements			Effects
A. Chromatophores	1. Melanophores	BLACK	By absorption (all wavelengths)
	2. Yellow through red chromatophores	Red, Orange, Yellow	By differential absorption
B. "Iridocytes"	1. Iridocytes	Yellow, Green, Blue	By differential reflection (constructive interference)
	2. Leucophores	WHITE	By reflection and scattering (all wavelengths)

present and the degree of expansion of the sac that contains them; i.e. of how thinly the *pigment body* is spread. Because these effects are combined there is some confusion about the number of different classes of chromatophores present. The following may help to clear this up, bearing in mind that similar considerations hold for the way pigments are used in the skin of a reptile or in painting a picture.

Chromatophores can only give colour when illuminated, subtracting those wavelengths that do not contribute to the final hue (from which it follows that they will always give off less energy than is present in the illuminating source). The light may either come from above the chromatophore or below it, by reflection from underlying structures, so that we see the colour by transparency. The hue that is perceived will depend on the proportion in which the visible wavelengths are present in the light that illuminates the pigment body. (Similarly, chromatophores of a particular hue illuminated by light deficient in that hue will appear black, as do the red chromatophores when seen at depths below 20–30 m.) Particularly when the chromatophore is seen by transparency its hue will alter if certain wavelengths have been filtered as the light passes through the tissues of the skin or another part of the body. Any structures such as *iridocytes* (Kawaguti & Ohgishi, 1962; Brocco, 1975) below the chromatophore will add their own colours and if these are complementary to those characteristic of the chromatophore pigment they will result in colours of intermediate hue.* The brightness of the colours seen by transparency will

* The iridocyte layer has been variously reported as consisting of "iridocytes", "iridophores", "iridocystes", "iridiocytes". A number of authors (e.g. Girod, 1883, and Schaeffer, 1938) have noted that they differ in form and size from one species to another. All of the names, however, (as also the German "Flitterzellen") refer to iridescent properties, though it is plain from their descriptions and our own observations that many of the so-called "iridocytes" scatter white light and so have optical properties very different from iridescent. Packard & Sanders (1971) called those structures that reflect white by scattering, and that are concentrated in the patches of the skin, "leucophores". This followed a personal observation by J. Bagnara who pointed out their similarity to amphibian leucophores. S. Brocco (1975) has since shown that there are two classes of reflecting cells in the deep dermal ("iridocyte") layer of the skin: iridocytes with stacks of electron-dense platelets and intact cytoplasm and nucleus, and a second class, identified as leucophores, with electron-dense sub-spherical and pear-shaped bodies ("leucosomes") surrounding a nucleus and cytoplasm whose membrane has broken down in places.

The problem here, as elsewhere, is that any attempt to classify structures on the basis of their visual effects leads to difficulties if we are simultaneously trying to classify them morphologically or histologically. Quite different structures produce similar effects (e.g. the different types of structure in the animal kingdom giving pure blue) and a slight modification affecting a particular cell may produce a new visual effect (e.g. when the pigment contained in a chromatophore changes colour owing to polymerization). The depend on how much light is being reflected through the chromatophore. This is partly a matter of the efficiency of light-scattering and reflecting structures below them—leucophores being more efficient than muscle in this respect—and partly a matter of how much light is available for reflection (aspects of reflection are discussed in some detail by Messenger, 1974). A screen of expanded chromatophores of the dark variety, familiar to most students as *melanophores*, will absorb most of the incident light of all wavelengths, and thus leave little light to be reflected by structures beneath. Such a screen will act as a *neutral density filter* (see later). The neutral density role of the melanophores is emphasized by their usual placing in the skin as the most superficial layer interposed between the surface—and the observer—and the rest of the colour-giving elements. Finally, the appearance of the chromatophores will depend on their orientation relative to the light, whether incident or reflected.

All these principles operate in painting, and as in painting they are not immediately obvious and their effects are rarely seen in isolation. We can recognize two extreme conditions. One in which dark chromatophores (i.e. melanophores), expanded or piled up as on the side walls of a groove, are embedded in transparent skin not underlain by any reflecting structures. This will give *black*, often as

problem does not go away by trying to improve on definitions so long as those definitions are based on visual effects (Fries, 1958; Mirow, 1972). And even the broadest of classifications as "pigment cell" or "chromatophore"—with pigment defined as "a substance that imparts colour *including black or white*" and chromatophore as "a pigment-containing cell" (*Encyclopaedia Britannica*, 15th Edn)—may not be adequate. It is not the substance *per se* within the iridocytes that imparts colour, but the way this substance is spatially arranged. If it is in the form of platelets of uniform thickness piled in successive layers with characteristic frequency they will give one colour, depending on the angle of incident light, and another if the layers are slightly closer together or further apart (Land, 1972) (i.e. the cell is called an "iridocyte" or "iridophore", etc.). It is a closely similar cell type (or the same cell at a later stage of differentiation) in which the substance is present not as platelets but as sub-spherical or pear-shaped droplets that scatter light giving white [i.e. the cell is called a "leucophore" (Brocco, 1975) or "autangophore" (Fries, 1958)]. In yet another cell, just described, associated with photophores (R. E. Young, his chapter, this volume), the platelets, still arranged in layers but no longer with regular spacing, have a common orientation and so act as light guides for the endogenously produced luminescence. What should these cells be called? It is hardly appropriate to call them chromatophores or pigment cells, though, like them, they do have effects on light, and have structural peculiarities similar to iridocytes.

So long as it is the cells that are being considered, it would seem logical to classify them according to histological criteria, much as pigments are classified according to chemical formulae and not according to the colours they give. As it appears from electron micrographs that the light-affecting substance of the platelets and droplets ("leucosomes") is the same in all of them, they could be defined in terms of this substance once it has been identified.

a sharp thin line or as a broader strip grading into grey at the edges. The other is the condition in which there are no chromatophores or the chromatophores present are fully contracted and the transparent skin, or layers deep to it, contains totally reflecting and diffusing structures, principally leucophores and body musculature. This gives *white*. Between the two extremes—extremes of intensity, not of hue—all variations, both of intensity and of hue, are possible, it being the job of the iridocytes to give pure hues of blue, green and yellow and of the pigment bodies of the chromatophores to give hues of yellow, orange, reds and browns of different intensity (i.e. tone) depending on their thickness and the amount of black melanins they contain. The optical properties of the skin are illustrated in Fig. 1a.

Blue, as the intense blue in the ocellus of some species of *Octopus* and on the eyeball of some of the cuttlefishes and squids, provides an illustration of the way in which melanins and reflecting and scattering elements combine to produce colours of a single hue (Fig. 1b). Using a dissecting microscope with top illumination, the blue can be traced directly to individual iridocytes. In areas of skin that appear blue these elements are present in large numbers. The physical origins of the particular hues they reflect are explained by Denton and Land (see Denton & Land, 1971 and Land, 1972) as due to thin-layer constructive interference. In the ocellus the

---

FIG. 1. (a) Schematic view of the skin to illustrate its optical properties with respect to an observer. Although generalized, the diagram may be regarded as covering the area of a single patch and its bounding grooves cut normal to the surface. Incident light passes successively through a transparent refracting layer 1 and neutral density and colour filters 2, 3 and 4 before being reflected or absorbed by layers 5, 6 and 7. Rays of light are shown as if coming from one direction only; refraction is not indicated. The variable filters are under nervous control. The neutral density filter 2 is formed of melanophores and may be tinted brown or red at the dark end of its range. Peak absorption in the colour filters 3 and 4 shifts towards the short end of the spectrum as they expand. The mirror-reflecting layer 5 reflects light of narrow band width (mostly towards the short end of the spectrum) in directions that depend on the orientation of the iridocyte platelets composing it (see b and c); a few orientations only are shown. The strongly reflecting white backing layer 6 scatters as well as reflects; it is composed of leucophores (see footnote on p. 194). (b) Schematic section through one of the grooves in the blue ring of the ocellus of *O. bimaculoides* indicating the unusual position of the mirror-reflecting iridocytes. The "blue" which they produce by constructive interference (see Land, 1972) is kept separate from other wavelengths by the absorbing melanophore layer (shown partly switched on). Total light flux increases, by reflection from underlying layers, when the ocellus is switched off swamping the "blue". (c) Diagram of an iridocyte cut in cross-section from an unpublished electron micrograph by S. Brocco to show the various orientations of its stacked platelets, one of which (seen as a black circle) has been cut in the plane of the platelet. Each stack is theoretically capable of reflecting pure hues by constructive interference in directions about the normal dependent on the angle of incident light (see Land, 1972).

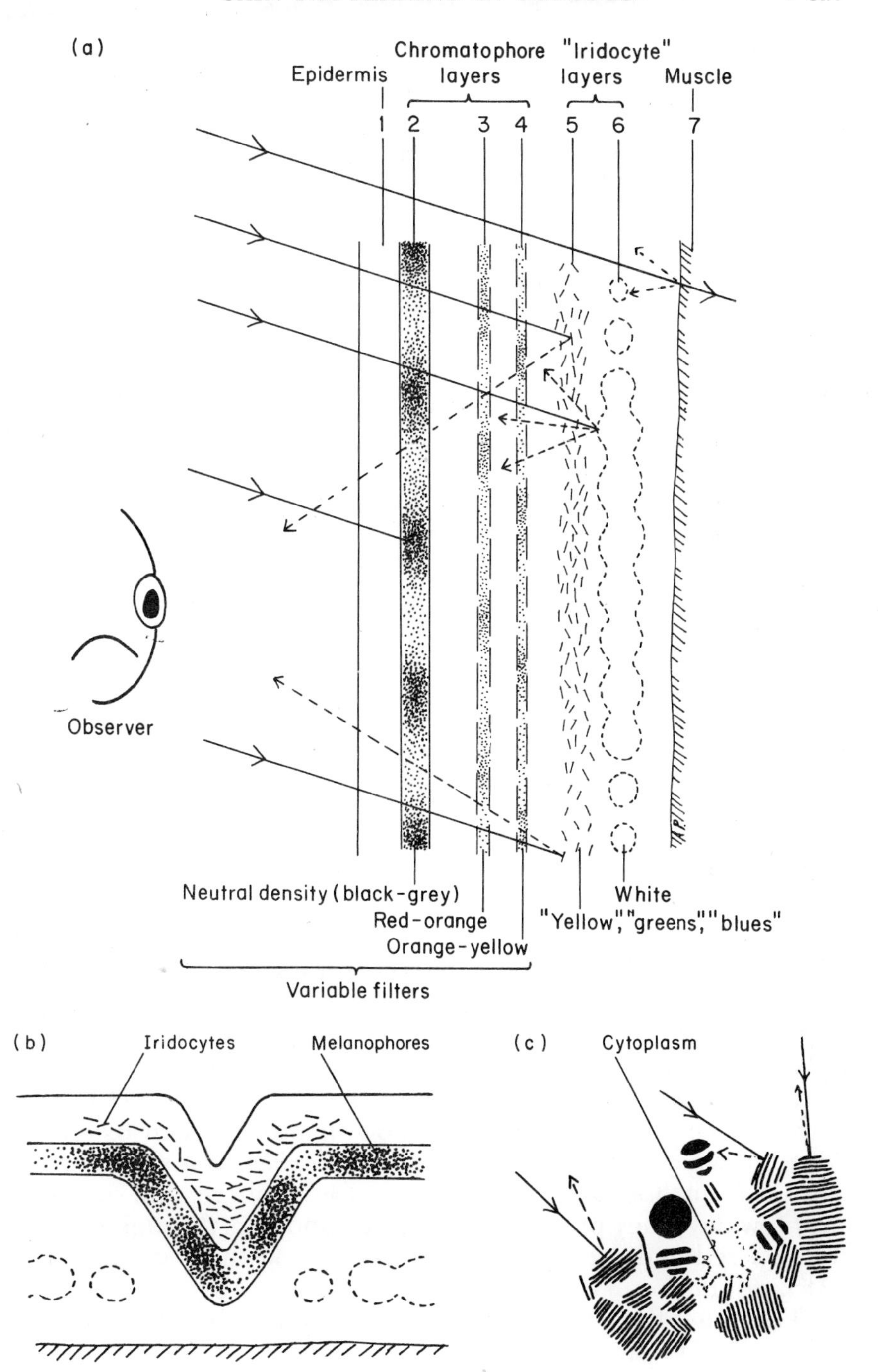
(a)
Epidermis
Chromatophore layers
"Iridocyte" layers
Muscle
1 2 3 4 5 6 7
Observer
Neutral density (black-grey)
Red-orange
Orange-yellow
Variable filters
White
"Yellow", "greens", "blues"
(b)
Iridocytes
Melanophores
(c)
Cytoplasm

iridocytes lie *above* the chromatophore layer (Fig. 1b), and the importance of this becomes clear as one watches the ocellus flashed on and off—by momentary expansion of the melanophores—and the blue come and go. Blue emerges only from those iridocytes directly above a melanophore, or projecting just beyond its edge. Presumably—as elsewhere in the animal kingdom—the melanins provide an absorbing screen to extract any rays of light that have passed through or between the iridocytes and that—if they had been allowed to re-emerge from the skin—would have swamped the pure colours reflected by the iridocytes (see also Fox, 1953: 25). To judge from published photographs (Voss & Sisson, 1971), the brilliant blue of the blue-spotted octopus is similarly supplied. The spots are surrounded and backed by black. Presumably iridocytes are also the source of blue in such pelagic species as *Tremoctopus*, *Ocythoë*, and the squid *Onychia*. Individual iridocytes refracting blue can also be observed in the anterior chamber of the eye. Here again they are underlain by the large melanophores in the iris.

How many classes of chromatophores?

The actual pigments present in the chromatophores may be relatively few. Indeed there is growing evidence that the orange, red, brown and black ones belong to a single biochemical series; coloured stages in the synthesis either of melanin or of ommochromes (Fox & Vevers, 1960). True melanin (eumelanin), identified in the skin by several authors (Fox & Crane, 1942), is black, but the various phaeomelanins formed from tyrosine—e.g. dopamelanins and 5,6-hydroxyindole-melanins—which arise as a deviation of the main melanin pathway (Prota & Thomson, 1976) are progressively coloured light straw-yellow, orange, red, and brown (Fox, 1973; Nicolaus, 1968). Ommochromes formed from the breakdown of tryptophan have been extracted from the skin of the cuttlefish, *Sepia* (Schwinck, 1956). These too can be dark but their colour depends on the size of the molecule, whether it is oxidized or reduced or is attached to a protein (Fox & Vevers, 1960). The extracted cuttlefish pigment is yellowish brown when oxidized, wine-red when reduced. In young *Octopus vulgaris* melanophores arise as a small, clear and almost colourless sphere, and over the next few days they progressively darken and pass through orange to deep red and eventually muddy brown on their way to black (Packard, in prep.) (see Fig. 2).

This raises the question of how many classes of chromatophore there are. Do all chromatophores eventually turn black? The

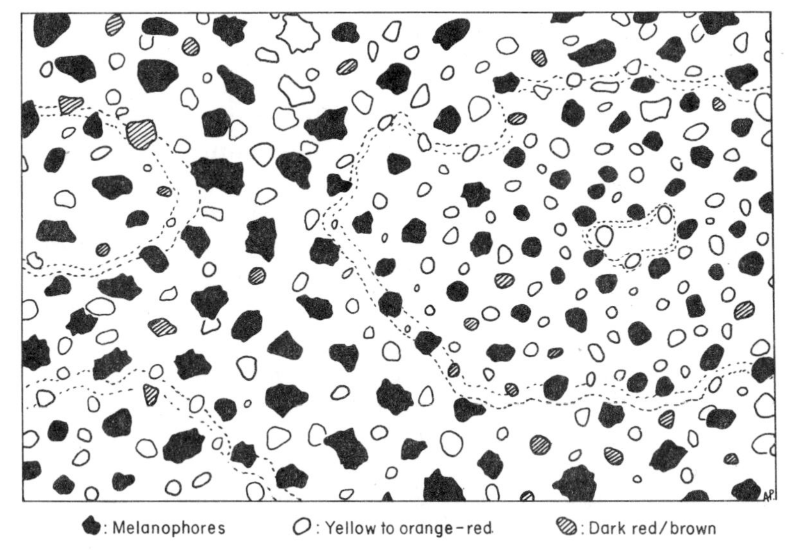

FIG. 2. Arrangement of the chromatophores in the dorsal surface of a young *O. vulgaris* (body weight 10 g). From a photomicrograph of incompletely relaxed fresh skin. The drawing covers an area of 1·1 mm^2 and includes the centre of a patch (right) and portions of two neighbouring patches separated by grooves. Chromatophores have been divided into three broad classes; red/brown ones are nascent melanophores. Local density of chromatophores falls off from the centre of a patch towards the grooves but average size increases. The same general arrangement holds in adult skin.

question is tied up with that of how many pigments there are and whether these are mixed or separate. If there are, say, three different pigments, a yellow, a red, and a black, then one might expect three classes of chromatophore. But if these pigments are mixed, then any classification based on colour alone would have to contend with a near-infinite series of hues depending on the proportions in which the pigments are mixed. The difficulty is a real one. Having looked at hundreds of chromatophores, we are still not sure whether there is a separate class "red" distinct from deep "orange-red" and are unable to decide whether "red" is in a different class from "brick-red". They look different and there are species of *Octopus*, e.g. *O. dofleini*, that have a range of "brick-red" chromatophores but no "red". Partly because of this difficulty we incline to the view that the orange, red, and brown series of chromatophores contain *mixtures of pigment*, either in separate granules or mixed in the same granule. Unpublished electron micrographs of the skin taken by S. Brocco show at least two main classes of pigment granule: large spherical uniformly dense granules and less dense granules of similar diameter and roughly spherical in shape, but with a blurred outline. Some chromatophores have only the first type of granule in their pigment bodies. These could be fully differentiated melanophores containing only melanin granules. Others contain both types of pigment granule.

Chromatophores are not randomly distributed (Fig. 2) but are fairly regularly spaced out over the skin with yellow-orange and brown ones filling in the spaces between melanophores. This is the first principle of patterning and may be stated as follows: *chromatophores of any one class are evenly distributed with respect to each other and occupy the spaces between chromatophores of other classes.* The picture is very similar to that of bristle and hair distribution in the integument of insects (Lawrence, 1973) and hair follicle distribution in mammals. Probably the underlying mechanism governing the location of the cells from which future chromatophores arise from a multipotent sheet of cells is similar, with the chromatophores already present inhibiting the differentiation of chromatophores in their immediate neighbourhood (Packard, in prep.).

Units: the static morphological array (Table II)

That the skin surface is broken up into units will be apparent to anyone who has looked closely at an octopus, whether dead or alive. It has the appearance of a loosely woven network, the strands of which are usually darker than the rest of the skin. Figures 4 to 21 and Fig. 26 are a set of low-power photographs showing examples of the network in different species of the genus *Octopus.* The individual "cells" of the net may be circular, oval or angular. They have a characteristic upper size in any one animal—equivalent to the mesh size of the net—and in some species there are also many "cells" of small diameter filling the nodes between the main "cells". Elsewhere (Packard & Sanders, 1971) we have referred to the lines of the net as a system of *grooves* and the areas which they bound as *patches.* A skin patch and its surrounding groove is one *unit.* Patches appear to retain their integrity throughout ontogeny, growing in size as the animal grows. The number of major patches remains constant from early in ontogeny (Packard & Sanders, 1971) though increasing numbers of smaller ones become distinguishable as the main network stretches with growth. The skin of a groove dips inwards; the depth of the infolding, and inversely the width of the groove, varies much in life, the grooves being deep and narrow in older animals, and when the muscles of the skin are contracted, and shallow and broad when the skin is relaxed.

Many of the units, especially the larger ones, may be seen to have tributary grooves running into the patch from the surrounding groove (Figs 13 and 16). Sometimes these tributary grooves form a short spiral—a sign that the patch contributes to the formation of a papilla (Figs 7 and 8). If these smaller grooves are taken into account, then the skin patches could be subdivided into smaller units. We have not considered it useful to do this. They are far less pronounced than the major grooves separating the patches, but they do contribute to the way in which the patches are used during patterning (see p. 220).

The appearance of the skin at any one moment depends to a large extent on the way in which the individual parts of the network are disposed. That is, the static morphological array of units begins to make itself felt as a component part of the various patterns at a level above the grain level of the picture. In all of the patterns that we have observed, the grain level—i.e. the level of the chromatophores and iridocytes—is too fine to be appreciated by the unaided eye, except for the huge larval chromatophores which reach up to 1 mm across when expanded. Not so the grooves and patches. Individual patches, or groups of a few patches, are used singly to produce visual effects. This is particularly evident in the case of the many white spots (see next section). They are underlain by a larger amount of reflecting material than are the neighbouring patches. The contribution of static white reflecting patches to patterning is even more marked in cuttlefish (Maynard, 1967).

At points on the skin from which major *papillae* arise, the patches are arranged concentrically with the smallest patches at the centre forming the peak of the papilla. The location of papillae does not change and a given papilla will always have the same patch at its peak. Papillae vary in height, however, and some of the larger ones—for instance the four large mantle papillae, or the long papillae above the eyes—carry smaller second- and third-order papillae up onto their lower slopes during full expression [e.g. in the acute flamboyant pattern (Fig. 37): Packard & Sanders, 1971].

Components

Physiological basis

This is a suitable point at which to introduce the idea of physiological components of patterning. Papillae come and go leaving little or

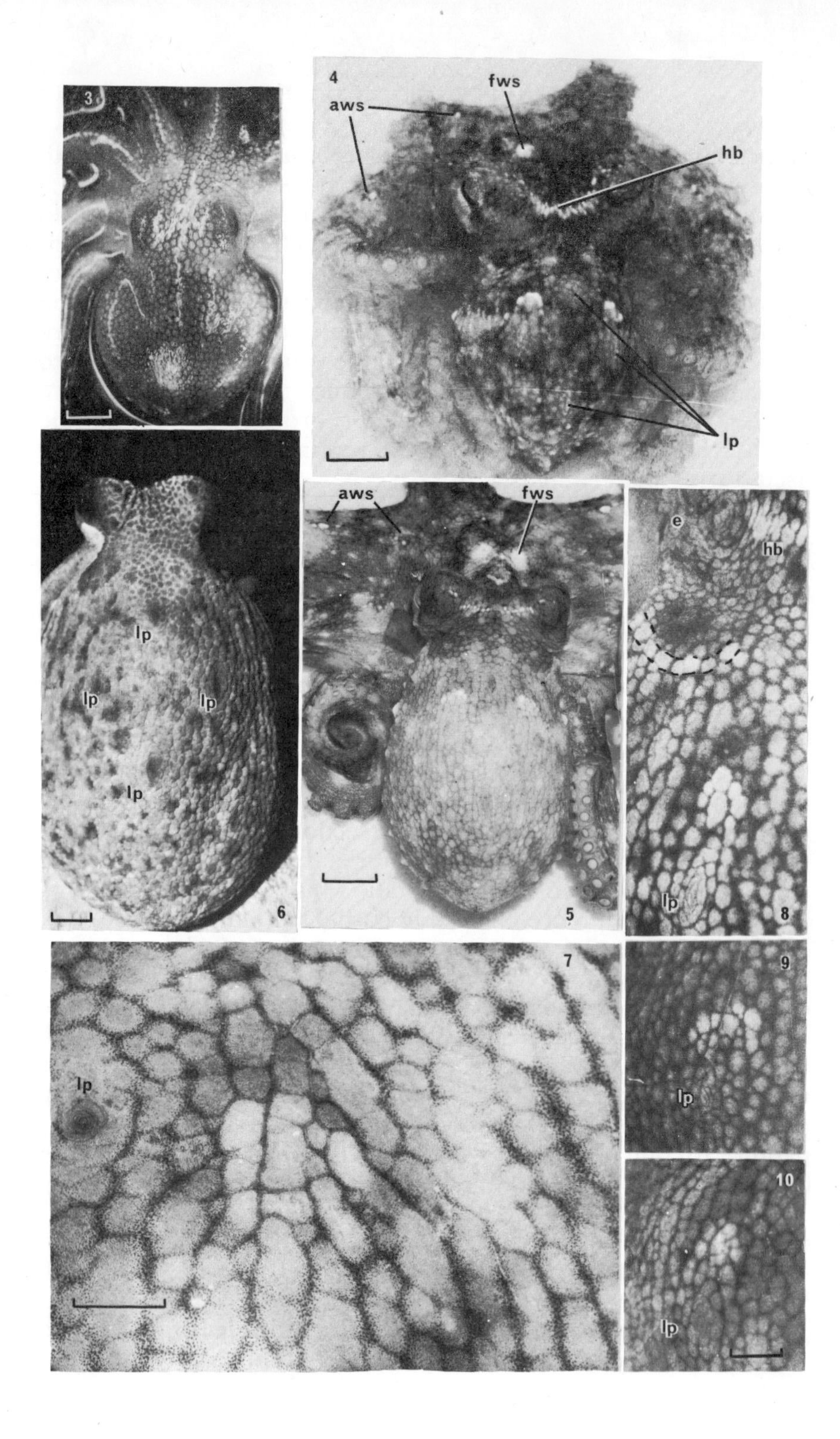
3
4
aws
fws
hb
lp
6
lp
lp
lp
lp
aws
fws
5
e
hb
lp
8
7
lp
9
lp
10
lp

no trace in the static morphological array once they have gone, even though, by careful observation, one can see that they always arise at the same point in the skin. They are physiological rather than morphological, though presumably, if one could see them, they would be found to involve contractions in particular sets of subcutaneous muscles supplied by particular nerve fibres. If one uses a stimulating silver wire electrode (delivering shocks of a few volts a.c. at 2–10 shocks per second) one can, in fact, evoke papilla-raising from the skin of a freshly dead or recently anaesthetized octopus. With continued stimulation the larger papillae will increase their height and may lean towards the side of the electrode. That electrical stimulation gives rise to a true papilla, and not just to local contraction of the skin around the point of stimulation, demonstrates that papilla-raising is a function of nerve

---

FIGS 3–10. *O. californicus* and *O. vulgaris*, skin network and white spots etc. aws, arm white spots; fws, frontal white spots; e, eye; hb, head-bar; lp, long mantle papillae.

FIG. 3. Male *O. californicus* (body weight 166 g): patches present over whole surface, grooves broad, chromatophores in grooves expanded. No differentiation of patches into special white spots apparent in this freshly dead specimen. (Scale 2 cm.)

FIGS 4–6. Different ontogenetic stages of Mediterranean common octopus, *O. vulgaris*.

FIG. 4. Young specimen (body weight approximately 2 g) exhibiting conflict mottle. Patches present, grooves broad, chromatophores in grooves expanded in certain areas. Mantle white spots, head-bar etc. clearly differentiated. (Scale 4 mm.)

FIG. 5. Juvenile octopus (body weight 50 g) exhibiting same white components as Fig. 4. Grooves narrow at this stage, head-bar formed by single line of spots. Note that the head, innervated by separate nerves, is darker than the skin of the mantle, though both are exhibiting the same pattern. The boundary between the two areas is formed by the back of the "hood". (Scale 8 mm.)

FIG. 6. Old octopus (body weight >5 kg). Grooves deep and narrow, light in colour. Mantle white spots and head-bar lost. Back of hood and long mantle papillae still apparent. (Despite increase in size, the number of major patches has not greatly increased.) (Scale approx. 4 cm.)

FIG. 7. Detail of part of the mantle including right mantle white spot and anterior long mantle papilla of lightly anaesthetized *O. vulgaris* (body weight *c.* 300 g) showing differentiation of skin patches and grooves. Individual chromatophores (partly expanded in the grooves) are resolvable as dots. (Scale 5 mm.)

FIGS 8, 9 and 10. Details of the skin in life showing variations in form of the left mantle white spot of three *O. vulgaris* [all to same scale (5 mm), body weight similar to Fig. 7]. Figure 8 includes white head-bar, the line of patches forming the boundary between head and mantle (dotted lines), and the area of the dark "horseshoe" (above mantle white spot). (The right-hand edge of these photographs corresponds to the midline.)

connections in the skin. It has even been found that stimulation of one papilla, for example the long mantle papilla forming the front of the quadrangle of papillae on the dorsal surface of the mantle (Figs 4, 5 and 6), will evoke erection of a second papilla at a distance, for example the long one forming the lateral member of the quadrangle. Presumably this group of papillae is a physiologically stable component.

And so too with other components. If the same physiological preparation is used (Packard, 1974) and shocks are shortened to 1 msec or less at a strength just above threshold, chromatophores alone can be made to expand causing little or no change in shape of the skin. The pattern of response to electrical stimulation varies in different parts of the skin, but there are a number of consistent features (Figs 14 and 23). One is that many chromatophores expand over a definite region of the skin, there being usually quite

---

FIGS 11–17. *O. bimaculoides* and *O. dofleini*. Skin network and white spots, etc.

FIG. 11. Adult female spotted octopus, *O. bimaculoides* (body weight 850 g) showing ocelli (o), frontal white spots (fws) and mantle white spots (also scattered white spots in head-bar region) during life. (Scale 1 cm.)

FIG. 12. Similar view of juvenile (body weight 45 g). (Scale 1 cm.) Long mantle papilla (lp), ocelli (o). Compare the region between right mantle white spot and eye in these two specimens.

FIG. 13. Detail of right side of mantle including right mantle papilla (lp) of octopus in Fig. 11, showing the system of patches and grooves and mantle white spot area (centre), which is relatively undifferentiated in this species (compare Fig. 7). (Scale 5 mm.)

FIG. 14a, b and c. Detail of the skin of an adult *O. bimaculoides* (body weight 440 g) showing the progress of a stimulating silver wire electrode (white at the tip) across an area known to contain the right mantle white spot (outlined). The stimulus (5V shocks at 20 per sec, shock duration 1 msec) is above threshold strength for the nerve fibres innervating the chromatophores that serve to mask the white spot (dark area in b), but below threshold for the patches (a) and grooves (b) on either side of this area. (Scale 5 mm.)

FIG. 15. Sub-adult Pacific giant octopus, *O. dofleini* (body weight 425 g), exhibiting dark eye-bar and mottle (groove chromatophores expanded). (Scale 1 cm.)

FIG. 16. Same individual showing details of skin network, mantle and head. Patches long, grooves broad, chromatophores in grooves expanded, mantle white spots (each consisting of one patch) exposed. Note transverse bar equivalent in position to back of hood. (Compare Figs 7 and 13.) (Scale 5 mm.)

FIG. 17. (a) Adult *O. dofleini* (body weight 4–5 kg) exhibiting uniform light phase and frontal white spot. (b) Same individual during darker phase exhibiting frontal white spot, arm-bars and spots, head-bar, raised mantle papillae (lp) etc. (N.B. The white area on the right of the mantle is due to skin damage.) (Scale approx. 3 cm.)

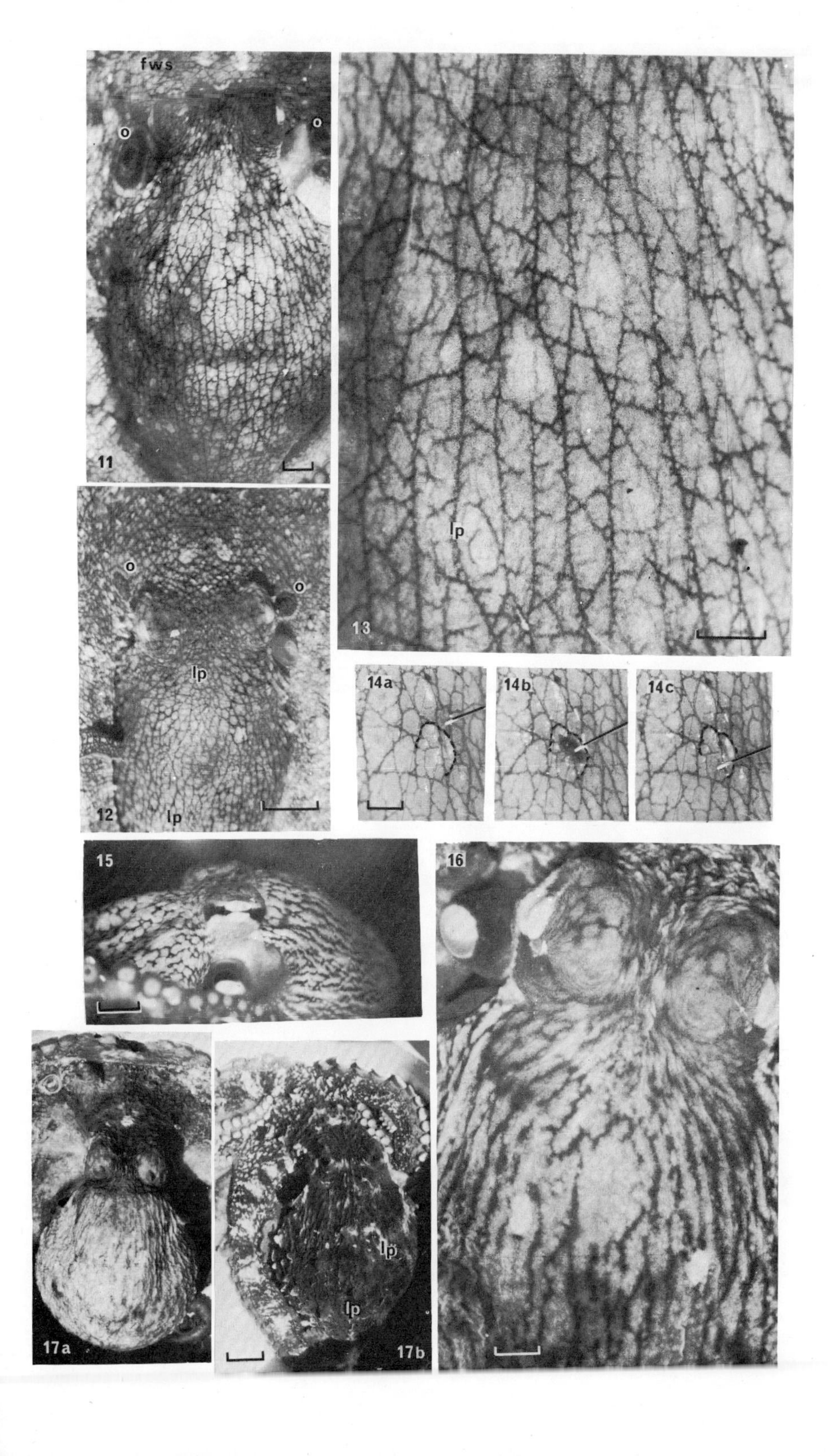

fws
o
o
11
13
lp
12
o
o
lp
lp
14a
14b
14c
15
16
17a
lp
lp
17b

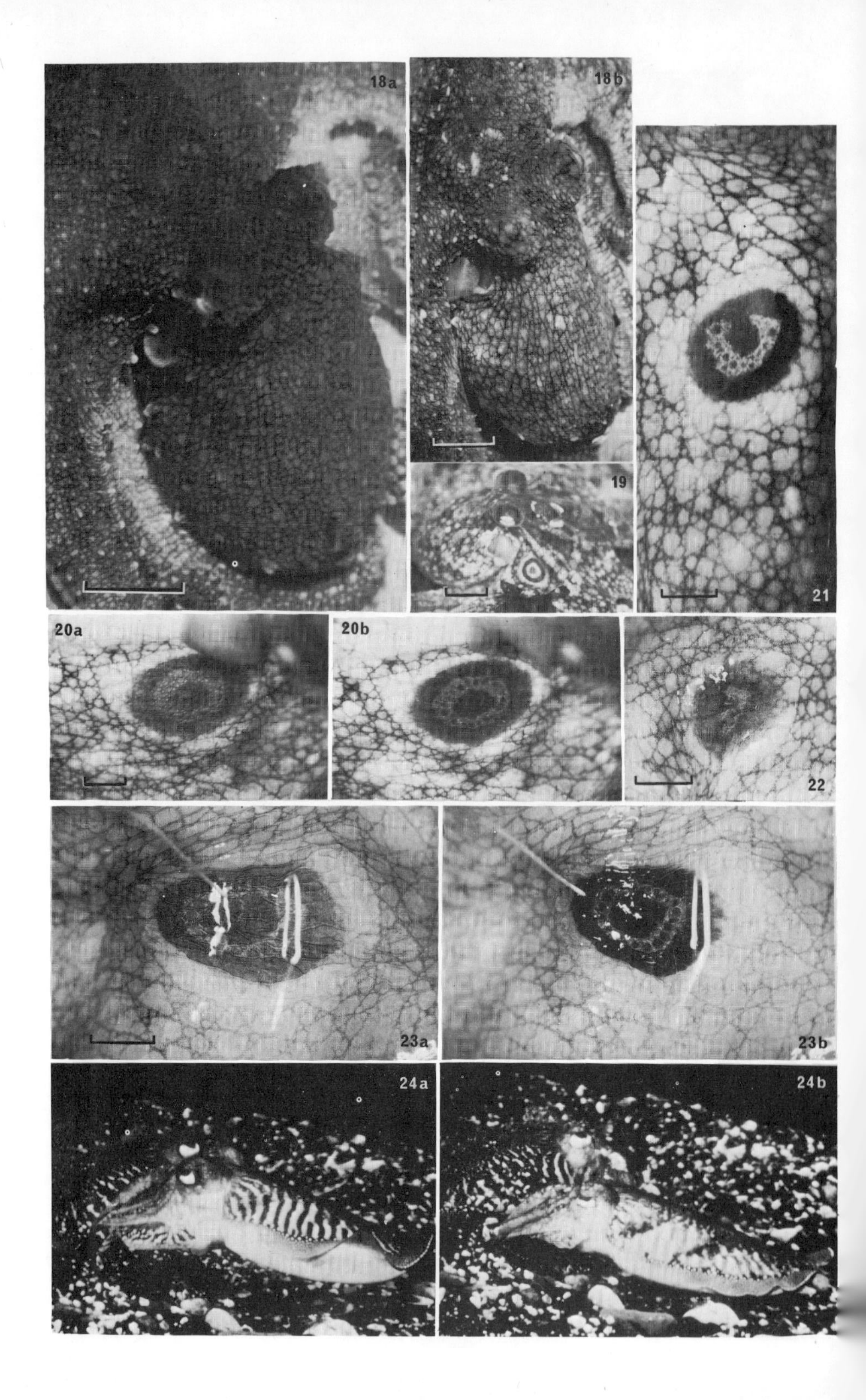
18a
18b
19
21
20a
20b
22
23a
23b
24a
24b

a sharp boundary between them and other chromatophores not involved in the response. The other is that movement of the stimulating electrode within the area of responding chromatophores does rather little to change the pattern of response, though additional groups of chromatophores near to the electrode may be recruited while others further away drop out. Immediately the stimulating electrode crosses the boundary between responding chromatophores and non-responding ones, however, the pattern of expansion can change abruptly. There may be no further response at that stimulus strength, or an

---

FIGS 18–24

FIG. 18. (a) Sub-adult *O. bimaculoides* exhibiting nearly uniform phase (chromatophores of patches and grooves equally expanded). Frontal white spots (upper left) just visible. (b) Same octopus some minutes later showing differential expansion of chromatophores in grooves (forming a black trellis in certain areas) and contraction over the patches exposing frontal, mantle and arm white spots. (Scales 1 cm.)

FIG. 19. Juvenile *O. bimaculoides* (body weight 35 g) displaying ocellus (eye-spot). Note large frontal white spot complex, small mantle white spot and acute mottle (possibly equivalent to conflict mottle of other species, see Figs 4 and 33). (Scale 1 cm.)

FIG. 20. Close-up of right ocellus of large *O. bimaculoides* shown in Fig. 11. (a) With eye-spot switched off (little contrast between ocellus and surrounding skin), (b) with eye-spot switched on.

Note that differential expansion of chromatophores (melanophores) takes place not only between the eye-spot and its immediate surround, but also in other areas enhancing the dark trellis. Note also that the overall intensity of light reflected from the inner necklace of iridocytes is less in (b) than in (a) but that the contrast between it and the inner and outer parts of the eye-spot is rendered greater by melanophore expansion. In life the inner necklace appears pure blue when the eye-spot is displayed (see text). (Scale 5 mm.)

FIG. 21. Unusual left eye-spot displayed by sub-adult *O. bimaculoides*. The iridocyte necklace of the ocellus (bright blue in life) is incomplete. The dark trellis formed by differential expansion of groove chromatophores, contributing to the acute mottle, is well illustrated in this photograph. (Scale 5 mm.)

FIG. 22. Incomplete ocellus of adult *O. bimaculoides* (body weight 440 g): iridocyte necklace rudimentary. (Scale 5 mm.)

FIG. 23. Close-up of lightly anaesthetized sub-adult *O. bimaculoides* (body weight 166 g) showing the eye-spot being switched on by electrical stimulation of the skin; (a) electrode in position (no stimulating current), individual chromatophores (melanophores) can just be resolved; (b) stimulation with brief multiple shocks produces expansion of all melanophores in the dark area of the ocellus which responds as a single motor unit. (Scale 5 mm.)

FIG. 24. Cuttlefish *Sepia officinalis* exhibiting (a) zebra display; (b) oblique stripes. Note that parts of the vertical stripes in (a) contribute to the formation of the obliques in (b). (A second cuttlefish can be seen in the background.) From a colour film by D. M. Maynard. Distance between frames approximately 1 sec.

altogether new set of chromatophores may come into play while those previously responding drop out. So far only the responses of the melanophores have been followed in this way; because of the screening melanophores, it is difficult to see what is happening to the other classes of chromatophores immediately beneath them. Florey (1969) states that in the squid, orange chromatophores are separately innervated from melanophores, and we have reason to believe that in the adult skin the same is true of *Octopus*. It turns out that the sets of chromatophores that do expand—usually many hundreds at a time—follow the patch and groove structure of the skin and form parts of the dark components seen under natural conditions. Examples of these are the dark arm-bars (Figs 17b, 26a, 27, 30 and 32d), the transverse zigzag (or "chevron") stripes on the mantle (Figs 27 and 32d), the dorsal trellis (involving expansion of groove chromatophores only, Figs 13, 18b and 21), the eyespots (Figs 19, 20a,b, 21, 23a,b, 26b and 27), and screening chromatophores masking out white spots (Fig. 14). All of these can be evoked in whole or in part by suitable electrical stimulation of the skin even many hours after death of the animal or separation of the mantle nerve from its connection with the brain. Some of the components appear to arise from a number of overlapping but staggered motor units; others, notably the ocellar eyespot of *O. bimaculoides* (Fig. 23) and the masking chromatophores of mantle and arm white spots on *O. dofleini* and *O. bimaculoides* (Fig. 14) appear to be single motor units.

Light and dark components (Tables III and IV)

In Packard & Sanders (1971) there is a list of the components of patterning seen in *O. vulgaris*. Most of these components are present in the other species we have studied, but there are differences in emphasis. For instance, the two white spots on the middle of the back (of the mantle) are relatively small in *O. dofleini* and *O. bimaculoides* but well developed in *O. vulgaris* and *O. rubescens*, whilst the frontal white spots are well developed in *O. bimaculoides*. Table III provides a comparative guide to the white components in these four species. Both in the octopus and the cuttlefish (Maynard, 1967) much of the "whiteness" is due to the complete contraction of all chromatophores over the spots, enhanced by the expansion of surrounding chromatophores (fig. 6 in Maynard, 1967). Conversely, the white spots disappear completely when the masking chromatophores are expanded. In *O. vulgaris* the density of all chromatophores over the mantle white spots is about half the

TABLE II

Skin network of some adult and sub-adult Octopus

	O. vulgaris	*O. dofleini*	*O. bimaculoides*	*O. rubescens*	*O. californicus*
Mantle patches	Round and polygonal; clearly separated one from another; tend to line up longitudinally	Long with major axis running longitudinally, boundaries irregular and often indistinct; incompletely dissected by tributary grooves. Longitudinal grooves broad	Round and polygonal often incompletely separated one from another, especially in longitudinal direction. Transverse grooves often shallower than longitudinal. All nodes between large patches filled by smaller patches	Round and polygonal, clearly separated one from another. Large patches often surrounded by smaller	Round and polygonal. Surrounding grooves broad
Number of patches touching line CDFE[a] (weights indicate sizes of individual specimens)	46 (550 g)[b] 54 (400 g)	(420 g) 15	30 (65 g) 36 (125 g) 30 (880 g)	56 (~25 g)	

[a] This is the line bounding the four long mantle papillae. Counts made from photographs of living specimens; they do not include the patch or patches at the centre of the papilla.
[b] For other counts see Packard & Sanders (1971).

density in the area that lies immediately anterior to them and that contributes part of a dark component, the dark "horseshoe" (Figs 4 and 8).

Table III is only a rough guide. Even in the few animals we have photographed there is considerable variation in the form of the white spots; and they are more emphasized in small animals than in large ones. Details of some mantle white spots in *O. vulgaris* are shown in Figs 7 to 10. The particular arrangement of the patches is probably as individual a matter as the details of a finger-print and we have found it a useful marker for identifying photographs of individuals that may be exhibiting quite different patterns (see e.g. Figs 32 and 33).

We are not in a position to give a comprehensive list of the dark components observable in the species we have photographed, but Table IV is a guide to some of the commoner components. The well known *ocellus* of *O. bimaculoides* consists usually of a small group of central patches surrounded by two concentric circles; these are the areas which give a uniform dark response. There is also a third concentric ring of pale skin in which the grooves are little evident (Fig. 23a) and which serves to emphasize the ocellus (Figs 20b and 21). This ring has a high threshold to electrical stimulation. Beyond it is a further area in which the grooves have a low threshold. The first concentric ring is the blue one. It consists of a necklace (Figs 20b and 21) of small patches strung together by grooves rich in iridocytes. As explained on p. 198 the relative intensity of the blue increases as the melanophores below them expand.

In the 11 *O. bimaculoides* that we have studied we found as many as four incomplete ocelli. In one individual both ocelli were aberrant, there being only a rudimentary trace of the blue ring (Fig. 22), and in two others the ring was incomplete (Fig. 21). We also saw a rudimentary third ocellus in one animal.

During the electrophysiological studies we had occasion to observe that the extended eye-bar varies in a systematic way. The eye-bar (Fig. 15) continues the line of the horizontal pupil forwards and backwards. Its backwards extension continues onto the back of the mantle (Fig. 30). Figure 25 shows how this varies in *O. vulgaris*, *O. dofleini* and *O. bimaculoides*. The dark line persists even after severing the pallial nerve connections, which carries most of the chromatophore nerves to the mantle (Froesch, 1973a), suggesting that this particular component is innervated by nerve fibres from the head (probably in the collar nerve). In *O. bimaculoides* the

TABLE III

White components in species of Octopus

	O. vulgaris	*O. dofleini*	*O. bimaculoides*	*O. rubescens*	*O. californicus*
White mantle spots	Conspicuous (especially in sub-adults). Circular or horseshoe-shaped. Several patches to each spot: arrangement variable. Innervated	Relatively inconspicuous in adults and sub-adults. One patch only involved. Innervated	Smaller than frontal white spots and large arm spots. One or two patches only involved. Innervated	Large oval complex, with conspicuous anterior horseshoe-shaped border. Innervated. Individual patches separately masked	Diffuse accumulation of white reflecting substance in many patches especially frontal white spot region. No signs of separate innervation
Head-bar	Row or rows of white patches appearing as single or double line	Present	Absent (isolated spots only in this region)	Present (single line of spots)	
Frontal white spots	Small proximal inverted V above conspicuous paired spots or dumb-bell traversing midline.	Present (single round spot in proximal position)	Conspicuous. Arranged in circle. Similar to *O. vulgaris*	Similar to *O. vulgaris*. Less conspicuous than mantle white spots	

continued

TABLE III—*continued*

	O. vulgaris	*O. dofleini*	*O. bimaculoides*	*O. rubescens*	*O. californicus*
Arm white spots	Serially arranged usually with one spot larger than rest at level of inner margin of interbrachial web on arms 1–3	Present	Similar to *O. vulgaris*. Large spots particularly conspicuous	Similar to *O. bimaculoides*	
Other		Edges of spots less clearly defined than in *O. vulgaris*, *bimaculoides* and *rubescens*		Numerous other white spots on mantle head and arms. Each spot one patch	

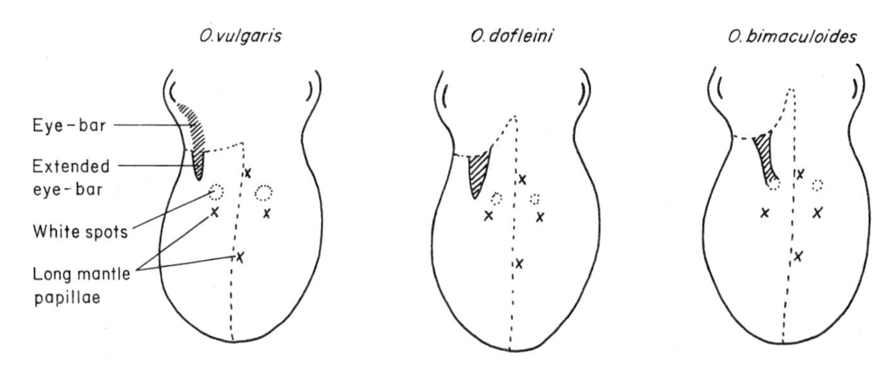

FIG. 25. Systematic differences in the form of a dark component (the extended eye-bar) in three species of *Octopus*. The dotted line indicates the area of skin appearing denervated after section of the left pallial nerve. (From electrophysiological findings on not less than two specimens for each species: see text.)

extended eye-bar reaches the mantle white spot and so encroaches upon the territory of the dark horseshoe (see later, p. 217) (Table IV).

TABLE IV

Other components observed

	O. vulgaris	*O. dofleini*	*O. bimaculoides*	*O. rubescens*
Dark arm-bars	+	+	+	+
Eye-bar and extended eye-bar	+	+	+	+
Dark eye-ring	+	−	−	+
Eye-spot (ocellus)	−	−	+	−
Four long mantle papillae	+	+	+	+

PATTERNS—COMPLETING THE PICTURE

These and other light and dark components together with changes in skin texture, body shape and movements combine as complete patterns: for instance, the various cryptic patterns ranging from uniform tone and colouring (Fig. 18a) through chronic general mottle (Figs 5, 7, 11, 12 and 20a) to acute mottling (Figs 15, 16, 19,

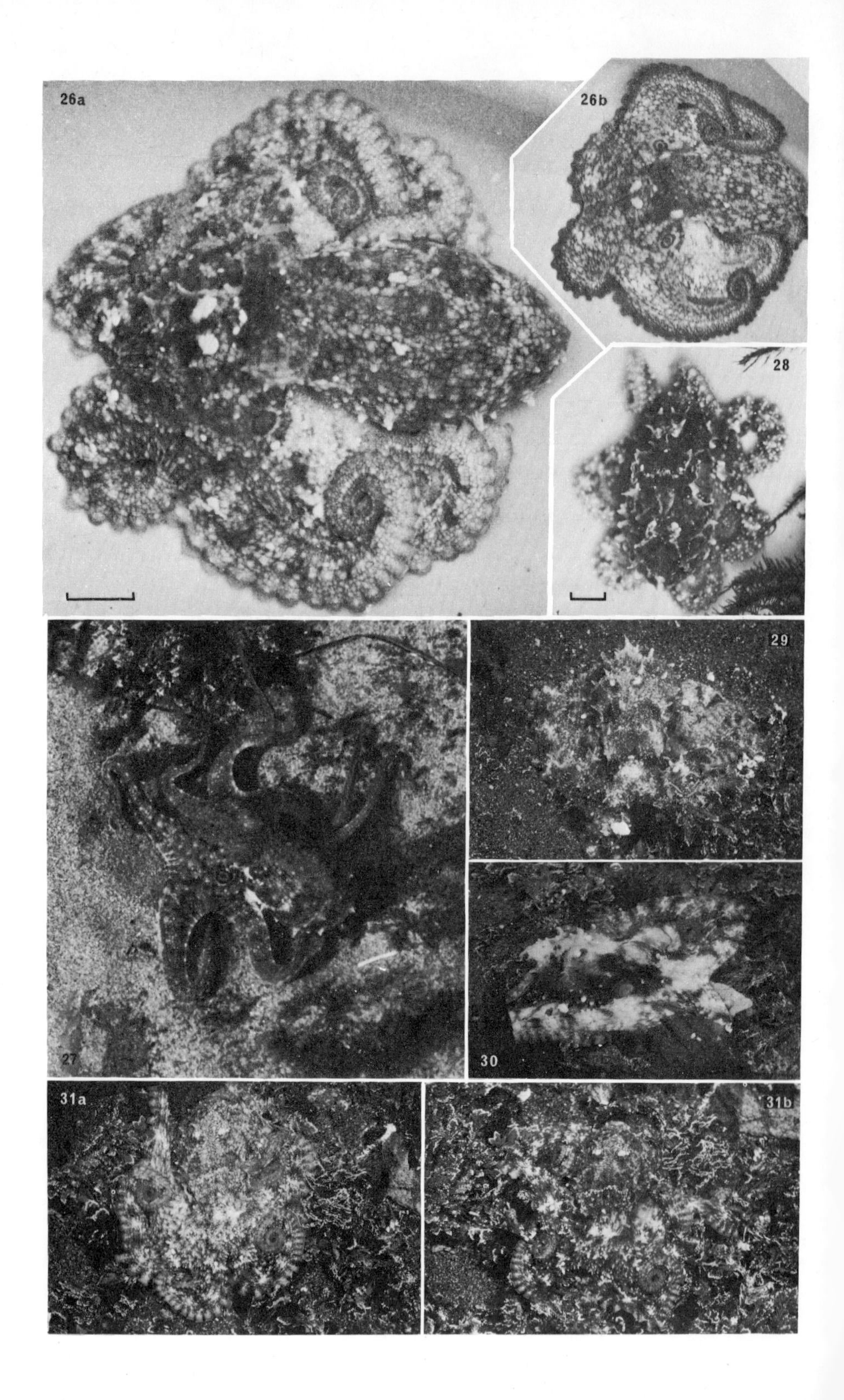
26a
26b
28
27
29
30
31a
31b

26a), and the various stereotyped displays such as the dymantic and the zebra (Figs 26b, 32 and 33). It is unlikely that we have seen the full range of patterns and we will not attempt to make a list of their occurrence in the different species. Instead, we will refer to the way in which some of the commoner ones illustrate what appear to be general principles of patterning.

The first is that components vary in the extent of their expression from barely perceptible to fully expressed. Thus an eye-bar may be just distinguishable from the pale skin above and below it, or increase in intensity until it is equal in "blackness" to the pupil itself. A papilla may be just discernible as a single patch raised above the level of the rest of the skin or increase in height until it has gathered onto its lower slopes many of the patches and grooves between it and the next large papilla (see above).

The second is that single components—for instance a white spot, a dark "horseshoe", a raised papilla—do not appear in

---

FIGS 26–31. Cryptic patterns and displays. (Scales 1 cm.)

FIG. 26. Juvenile *O. bimaculoides* (45 g) exhibiting (a) acute general mottle with frontal, mantle and arm white spots, arm-bars etc. and raised papillae (eye-spot not noticeably expressed), (b) "dymantic" pattern achieved by spreading arms which become dark at the edges, displaying eye-spots, darkening central "hood" area and suppressing arm-bars and certain areas of expanded chromatophores (e.g. around each long mantle papilla).

FIG. 27. *O. bimaculoides* photographed in shallow tide pool in sunlight, displaying eye spots with "zebra" stripes. The rest of the skin tone matches with the sandy background. Notice that there is a certain similarity between the "hood" region of expanded chromatophores behind the eye and the ocellus to the side of the eye suggesting a pathway for the evolution of such features as eye spots.

FIG. 28. *O. rubescens* (same individual as Figs 32 and 33) exhibiting fully raised papillae and partially extended, partially curled, arms characteristic of the "flamboyant" pattern seen when the octopus is hiding in seaweed.

FIG. 29. Similar pattern involving papillae, white spots, arm-bars etc. exhibited by juvenile *O. vulgaris* hiding amongst sand and coralline algae.

FIG. 30. Juvenile *O. vulgaris* exhibiting a complex pattern in which there is an interplay of longitudinal dark components (extended eye-bar) and transverse one (arm-bars etc.).

FIG. 31. Tone-matching by juvenile *O. vulgaris*. In (a) the animal is lighter than the background. In (b) it has succeeded in matching the reflectance of the background by adjusting (increasing) the intensity of expression of the dark components all over the body (increasing the absorbance of the neutral density filter, see Fig. 1). The effect appears to be achieved by enhancing the pre-existing firing pattern of nerve fibres supplying the skin [further expanding chromatophores that are already partly expanded in (a)] and to a less extent by bringing in other nerve fibres (enlarging the skin areas occupied by dark components).

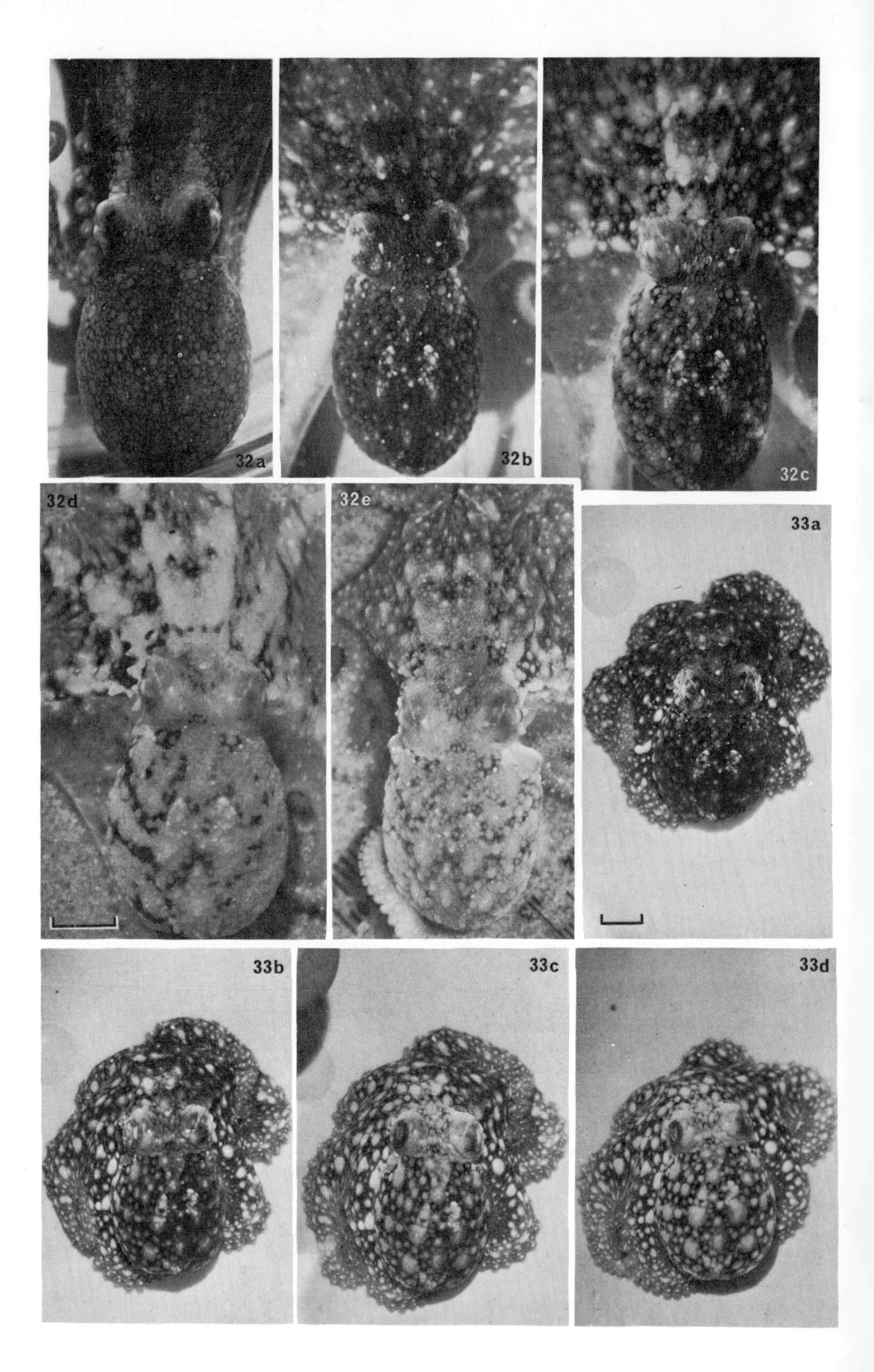
32a
32b
32c
32d
32e
33a
33b
33c
33d

isolation, but appear along with other components of the same category (i.e. in series). Thus if one dark arm-bar appears so will others, both along the length of the same arm and upon other arms. If a dark "horseshoe" appears ahead of the mantle white spot, areas of skin of similar size or shape will go dark elsewhere on the mantle. Papillae in one part of the skin appear at the same time as papillae over the rest of the skin. When such serially repeating components are expressed with less than maximal intensity, a proximo-distal gradient of intensity can be observed along the arms at any one moment. Bars furthest from the head are faintest. There are exceptions to the second rule. One is that components may express themselves unilaterally with little or no sign of the same components on the other side of the body. More usually unilateral expression turns out to be a difference in intensity of expression of the components on the two sides (Fig. 32d). Second, some components, the eye-bar, the eye-ring (Fig. 33c), the eye-spot, are by nature single components and can be expressed in apparent isolation from others. Differentiation of patterning has probably consisted in part in isolating components from other components. It may be that the frontal white spots can operate independently from other white spots and an example of increased independence of parts is seen in *O. rubescens*, which can differentially unmask individual patches of its large mantle white spots.

FIGS 32 and 33. Lateral inhibition in *Octopus rubescens*. All the figures in this plate are of the same individual exhibiting:

FIG. 32. (a) Nearly uniform dark phase (compare Fig. 18a); (b) general mottle with frontal, head and mantle white spots unmasked; (c) differential expansion and contraction of chromatophores in certain grooves and patches in a basically proximo-distal sequence with superimposed transverse differences; (d) enhancement of process seen in (c) to produce "zebra" stripe pattern displayed to a second octopus to the left; (e) same pattern as (d) intensity reduced.

FIG. 33. Another programme, this time involving greater expression of dark components in the longitudinal direction (i.e. differential expansion of chromatophores of grooves and patches in transverse sequence). (a) Similar stage to Fig. 32b but mantle rounded and arms curled. (b) and (c) Progressive enhancement of conflict mottle achieved by increasing the sizes of the light areas of skin. Chromatophores are contracted in these areas (i.e. nerve fibre activity is inhibited), (d) same as (c), intensity of expression slightly reduced. In (c) the octopus is displaying the dark eye-ring (part of the "dymantic" display) towards an intruding rubber ball (upper left).

To be convinced that Figs 32 and 33 are of the same animal, concentrate on the form of individual spots and patches, e.g. the mantle white spots and white spot above the right eye. That the central nervous programmes giving the two patterns employ some of the same components can be seen by studying the arrangement of individual patches, e.g. at the back of the "hood" and in the conspicuous chevron on the left of the mantle in the two sets of figures. (Scales 1 cm.)

The third principle is that different components combine together (i.e. in parallel) to give distinct patterns. Singularly striking examples of this are the dymantic pattern (Fig. 26b) and the flamboyant (Packard & Sanders, 1969, 1971). In its fullest expression in *O. vulgaris* the dymantic display involves spreading of the arms and web, distension of the mantle, head flattening, dilation of the pupil, the dark eye-bar and eye-ring, darkening of the edges of the arms and paling of the rest of the skin. Not only do these components commonly occur together (Packard & Sanders, 1971: fig. 21) but others, for example papilla-raising over the whole surface of the body, seem at the same time to be excluded. In the flamboyant pattern (Packard & Sanders, 1969, 1971) in which the arms are twisted and held away from the substratum, the web is contracted, papillae are maximally extended (Fig. 28), the mantle is constricted and the skin is uniformly dark or dark on the sides with a light strip down the middle. (Some of these components, it will be noticed, are diametrically opposite to those of the dymantic. See later.) Less easily analysed are the various patterns illustrated in Figs 30 and 31 in which certain components will be observed in combination with others.

In illustrating the principle that patterns consist in the combination of different sets of components, we have introduced postural components (shape of body, position of arms) that go together with certain appearances of the skin. This brings us to the final principle in this section: that patterns are reflections of the whole behaviour of the animal. Put another way, we can say that body patterns are themselves components of particular behaviour sequences. This was illustrated by Packard in 1963 with regard to the approach and withdrawal responses of *O. vulgaris*; systematic evidence has now been provided by Warren, Scheier and Riley (1974) who recorded the occurrence of certain body patterns in *O. rubescens* being trained to attack crabs and conditioned stimuli.

Several questions arise at this point. How stereotyped are the patterns? Should a pattern consisting of the combination of certain components be called by one name and given another name if one of the sets of components changes? Do patterns grade into each other? The questions are obviously related. The answers have to do with the difficult matter of pattern recognition, but for an exposition of the problems this raises for humans in general, see Casey & Nagy (1971). Individual octopuses—and the same is true of other cephalopods—show patterns that appear to be quite stereotyped. Time and again the same pattern, with individual components involving the same set of skin patches each time, will be exhibited under the same set of conditions. But the combinations of components both within individuals and between individuals are not entirely fixed. The various postural components described above for the dymantic pattern may be exhibited with a smooth skin showing the broad conflict mottle (Fig. 26b; Packard & Sanders, 1969: fig. 14). The flamboyant pattern, described above as displayed with a uniform dark colouration, may also be seen with a light band along the middle of the body (i.e. extended hood) or with the transversely orientated chevron stripes.

Patterns also grade into each other. Examples are shown in Figs 32 and 33 of *O. rubescens*. Photographs such as these provide perhaps the most important single set of data for understanding the neural basis of patterning (see next section).

WHAT DO PATTERNS TELL US ABOUT THE NERVOUS SYSTEM BEHIND THEM?

The principles of patterning just enunciated have a common thread. They amount to the single principle that parts are subordinated to the whole. The skin of the octopus provides, in fact, an extraordinarily vivid pictorial demonstration of this universal biological principle in action. We have suggested four levels at which this can be seen: the first, the chromatophore level, with chromatophores regularly spaced with respect to each other; the second, the level of the patches, groups of chromatophores being organized into patches and surrounding grooves; the third, the level of components, many patches or grooves usually contributing to one component; and the fourth, the level of the whole skin in which components of one kind combine with components of another to give a single pattern, which in turn is subordinate to a fifth level: that of the animal's whole behaviour.

The four levels of patterning can be represented as a hierarchy (Fig. 34). It turns out that the branches of this hierarchy represent real branchings in the peripheral nervous system (Fig. 35)—possibly of single nerve fibres—from the component level downwards. For instance (Fig. 35a) the groups of patches and intervening grooves that go to form the ocellus in *O. bimaculoides* appear to have all their melanophores under the control either of a single motor unit or several closely overlapping motor units (see p. 210). Elsewhere (see p. 208) grooves and patches are separately innervated and the situation would be as in Fig. 35c.

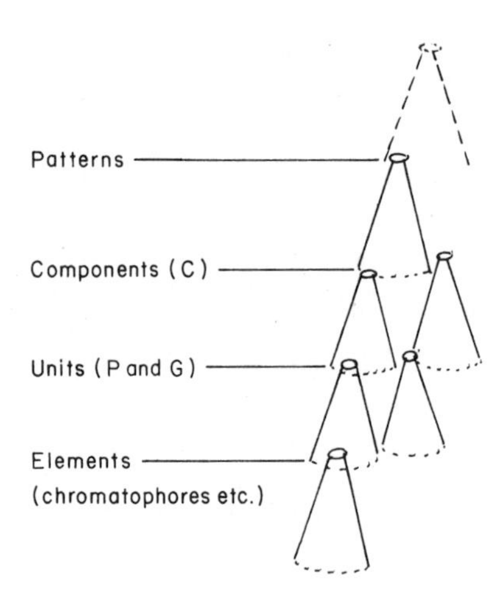

FIG. 34. The hierarchical scheme of patterning.

But are there really four levels of pattern organization? Why four?

A close look at some of the major patches of *O. vulgaris* reveals that there is considerable differentiation within them (Packard, in prep.) into a central zone of small densely packed chromatophores and a surrounding zone of less densely packed larger chromatophores. In some of these patches the inner and outer zones are separately innervated (Fig. 35d). In terms of hierarchical organization this is tantamount to saying that the patch level is subdivided into more than one level, and the same turns out to be true of other levels described. Only in the least differentiated species, such as *O. californicus* (Fig. 3), are there no obvious groupings of the different classes of chromatophores within a patch. At the component level we observe that patches of skin which contribute an isolated dark component to one pattern, combine in other groupings to form a different dark component of another pattern. This is nowhere better illustrated than in D. M. Maynard's film of the cuttlefish (*Sepia officinalis*) as it changes from the vertical stripes of the zebra pattern (Fig. 24a) to the oblique stripes of the saddle (Fig. 24b) seen in the white square pattern (Holmes, 1940). Two possible neurophysiological interpretations of this are given in Fig. 36.

The highest level of patterning—involving the combination of several different components—evidently requires the brain (see Young, 1971: chap. 15) and cannot be elicited by stimulation of the

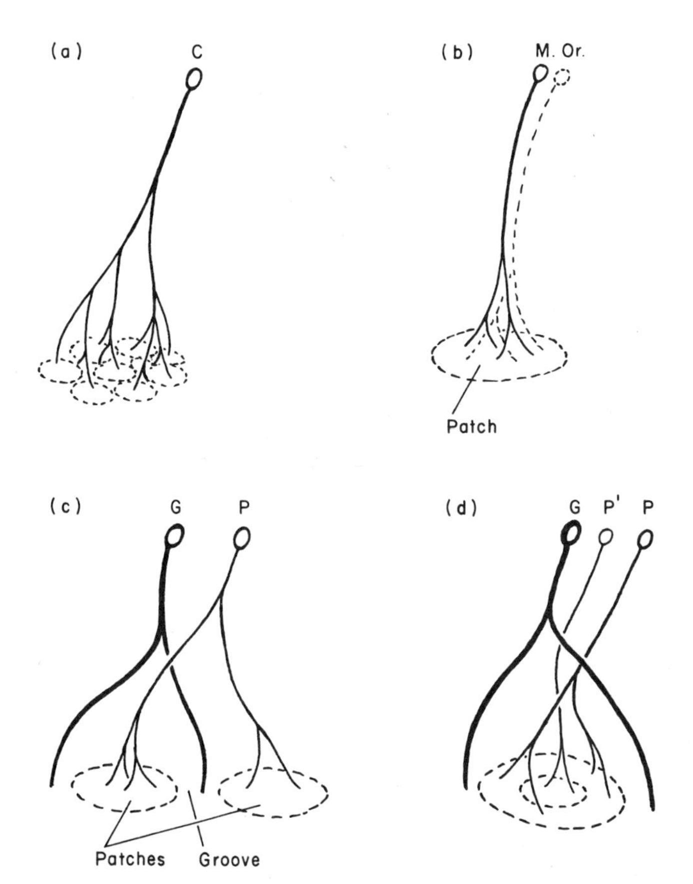

FIG. 35. Various manners of innervation indicated by electrical stimulation of the skin (see text). (a) A motor unit covering several patches and grooves constituting a whole component (C), as in the ocellus of *O. bimaculoides* (see Fig. 23). (b) Melanophores and orange-red chromatophores within a patch separately innervated by fibres M and Or. (c) Grooves and patches separately innervated by fibres G and P. (d) Differentiation within a patch; chromatophores in the central region innervated by a separate fibre P′.

periphery. But, again, we do not know exactly what level or levels within the brain. Boycott (1961) obtained whole patterns from electrical stimulation of the optic lobes, and one might conclude that the programmes for the patterns reside in these lobes. But there is at least one intermediate lobe lying between them and the chromatophore lobes of the suboesophageal mass that contain the motoneurones supplying the muscle fibres of the skin and chromatophores (Boycott, 1953). It may be that some patterns require the optic lobes while others reside in the intermediate

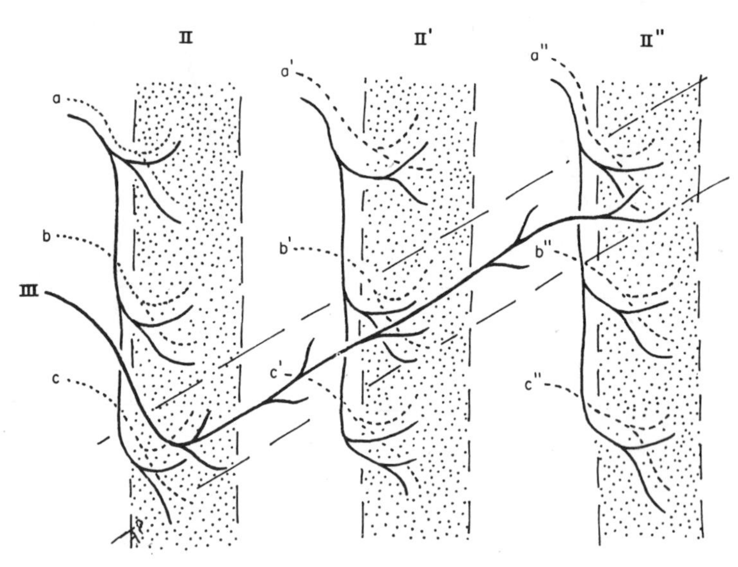

FIG. 36. Theoretical schemes of innervation of the mantle of the cuttlefish *Sepia officinalis* seen from the left side. (Compare Fig. 24a,b.) Part of an oblique stripe is seen superimposed upon three dark "zebra" stripes. In scheme 1 (continuous lines) vertical chromatophore fields are innervated by fibres II, II′ and II″ and the oblique field by fibre III. In scheme 2 (broken lines) the vertical fields are innervated by fibres a, b, c; a′, b′, c′ and a″, b″, c″ firing synchronously, and the oblique field by c, b′ and a″ firing together with two intermediate fields (not shown). Scheme 1 is an example of "position-pattern separation" proposed by Maynard (1967). Not shown in this scheme are class I fibres responsible for uniform darkening of the whole area covered by the diagram.

(lateral basal) lobes, and are merely evoked by stimulation of the optic lobes; while the simplest of all patterns—uniform darkening—may require only the chromatophore lobes (Boycott, 1961).

It does not require electrical stimulation to show that higher centres are involved in patterning. All the patterns that are part of behavioural responses (see next section) demonstrate this, especially the visual ones. Particularly interesting are the patterns that regularly appear during learning experiments and situations of behavioural conflict (Boycott & Young, 1958; Packard, 1963). The smooth forward movement of a fully developed attack upon a crab or upon a positively conditioned stimulus is associated with uniform darkening of the skin. Upon landing, the pattern changes. The changes have been followed in detail in *O. rubescens* by Warren *et al.* (1974). At first the skin goes light, then, on seizing the crab or object a mottle of light on dark, equivalent to the conflict mottle (see next section) appears.

Now "dark" represents activity in chromatophore nerve fibres and "light" represents no activity. All of the patterns illustrated in this paper are, in fact, achieved by a change in the balance of these two conditions, i.e. of central excitation and inhibition. There is nothing new in this. The same is true of all motor patterns, but only in cephalopods is it so vividly brought to view. The photographs in Figs 32 and 33 show a stepwise increase in the proportion of light skin, as if some process of lateral inhibition were building up (Packard, in prep.).

The conflict mottle (Fig. 33) represents a neat balance between dark and light, as if the processes of excitation and inhibition were in equilibrium. That this pattern appears when there is psychological conflict between a "positive" learned response and a "negative" learned response is a dramatic demonstration of the fact that overall conditions in the highest centres of the brain are reflected in their lowest and most distal connections. (These behavioural aspects of certain kinds of conditioning experiment were discussed briefly by Boycott & Young, 1958, and again by Packard, 1963.)

In the dymantic display of *O. vulgaris*, with its extensive area of pale skin bounded by dark peripherally and centrally, we see a suppression of nerve activity in all chromatophore fibres except those supplying the edges of the arms and the region of the eyes. Such extensive inhibition of chromatophore nerve activity bounded by a narrow region of excitation apparently develops only late in ontogeny, since the dymantic display is seen only in incomplete form in juveniles and not at all in early settling stages (Wells, 1962; Packard & Sanders, 1971).

Sanders & Young (1974) have shown that nerves grow back to the skin and apparently re-establish their original connections months after being severed. A striking feature of this regeneration is the finding that whole components return.

DISPLAYS AND CONCEALING PATTERNS

Displays

These are the visually simpler patterns (see Packard, 1972: 277). They consist of a few broad bands, stripes or circles of dark contrasting with lighter skin, and it is the high contrast and rapidity with which they appear that makes them conspicuous. Such are the black hood, the zebra, the dymantic and the dark flush (or passing cloud). They are seen only in the presence of other animals either of the same species, or of other species when the wearer has been observed. Included in this category are the courtship displays described by various authors (Tinbergen, 1939; Tardent, 1962; Young, 1962; Arnold, 1962, 1965; Packard, 1963; Cousteau, 1969, 1971; Wells & Wells, 1972). In some of these patterns dark vertical components are a dominant feature. Presumably any communication function will be intraspecific.

The dymantic is an example of a display serving an interspecific function. It means "boo!"—i.e. "go away: don't you see how frightening I am!"—and is used when other ruses have failed. The dark flush, or passing cloud (Packard & Sanders, 1969, 1971: fig. 15) also apparently serves as an attention-catching device in an interspecific context—perhaps to make an animal move slightly and so become visible (but see Cott, 1940). It has been observed both in the cuttlefish *Sepia officinalis* (Holmes, 1940) and in *O. vulgaris* when they are lined up for an attack upon prey. It consists of a pulse of dark that spreads out from the head along the first and second pair of arms. It may be that the frontal white spots are used in a similar manner to the dark flush. They are sometimes winked off and on.

Cryptic patterns

Much more difficult to categorize are the various concealing patterns. A large part of their effect is achieved by the shape of the outline taken on by the body and the texture of the skin which changes from smooth to rough by the raising of papillae (Figs 17b, 18a, 26a, 28, 29 and 30). Conversely colour—i.e. hue—may play only a minor part. The most important single function of the chromatophores seems to be in enabling cephalopods to match the reflectance of the background. The manner in which this is done through the aid of the white reflecting layers deep to the chromatophores (see Fig. 1) has been examined photographically by Messenger (1974). The white material, particularly the leucophores, succeeds in reflecting the different wavelengths of visible light in proportions similar to those in the incident light. The total amount of light reflected is, however, determined by the darkness and extent of the neutral density filter (see p. 195) provided by the dark components above the reflecting layers. An example of how this can be adjusted in time after settling on a particular background is shown in Figs 34a and b (see also Packard & Sanders, 1969: 99, figs 18, 19). Does the eye of an octopus measure the albedo of its surroundings and alter the degree of expression of its dark components according to the proportion and distribution of reflected light present? Presumably: a few simple experiments could be designed to check this, but so far they have not been done. What we do know with reasonable certainty is that octopuses are not capable of learning to discriminate objects on the basis of hue alone (Messenger, Wilson & Hedge, 1973)—i.e. they appear to have no colour vision. We also know that there is a limited repertoire of cryptic patterns. Placed on a background which does not form part of its natural environment—for instance a sheet of coloured formica—a cephalopod can do no more than call on this repertoire through its central nervous system and adjust the intensity of expression of the several components to match the reflectance. Long-term changes in the static morphological array of elements present in the skin—for instance the amount of leucophore material or proportion of orange-red chromatophores—should not, however, be ruled out. Is there a paucity of reflecting material in cephalopods reared in the dark? Do octopuses reared in an environment of orange sponges, ascidians and bryozoans—such as is met with in the rock pools of Northern California—grow up to become more orange than siblings reared in an environment deprived of these? The two-spotted octopus (*O. bimaculoides*) is greener than *O. vulgaris* and *O. rubescens*; is this partly due to an environmentally induced increase in the amount of iridocyte material present in the skin of animals that have grown up on the intertidal flats of Southern California where large amounts of green sea lettuce (*Ulva*) and eelgrass grow? Experiments on other animal groups would lead us to expect positive answers to all of these questions.

EVOLUTIONARY CONSIDERATIONS

Homology

There are a number of patterns that appear separately in quite different cephalopods. The dymantic, the zebra and the flamboyant are seen not only in octopods but also in decapods (both sepioids and teuthoids). Three expressions of the dymantic pattern are illustrated in Packard (1972: fig. 24). Figure 37 shows the flamboyant in a young octopus and a young cuttlefish. If these patterns are considered homologous one with another, then what is the status of the homology? It is not strictly a morphological one. The dark border of the dymantic pattern is provided in *O. vulgaris* by the edge of the arms and suckers; in the cuttlefish *Sepia officinalis* by the edge of the fins running round the mantle. Different species

FIG. 37. Homologous patterns in an octopod and sepiid. Drawn from photographs of (a) young *Octopus vulgaris* (see Packard & Sanders, 1969, 1971) and (b) young *Metasepia tullbergi* (Y. Natsukari, unpublished) to show the strange postural and skin textural components characteristic of the flamboyant display. The display is generally cryptic and is seen as a response to disturbance especially in young cephalopods.

of *Octopus* produce the dark eye-ring from different morphological parts (Table III): *O. vulgaris* and *O. rubescens* through expansion of the chromatophores around the eye itself, *O. bimaculoides* by expansion of the ocellus just below the eye. And yet these patterns have enough in common for them to be given the same name in the different species. Perhaps we should look upon them as behaviourally homologous—i.e. that they have the same behavioural origins. Judging by the accounts in the literature (Holmes, 1940; Young, 1950; Wells, 1962; Boycott, 1965), behaviour contexts that evoke the dymantic in one species, evoke the dymantic in another (see below). This is an area where more comparative ethology is required.

On the basis of the data in this paper we are inclined to place *O. bimaculoides* nearer to *O. dofleini* than to *O. vulgaris*, and *O. rubescens* nearer to *O. vulgaris* than to *O. bimaculoides* (see Tables II-IV).

The role of the skin in evolution

Packard (1972) has argued that the evolution of modern cephalopods should be regarded as taking place in behaviour space. Once the chief physiological adaptations had arisen—and this presumably took place early in the Palaeozoic, as most of them are little different from those of other molluscs—subsequent evolution will have been mainly of systems involved in behaviour and psychology. The skin and the brain are important organs in this respect. Many authors have pointed out that in losing the external protective shell, into which their ancestors withdrew, coleoids came to rely on the skin: protecting themselves by pictorial means.

The survival value of cryptic and warning patterns is indeed so obvious as almost to be taken for granted and their relevance missed. It is still not generally recognized that cephalopods are a favourite food of many vertebrate animals. Up and down the west coast of North America octopuses are eaten by nearly all of the reef-living fish species whose stomach contents have been analysed (see Carlisle, Turner & Ebert, 1964; Taylor & Chen, 1969; also the unpublished observations of other diving biologists: L. Moulton, K. McKay and L. Hallacher). These teleosts have colour vision and good acuity and it is they that provide the selection pressures for many of the cryptic patterns that have evolved. The widespread occurrence of the dymantic display suggests also that fish in general are startled by the sudden appearance of a pale mask with its dark rings and eye-spots, and that there would be sustained selection for this pattern in coastal rocky habitats. It may be that the ocellus of *O. bimaculoides* and *O. bimaculatus*, which is placed at the base of the arms and so faces upwards when they are spread, and not sideways like the ring round the eye, has evolved as a variant of the dymantic pattern directed against mammals and birds hunting in shallow water (Fig. 27).

Here then is one way of answering the homology question. If the selective advantage is in the behaviour itself, then natural selection may be relatively indifferent to the details of the means by which it is produced—so long as the means are effective.

The chromatophores of coleoids are unique in the animal kingdom, both in their form and the way in which they are controlled. However, we have made it our business to point out that patterns are not a matter of a single set of elements or structures, but of all of them working together—in the skin, in locomotory structures and in modes of action—and given coherence by the brain. It is the repertoires read out from the central nervous programmes that have been selected in evolution, and the chromatophores in the context of their use by the brain. In honouring J. Z. Young we are honouring the man who has done more than any other to fashion the key to unlock these central nervous programmes and with it to understand the brain's style of working.

Acknowledgements

We wish to thank Dr Edith Maynard for making available unpublished material of D. M. Maynard and for her permission to publish. The work was done at the Naples Zoological Station, Italy, the Friday Harbor Marine Laboratory of the University of Washington, and the marine laboratory of the University of California at Santa Barbara. We thank the directors and staff of these laboratories for their co-operation and use of facilities; also the director and staff of the Vancouver Aquarium who provided octopuses free of charge, and Dr Case, M. Fawcett, Steve Brocco, Elsie Dorsie and Donata Oertel who arranged for us to make observations in the field and the laboratory. We thank Y. Natsukari of the University of Nagasaki for permission to publish Fig. 37b.

The investigation was supported by a scientific grant-in-aid from the Royal Society to one of us (A.P.).

References

Arnold, J. M. (1962). Mating behavior and social structure in *Loligo pealii*. *Biol. Bull. mar. biol. Lab. Woods Hole* **123**: 53–57.

Arnold, J. M. (1965). Observations on the mating behaviour of the squid *Sepioteuthis sepioidea*. *Bull. Mar. Sci.* **15**: 216–222.

Berry, S. S. (1911). Preliminary notices of some new Pacific cephalopods. *Proc. U.S. natn. Mus.* **40**: 589–592.

Berry, S. S. (1912). A review of the cephalopods of western North America. *Bull. Bur. Fish., Wash.* **30**: 263–336.

Berry, S. S. (1953). Preliminary diagnoses of six West American species of *Octopus*. *Leafl. Malac.* **1** (10): 51–58.

Boycott, B. B. (1953). The chromatophore system of cephalopods. *Proc. Linn. Soc. Lond.* **164**: 235–240.

Boycott, B. B. (1954). Learning in *Octopus vulgaris* and other cephalopods. *Pubbl. Staz. zool. Napoli* **25**: 67–93.

Boycott, B. B. (1961). The functional organization of the brain of the cuttlefish *Sepia officinalis*. *Proc. R. Soc.* (B) **153**: 503–534.

Boycott, B. B. (1965). A comparison of living *Sepioteuthis sepioidea* and *Doryteuthis plei* with other squids, and with *Sepia officinalis*. *J. Zool., Lond.* **147**: 344–351.

Boycott, B. B. & Young, J. Z. (1958). Reversal of learned responses in *Octopus vulgaris*. *Anim. Behav.* **6**: 45–52.

Brocco, S. (1975). The fine structure of the frontal and mantle white spots of *Octopus dofleini*. *Am. Zool.* **9**: 782.

Carlisle, J. G., Turner, C. H. & Ebert, E. E. (1964). Artificial habitat in the marine environment. *Fish Bull. Calif.* **124**: 1–93.

Casey, R. G. & Nagy, G. (1971). Advances in pattern recognition. *Scient. Am.* **224**: 56–57.

Cloney, R. A. & Florey, E. (1968). Ultrastructure of cephalopod chromatophore organs. *Z. Zellforsch. mikrosk. Anat.* **89**: 250–280.

Cott, H. B. (1940). *Adaptive colouration in animals*. London: Methuen.

Cousteau, J.-Y. (1969). The night of the squid. *The undersea world of Jacques Cousteau*. Television film, general release. Metromedia: Hollywood.

Cousteau, J.-Y. (1971). *Octopus, octopus*. Television film, general release. Metromedia: Hollywood.

Cousteau, J.-Y. & Diolé, P. (1973). *Octopus and squid: the soft intelligence*. London: Cassell.

Cowdry, E. V. (1911). The colour changes of *Octopus vulgaris*. *Univ. Toronto Stud. biol. Ser.* No. 10: 1–53: *Contr. Bermuda biol. Stud.* **2** (22).

Denton, E. J. & Land, M. F. (1971). Mechanism of reflexion in silvery layers of fish and cephalopods. *Proc. R. Soc. Lond.* (B) **178**: 43–61.

Florey, E. (1969). Ultrastructure and function of cephalopod chromatophores. *Am. Zool.* **9**: 429–442.

Florey, E. & Kriebel, M. E. (1969). Electrical and mechanical responses of chromatophore muscle fibres of the squid, *Loligo opalescens*, to nerve stimulation and drugs. *Z. Vergl. Physiol.* **65**: 98–130.

Fox, D. L. (1953). *Animal biochromes and structural colours*. Cambridge: University Press.

Fox, D. L. (1973). Colouration, biological. In *Encyclopaedia Britannica*, 15th Edition.

Fox, D. L. & Crane, S. C. (1942). The pigment of the two-spotted octopus and the opalescent squid. *Biol. Bull. mar. biol. Lab. Woods Hole* **82**: 284–291.

Fox, H. M. & Vevers, H. G. (1960). *The nature of animal colours*. London: Sidgwick & Jackson.

Fries, E. F. B. (1958). Iridescent white reflecting chromatophores in certain teleost fishes, particularly in *Bathygobius*. *J. Morph.* **103**: 203–242.

Froesch, D. (1973a). Projection of chromatophore nerves on the body surface of *Octopus vulgaris*. *Mar. Biol.* **19**: 153–155.

Froesch, D. (1973b). On the fine structure of the *Octopus* iris. *Z. Zellforsch. mikrosk. Anat.* **145**: 119–129.

Girod, P. (1883). Recherches sur la peau des Céphalopodes. *Archs Zool. exp. gen.* (2) **1**: 225–266.

Holmes, W. (1940). The colour changes and colour patterns of *Sepia officinalis* L. *Proc. zool. Soc. Lond.* (A). **110**: 17–36.

Holmes, W. (1955). The colour changes of cephalopods. *Endeavour* **14**: 78–82.

Kawaguti, S. & Ohgishi, S. (1962). Electron microscopic study on iridophores of a cuttlefish *Sepia esculenta*. *Biol. J. Okayama Univ.* **12**: 57–60.

Land, M. F. (1972). The physics and biology of animal reflectors. In *Progress in biophysics and molecular biology* **24**. Butler, J. A. V. & Noble, D. (eds).

Lane, F. W. (1957). *Kingdom of the octopus.* London: Jarrolds.
Lawrence, P. A. (1973). The development of spatial patterns in the integument of insects. In *Developmental systems: Insects.* Chap. 2: 157–209. Counce, S. J. & Waddington, C. H. (eds). London and New York: Academic Press.
Maynard, D. M. (1967). Organization of central ganglia. In *Invertebrate nervous systems.* Wiersma, C. G. A. (ed.). Chicago: University Press.
Messenger, J. B. (1974). Reflecting elements in cephalopod skin and their importance for camouflage. *J. Zool., Lond.* **174**: 387–395.
Messenger, J. B., Wilson, A. P. & Hedge, A. (1973). Some evidence for colour blindness in *Octopus. J. exp. Biol.* **59**: 77–94.
Mirow, S. (1972). Skin colour in the squids *Loligo pealii* and *Loligo opalescens* II Iridophores. *Z. Zellforsch. mikrosk. Anat.* **125**: 176–190.
Nicolaus, R. A. (1968). *Melanins.* Paris: Hermann.
Packard, A. (1963). The behaviour of *Octopus vulgaris. Bull. Inst. oceanogr. Monaco* No. spéc. **1D**: 35–49.
Packard, A. (1972). Cephalopods and fish: the limits of convergence. *Biol. Rev.* **47**: 241–307.
Packard, A. (1974). Chromatophore fields in the skin of the octopus. *J. Physiol., Lond.* **238**: 38–40P.
Packard, A. (in prep.). *Spatio-temporal organization of chromatophore patterning in the octopus.*
Packard, A. & Sanders, G. D. (1969). What the octopus shows to the world. *Endeavour* **28**: 92–99.
Packard, A. & Sanders, G. D. (1971). Body patterns of *Octopus vulgaris* and maturation of the response to disturbance. *Anim. Behav.* **19**: 780–790.
Pickford, G. E. (1964). *Octopus dofleini* (Wülker). *Bull. Bingham oceanogr. Coll.* **19**: 1–70.
Pickford, G. E. & McConnaughey, B. H. (1949). The *Octopus bimaculatus* problem: a study in sibling species. *Bull. Bingham oceangr. Coll.* **12** (4): 1–66.
Prota, G. & Thomson, R. H. (1976). Melanin pigmentation in mammals. *Endeavour* **35**: 32–38.
Sanders, G. D. & Young, J. Z. (1974). Reappearance of specific colour patterns after nerve regeneration in *Octopus. Proc. R. Soc. Lond.* (B). **186**: 1–11.
Schaeffer, W. (1938). Bau, Entwicklung und Farbentstehung bei den Flitterzellen von *Sepia officinalis. Z. Zellforsch. mikrosk. Anat.* **27**: 222–245.
Schwinck, L. (1956). Vergleich des Redox-Pigmentes aus Chromatophoren und Retina von *Sepia officinalis* mit Insektenpigmenten der Ommochromgruppe. *Zool. Anz.* [*Verh. dtsch. zool. Ges.*] *Supplementbd.* **19**: 71–75.
Tardent, P. (1962). Keeping *Loligo vulgaris* L. in the Naples aquarium. *Bull. Inst. océanogr. Monaco* No. spec. **1A**: 41–46.
Taylor, B. & Chen, L.-C. (1969). The predator prey relationship between the octopus (*Octopus bimaculoides*) and the California scorpion fish (*Scorpaena guttata*). *Pacif. Sci.* **23**: 311–316.
Tinbergen, L. (1939). Zur Fortplanzungsethologie von *Sepia officinalis. Arch. néerl. Zool.* **3**: 323–364.
Voss, G. L. & Sisson, R. F. (1971). Shy monster, the octopus. *Natn. geogr. Mag.* **140**: 776–799.
Warren, L. R., Scheier, M. F. & Riley, D. A. (1974). Colour changes of *Octopus rubescens* during attacks on unconditioned and conditioned stimuli. *Anim. Behav.* **22**: 211–219.

Wells, M. J. (1962). *Brain and behaviour in cephalopods.* London: Heinemann.
Wells, M. J. & Wells, J. (1972). Sexual displays and mating of *Octopus vulgaris* Cuvier and *O. cyanea* Gray and attempts to alter performance by manipulating the glandular condition of the animals. *Anim. Behav.* **20**: 293–308.
Young, J. Z. (1950). *The life of vertebrates.* Oxford: Clarendon Press.
Young, J. Z. (1962). Courtship and mating by a coral reef octopus (*O. horridus*). *Proc. zool. Soc. Lond.* **138**: 157–162.
Young, J. Z. (1971). *The anatomy of the nervous system of* Octopus vulgaris. Oxford: Clarendon Press.

Symp. zool. Soc. Lond. (1977) No. 38, 233–241.

THE SQUID AND ITS GIANT NERVE FIBRE

J. B. GILPIN-BROWN

The Laboratory, Plymouth, England

SYNOPSIS

An explanation of the importance of the squid's giant nerve fibre to the squid, and to man.

The following paper is a transcript of the commentary of a 16 mm film of the same title which was made for the Symposium. It is produced here because it gives a brief account of Professor Young's investigations of the giant nerve fibres of squid in the 1930s and describes some of the techniques that different scientists have developed subsequently for their study. Although obviously incomplete without the accompanying film, it is hoped that its usefulness has been increased by the inclusion of additional references.

COMMENTARY

26th October 1973

. . . And the research vessel SULA sails out of Sutton Harbour to spend yet another winter's day trawling for squid over the inshore fishing grounds near Plymouth. Weather permitting, the ships of the Plymouth Marine Laboratory have been doing this for at least 40 years. This film explains why these animals have attracted so much attention for so long.

The first reason is that they are intensely interesting animals in themselves. They are jet-propelled predators—a habit which demands keen observation and immediate, precise, responses. As a result, they have many of the characteristics which we normally associate only with the vertebrates. But squid are not vertebrates. They belong to a far older group of animals—the cephalopods—whose ancestors once dominated the ancient Palaeozoic and Mesozoic seas. Cephalopods are molluscs—which means that these active creatures are allied to the sedentary seashells of our shores.

In modern forms the lower half of the body consists almost entirely of the mantle cavity; and it is the contraction of its muscular wall which expels seawater out through the funnel, or siphon, below the head. About 400 species are probably alive today, and many of them are numerous enough to provide important fisheries—especially in Japan, where about half-a-million tons are landed annually. Around Britain, large specimens of *Loligo forbesi* and *Loligo vulgaris*, a foot or more long, are only found in the English Channel in the autumn. The Laboratory at Plymouth then, is particularly fortunate as it is one of the few research centres where they can be closely studied. The head is distinct from the body and bears conspicuous eyes—and arms set with rows of suckers. On each side of the tapered body there are large lateral fins, used when the squid is cruising gently along. Jet propulsion is used for escape—or attack—sea-water being drawn into the mantle cavity through openings just behind the eyes and then forced out through the siphon when all the mantle muscles contract in unison. A 350g squid can squirt about 200ml of water (about half-a-pint) in less than a fifth of a second (Trueman & Packard, 1968). To achieve this, the nervous system must ensure that all the different mantle muscles involved must—even in a squid several feet in length—receive the stimulus rapidly and simultaneously.

Professor J. Z. Young began his detailed studies of the squid's nervous system in 1933 (Young, 1944). Opening the mantle cavity in the mid-ventral line exposes the squid's true body wall (through which the viscera can be seen); the muscular siphon itself; and, on each side close to the inhalant openings of the mantle cavity, a large ganglion. This is called the stellate ganglion and (when the gill is removed) the reason is clear: numerous nerves, just below the skin, radiate from it in a star-like pattern. Professor Young showed that each stellate ganglion has about ten of these nerves, and that they fan out to all parts of the mantle wall. He also showed that each of these stellar nerves contained a large transparent tubular structure, about a millimetre in diameter, which Young at first took to be a blood vessel. In transverse sections, though, he found that this structure was—apart from its size—little different from the numerous small nerve fibres which surrounded it. Moreover, unlike a small arteriole, it never contained blood or amoebocytes. Examination of the stellate ganglion—particularly in the small transparent squid *Alloteuthis*—showed that all these tubular structures arose within the ganglion itself. Closer examination of the dissected ganglion confirmed this and detailed studies of its histology convinced Young in 1936, that each of these tubular structures must be a single nerve fibre—a giant axon—formed by the fusion of very many smaller ones. Young confirmed this hypothesis when, using a pair of simple electrodes, he stimulated a stellar nerve (Young, 1938). Large contractions of that part of the mantle served by the stellar nerve, were only obtained if the tubular structure within it remained intact. These structures therefore, must be single nerve axons—each almost a millimetre in diameter.

The next step, of course, was the removal of a living stellar nerve; and this has now become a routine procedure at those Marine Laboratories where squid are studied. After the head and viscera have been removed, the mantle is cut down the middle and one-half pinned out in a transparent dissecting dish. Very little magnification is required to see the nerve and to cut away the overlying tissue. A simple copper sulphate heat filter cools the light; and the sea-water bathing the mantle is chilled to about 4°C. Different scientists use slightly different procedures for the removal of the nerve. At Plymouth, Dr Meves first lifts the stellate ganglion—and then progressively frees the nerve from the mantle. Apart from removing any superfluous tissue, no further preparation of the nerve is required before simple physiological experiments can be made.

It has long been known that, as an impulse travels along a nerve fibre, the active region becomes electrically negative to all the neighbouring regions. Two electrodes, connected to suitable amplifying and recording devices, will, therefore, register a diphasic change as the impulse passes each electrode in turn. The change is only produced though if the stimulus is large enough, because (like all nerve fibres) the giant axon has an "all or none" response. This is the action potential—here continuously displayed because the frequency of stimulation is high. In 1938 Pumphrey & Young found that the conduction velocity of these axons increases with the square root of their diameter—the giant axons being about five times faster than the small fibres. Moreover, they are graded in size. The largest—and hence the fastest—supply the most distant parts of the mantle. These giant axons thus play a vital role in the squid's life. They not only ensure that the escape response is as fast as possible; but also that all parts of the mantle contract simultaneously—an essential requirement for efficient jet propulsion.

Of course they are also one of the largest animal cells known to man; and many techniques have been developed for their study.

In order to expose the giant axon itself, all the small fibres surrounding it have first to be removed. This is called "cleaning" the axon—one end of which is often left with the small fibres attached for handling purposes while the axon is prepared for the experiment. It usually takes about half-an-hour to clean an axon, and to leave it bare of all other tissue—often over a distance of about 5 cm. When finished, the giant axon alone remains—a single animal cell several centimetres in length and (in this case) about 650μm in diameter (each small division of the scale is 100μm).

A cell this size weighs about 20mg, so that it is not surprising that the giant axon was first used as a source of cytoplasm. In 1937 Bear, Schmitt & Young, working at Wood's Hole, studied the protein constituents from samples which they obtained simply by squeezing the cytoplasm out of the cut end. Later workers (e.g. Webb & Young, 1940; Steinbach & Spiegelmann, 1943) made precise measurements of the electrolytes and found that the concentration of potassium was much higher, and that of sodium much lower, than in the surrounding body fluids.

It was for studies of the viscosity of the cytoplasm that a cannula was first inserted into the axon—and it has since proved such a valuable tool that it is now used routinely. A thread is attached at each end of the axon and pinned down to hold it taut. In addition, two threads, which will later secure the cannula, are slackly tied round it. The cannula itself is a carefully drawn piece of glass tubing, the end of which has been ground down to form a smooth, oblique, tip—slightly smaller than the diameter of the axon. The next step is to cut—without severing the axon—a notch in its wall large enough to accommodate the cannula. A micro-manipulator provides the fine adjustment needed to pass the cannula through the notch and some distance along the axon. Once in place, the loop of thread is brought up and the cannula secured. Finally, the threads are trimmed, and the uncleaned end of the axon removed. This is the preparation with which the basic mechanism of nervous conduction has been studied.

An electrode was first placed inside the axon by Hodgkin & Huxley in 1939 at Plymouth, and by Curtis & Cole in 1940 at Woods Hole. Hodgkin & Huxley first made a plastic cell, and mounted it on a platform, which could be raised and lowered like a lift. As Professor Baker shows, the axon, held by the cannula, was then hung in the cell, which was filled with sea-water and connected to an external electrode. The internal electrode—which was not attached to the lift—was then placed vertically above the axon, with its tip in the cannula. It was centred, with respect to the cannula and axon, by placing a small mirror beside the axon, arranged so that a second image, at right angles to the first, was seen through the microscope. By adjusting the position of the cell—both horizontally and vertically—the cannula and axon were now raised up, over the tip of the electrode—which always remained in the same line of sight. Hodgkin & Huxley found that, as the electrode entered the axon, a negative potential (with respect to the external sea-water) of about 65 mV was obtained. This was the resting potential of the axon and, although its existence had long been suspected, this was the first time it had been directly measured. Moreover, when the axon was stimulated the potential did not simply fall to zero during the impulse but became positive with respect to the outside—shown by the overshoot of the action potential. This important discovery suggested that the nerve membrane which, at rest, is mainly permeable to potassium, becomes primarily permeable to another ion during excitation. This other ion is sodium; since, if its concentration in the external solution was lowered, the action potential immediately became smaller—by an amount depending on the sodium concentration.

If (as these experiments suggest) the action potential was dependent on the passage of ions across the membrane, it was obviously important to measure the currents carried by these ions. To do this it is necessary to hold the membrane potential at a chosen value—this is the powerful voltage-clamp technique, originally developed by Cole (1949) in America, and applied by Hodgkin, Huxley & Katz (1952). The technique required that an extra electrode—the current electrode—be inserted into the axon. For this purpose a double electrode was made by winding very fine silver wires around a thin glass capillary. As the wire, which was only 20μm in diameter, was wound on it was kept taut (as Sir Alan Hodgkin shows) by dangling a small piece of plasticine on the free end. The finished electrode consists of two entirely separate spirals—insulated where necessary, by shellac varnish. While it is inserted in the same way as the simple electrode, the information it gives is quite different. As a change of potential is imposed on the axon, as seen in the top trace, the currents flowing across the membrane are revealed in the bottom trace. As the membrane is clamped by successively larger voltage pulses, so the direction of the currents across the membrane changes. The early downward dip, seen on the left, is the transient current carried by the influx of sodium ions; and it is superseded by an opposite and persistent current, attributed to the outward flow of potassium ions. This suggestion was later confirmed by changing the potassium concentration within the fibre.

In 1956, Hodgkin & Keynes made a micro-injector which could be inserted down the axon, in the same way as the simple electrode. During tests, repeated here by Professor Keynes, the fine glass capillary contained a column of dye a few millimetres long. The syringe was mounted so that, during injection, the barrel moved over the stationary plunger—thus the capillary withdraws, leaving the injected solution behind. Of course, during an experiment the solution was colourless—so its limits were marked by two small bubbles. This technique has proved particularly useful when used with radioactive isotopes. As Dr Caldwell demonstrates, the external solution was sampled at known times after the injection of the isotope, and its activity measured. These experiments showed the presence of a "sodium pump". This maintains the difference between the internal and external sodium concentrations, and is driven by the breakdown of energy-rich phosphate (Caldwell, Hodgkin, Keynes & Shaw, 1960). But the micro-injection technique could only add substances to the axoplasm; there was no opportunity of removing them.

In 1961, Baker, Hodgkin & Shaw succeeded in doing just this (Baker, Hodgkin & Shaw, 1961, 1962). In this short sequence (made at the time) Professor Shaw shows how, with a device rather like a miniature garden roller, the axoplasm was gradually squeezed out. An uncleaned axon was used, so that the small fibres surrounding it would give the membrane some protection. It was then possible to reinflate the axon, by forcing an artificial solution through it—using (in these early experiments) an old gramophone motor connected, through gears, to a caliper syringe. During the rolling out process, some of the axoplasm is forced back, up into the tip of the cannula. As the pressure builds up, this is forced out of the cannula and travels down the length of the axon, as a small plug, until finally ejected. The lumen of the axon is now completely clear; and solutions flow through it with ease. Previously, such an isolated membrane was thought to be completely dead. However, the striking fact was that this supposedly lifeless tissue still displayed all the essential behaviour of a living nerve. It could still conduct up to half-a-million impulses—even though large volumes of artificial solutions were streamed through it. This cannulated and perfused axon can, of course, be treated in exactly the same way as an intact one (Shaw, 1966). A simple internal electrode will, as before, give the resting potential (here about 80 mV). But now, unlike the earlier experiments, it is a simple matter to change the internal solution. If the new solution—which rapidly displaces the old—contains little potassium, a sudden change occurs. The resting potential drops to a small fraction of its former value. It is equally simple to replace the original solution—and to restore the resting potential. The resting potential, therefore, is dependent upon the high internal concentration of potassium ions. This surprising technique, then, has been extremely successful.

At about the same time, Doctor Tasaki and his colleagues (Oikawa, Spyropoulos, Tasaki & Teorelli, 1961) found that they could drill out a central core of axoplasm. Their technique, which is very different to those previously used at Plymouth, is rather less drastic, and suitable for smaller fibres. A cleaned axon is laid out horizontally in a pool of liquid and held in place by a thread attached at each end. When the position of the axon is correct the usual small notch is cut in it to receive the cannula. In this method the cannula has to be guided along the whole length of the axon, without damaging its membrane, and the double-image technique is again used, but with a prism instead of a mirror. The cannula itself is a long, thin, hollow, glass tube which (during its passage along the axon) removes the central core of axoplasm. This is achieved by the experimenter (here John Kimura) gently sucking on the cannula so that, as it passes down the axon, axoplasm is drawn into it. Another notch is cut in the far end and the cannula passed through it. A positive pressure is now applied to the perfusion liquid within the cannula, and the viscous axoplasm rinsed out. The internal electrode, which is on exactly the same axis, is now brought up and its tip inserted into the end of the cannula. While the electrode is within the cannula both are moved, in unison, to the middle of the axon. Here the internal electrode remains, but the cannula is brought back to its starting point and secured with thread. All that remains now is to remove the prism, and to bring into position the external electrodes. The particular value of this elaborate and powerful technique is that it permits a cleaned axon to be perfused. It is one of the many techniques which physiologists have used since the importance of research on the squid's giant axon was first recognized.

Many scientists have contributed their results, which have greatly increased our understanding of the basic mechanism of nervous conduction in all animals—including man—and have provided the basis of modern neurophysiology (*vide* Hodgkin, 1964; Tasaki, 1968).

But, as the winter days draw in, the Laboratory's ships still return with their cargoes of squid, for much has yet to be learnt. How, for instance, does the membrane distinguish between different ions? And what is the mechanism which brings about the changes in permeability?

Small wonder then, that in fair weather or foul, the work of the ships—and their crews—continues every winter.

Small wonder then, that each day's haul is eagerly awaited, that results are discussed and experiments made, often far into the night.

Small wonder then, that at Marine Laboratories like those at Plymouth in England and Woods Hole in America—the squid is a very important animal.

Acknowledgements

The following scientists planned and demonstrated the observations and experiments on which this film depends. Their generous help and encouragement is gratefully acknowledged. Professor P. F. Baker FRS, Dr P. C. Caldwell FRS, Sir Alan L. Hodgkin OM, KBE, PRS, Professor R. D. Keynes FRS, J. E. Kimura, Dr P. McNaughton, Dr H. Meves, Professor T. I. Shaw FRS, Professor J. Z. Young FRS. Filmed, with the help of the Director and Staff, at the Plymouth Laboratory of the Marine Biological Association.

References

Baker, P. F., Hodgkin, A. L. & Shaw, T. I. (1961). Replacement of the protoplasm of a giant nerve fibre with artificial solutions. *Nature, Lond.* **190**: 885–887.

Baker, P. F., Hodgkin, A. L. & Shaw, T. I. (1962). Replacement of the axoplasm of giant nerve fibres with artificial solutions. *J. Physiol., Lond.* **164**: 330–354.

Bear, H. B., Schmitt, F. O. & Young, J. Z. (1937). Investigations on the protein constituents of nerve axoplasm. *Proc. R. Soc. (B.)* **123**: 520–529.

Caldwell, P. C., Hodgkin, A. L., Keynes, R. D. & Shaw, T. I. (1960). The effects of injecting "energy-rich" phosphate compounds on the active transport of ions in the giant axons of *Loligo. J. Physiol., Lond.* **152**: 561–590.

Cole, K. S. (1949). Dynamic electrical characteristics of the squid axon membrane. *Archs Sci. physiol.* **3**: 253–258.

Curtis, H. J. & Cole, K. S. (1940). Membrane action potentials from the squid giant axon. *J. cell. comp. Physiol.* **15**: 147–157.

Hodgkin, A. L. (1964). *The conduction of the nervous impulse.* Liverpool: University Press.

Hodgkin, A. L. & Huxley, A. F. (1939). Action potentials recorded from inside a nerve fibre. *Nature, Lond.* **144**: 710–711.

Hodgkin, A. L., Huxley, A. F. & Katz, B. (1952). Measurement of current-voltage relations in the membrane of the giant axon of *Loligo. J. Physiol., Lond.* **116**: 424–448.

Hodgkin, A. L. & Keynes, R. D. (1956). Experiments on the injection of substances into squid giant axons by means of a microsyringe. *J. Physiol., Lond.* **131**: 592–616.

Oikawa, T., Spyropoulos, C. S., Tasaki, I. & Teorelli, T. (1961). Methods for perfusing the giant axon of *Loligo pealii. Acta physiol. scand.* **52**: 195–196.

Pumphrey, R. J. & Young, J. Z. (1938). The rates of conduction of nerve fibres of various diameters in cephalopods. *J. exp. Biol.* **15**: 453–466.

Shaw, T. I. (1966). New aspects of nerve behaviour. *Discovery, Lond.* **27**(3): 38–43.

Steinbach, H. B. & Spiegelmann, S. (1943). The sodium and potassium balance in squid nerve axoplasm. *J. cell comp. Physiol.* **22**: 187–196.

Tasaki, I. (1968). *Nerve excitation.* Illinois: Charles C. Thomas.

Trueman, E. R. & Packard, A. (1968). Motor performances of some cephalopods. *J. exp. Biol.* **49**: 495–507.

Webb, D. A. & Young, J. Z. (1940). Electrolytic content and action potential of the giant nerve fibres of *Loligo. J. Physiol., Lond.* **98**: 299–313.

Young, J. Z. (1936). The giant nerve fibres and epistellar body of cephalopods. *Q. Jl microsc. Sci.* **78**: 367–386.
Young, J. Z. (1938). The functioning of the giant nerve fibres of the squid. *J. exp. biol.* **15**: 170–185.
Young, J. Z. (1944). Giant nerve fibres. *Endeavour* **3**: 108–113.

Note Added in Proof

The film of which the commentary is given in this chapter is now available. Title: *The squid and its giant nerve fibre.* Scientific Advisor: Hans Meves. Written, photographed and produced by J. B. Gilpin-Brown. British Film Institute classification: 591·18. 28 mins., sd., col., 16 mm. Prints available from: Dr J. B. Gilpin-Brown, The Laboratory, Citadel Hill, Plymouth.

Symp. zool. Soc. Lond. (1977) No. 38, 243–260.

THE SQUID GIANT AXON: A MODEL FOR STUDIES OF NEURONAL CALCIUM METABOLISM

P. F. BAKER

Department of Physiology, King's College, London, England

SYNOPSIS

The squid giant axon has proved an ideal preparation for studying the basic mechanisms underlying nervous function. The analysis of the mechanism of nervous conduction by Bear, Schmitt and Young, Cole and Marmont, Hodgkin, Huxley, Katz and Keynes and many others now occupies a central position in the classical framework of neurobiology, and principles elucidated from experiments on the squid giant axon have proved of general significance and to be applicable to all nerve cells. More recently, the squid giant axon has been used as a model for studying mechanisms that may be relevant to synaptic function. These studies can be related very directly to the work of Katz and Miledi on the squid giant synapse; but as with studies of nervous conduction, they seem to be of general relevance.

These studies have centred on the transport and metabolism of calcium in the squid giant axon. Calcium is known to be required for transmitter release and studies on the giant axon have shown how the concentration of ionized calcium inside the nerve is regulated and have permitted characterization of the mechanisms responsible for entry and exit of calcium across the surface membrane. Extruded axoplasm contains about 500 μM calcium, the bulk of which is bound. The intracellular ionized calcium is less than 1 μM, whereas that in the external solution is 10 000 μM. Calcium can enter the cell by a variety of routes including two that open and close in response to changes in membrane potential. Depolarizing the cell permits calcium ions to enter and it seems to be calcium entry that evokes the release of transmitter. Calcium that enters is ultimately removed from the cell by calcium pumping systems located in the surface membrane. Alterations in either intracellular binding or membrane pumping can lead to changes in intracellular calcium and it seems likely that the level of intracellular ionized calcium determines the resting, so-called "spontaneous", rate of transmitter release.

INTRODUCTION

As John Gilpin-Brown's magnificent film so clearly showed at the Symposium (see previous chapter), the squid giant axon is peculiarly suitable for studying many aspects of cell physiology, in particular those that relate to the conduction of the nervous impulse. Every neurophysiologist owes a debt of gratitude to J. Z. Young for bringing this extraordinary preparation to the attention of physiologists. The large size of the axon has enabled investigators to probe the mechanism of nervous conduction in ever increasing detail—carrying out experiments that would be extremely difficult, if not impossible, on nerves of more conventional diameter. Fortunately, results obtained with squid axons seem to have a general relevance to nervous function in other phyla and the mechanism of nervous conduction that was worked out on the squid giant axon by Bear, Schmitt and Young, Cole, Curtis and Marmont, Hodgkin, Huxley, Katz, Keynes and many others is now a corner stone of modern neurobiology. For a description of this classical work, the reader should consult the excellent monographs of Hodgkin (1964) and Katz (1966).

More recently attention has shifted from the mechanism of conduction to the contribution that the squid axon and its giant synapse can make to our understanding of the mechanism of transmission of the nerve impulse from one cell to another. As Katz and Miledi, in particular, have shown, transmission at the squid giant synapse, as at most other chemical synapses, involves the release of specific chemicals from the nerve terminals in a reaction that requires calcium ions. Studies of the metabolism of Ca in the squid giant axon have provided much new information that seems to have considerable relevance to the control of transmitter release and neurosecretory processes in general (Baker, 1972). It is these experiments on the metabolism of calcium and other divalent cations in squid axons that I would like to discuss in this paper.

THE CALCIUM GRADIENT

As Bear, Schmitt & Young showed in 1937, it is a fairly simple matter to obtain a relatively pure sample of protoplasm from an intact giant axon. All that is necessary is to cut open one end of the axon and gentle squeeze the axoplasm out of the cut end rather like squeezing toothpaste out of a tube. In order to avoid contamination of the extruded axoplasm, it is usual to clean the cut end of the giant axon of adhering small nerve fibres before extruding the axoplasm on to a clean microscope coverslip or piece of parafilm. Axoplasm obtained in this way has the consistency of a stiff gel, and examination in the electron microscope shows it to be free of surface membrane (axolemmal) contamination. Provided the extruded axoplasm is protected from drying out, it can retain its rod-like appearance and normal ATP content for many hours (Baker & Saw, 1965).

It is instructive to compare the Ca and Mg contents of extruded axoplasm with that of the solution in which the nerve is usually immersed. Measurement of axoplasmic Ca and Mg by atomic absorption spectroscopy shows the interior of the cell to contain much less Ca and Mg than squid blood. Axoplasm contains about 0·5 mmol Ca/kg and about 7 mmol Mg/kg whereas the concentration of Ca and Mg in squid blood is about 10 mM and 55 mM respectively.

Atomic absorption measures the total Ca and Mg contents of axoplasm; but it provides no information on what fraction of the axoplasmic Ca and Mg is bound and what fraction is ionized. Information on the state of axoplasmic Ca and Mg can be obtained by introducing a small patch of radioactive Ca and Mg into the axon by micro-injection. The extent to which the patch broadens provides information on the ease with which the ions can diffuse in axoplasm. Ca and Mg behave quite differently in this kind of experiment (Fig. 1). ^{45}Ca remains at the site of injection whereas

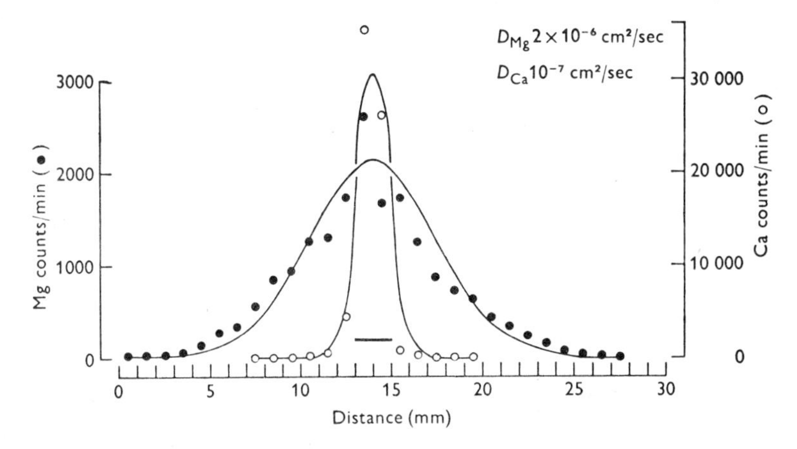

FIG. 1. Self-diffusion of ^{28}Mg and ^{45}Ca in axoplasm. Squid giant axon injected with 2 mm ^{28}Mg, ^{45}Ca and left in artificial sea-water for 510 min at 21°C. The injector entered from the left-hand side and it is likely that some of the tracer moved up the track left behind as the injector was withdrawn. Axon diameter 750 μm. The diffusion constants listed were used to calculate the smooth curves. (From Baker & Crawford, 1972.)

the patch of ^{28}Mg broadens considerably. The data suggest that the bulk of the injected ^{45}Ca is rapidly bound whereas about one-half to one-third of the ^{28}Mg is free to diffuse. If the divalent cations in axoplasm are ionized, they should move towards the cathode if the axon is subjected to a longitudinal electric field. Again there is a marked difference in the behaviour of ^{45}Ca and ^{28}Mg in this kind of experiment (Fig. 2). The patch of injected Ca remains at the site of injection whereas the patch of Mg moves towards the cathode, broadening as it does so. This experiment indicates that the concentration of ionized Ca in axoplasm is probably very low (less than

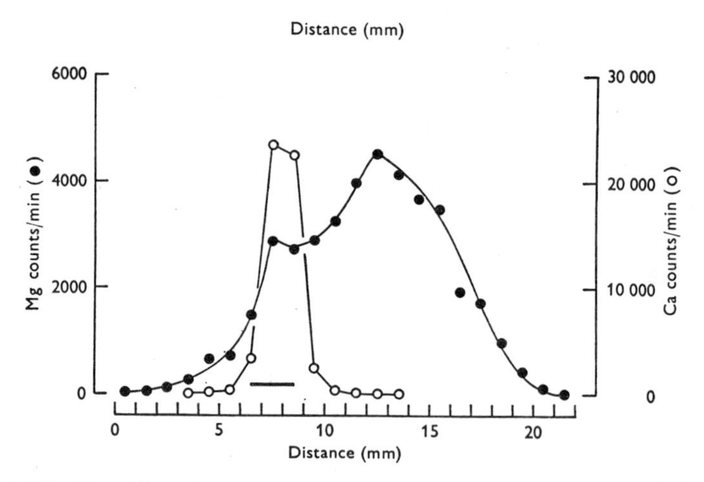

FIG. 2. Mobility of Ca and Mg in axoplasm. The injector was inserted from the left-hand side. The short horizontal bar marks the site of injection. 1·025 V was applied over 4·3 cm for 240 min. Temp. 19°C. Axon diameter 790 μm. If the loss of ^{28}Mg from the axon is neglected and if Mg has the same mobility in axoplasm as in free solution, the peak should have moved 13·7 mm. (From Baker & Crawford, 1972.)

10 μM) whereas one-half to one-third the Mg in axoplasm seems to be ionized.

Another more sensitive technique is obviously needed to determine the concentration of ionized calcium. The experimental solution to this problem is to be found in many coelenterates! A number of animals in this phylum contain proteins that react with Ca ions to generate light. The first example of this class of Ca-sensitive proteins was isolated in 1963 by Shimomura, Johnson & Saiga from the jelly-fish *Aequorea forskalea* and has been called aequorin. Analogous substances are present in other species; for example *Obelia* yields the protein obelin (Morin & Hastings, 1971; Campbell, 1974). The great value of these so-called photoproteins is that they react with Ca ions to generate light. No other cofactors are required and they respond to Ca concentrations as low as 10^{-8} M.

Aequorin can be purified and injected into a squid axon where it acts as a reporter molecule for the level of ionized calcium. In a resting axon, the rate of light emission is very low, which is fully consistent with the conclusions drawn from the diffusion and mobility experiments already described. In order to say how low, it is necessary to calibrate the light output from the axon and this can be done by introducing into the axon a series of known calcium concentrations. As very low calcium concentrations are involved it is necessary to use calcium buffers that can maintain within the axon a stable, low concentration of ionized calcium. The best calcium buffer is the substance EGTA and by injecting various Ca-EGTA buffers it is possible to find a buffer that causes no change in light emission. By arguing that the ionized Ca stabilized by this buffer must be the same or very close to that which normally exists inside the axon, it is a relatively simple matter to estimate the intracellular concentration of ionized calcium. The value obtained is very low: 0·1–0·3 μM. The squid axon is not unique in this respect; and wherever measurements of intracellular Ca have proved possible, the value has always been less than 1 μM.

As the Ca concentration that usually bathes squid axons is about 10 mM or 10 000 μM, there is an inwardly directed calcium gradient across the axon membrane of at least 10 000 : 1. In absolute terms, as the intracellular concentration of ionized Ca is so low, a very small increment in Ca in the cytosol can cause a large percentage change in ionized calcium. For instance, the additon of 1 μM Ca will more than double the concentration of ionized Ca in the cytosol, whereas to achieve a similar alteration in Na would require the addition of about 30 000 μmol of sodium. This ease with which Ca_i can be varied underlies the well known importance of changes in intracellular Ca as a trigger for many aspects of cell behaviour and function, including muscular contraction, secretion, membrane permeability, cell division and the control of locomotion in *Paramecium* (see Baker & Glitsch, 1975).

MAINTENANCE OF THE CALCIUM GRADIENT

Having established the existence of a very steep inward gradient of calcium ions, it is necessary to enquire how this gradient is maintained. Chemical gradients do not arise spontaneously: they always reflect the existence of some other driving force providing a source of energy. One such possibility in a nerve cell would be a difference in potential difference, but to achieve the observed chemical gradient of calcium the interior of the axon would have to be positive with respect to the exterior by at least 120 mV. In fact the opposite is true: the interior of the axon is negative with respect to the outside by about 60 mV. It follows that the low intracellular concentrations of ionized calcium are maintained against both a steep chemical and adverse electrical gradient. As the axolemma is permeable to Ca ions, there is no reason why the Ca gradient should persist unless mechanisms exist in the cell for expelling Ca against both a chemical and electrical gradient.

Such mechanisms, called calcium pumps, do exist as we shall see; but the regulation of intracellular Ca is complicated by the existence within the cell of a number of Ca-binding systems, the evidence for which has already been mentioned. These intracellular binding systems serve to buffer the ionized Ca in the cytosol. Less than 0·2% of the total intracellular Ca is ionized and the rest is bound to these intracellular buffer systems. In the short-term regulation of intracellular ionized Ca these systems must be very important, for instance in switching off contraction or secretion after a period of activity; but in the long term, the cells do not continue to gain Ca, which must mean that any calcium that enters the cell is ultimately expelled across the surface membrane by some form of calcium pump.

Even if the intracellular binding systems are not in the long term responsible for setting up the steep calcium gradient, they undoubtedly have an important short-term regulatory function and it is important to know what kinds of systems are involved and what are their respective properties.

Once again, the squid is an ideal preparation. It is possible to follow changes in Ca binding in intact axons by the use of intracellular aequorin. In a normal resting axon the ionized Ca is very low: but application of metabolic poisons such as cyanide or dinitrophenol results in a dramatic increase in light emission indicating that intracellular Ca binding requires energy. The only recognized Ca-binding systems in nerve are mitochondria, and Ca accumulated in mitochondria is known to be released following treatment with cyanide and dinitrophenol. It follows that the mitochondria are the most likely source of the extra ionized Ca seen in poisoned axons (Baker, Hodgkin & Ridgway, 1971).

Rather more direct evidence can be obtained using isolated axoplasm. Axoplasm extruded from squid axons can be drawn into a glass capillary tube of diameter roughly equal to that of the original axon. Axoplasm treated in this way can maintain an ATP concentration in the normal range for many hours, and injection of aequorin reveals that the ionized Ca is close to its level in an intact axon. Subsequent injection of Ca only produces a transient rise in light output, whereas application of cyanide or dinitrophenol leads to a maintained rise in ionized Ca. These observations suggest that the main intracellular Ca binding systems all survive extrusion. More information about these binding systems can be obtained by putting the isolated axoplasm in a small dialysis tube and measuring the accumulation of ^{45}Ca by axoplasm immersed in different media (Baker & Schlaepfer, 1975).

When axoplasm is exposed to an isotonic solution, rich in K ions, containing 10 μM Ca, the axoplasm accumulates calcium. Some accumulation occurs without any added substrate; but addition of succinate and P_i or ATP and P_i, both known to activate mitochondrial Ca binding, leads to a much increased uptake (Fig. 3). Careful analysis of the uptake confirms the presence of at least two components that can be characterized in operational terms: one that requires either ATP and P_i or succinate and P_i, which is blocked by DNP and has a rather low affinity for Ca, requiring about 30 μM Ca for half-maximal activation, and a second that persists in the absence of substrate, is not affected by concentrations of DNP that block mitochondrial uptake of Ca and has a very high affinity for Ca, the apparent K_m for Ca, $<0{\cdot}5$ μM (Baker & Schlaepfer, 1975). This data is entirely consistent with the existence in axoplasm of both mitochondrial and non-mitochondrial Ca-binding systems; but further evidence is needed to prove the point.

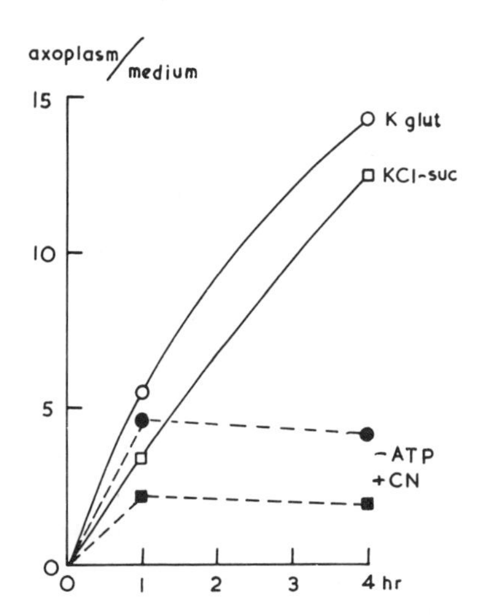

FIG. 3. Calcium accumulation by isolated axoplasm. Extruded axoplasm enclosed in small dialysis sacs was exposed to different isotonic "artificial axoplasms" containing 10 μM Ca labelled with ^{45}Ca either in the presence of ATP and P_i (solid lines) or in the absence of ATP and the presence of cyanide, a mitochondrial poison (dashed lines). K glut refers to an axoplasm containing 0·6 M K glutamate; and KCl-suc to one containing a mixture of 300 mM KCl and 600 mM sucrose. The ordinate shows the accumulation of counts in the axoplasm expressed relative to that in an equal weight of the medium; abscissa time in hours. Temp., 21°C. (Unpublished data of Baker & Schlaepfer.)

Two particularly interesting observations are that Ca binding by isolated axoplasm is lower in media rich in sodium than in the more physiological potassium media and is markedly inhibited by the presence of other polyvalent cations including Mn^{2+} and La^{3+}. Bearing in mind the very low concentration of ionized Ca inside cells, these observations suggest that a small fall in the ratio of intracellular K/Na, or the presence of foreign polyvalent cations, may lead to a reduction in Ca binding and a concomitant rise in ionized calcium with attendant effects on any Ca-sensitive systems inside the cell.

To return to the long-term maintenance of the Ca gradient. As we have already seen, if the axon is not to gain Ca steadily, mechanisms must exist in the cell surface to expel Ca ions from the interior of the axon against a steep electrochemical gradient. The transport of calcium across the axolemma is most easily studied by the use of radioactive calcium (Hodgkin & Keynes, 1957). By including ^{45}Ca in the external medium and subsequently extruding and counting the axoplasm it is possible to measure the unidirectional influx of ^{45}Ca, and by injecting ^{45}Ca into the interior of the axon and following its rate of appearance in the external solution it is possible to measure the unidirectional efflux of ^{45}Ca. An axon immersed in sea-water seems to be in a steady state and both Ca efflux and influx are roughly equal, the unidirectional fluxes amounting to 0·2 $pmol/cm^2 \ sec^{-1}$. Analysis of the dependence of these fluxes on the ionic environment of the axon and on its metabolic state provides information about the processes responsible for maintaining the very steep Ca gradient. Four particularly pertinent observations are:

(1) The Ca efflux is increased in fully poisoned axons (Blaustein & Hodgkin, 1969). This is surprising as under comparable conditions the Na efflux due to the Na pump is completely inhibited. The rise in Ca efflux comes about in part because the intracellular ionized Ca increases dramatically in poisoned axons because of release of Ca from presumed mitochondrial binding sites and in part because the extrusion of Ca does not seem to be as dependent on ATP as is the Na–K exchange pump.

(2) In both unpoisoned and poisoned axons part of the Ca efflux is dependent on the presence of Ca in the external medium (Fig. 4) and seems to reflect an exchange of internal Ca for external Ca. This has no obvious physiological significance; but its existence greatly complicates the analysis of flux data.

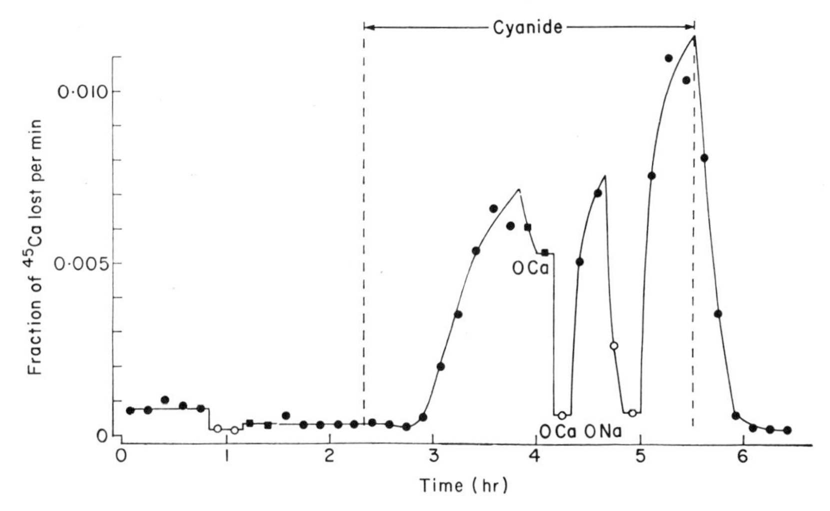

FIG. 4. Effect of external Ca and Na on the rate constant of Ca efflux from an intact squid axon. During the period indicated all solutions contained 2 mM cyanide. ●, Na-ASW; ■, Ca-free Na-ASW; ○, Ca-free, Na-free, Li-ASW. Axon diameter 800 μm. Temp., 20°C. (Unpublished data of Baker & Crawford.)

(3) In both unpoisoned and poisoned axons part of the Ca efflux is dependent on the presence of Na ions in the external medium (Baker, Blaustein, Hodgkin & Steinhardt, 1967). K, Li, Tris, Mg and choline cannot replace Na. In Na-free media, the Ca influx is increased and Ca enters the axon in exchange for internal Na. Under these conditions Ca influx is roughly proportional to the square of the internal sodium concentration.

(4) In both unpoisoned and poisoned axons Ca efflux and influx are unaffected by concentrations of cardiac glycosides that completely inhibit the Na–K exchange pump.

These observations suggest that calcium gradients may be generated by exchange of Na ions for Ca ions and only be dependent on ATP indirectly for the maintenance of the Na gradient by the ATP-dependent Na–K exchange pump. According to this model, under physiological conditions Na ions enter the axon in exchange for intracellular Ca ions. Lowering Na_o should reduce Ca efflux and favour reversal of the exchange, calcium ions now entering the axon in exchange for intracellular sodium. This reversal of Na–Ca exchange is observed (see Fig. 5). The Ca gradient that can be generated will depend on the number of Na ions that exchange for one calcium. If exchange is electroneutral

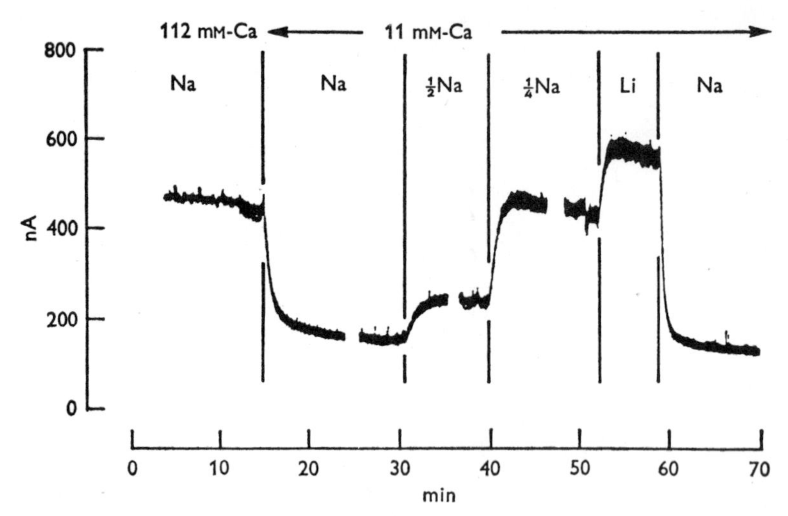

FIG. 5. Changes in ionized calcium in squid giant axon in response to alteration in the external Ca and Na concentrations. The axon was injected with aequorin and an increase in light output (ordinate) reflects a rise in intracellular ionized Ca. Axon diameter 700 μm. Temp., 20°C. (From Baker, Hodgkin & Ridgway, 1971.)

with two Na ions exchanging for one Ca, the Ca gradient will approach

$$\frac{[Ca]_i}{[Ca]_o} = \frac{[Na]_i^2}{[Na]_o^2}.$$

As Na_i/Na_o is roughly 1/10, the Ca gradient should approach 1/100. As the Ca concentration external to a squid axon is probably in the region of 10 000 μM, electroneutral exchange could only give Ca_i of 100 μM which is much higher than is indicated by experiments with aequorin. In order to approach the observed Ca_i^{2+}, it is necessary for three Na ions to exchange for each Ca ion. Under these conditions either exchange is no longer electroneutral and one Na ion enters the axon down both a chemical and electrical gradient in which case the Ca gradient will approach

$$\frac{[Ca]_i}{[Ca]_o} = \frac{[Na]_i^3}{[Na]_o^3} e^{VF/RT}$$

or electroneutrality is maintained by exchange of three Na ions for one K ion plus one Ca ion in which case the gradient will approach

$$\frac{[Ca]_i}{[Ca]_o} = \frac{[Na]_i^3}{[Na]_o^3} \frac{[K]_o}{[K]_i}.$$

The Na-dependent Ca efflux is reduced in depolarized axons (Brinley & Mullins, 1975), which suggests that Na–Ca counter transport may not be electroneutral and is consistent with the entry of three Na ions for each calcium ion. Clear proof requires that a Ca_i-dependent Na entry be demonstrated and that the stoichiometry of this entry be shown to be 3 Na : 1 Ca. The small size of the fluxes makes this a very difficult point to settle experimentally.

Another approach is to look at the kinetics of the Ca efflux. If more than one Na ion enters the axon in exchange for each Ca that leaves, the form of the activation of the Ca efflux by external Na should be sigmoidal. This is quite definitely the case in poisoned axons, but is not obviously so in unpoisoned nerves (Baker & Glitsch, 1973). At this point we begin to touch on an extremely complex and as yet ill understood feature of the Ca efflux system which I cannot examine in any detail here. But there is growing evidence that the kinetic properties of the Ca efflux system depend on the metabolic state of the cell. In unpoisoned axons the affinity of the Ca_o-dependent Ca efflux for external Ca is much higher than in poisoned axons, and a similar pattern is found for the affinity of the Na_o-dependent Ca efflux for external Na. Provided Ca_i is kept constant, these changes can be reversed in poisoned axons by injecting ATP but not AMP or cAMP (Baker & McNaughton, unpublished data). The emerging picture is that although the energy for Ca efflux may be derived from the inward gradient of Na, the affinity with which the Ca efflux mechanism is coupled into the Na gradient may be dependent on ATP or some derivative of it. The precise role of ATP in this system is difficult to elucidate in an intact axon; but a new technique developed by Brinley & Mullins (1967) for the internal dialysis of squid axons may facilitate further analysis (Di Polo, 1973, 1974).

In view of the dependence of transmitter release on calcium and the growing evidence that this Ca is most probably required at an intracellular site, it is of interest to find that conditions associated with a rise in Ca, either through changes in axoplasmic binding or membrane pumping, all lead to an increased spontaneous release of transmitter from nerve terminals (see Baker, 1974). The release of transmitter evoked by a nerve impulse is absolutely dependent on external Ca and seems to require Ca entry into the cell. Again the squid giant axon provides a convenient model because it possesses voltage-sensitive Ca entry systems that have many features in common with the Ca entry that is presumed to evoke release of transmitter at the nerve terminal.

VOLTAGE-SENSITIVE ENTRY OF CALCIUM

The permeability of nerve cell membranes to calcium is not constant but can alter in response to changes in membrane potential, and the squid giant axon provides an ideal preparation for studying the properties of these voltage-sensitive changes in permeability. An increased uptake of calcium during nervous activity was first reported by Hodgkin & Keynes (1957). They measured the influx of ^{45}Ca into pairs of squid axons and it was necessary to cause one axon of the pair to carry a large number of impulses in order to obtain a measurably greater Ca uptake in stimulated axons as compared to their unstimulated controls. Nevertheless, the results show quite clearly that nervous activity increases calcium uptake and this uptake increases in a roughly linear manner with increasing external Ca concentrations. Stimulation had no effect on the efflux of ^{45}Ca. In an axon immersed in sea-water at 20°C, the extra entry of calcium averaged 0.01 pmol/cm^2 impulse or about 1/400 of the net entry of Na under the same conditions.

Intracellular aequorin provides a much more sensitive method for detecting calcium entry and under suitable conditions it is possible to detect a rise in internal Ca during a single nerve impulse (Baker, Hodgkin & Ridgway, 1971; Hallett & Carbone, 1972; Stinnakre & Tauc, 1973). A feature of the records that are obtained with aequorin is that the light output from the axon increases during stimulation; but falls back close to its initial value when stimulation ceases (Fig. 6). The increased light output seems to result from entry of calcium into the cell because it is not seen when axons conduct impulses in solutions lacking Ca ions. The return after stimulation to the initial rate of light emission probably reflects uptake of Ca by intracellular binding systems.

Depolarization is known to open channels through which sodium and potassium ions can cross the nerve cell membrane, the well-known Na and K channels that underlie the action potential, and it is pertinent to enquire whether the observed entry of Ca occurs through either or both of these channels or whether Ca enters by a separate route. The aequorin reaction is too slow to enable Ca entry to be followed directly: but by measuring the light output from intracellular aequorin in response to voltage clamp pulses of different duration, Baker, Hodgkin & Ridgway (1971) were able to demonstrate two phases of Ca entry: an early phase that occurred at roughly the same time as the increase in sodium conductance and a later phase that roughly paralleled the increase

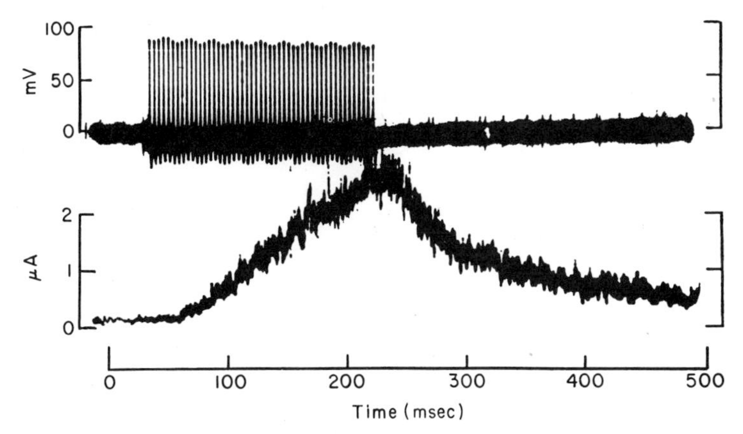

FIG. 6. Increase in intracellular ionized Ca in response to a brief period of repetitive stimulation. Upper trace, action potentials; lower trace, light output from injected aequorin. The light output was detected by a photomultiplier and displayed in an oscilloscope. An increase in light reflects a rise in ionized Ca. The axon was immersed in Na-ASW containing 112 mM Ca. Temp., 20°C. (Unpublished data of Baker, Hodgkin & Ridgway.)

in potassium conductance (Fig. 7). Although the time course is consistent with Ca entry occurring through both the Na and K channels, other evidence suggests that this explanation is only

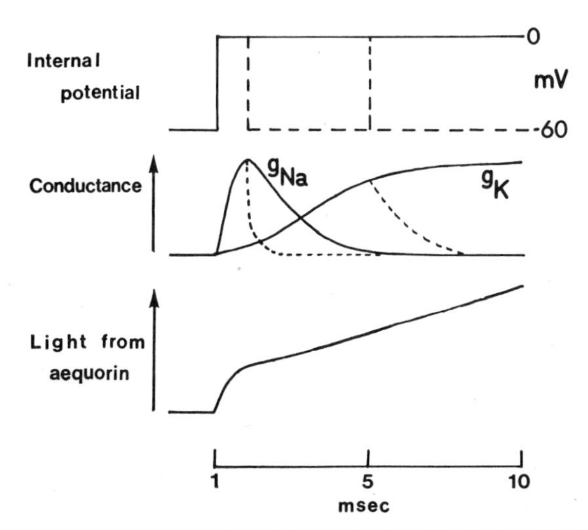

FIG. 7. Diagrammatic representation of the method used by Baker, Hodgkin & Ridgway (1971) to elucidate the time course of Ca entry in response to depolarization. Changes in the light output from intracellular aequorin are shown in response to repetitive depolarizations from −60 mV to zero potential for different lengths of time. The time courses of the Na and K conductances in response to a maintained depolarization of the same magnitude are shown for comparison.

partly correct. This evidence rests heavily on pharmacological methods of detecting the calcium entry. The sodium conductance is blocked very specifically by low concentrations of tetrodotoxin (TTX) and application of TTX in concentrations adequate to block the Na conductance also blocks the early phase of Ca entry suggesting that this component of uptake probably reflects Ca entry through the sodium channels. The late phase of Ca entry cannot be explained so easily. Agents such as tetraethylammoniun ions (TEA) that block the K channel have no effect on the late phase of Ca entry whereas various divalent cations including Mg, Mn, Ni and Co, and the organic calcium antagonists iproveratril and D600, block the late phase of calcium entry without affecting the K channels appreciably. These experiments are most consistent with the existence of a separate Ca channel which has been called the "late Ca channel" to distinguish it from the early TTX-sensitive entry of Ca through the Na channel. Another difference between the late Ca channel and the K channel is their response to maintained depolarization. Both channels ultimately cease to conduct ions, a process known as inactivation, but the time courses of onset and especially of recovery from inactivation are quite different.

The main criticism of these conclusions is that they rely entirely on the use of aequorin and it would be more satisfactory if they could be corroborated by some other technique. It might, for instance, be possible to detect a Ca current associated with Ca entry. This has not so far proved possible in intact squid giant axons, although under certain conditions calcium action potentials can be detected at the squid giant synapse and in a variety of other nerve cell preparations. These action potentials are blocked or reduced by Co, Mn and D600, and calcium entry seems to take place through a channel that exhibits many of the properties of the "late Ca channel" (Reuter, 1973). The only Ca currents that have been characterized in the squid giant axon have been studied in perfused axons that have been specially treated to minimize the currents of Na and K. Under these conditions Meves & Vogel (1973) were able to detect a small inward current carried by Ca ions. This inward current showed many features in common with the Na channel including being blocked by TTX and it seems most likely that this current reflects the early phase of Ca entry. Failure to detect any current associated with the late phase may simply reflect the fact that Ca entry during the second phase occurs over a much longer period of time than the early phase. This would make detection of a "late Ca current" very difficult. It is of course possible that the late entry of Ca may occur in an electroneutral fashion; but this seems unlikely in view of the clear evidence in other tissues for a Ca current with remarkably similar properties to the late phase of Ca entry.

Some independent evidence for the late Ca channel was obtained using a combined voltage clamp electrode and internal glass scintillator (Baker, Cook & Glitsch, 1974). The principle of this method is to thread down the long axis of the axon a 100–150 μM diameter rod of scintillation glass. Immersion of the axon in sea-water containing ^{45}Ca produces very few scintillations until ^{45}Ca begins to enter the axon when the counts from the scintillator begin to rise in a linear fashion. By winding around the scintillator wires for passing current and measuring potential, it is possible to monitor ^{45}Ca uptake under voltage clamp conditions. Although the technique is very insensitive, it does show increased ^{45}Ca uptake in response to trains of depolarizing pulses and the increased uptake persists in the presence of enough TTX to block the increase in Na conductance. Hyperpolarizing pulses gave no extra uptake of ^{45}Ca.

Taken together these results suggest that in response to depolarization some Ca enters through the TTX-sensitive Na channel: but the rest enters through a channel that is insensitive to both TTX and TEA and seems to be separate from the well known Na and K channels of the action potential. The relative amount of Ca entering through the Na channel and "late Ca channel" varies widely. In some axons it is about equal while in others Ca entry through the "late Ca channel" predominates.

The "late Ca channel" is particularly interesting because its properties parallel very closely those of the mechanism that evokes release of transmitter substances and neurosecretory products from nerve terminals. Thus transmitter release appears to depend on Ca entry and can be activated by iontophoresis of Ca into the nerve terminal (Miledi, 1973). Transmitter release in response to depolarization is unaffected by both TTX and TEA but is blocked by Mg, Mn, Co and D600. The insensitivity of transmitter release to TTX does not imply that Ca entry through the Na channel is of no significance in release; but merely that Ca entry can occur by a route that persists in the presence of TTX. This route has all the properties of the late Ca channel.

The only area where there is some dispute about whether the properties of the "late Ca channel" and transmitter release do alter in parallel is in response to maintained depolarization. Under these conditions the "late Ca channel" inactivates (Baker, Meves & Ridgway, 1973) but there is rather conflicting evidence on what happens to transmitter release mechanisms. We have recently obtained evidence that the most likely explanation for the transient release of catecholamines from the depolarized adrenal medulla is that depolarization promotes first activation followed by inactivation of Ca entry in this tissue (Baker & Rink, 1974). Both Maddrell & Gee (1974) and Nordmann (1975) have interpreted their results on an insect neurosecretory system and the rat neurohypophysis respectively in a similar way. The available evidence suggests that inactivation of Ca entry probably does occur in most secretory systems but at widely different rates.

A particularly interesting and as yet unresolved question concerns the effects of metabolism on the voltage-sensitive entry of calcium. It is usually assumed that the voltage-sensitive passive permeability systems are rather insensitive to changes in the metabolic state of the cell. Thus the action potential persists in fully-poisoned squid axons and these axons continue to conduct impulses when perfused with media lacking ATP or substrate. Although the Na and K permeability systems may be quite stable, there is growing evidence that the "late Ca channel" may not be. The evidence is still tenuous but it seems a strong possibility that the "late calcium channel" may be subject to metabolic control. Perhaps the clearest example is in the heart, where Ca entry during the action potential can be increased by noradrenalin (Reuter, 1974). The cardiac calcium channel shares many properties with the "late Ca channel" of nerve and if modification can occur in one tissue it is a very real possibility that it may occur in others. One possible example in the nervous system is the well-known rapid failure of many synapses following exposure to anoxia. In the squid giant synapse, synaptic transmission fails without loss of the presynaptic action potential and it also occurs with little change in the frequency of miniature potentials. This suggests that transmission has failed with minimal alteration in resting potential and ion concentrations. Prolonged exposure to anoxia results in a rise in the frequency of miniature potentials presumably as a result of the release of Ca from internal stores. The rapid failure of evoked release is difficult to explain but could be rationalized if Ca entry into the presynaptic terminal is extremely sensitive to the metabolic state of the cell. As transmitter release depends on a high power of the Ca concentration, a relatively small reduction in Ca entry would result in a much larger reduction in transmitter release.

Regulation of calcium permeability systems would seem to offer considerable scope for effecting fine control of a number of aspects of cell behaviour and function, and further investigation of this possibility in both the squid giant axon and giant synapse may prove profitable.

Acknowledgement

I wish to thank the Medical Research Council for supporting this work.

References

Baker, P. F. (1972). Transport and metabolism of calcium ions in nerve. *Prog. Biophys. Molec. Biol.* **24**: 177–223.

Baker, P. F. (1974). Excitation-secretion coupling. In *Recent advances in physiology*: 51–86. Linden, R. J. (ed.). Edinburgh & London: Churchill Livingstone.

Baker, P. F., Blaustein, M. P., Hodgkin, A. L. & Steinhardt, R. A. (1967). The effect of sodium concentration on calcium movements in giant axons of *Loligo forbesi*. *J. Physiol., Lond.* **192**: 43–44*P*.

Baker, P. F., Cook, R. H. & Glitsch, H. G. (1974). Apparatus for the continuous measurement of ^{45}Ca influx into intact squid axons under voltage clamp conditions. *J. Physiol., Lond.* **242**: 48–50*P*.

Baker, P. F. & Crawford, A. C. (1972). Mobility and transport of magnesium in squid giant axons. *J. Physiol., Lond.* **227**: 855–874.

Baker, P. F. & Glitsch, H. G. (1973). Does metabolic energy participate directly in the Na^+-dependent extrusion of Ca^{2+} ions from squid giant axons? *J. Physiol., Lond.* **233**: 44–46*P*.

Baker, P. F. & Glitsch, H. G. (1975). Voltage-dependent changes in the permeability of nerve membranes to calcium and other divalent cations. *Phil. Trans. R. Soc. Lond.* (B) **270**: 389–409.

Baker, P. F., Hodgkin, A. L. & Ridgway, E. B. (1971). Depolarization and calcium entry in squid giant axons. *J. Physiol., Lond.* **218**: 709–755.

Baker, P. F., Meves, H. & Ridgway, E. B. (1973). Calcium entry in response to maintained depolarization of squid axons. *J. Physiol., Lond.* **231**: 527–548.

Baker, P. F. & Rink, T. J. (1974). Transient release of catecholamines in response to maintained depolarization of bovine adrenal medulla. *J. Physiol., Lond.* **241**: 107–109*P*.

Baker, P. F. & Schlaepfer, W. (1975). Calcium uptake by axoplasm extruded from giant axons of *Loligo*. *J. Physiol., Lond.* **249**: 37–38*P*.

Baker, P. F. & Shaw, T. I. (1965). A comparison of the phosphorus metabolism of intact squid nerve with that of the isolated axoplasm and sheath. *J. Physiol., Lond.* **180**: 424–438.

Bear, R. S., Schmitt, F. O. & Young, J. Z. (1937). Investigations on the protein constituents of nerve axoplasm. *Proc. R. Soc.* (B) **123**: 520–529.

Blaustein, M. P. & Hodgkin, A. L. (1969). The effect of cyanide on the efflux of calcium from squid axons. *J. Physiol., Lond.* **200**: 497–527.

Brinley, F. J. & Mullins, L. J. (1967). Sodium extrusion by internally dialysed squid axons. *J. gen. Physiol.* **50**: 2303–2331.

Brinley, F. J. & Mullins, L. J. (1975). The sensitivity of calcium efflux from squid axons to changes in membrane potential. *J. gen. Physiol.* **65**: 135.

Campbell, A. K. (1974). Extraction, partial purification and properties of obelin, the calcium activated luminescent protein from the hydroid *Obelia janiculata. Biochem. J.* **143**: 411–418.

Di Polo, R. (1973). Calcium efflux from internally dialysed squid axons. *J. gen. Physiol.* **63**: 5–36.

Di Polo, R. (1974). Effect of ATP on the calcium efflux in dialysed squid giant axons. *J. gen. Physiol.* **64**: 503–517.

Hallett, M. & Carbone, E. (1972). Studies of calcium influx into squid giant axons with aequorin. *J. Cell Physiol.* **80**: 219–226.

Hodgkin, A. L. (1964). *Conduction of the nervous impulse.* Liverpool: University Press.

Hodgkin, A. L. & Keynes, R. D. (1957). Movements of labelled calcium in squid giant axons. *J. Physiol., Lond.* **138**: 253–281.

Katz, B. (1966). *Nerve, muscle and synapse.* Maidenhead: McGraw-Hill.

Maddrell, S. H. P. & Gee, J. D. (1974). Potassium-induced release of the diuretic hormones of *Rhodnius prolixus* and *Glossina austeni*: Ca dependence, time course and localization of neurohaemal area. *J. exp. Biol.* **61**: 155–171.

Meves, H. & Vogel, W. (1973). Calcium inward currents in internally perfused giant axons. *J. Physiol., Lond.* **235**: 225–265.

Miledi, R. (1973). Transmitter release induced by injection of calcium ions into nerve terminals. *Proc. R. Soc.* (B) **183**: 421–425.

Morin, J. G. & Hastings, J. W. (1971). Biochemistry of the bioluminescence of colonial hydroids and other coelenterates. *J. Cell Physiol.* **77**: 305–312.

Nordmann, J. J. (1975). Hormone release in relation to inactivation of Ca-entry in the rat neurohypophysis. *J. Physiol., Lond.* **249**: 38–39*P*.

Reuter, H. (1973). Divalent cations as charge carriers in excitable membranes. *Prog. Biophys. Mol. Biol.* **26**: 1–43.

Reuter, H. (1974). Localization of *beta* adrenergic receptors, and effects of noradrenaline and cyclic nucleotides on action potentials, ionic currents and tension in mammalian cardiac muscle. *J. Physiol., Lond.* **242**: 429–451.

Shimomura, O., Johnson, F. H. & Saiga, Y. (1963). Microdetermination of calcium by aequorin luminescence. *Science, N.Y.* **140**: 1339–1340.

Stinnakre, J. & Tauc, L. (1973). Calcium influx in active *Aplysia* neurones detected by injected aequorin. *Nature, New Biol.* **242**: 113–115.

Symp. zool. Soc. Lond. (1977) No. 38, 261–275

THE GIANT NERVE FIBRE SYSTEM OF CEPHALOPODS. RECENT STRUCTURAL FINDINGS

R. MARTIN

*Stazione Zoologica di Napoli, Naples, Italy**

SYNOPSIS

The first-, second- and third-order giant axons of the giant fibre system of decapods can already be identified by their large size in the freshly hatched larva and in late embryonic stages. *Sepia* embryos have several first-order cells and axons; at hatching these are reduced to two, one from the right, the other from the left. In *Loligo* larvae the crossing first-order axons are not yet fused, but possibly connected by synapses.

As shown by Young (1939), in postlarval *Loligo* the first-order giant axons are fused at the chiasm. In *Sepia* and *Illex* the two axons are unfused; possibly there are two-way synapses. It appears that each systematic group has its own mechanism at the chiasm for obtaining functional bilaterality.

The "giant synapse" between second- and third-order axon is formed by 15 000 to 20 000 single contacts of about 1 μm diameter. Iontophoretically-injected cobalt ions render visible the processes of the third-order axon that form these contacts. In the "giant synapse" of *Sepia* and *Todarodes* collaterals of the second-order axon meet processes from the third-order axon in the sheath that separates the two axons. After the stellate ganglion of *Loligo* has been incubated with lanthanum ions, or after calcium ions have been injected iontophoretically into the presynaptic axon, the "giant synapse" contains fewer synaptic vesicles than it does when fixed by perfusion. The number of vesicles is higher when the ganglion is fixed by perfusion in the presence of large amounts of magnesium ions.

The populations of synaptic vesicles in synapses at five different levels of the chain of giant neurons are all electron transparent. They differ in size and in appearance when the synapses are incubated with zinc iodide-osmium reagent.

The synaptic contacts between the accessory giant axon and the unfused third-order axons are very similar in fine structure to the contacts of the giant synapse between second- and third-order axons.

INTRODUCTION

Forty years ago, a note of Young (1934) reported the discovery of the giant nerve fibres in the cuttlefish, *Sepia*. As he pointed out in his comprehensive description of the system (Young, 1939) it was really a rediscovery, since Williams (1909) had already seen the two first-order giant cells, the chiasm of their processes, and the giant motor axons in the squid. Williams' observations had remained unnoticed. After Young's (1939) classical study, once the cathode ray tube and electrolyte-filled electrodes were available as research tools, an ever growing number of physiological and biochemical papers appeared that made use of the "squid axon" or the "squid synapse". Basic models in neurophysiology, elucidating the processes involved in impulse conduction (Hodgkin, 1957) and impulse

* Present address: *Universität Ulm, Ulm, Germany.*

transmission (Miledi, 1967; Kusano, Livengood & Werkman, 1967; Katz & Miledi, 1969) have been elaborated through the study of cephalopod axons.

This paper is a summary of complementary results on the construction of the giant fibre system. Interest lies in the fact that in the last 15 years the transmission electron microscope has allowed ultrastructural details to be correlated with functional events. By iontophoretic injection of cobalt ions or dyes it has been possible to trace and identify single axons and branches. Some of the results reviewed are taken from a comparative study of the giant synapse by Martin & Miledi (in prep. a, b). The chain of giant neurons forming the giant fibre system in *Loligo* is represented schematically in Fig. 1.

SYNAPTIC CONNECTIONS ALONG THE CHAIN OF THE GIANT NEURONS

Six synaptic contact areas follow each other along the chain of giant neurons in *Loligo*: (A) (Fig. 1), afferent boutons of unknown origin that cover the cell body and the dendrites of the two first-order giant cells in the magnocellular lobes of the brain; (B) masses of synaptic boutons, also of unknown origin, that end on the first-order axons shortly before contacts C; (C) contacts between the first- and second-order giant axons in the palliovisceral lobe; (D) contacts between the accessory giant axon and the unfused third-order axons near to their cell bodies in the stellate ganglion; (E) the contacts between the second- and third-order giant axons in the stellate ganglion—the "giant synapse" of physiologists; (F) the neuromuscular junctions between third-order axons and mantle muscles.

In contrast to synapses in other giant fibre systems—for instance annelids (Hama, 1966), crustaceans (Furshpan & Potter, 1959) and teleosts (Furukawa, 1966), in which electrotonic transmission has been demonstrated—the electron microscopic picture suggests that the giant fibre system of cephalopods in contacts A, B, D, E, F contains chemically transmitting synapses. However, in contact C of *Loligo* (Gervasio, Martin & Miralto, 1971) and in the chiasm of the first-order axons of *Sepia* (Martin, 1969) the cleft width between pre- and postsynaptic membranes is reduced, as it is in electrical contacts (Hama, 1966). For none of the squid synapses has the chemical nature of the transmitter substance been identified (see Miledi, 1969, for contact E).

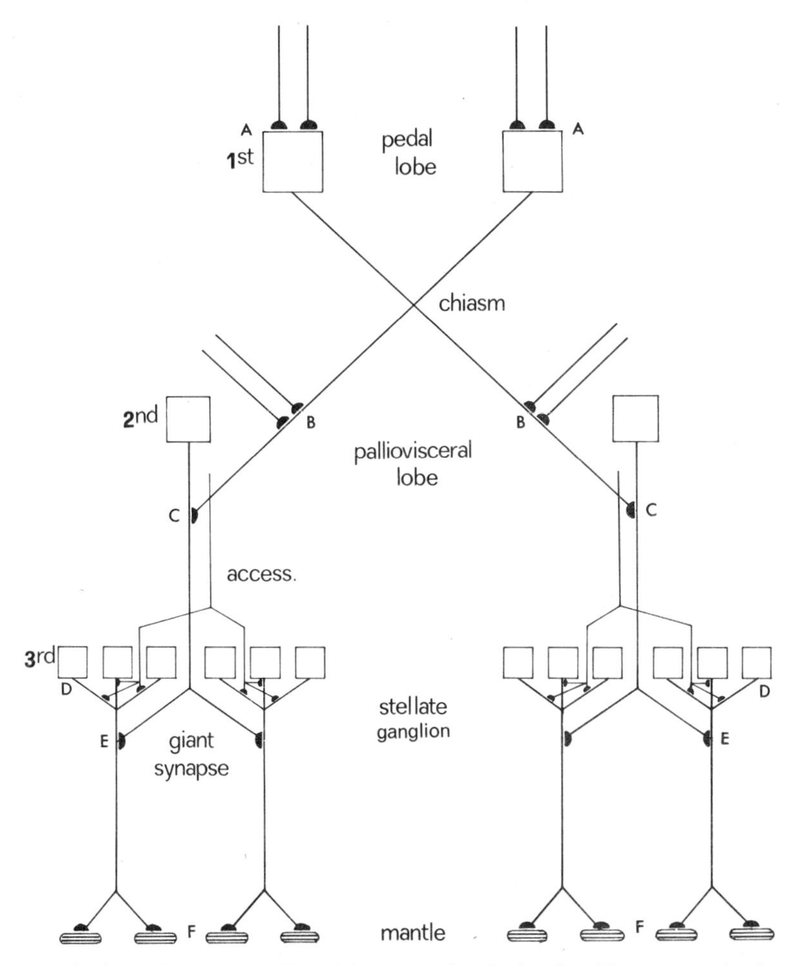

FIG. 1. A schematic representation of the connections in the giant fibre system of *Loligo*.

In view of the great diversity in synapse structure in the cephalopod brain (Barlow & Martin, 1971) it is surprising how homogeneous the different contacts are in the giant fibre system. With the exception of very rare boutons on the first-order cell that contain vesicles with electron-dense cores (Gervasio, Martin & Miralto, 1971), all synapses are provided with clear vesicles. Only by statistical evaluation of the diameters of vesicles and by incubation with zinc iodide-osmium did it become evident that no two successive vesicle populations are identical (Martin, Barlow & Miralto, 1969; Froesch & Martin, 1972).

THE CHIASM OF THE FIRST-ORDER GIANT FIBRES

Complete syncytial fusion of the two crossing first-order giant axons (Young, 1939) in the palliovisceral lobe of the postlarval *Loligo* has been confirmed by transmission electron microscopy. The fusion must take place in postembryonic development, since at hatching the two axons are still separate (Martin, 1969). Possibly in the freshly hatched squid the two crossing axons transmit impulses from one to the other since, in both fibres, vesicles have been observed at the opposed membranes. We could not study the process of fusion since we were unable to raise squids in the laboratory.

The chiasm of the first-order axons guarantees bilaterally symmetrical spread of impulses from the chiasm downward, resulting in simultaneous contraction of the circular muscles of the mantle. A comparative study of the chiasm in *Loligo*, *Sepia* and *Illex* revealed that the construction in each cephalopod group is different (Fig. 2). In *Sepia* both an ipsilateral and a contralateral branch

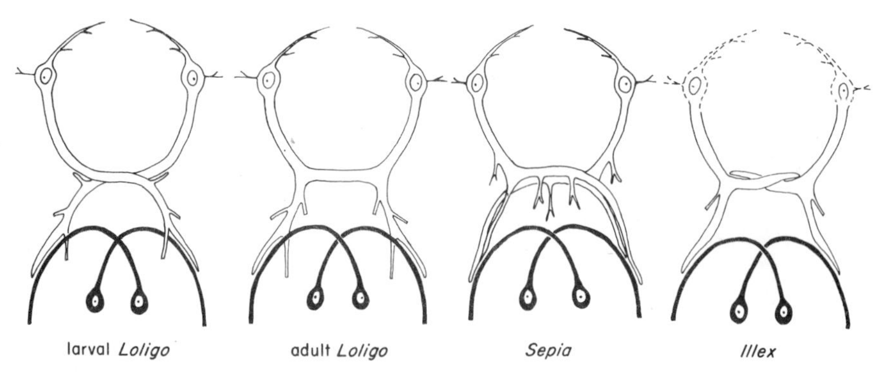

FIG. 2. Schematic drawings of the course of fibres in the chiasm of the first-order axons in the larval *Loligo*, the adult Loligo, in *Sepia* and in *Illex*. The first-order cells in *Illex* (broken lines) have not been studied. The black fibres are the second-order axons.

of the two first-order axons make contact with the second-order axon. At the cross-over point the two fibres are unfused, and perfectly symmetrical synapses occur with a single row of vesicles on each side. This structural arrangement suggests a two-way passage of impulses at the contacts. In *Illex* only ipsilateral branches of the first-order axons contact the second-order axon; in the chiasm the two crossing axons end blindly forming synaptic contacts (Martin, 1969).

So in the three systematic groups three different constructions of the chiasm have evolved that presumably fulfil the same function. Oegopsid and sepiid species have adopted similar solutions: bilateral synaptic spread of impulses in the chiasm—while the most simple and elegant solution has evolved in myopsids by complete fusion of the crossing axons.

THE LOCALIZATION OF THE CELL BODIES OF THE THIRD-ORDER AXONS

By means of dye and cobalt injections (Pitman, Tweedle & Cohen, 1972) it is possible to follow the many axons that fuse to form the third-order axons back to their cell bodies in the giant fibre lobe of the stellate ganglion. In *Loligo pealeii* Staub (1954) described discrete areas in which the neurons of the different third-order axons are located. When we injected these axons in *Loligo vulgaris* with cobalt we could not confirm the existence of distinct localizations of the different neuron populations (Martin & Miledi, in prep. b). In general the larger and more distal axons were formed by neurons distributed over the whole giant fibre lobe, while the smaller proximal axons had their cell bodies only in the proximal part of the lobe (Fig. 3). The different neuron populations overlapped; they also reached from the dorsal to the ventral side of the lobe.

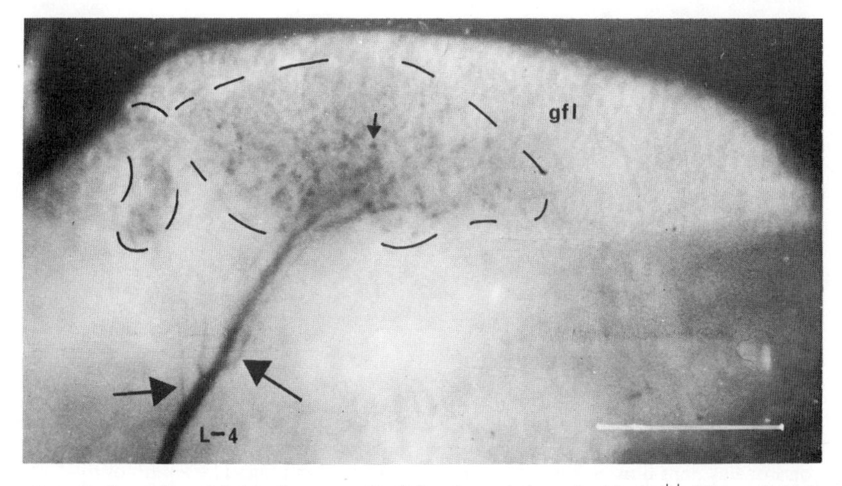

FIG. 3. A smaller third-order axon (L-4) has been injected with Co^{++}. The cell bodies (small arrow) in the giant fibre lobe (gfl), the axons of which fuse to form the giant axon, are distributed in two areas of the lobe. The large arrows point to lateral axon bundles. Scale line = 1 mm (×25).

The differences between Staub's and our findings may be due to species difference. It also appears that iontophoretic injection of cobalt is a better technique for discriminating the finer ramifications and displays the neuron populations more completely than the pressure injection of dyes used by Staub.

THE DIMENSIONS OF THE GIANT SYNAPSE OF *LOLIGO*

The giant synapse *sensu strictu* of *Loligo* that is usually studied by physiologists, is formed by the most distal branch of the second-order axon and the largest third-order axon in the stellate ganglion. The second-order axon (Fig. 4) is presynaptic, conducting impulses from the brain to the stellate ganglion; the third-order axon is postsynaptic, innervating the musculature of the mantle. A description of how to dissect and handle the ganglion and how to mount the giant synapse for work with microelectrodes is given by Miledi (1967) and by Katz & Miledi (1967). For work on the effect

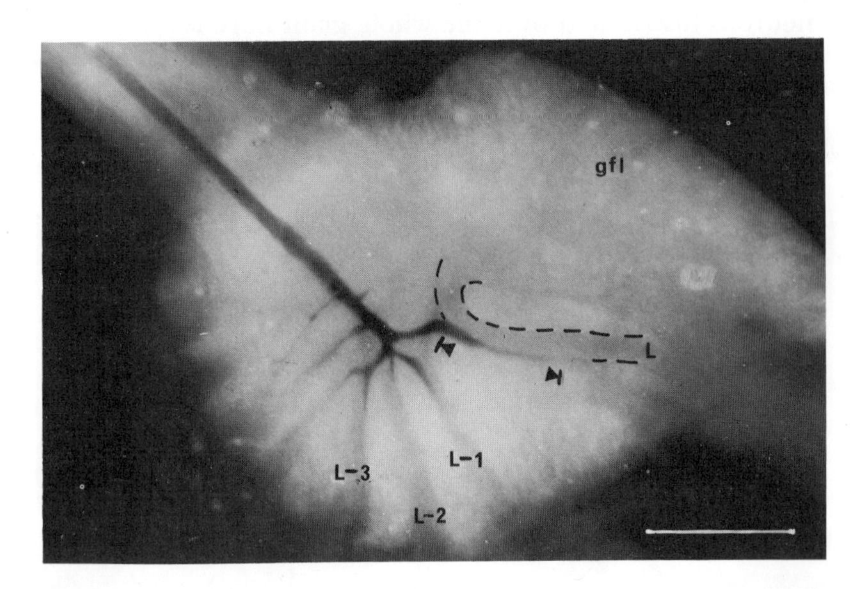

FIG. 4. A stellate ganglion of *Loligo vulgaris* that has been soaked in glycerine and photographed in reflected light. Co^{++} ions have been injected and precipitated in a second-order giant fibre to show its branchings in the ganglion. The length of the largest giant synapse is indicated by two triangles. The broken lines indicate the most distal (L) and largest third-order axon, the subsequent smaller third-order axons are indicated L-1, L-2, L-3 etc. The giant fibre lobe (gfl) contains the neurons of the third-order axons. Scale line = 1 mm (×20).

of pharmacologically-active substances on synaptic transmission, or for the determination of functional parameters, it is important to know the dimensions of the axons and of their contacts.

The presynaptic terminal in the centre of the synapse has a diameter of 30 to 40 μm in an animal of 170 mm mantle length, and 25 μm in an animal of 40 mm mantle length. The diameter of the postsynaptic axon in the former animal measures about 300 μm, in the latter 100 μm. The lengths of the synapses in 12 *Loligo vulgaris* of 40 to 160 mm mantle length reached from 0·5 to 1·2 mm. Tree-like processes from the postsynaptic axon (and not from the presynaptic, as erroneously reported by Casteion & Villegas, 1964) penetrate the thick glia sheath between the two axons, and end in round synaptic contacts at the presynaptic membrane each about 1·15 μm^2 in area (Figs 5, 6). The number of processes counted by Young (1973) in histological sections of a small *Loligo pealeii* was 181. After injecting cobalt into the postsynaptic fibre of a medium-sized *Loligo vulgaris* (180 mm mantle length) we counted 323 processes. Young estimated that there were 20 000 or more synaptic contacts in his specimen; from electron micrographs taken at 12 levels of the synapse we calculated 14 184 contacts (Martin & Miledi, in prep. b).

The synapse of *Loligo* transmits impulses by releasing a transmitter substance in quantal form (Miledi, 1967; Katz & Miledi, 1969), and, as would be expected, the giant synapse has the structural characteristic of chemical synapses (Hama, 1962). Electron-transparent synaptic vesicles with an average diameter of 54 nm (Froesch & Martin, 1972) are accumulated at the presynaptic membrane; the cleft between pre- and postsynaptic membrane is 15 nm wide. The number of vesicles, in the order of $2{\cdot}5 \times 10^7$ in the synapse of an adult squid fixed by perfusion, is considerably higher after perfusion fixation in the presence of high amounts of magnesium ions (Martin & Miledi, in prep. b). Synaptic vesicles are much reduced in number after incubation with lanthanum ions (Martin & Miledi, 1971), or after injection of calcium ions into the axons (Martin & Miledi, in prep. a). A problem in these studies is the irregular way in which the different populations of vesicles are distributed over the length of the presynaptic terminal.

A presynaptic complex of large lamellae associated with vesicles of the size of synaptic vesicles, similar in some respect to synaptic ribbons in sensory cells, has been observed very frequently in terminals of the second-order giant axon of *Loligo* (Martin & Miledi, 1975).

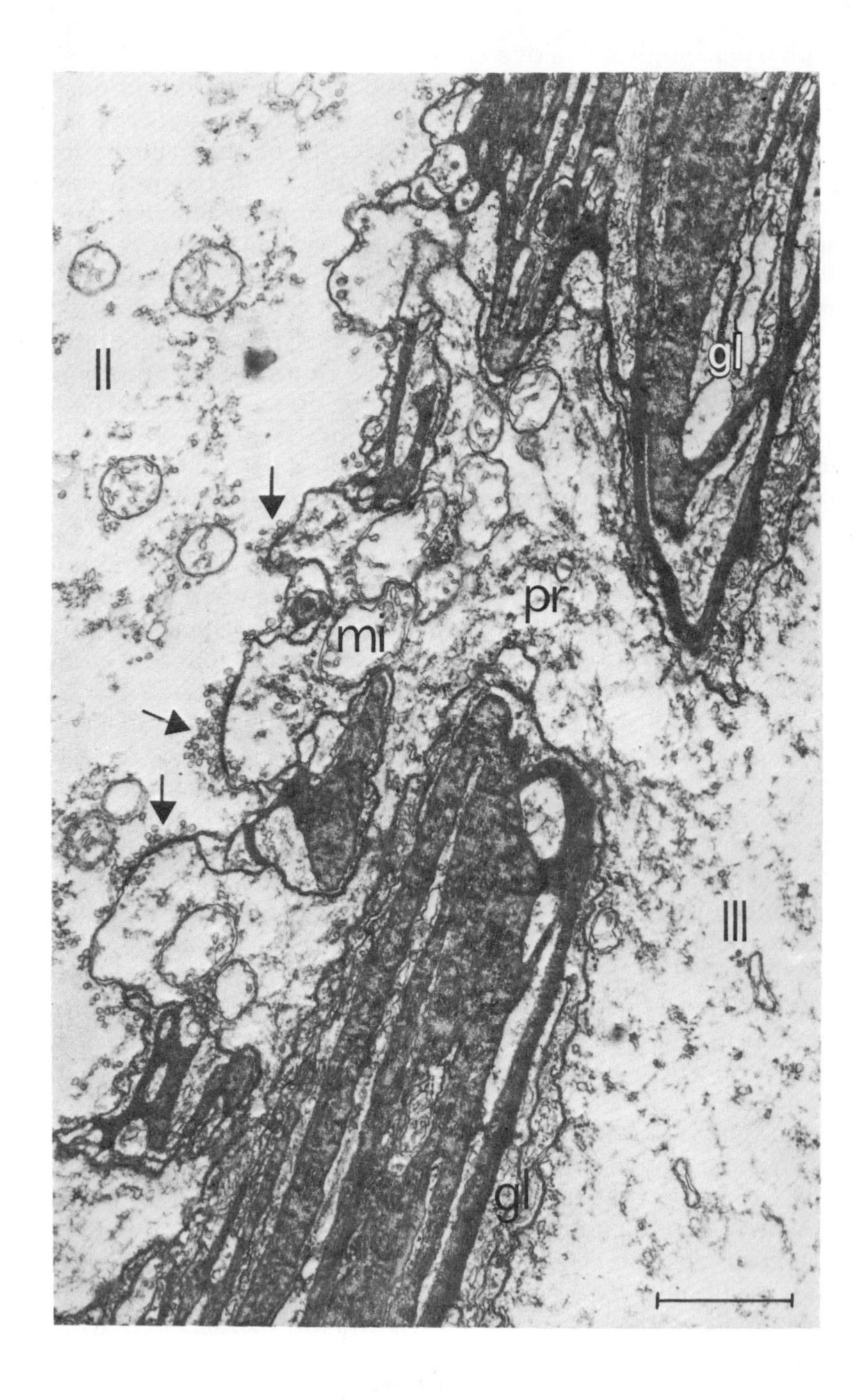
II
gl
pr
mi
III
gl

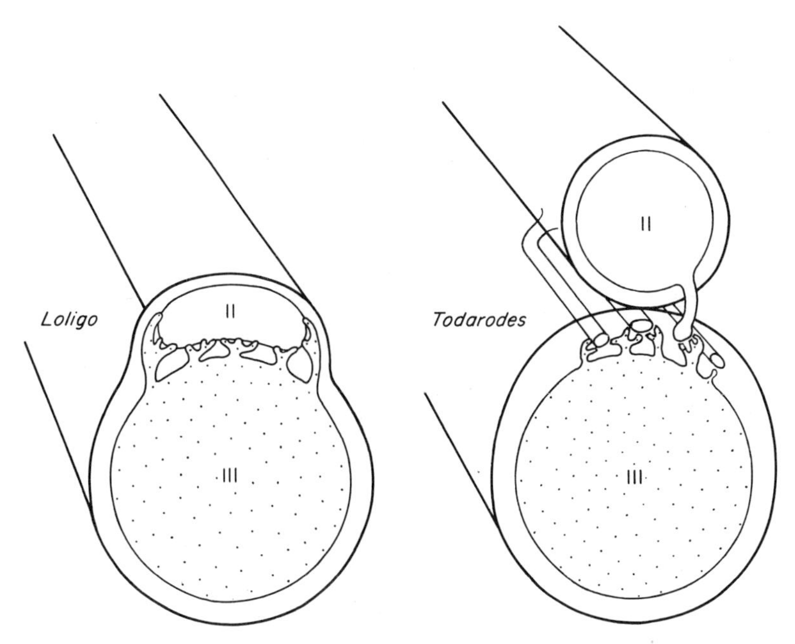

FIG. 6. Schematic drawings of the cross-cut "giant synapse" of *Loligo* and of *Todarodes* or *Sepia*. In *Loligo* processes of the third-order axon (III) reach the second-order axon (II). In *Todarodes* collaterals of the second-order axon meet processes of the third-order axon in many contacts that are embedded in the glia sheath.

THE "GIANT SYNAPSE" OF SEPIIDS AND OEGOPSIDS

Amongst the three decapod groups with giant fibre systems sepiid and oegopsid species differ in two respects from myopsid squids. In species of the former groups the neurons of the third-order axons are not concentrated in a giant fibre lobe, as they are in *Loligo* (see Young, 1939, for *Loligo* and *Sepia*; Martin & Miledi, in prep. b, for *Todarodes*), and in the giant synapse both the pre- and postsynaptic axon (second- and third-order) send out collaterals and processes that meet in a large number of contacts in the sheath between the two axons (Fig. 6) (Martin & Miledi, in prep. b). In *Loligo*, processes of the postsynaptic axon penetrate the sheath and end at the

---

FIG. 5. Electron micrograph of a small part of the giant synapse showing the second (II)- and third (III)-order axons. The sheath between the two axons is composed of collagen-filled extracellular layers (electron-dense) and glial processes (gl). A process (pr) of the third-order axon with five branches contacts the second-order axon. Synaptic vesicles are accumulated at the presynaptic membrane (arrows). mi = Mitochondria. Scale line = 1 μm (×18 000).

membrane of the presynaptic axon (Fig. 7). It appears that neither the synapse of *Sepia* nor that of *Todarodes* has been studied physiologically.

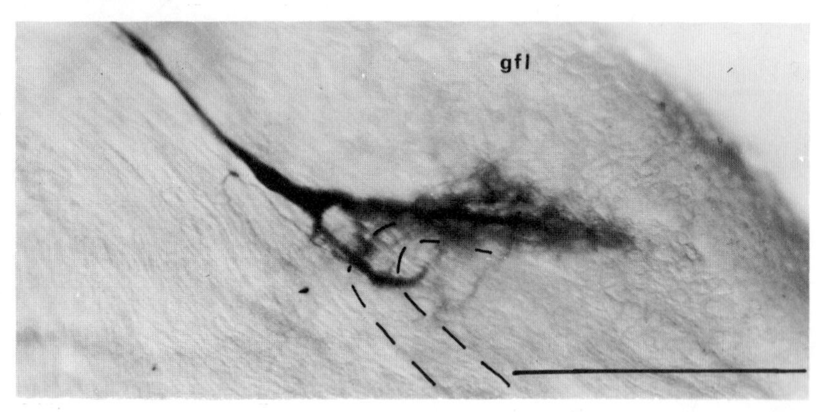

FIG. 7. An accessory axon injected with Co^{++} ions. A shorter branch proceeds to the bend of the third-order axon, a longer branch with numerous fine collaterals towards the neurons of the giant fibre lobe (gfl). The approximate course of the largest third-order axon is indicated by broken lines. Scale line = 1 mm (×40).

THE ACCESSORY SYSTEM

An accessory giant axon nearly as large in diameter as the second-order axon accompanies the latter from the brain to the stellate ganglion (Fig. 7). It synapses with the third-order axons through a large number of boutons on the unfused fibres near to their cell bodies in the giant fibre lobe (D, Fig. 1). These are the proximal synapses of Young (1939). The accessory axon is excitatory as is the second-order axon (Bryant, 1959). Casteion & Villegas (1964) studied the fine structure of the proximal synapses. The unequivocal identification of these contacts was possible after iontophoretic injection of cobalt ions into the accessory axon (Fig. 7). After injection and precipitation (Pitman, Tweedle & Cohen, 1972) of cobalt, electron-dense precipitates were found included in lysosome-like organelles in the terminals of the accessory axon (Martin & Miledi, in prep. b). As in the distal synapses, processes from the third-order axons that penetrate the sheath, contacted the branches of the accessory terminals (Fig. 8). This

FIG. 8. Cross-section of three unfused third-order axons (III) and a branch of the accessory axon (acc). A process (pr) of a third-order axon penetrates the sheath (sh) and contacts the accessory axon. Synaptic vesicles are accumulated at the presynaptic membrane (arrows). Scale line = 1 μm (×25 000).

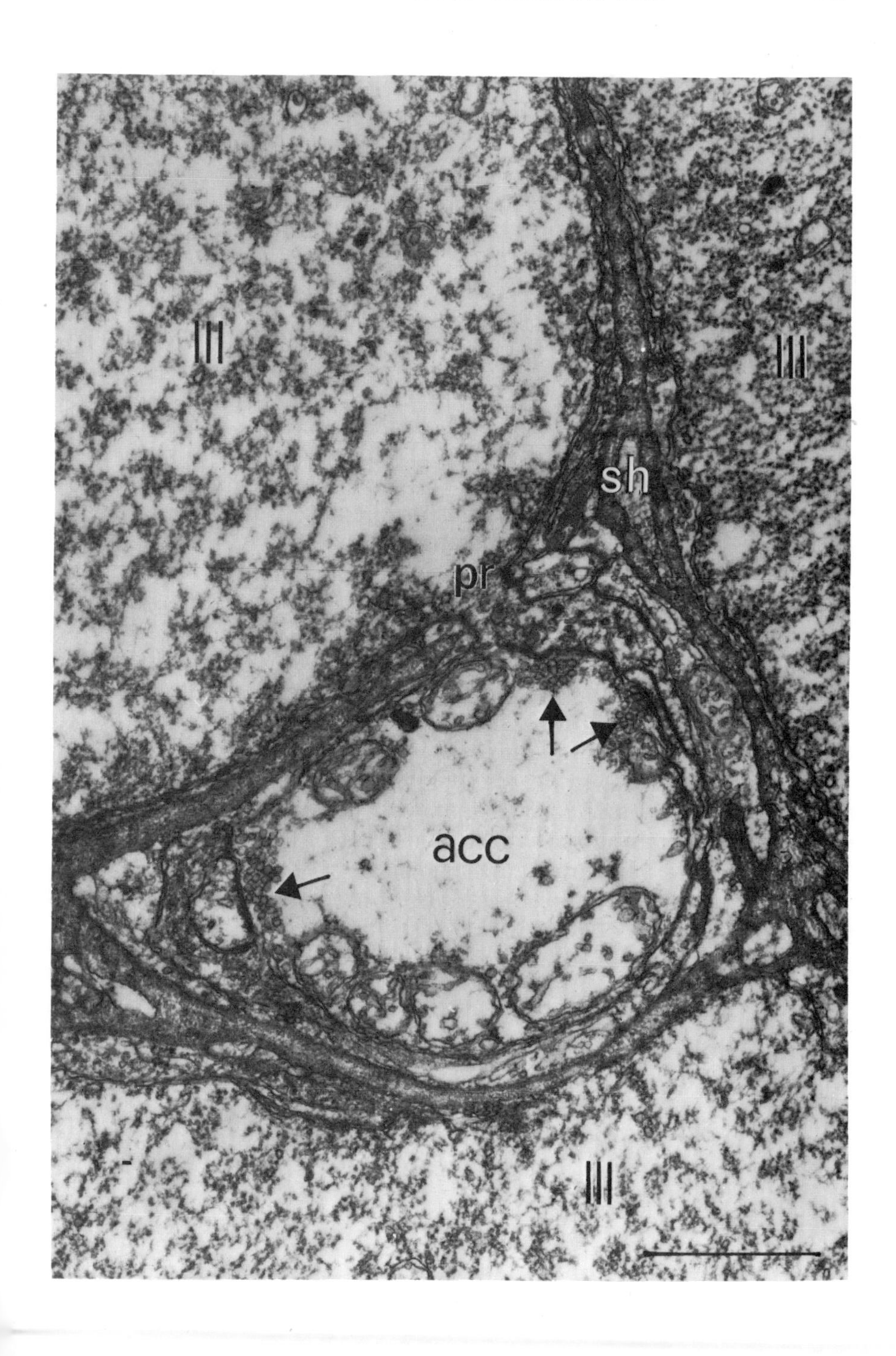
III
III
sh
pr
acc
III

synapse, too, has the structural characteristics of a chemical synapse. Electron-transparent synaptic vesicles of an average diameter of 52 nm are accumulated at the presynaptic membrane. The cleft between the pre- and postsynaptic membrane measures about 15 nm. So the distal and the proximal synapses on the giant motor axons are very similar in structure and function. The origin of the accessory axon and its input connections are not yet known.

EMBRYOLOGY OF THE GIANT FIBRE SYSTEM

The beautifully differentiated first-order giant neurons amongst masses of undifferentiated neuroblasts in the brain of the freshly hatched *Loligo* are an extraordinary sight (Martin, 1965). The crossing first-order axons are easily identified by their size. Moreover the neurons of the second-order giant axons in the roof of the palliovisceral lobe and their processes in the mantle connective also stand out, at hatching, from the accompanying cells and fibres by their large dimensions. But the accessory giant axon is not yet identifiable in the mantle connective. The synaptic contacts between second- and third-order axons in the stellate ganglion appear well differentiated at hatching. At this stage of development second- and third-order axons in the giant synapse are about the same size (Martin & Miledi, in prep. b).

An interesting indirect development of the first-order giant fibres has been observed in *Sepia* embryos. In stages some days before hatching two first-order giant neurons on each side of the brain are formed and four axons, two from each side, cross in the chiasm. At hatching the adult-like situation is found, only one first-order giant cell and one crossing axon from each side of the brain. During late embryonic development on each side of the brain one neuron degenerates (Martin & Rungger, 1966).

This indirect differentiation of the system may reflect steps in the evolutionary elaboration of the giant fibre system. Hypothetically, one may assume that in early forms a group of neurons in the magnocellular lobe controlled the motor fibres of the mantle muscles. This arrangement is found in recent octopods. Later on, however, to promote a single response in the system all afferent sensory stimuli appear to have become concentrated on a single neuron that reaches very large dimensions.

OPEN PROBLEMS

The giant synapse of the squid offers great opportunities for studying the effect of cations or pharmacologically active substances on synaptic transmission. It is one of the few preparations in which both the pre- and the postsynaptic axon are accessible for recording, stimulating or injecting electrodes. One obvious disadvantage, however, is the fact that the chemical nature of the transmitter substance is not yet known (Miledi, 1969).

On the structural side it has not yet been possible to identify in the mantle the neuromuscular junctions belonging to the giant motor fibres. Possibly the newly available methods of iontophoretic injection of cobalt ions may lead to a localization of the endings. Also the origin and input connections of the accessory giant axons may now be studied by injecting cobalt or dyes.

It is surprising that in the whole system an inhibitory control system has not been found. In other giant fibre systems, for example the Mauthner system of fishes, the very fast, startle response that is generated in this system is inhibited immediately afterwards by a combination of inhibitory systems (Furukawa, 1966). In the squid, since the endings of the second order and of the accessory axon (contacts D and E, Fig. 1) on the motor axons are excitatory, one might begin to look for inhibitory synapses at the level of the afferent boutons on the soma of the first-order giant neurons (contact A, Fig. 1). The masses of synaptic boutons on the terminal of the first-order axon in the palliovisceral lobe of the brain (contact B, Fig. 1) are also possible candidates. Unfortunately, owing to the delicacy of the tissues and the inaccessibility of this part of the brain, electrophysiological work in the palliovisceral or magnocellular lobe appears to be technically very difficult at the present time.

Acknowledgement

I should like to thank Dr A. Packard for revision of the manuscript and suggestions.

References

Barlow, J. & Martin, R. (1971). Structural identification and distribution of synaptic profiles in the *Octopus* brain using the zinc iodide-osmium method. *Brain Res.* **25**: 241–253.

Bryant, S. H. (1959). The function of the proximal synapses of the squid stellate ganglion. *J. gen. Physiol.* **42**: 609–616.

Casteion, O. J. & Villegas, G. M. (1964). Fine structure of the synaptic contacts in the stellate ganglion of the squid. *J. Ultrastr. Res.* **10**: 585–598.

Froesch, D. & Martin, R. (1972). Heterogeneity of synaptic vesicles in the squid giant fibre system. *Brain Res.* **43**: 573–579.

Furshpan, E. J. & Potter, D. D. (1959). Transmission at the giant motor synapses of the crayfish. *J. Physiol., Lond.* **145**: 289–325.

Furukawa, T. (1966). Synaptic interaction at the Mauthner cell of goldfish. *Progr. Brain Res.* **21A**: 44–70.

Gervasio, A., Martin, R. & Miralto, A. (1971). Fine structure of synaptic contacts in the first-order giant fibre system of the squid. *Z. Zellforsch. mikrosk. Anat.* **112**: 85–96.

Hama, K. (1962). Some observations on the fine structure in the stellate ganglion of the squid *Doryteuthis bleekeri. Z. Zellforsch. mikrosk. Anat.* **56**: 437–444.

Hama, K. (1966). Studies on fine structure and function of synapses. *Progr. Brain Res.* **21A**: 251–267.

Hodgkin, A. L. (1957). Ionic movements and electrical activity in giant nerve fibres. *Proc. R. Soc.* (*B*) **148**: 1–37.

Katz, B. & Miledi, R. (1967). A study of synaptic transmission in the absence of nerve impulses. *J. Physiol., Lond.* **192**: 407–436.

Katz, B. & Miledi, R. (1969). Tetrodotoxin-resistant electric activity in presynaptic terminals. *J. Physiol., Lond.* **203**: 459–487.

Kusano, K., Livengood, D. R. & Werkman, R. (1967). Correlation of transmitter release with membrane properties of the presynaptic fiber of the squid giant synapse. *J. gen. Physiol.* **50**: 2579–2602.

Martin, R. (1965). On the structure and development of the giant fibre system of the squid *Loligo vulgaris. Z. Zellforsch. mikrosk. Anat.* **67**: 77–85.

Martin, R. (1969). The structural organization of the intracerebral giant fibre system of cephalopods. The chiasma of the first order giant axons. *Z. Zellforsch. mikrosk. Anat.* **97**: 50–68.

Martin, R., Barlow, J. & Miralto, A. (1969). Application of the zinc iodide-osmium tetroxide impregnation of synaptic vesicles in cephalopod nerves. *Brain Res.* **15**: 1–16.

Martin, R. & Miledi, R. (1971). Lanthanum ions on the giant synapse of the squid. *Proc. Int. Un. Physiol. Sci.* **9** (Abstract).

Martin, R. & Miledi, R. (1975). A presynaptic complex in the giant synapse of the squid. *J. Neurocytol.* **4**: 121–129.

Martin, R. & Miledi, R. (in prep. a). *Fine structural changes in the giant synapse of the squid after intraaxonal injection of calcium ions.*

Martin, R. & Miledi, R. (in prep. b). *The form and dimensions of the giant synapse of cephalopods.*

Martin, R. & Rungger, D. (1966). Zur Struktur und Entwicklung des Riesenfaser-systems erster Ordnung von *Sepia officinalis* L. (Cephalopoda). *Z. Zellforsch. mikrosk. Anat.* **74**: 454–463.

Miledi, R. (1967). Spontaneous synaptic potentials and quantal release of transmitter in the stellate ganglion of the squid. *J. Physiol., Lond.* **192**: 379–406.

Miledi, R. (1969). Transmitter action in the giant synapse of the squid. *Nature, Lond.* **223**: 1284–1286.

Pitman, R. M., Tweedle, C. D. & Cohen, M. J. (1972). Branching of central neurons: Intracellular cobalt injection for light and electron microscopy. *Science, N.Y.* **176**: 412–414.

Staub, N. C. (1954). Demonstration of anatomy of the giant fiber system of the squid by microinjection. *Proc. Soc. exp. Biol. N.Y.* **86**: 854–855.

Williams, L. W. (1909). *The anatomy of the Common squid* Loligo pealii, *Lesueur.* Leiden, Holland: Brill.

Young, J. Z. (1934). Structure of nerve fibres in *Sepia*. *J. Physiol., Lond.* **83**: 27*P*–28*P*.
Young, J. Z. (1939). Fused neurons and synaptic contacts in the giant nerve fibres of cephalopods. *Phil. Trans. R. Soc.* (B) **229**: 465–505.
Young, J. Z. (1973). The giant fibre synapse of *Loligo*. *Brain Res.* **57**: 457–460.

Symp. zool. Soc. Lond. (1977) No. 38, 277–285.

PUPILLARY RESPONSE OF CEPHALOPODS

W. R. A. MUNTZ

Department of Biology, Stirling University, Stirling, Scotland

SYNOPSIS

The cephalopod pupil may have many shapes. In *Octopus vulgaris,* for example, it contracts to an elongated slit, and in *Sepia officinalis,* it has the form of an elongated letter W, with a horizontal middle section and vertical segments at each end. When the pupil of *Sepia* is contracted these vertical segments may remain relatively large compared with the horizontal section, giving an effect of a double pupil.

In eyes that obey Matthiessen's ratio a contracted pupil affects the quality of the image formed on different parts of the retina, by affecting both brightness and the angle at which the image-forming rays of light impinge on the receptors. The pupil shapes of *Octopus* and *Sepia* may be correlated with the regional variations in retinal structure described by Young (1963), for the horizontal strip of *Octopus* and the anterior and posterior areas of *Sepia,* which appear on anatomical grounds to be regions of visual specialization, are also the regions where the pupils of these two species should form an image of high quality.

Measurements of the pupil response of *Sepia* to a 3 log unit decrease in illumination showed that the response was slow, with the pupil taking about 30 sec to reach its maximum size. After this the pupil starts to contract again, and by about 10 min is only a little larger than its original size. It is suggested that the second phase of the response is correlated with the retraction of the retinal screening pigment, which also takes about 10 min to become complete.

INTRODUCTION

The pupil of cephalopods may have one of several shapes. In *Eledone* and *Octopus* it has the form of a rectangle, elongated in the horizontal direction, while in *Sepia* it has the form of an elongated letter W (see Fig. 1). The present paper consists of two parts. The first of these concerns the response of the pupil to light, and the second considers the effect that the shape of the pupil will have on the quality of the image formed on the retina. In both cases the migration of the screening pigment between the receptors must also be considered, for this is known to be affected by the illumination of the eye, migrating out between the receptor cells in the light and retracting in darkness (Hagins & Liebman, 1962; Young, 1963; Daw & Pearlman, 1974). Movement of this pigment will presumably affect the sensitivity of the eye by screening the receptors from light, and may well also affect the quality of the image formed by acting as a collimator and restricting the spread of light between the receptors.

THE RESPONSE OF THE PUPIL TO LIGHT

The pupillary response of *Sepia officinalis* was measured photographically. Adult *Sepia* were constrained in one corner of an aquarium, in about 8 cm of sea-water, by means of a moveable partition. They faced a white screen illuminated by a 150 W lamp, which gave a screen luminance of approximately 10 millilamberts. The pupil was photographed through a small hole in the screen, using an electronic flash. The following procedure was used. The animal was adapted to the screen at full luminance for 3 min, and the intensity of the light then decreased by 3 log units by inserting a neutral density filter (3ND) in front of the light source. The pupil was photographed at varying intervals after this decrease in illumination. Since the electronic flash caused the pupil to contract, only one photograph could be taken on each occasion, after which the animal was readapted to the full intensity for a further 3 min and the procedure repeated. Since the pupil of cephalopods dilates if the animal is excited or frightened, even if the illumination is constant, results were only used when the photographs showed the animal resting quietly on the floor of the aquarium.

Figure 1 shows tracings from photographs obtained during one experiment. It is not easy to estimate the total amount of light that will reach the retina at different times from such two-dimensional pictures, since the pupil extends around the projecting lens of the eye, covering almost half of a spherical surface. Furthermore, because of the pupil's shape, the effective aperture will be different for different directions of view. The area of the pupil's image on such photographs will, however, provide a measure of the effective pupillary aperture for light reaching the eye from the direction of the camera, so that although we cannot estimate the total amount of light reaching the retina, we can estimate the amount that will reach it from this particular direction. In Fig. 1 the change in the area of the pupil's image between 0 and 30 sec amounts to 0·7 log

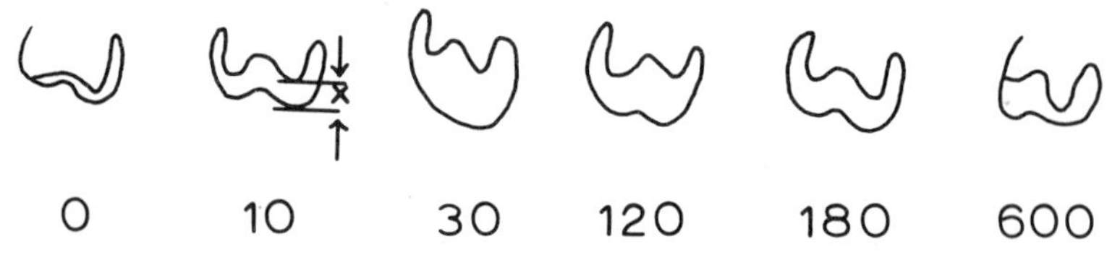

FIG. 1. Tracings from photographs of the pupil of *Sepia*, taken at various times (in sec) after a 3 log unit decrease in stimulus intensity. The distance x is the measure of pupil width used in Fig. 2.

units, which is considerably less than the 3 log unit decrease in illumination used to produce it.

In order to compare experiments performed on different animals, the size of the pupil was quantified in arbitrary units, by measuring in each case the distance labelled x in Fig. 1. Measurements for six *Sepia*, from which relatively complete results were obtained, are shown in Fig. 2. It is apparent that the pupil dilates slowly in response to a decrease in illumination, taking half-a-minute or more to reach its full size. Records for longer intervals were much more difficult to obtain, since the animals often moved during the interval so that the pupil was no longer in the field of view of the camera. Three animals for which such records were obtained all showed a subsequent decrease in pupil size with time, although in no case did the pupil regain its initial size.

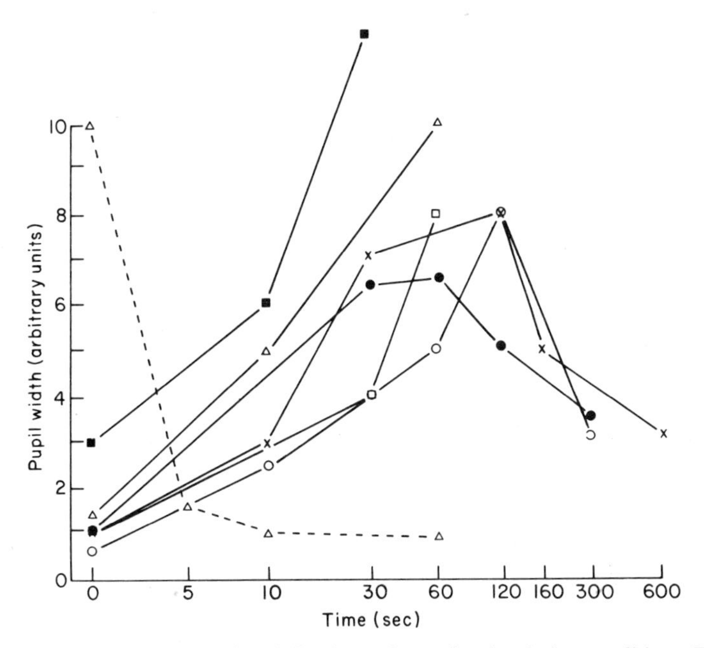

FIG. 2. Pupil width against time following a change in stimulating conditions. The solid lines and different symbols show the results for six different animals following a 3 log unit decrease in the stimulus intensity. The upright triangles and broken line show the results for one animal following a 3 log unit increase in stimulus intensity.

This latter result suggests that when the illumination is decreased an initial increase in sensitivity is obtained by pupil dilation, but that subsequently another mechanism takes over. The migration of the screening pigment is one obvious possibility, since this has been shown to take 5–15 min to become complete in both *Octopus vulgaris* and *Loligo pealeii* (Young, 1963; Daw & Pearlman, 1974). This means that, if the time course of pigment migration is more or less the same in *Sepia*, it will become complete at about the same time as the pupil reverts to a smaller size. Furthermore, Daw & Pearlman (1974) have estimated that the retraction of the screening pigment will increase the sensitivity by approximately 0·6 log units, which is roughly the same as the increase that can be achieved by pupil dilation (see above).

It is puzzling that the change in pupil size following a decrease in illumination is so slow, for, if the animal relies on changes in pupil size for accurate vision, a sudden decrease in illumination will result in a substantial time during which vision is defective. Under other circumstances the pupil can dilate rapidly: this happens, for example, when the animal attacks its prey or is frightened. This shows that the slow time course shown in Figs 1 and 2 is not due to any physiological limitations of the system.

The response to an increase in illumination was much quicker, and it was not possible to measure its time course accurately with the present method. Figure 2 shows one set of results obtained with an animal that was adapted for one minute to the dim illumination obtained by placing a 3 log unit neutral density filter in front of the light source. The pupil was photographed at different intervals after the filter was removed. It can be seen that after 5 sec the pupil was already close to its smallest size.

PUPIL SHAPE AND THE RETINAL IMAGE

In *Sepia* and *Octopus* the lens is spherical, and projects considerably beyond the front surface of the eye. The iris covers the surface of the lens, and so is curved. The eye obeys Matthiessen's ratio: that is, the centre of the lens is located a distance from the retina of about 2·5 times the lens radius (Pumphrey, 1961). These general relationships are shown in Fig. 3.

In the absence of a pupil, light passing through such a lens and coming to focus at the retina will form a cone of light, as shown in Fig. 3. In adult *Sepia* and *Octopus* the retinal receptors may be over 200 μm long, which means that a cone of light of these dimensions, if it is imaged at the surface of the retina, will cause a blur circle 166 μm in diameter by the time it reaches the base of the receptors. This is equivalent to a visual angle of 54′ in an eye having a focal

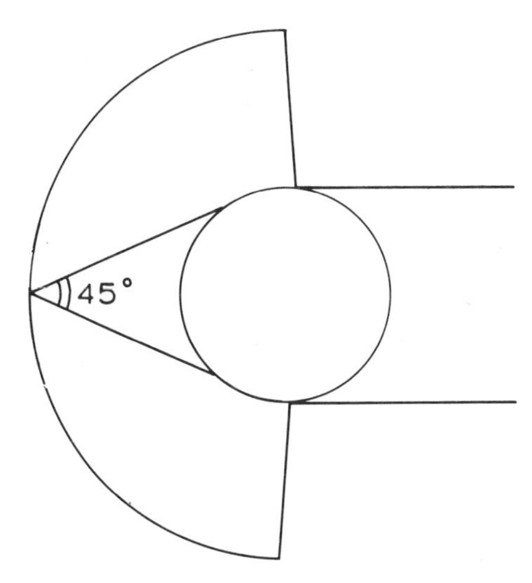

FIG. 3. The cone of light impinging on the retina in an eye obeying Matthiessen's ratio, in the absence of a pupil.

length of 10 mm, which is considerably worse than the value of 17′ for the visual acuity of the octopus, measured behaviourally (Sutherland, 1963).

Various possible mechanisms exist which might decrease this effect. For example, the image could be formed at the outer ends of the receptors, and the light trapped within the receptor by internal reflection, as occurs in some vertebrate rods (Denton & Nicol, 1964). Alternatively, the black screening pigment, which migrates out between the receptors in the light, could prevent the light from passing between the receptors. This has been demonstrated in the compound eye, and Lettvin and Maturana have found that the receptive field of single optic nerve fibres in *Octopus* increases greatly in the dark, a finding that they correlated with pigment migration (summarized in Lettvin & Pitts, 1962).

One consequence of a constricted pupil in an eye obeying Matthiessen's ratio is that although light from certain parts of the visual field will impinge on the retina perpendicularly, light from other parts of the field will not pass through the centre of the lens, and so will reach the retina at an angle (e.g. the pencil of light B in Fig. 4). Light rays of the former class will form a sharp image on the retina, but those that pass through the edge of the lens will only do so if there is some retinal mechanism, such as those discussed above, that prevents blurring. If this mechanism is the black

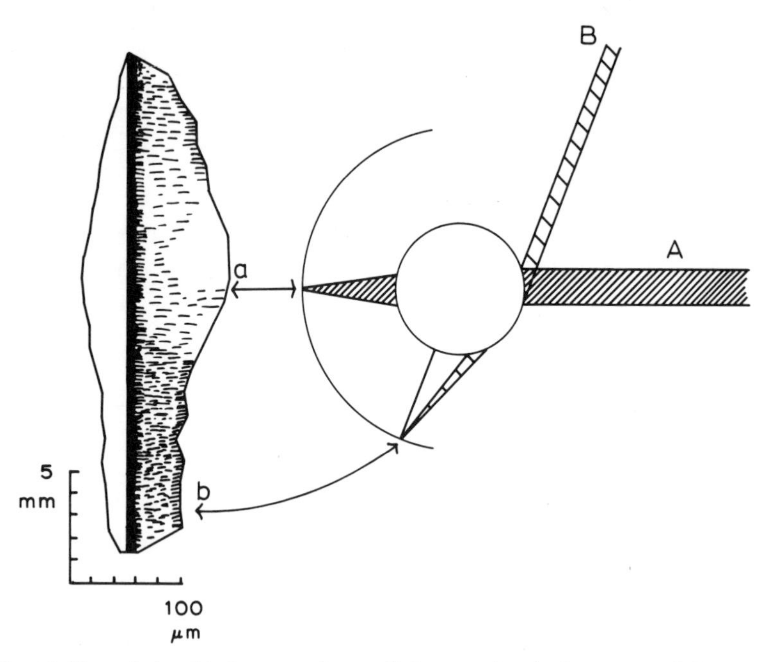

FIG. 4. The relationship between the pupil shape and retinal anatomy of *Octopus*. The left-hand half of the figure is taken from fig. 4 of Young (1963), and represents a straightened-out vertical section through the centre of a light-adapted *Octopus* retina. The figure shows the length of the basal segments (to the left of the central dark line) and rhabdomes (to the right) at different distances along the retina, and indicates the approximate position and density of the screening pigment granules. The right-hand half of the figure represents a vertical section through an octopus eye with a constricted pupil, and shows how light reaches parts of the retina such as (a) perpendicularly, but other parts such as (b) at a considerable angle.

screening pigment that migrates out between the receptors in the light, we might expect that the concentration of pigment would be greater in those parts of the retina that receive light rays at an angle. This appears to be the case. Thus in *Octopus* the concentrated pupil has the form of a horizontal slit. The right-hand side of Fig. 4 represents a vertical section through the *Octopus* eye. It can be seen that light from the horizontal plane will reach the equator of the retina perpendicularly, but light from above or below the horizontal plane must pass through the edge of the lens, and will reach the retina at an angle. This corresponds exactly with an area of the retina, the equatorial strip, where the rhabdomes are longer, and the pigment migration less pronounced (Fig. 4) (Young, 1963). Away from this strip the light reaches the retina at an angle, but in these areas the receptors are shorter and the screening pigment more pronounced, which will tend to preserve the accuracy of the image.

One obvious difficulty with this argument is that although the equatorial strip receives some light rays perpendicularly, it will receive a substantial amount of light at an angle as well. Thus Fig. 4 represents a vertical section through the eye, the plane in which the pupil is constricted, but a horizontal section through the eye would have an appearance closer to that shown in Fig. 3, with light entering along the whole length of the pupil. We might expect, therefore, that even in the equatorial strip blurring would occur in the horizontal plane unless pigment was present. However, in the first place it is at least possible under these circumstances to achieve a sharp effective image by "neural sharpening", mediated by inhibitory lateral interactions in the retina or optic lobes. Such neural sharpening can only take place if more light reaches the retina at the point of image formation than elsewhere, a circumstance that holds for the situation shown in Fig. 3, but not for the oblique rays in Fig. 4. Secondly, Sutherland (1963) has shown that narrow horizontal striations are more readily detected by octopuses than narrow vertical striations. Horizontal striations have their brightness differences in the vertical plane and *vice versa*, so that this result fits well with the horizontal pupil of the octopus, and suggests that there is indeed some blurring along the equatorial strip, as would be expected from Fig. 3. Chromatic and spherical aberration may, of course, also have contributed to this result, since, as Heidermanns (1928) pointed out, those will also have their main effect on the definition of horizontal differences.

The pupil of *Sepia* has a more complex shape (Fig. 1) than that of *Octopus*. There is a central section which is more or less horizontal, and the retina shows an equatorial strip like *Octopus*, which correlates with this. There are also conspicuous vertical segments at the front and rear of the pupil. On occasion these vertical segments may remain open, while the horizontal segment is closed, leading to the appearance of a double pupil. It seems likely that this double pupil can be correlated with two areas of specialization described by Young (1963) for the *Sepia* retina, which lie at the anterior and posterior poles of the retina, where the receptors are longer than elsewhere and the pigment migration less marked. Figure 5 shows how these two areas of retinal specialization may relate to the two vertical parts of the pupil. The light hits the retina perpendicularly at the specialized areas, so that collimation is not

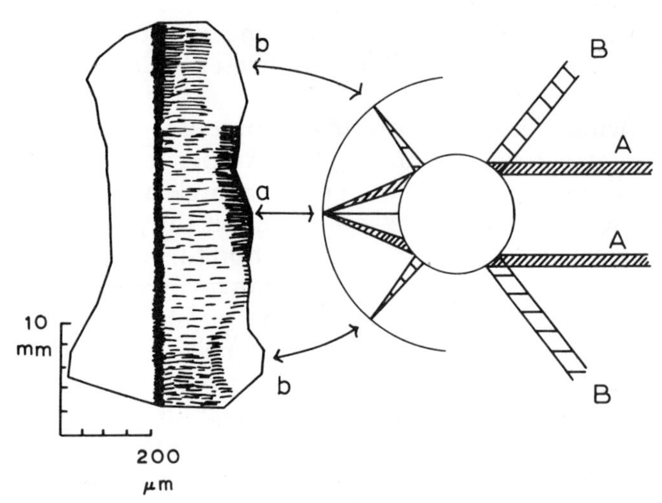

FIG. 5. The relationship between the pupil shape and retinal anatomy of *Sepia*. The left-hand half of the figure is taken from fig. 9 of Young (1963), and shows the distribution of screening pigment and rhabdome lengths in a *horizontal* section of a light-adapted *Sepia* retina. The right-hand half of the figure represents a horizontal section through *Sepia* eye and shows how pencils of light through the anterior and posterior vertical sections of the pupil will reach the retina.

needed, and each specialized area receives light through one of the pupil's vertical segments only.

The presence of a large spherical lens obeying Matthiessen's ratio is characteristic of aquatic animals, and has obvious advantages. Nevertheless, the combination of such a lens and a constricted pupil must result in vision being more efficient for some directions of view than others. The shape of the pupil can select which parts of the visual field will stimulate the retina most effectively. In the case of *Octopus* and *Sepia* the pupil shape and retinal structure can be correlated with the animals' habits. Both animals live on the bottom, and objects of the same horizontal level will be most important to them, which correlates with the presence of an equatorial strip. *Sepia* has in addition two areas of specialization, which may be correlated with swimming backwards, and feeding on prey directly in front of the animal (see Messenger, this Symposium, pp. 349–358), both of which habits are less developed in *Octopus*. Other cephalopods are free swimming; in one of these at least, *Loligo pealeii*, the central strip of the retina is lacking (Young, 1963).

Many other marine animals have eyes that obey Matthiessen's ratio. Denton & Nicol (1964) have shown that in some elasmobranchs the orientation of the reflecting layers of the tapetum correlates with the direction in which the light rays impinge upon the retina. The elasmobranchs studied by Denton & Nicol have circular pupils, but in other species a variety of pupil shapes occur, including in some cases (e.g. *Scyliorhinus caniculus*) the appearance of a double pupil rather similar to that found in *Sepia* (Franz, 1931; Walls, 1942). The shape of these pupils will affect the brightness and accuracy of the retinal image, but so far no corresponding regional differences of retina or tapetum have been described.

References

Daw, N. W. & Pearlman, A. L. (1974). Pigment migration and adaptation in the eye of the squid, *Loligo pealeii*. *J. gen. Physiol.* **63**: 22–36.

Denton, E. J. & Nicol, J. A. C. (1964). The choroidal tapeta of some cartilaginous fishes (*Chondrichthyes*). *J. mar. biol. Assoc. U.K.* **44**: 219–258.

Franz, V. (1931). Die Akkommodation des Selachienauges und seine Abblendungsapparate, nebst Befunden an der Retina. *Zool. Jb.* (Zool.) **49**: 323–462.

Hagins, W. A. & Liebman, P. A. (1962). Light-induced pigment migration in the squid retina. *Biol. Bull. mar. biol. Lab., Woods Hole* **123**: 498.

Heidermanns, C. (1928). Messende Untersuchungen über das Formensehen der Cephalopoden und ihre optische Orientierung im Raume. *Zool. Jb.* (Zool.) **45**: 609–650.

Lettvin, J. Y. & Pitts, W. H. (1962). Neapolitan studies. *M.I.T. Quart. Progr. Rep.* No. 64: 288–292.

Pumphrey, R. J. (1961). Concerning vision. In *The cell and the organism.* Ramsey, J. A. & Wigglesworth, V. B. (eds). Cambridge: University Press.

Sutherland, N. S. (1963). Visual acuity and discrimination of stripe widths in *Octopus vulgaris* Lamarck. *Pubbl. staz. zool. Napoli* **33**: 92–109.

Walls, G. L. (1942). *The vertebrate eye and its adaptive radiation.* Michigan: Cranbrook Inst. of Science.

Young, J. Z. (1963). Light- and dark-adaptation in the eyes of some cephalopods. *Proc. zool. Soc. Lond.* **140**: 255–272.

Symp. zool. Soc. Lond. (1977) No. 38, 287–308.

EXTRA-OCULAR PHOTORECEPTORS IN CEPHALOPODS

A. MAURO

The Rockefeller University, New York, USA

SYNOPSIS

Micro-anatomical, biochemical and electrophysiological data have established that both in octopods and decapods extra-ocular photoreceptors are present in vesicles—photosensitive vesicles. In octopods a single photosensitive vesicle is located at the hind end of each stellate ganglion. In decapods analogous photosensitive vesicles are located near the optic tract. In *Loligo vulgaris*, for example, they are distributed along the ventral side of the tract. In *Todarodes sagittatus*, for example, there are present, bilaterally, a dorsal set of vesicles at the base of the vertical lobe and a ventral set appearing as a grape-like structure consisting of many vesicles. The ventral set gives rise to a nerve which terminates in the peduncle lobe on the optic tract. That the extra-ocular photoreceptors in cephalopods do not serve a visual function but rather act as light detectors is indicated by the absence of a lens apparatus and any oriented array of radially extending rhabdomeric processes to form a "retina", as in the eye proper. The physiological role of such a system of light detectors may vary in different cephalopods; in any event it remains to be elucidated by future research.

HISTORICAL INTRODUCTION

Within the past decade it has been established both from anatomical and electrophysiological data that extra-ocular photoreceptors are present in cephalopods. In the case of octopods the photoreceptor cells reside in the "epistellar body", a small vesicle on the ventral surface at the hind end of both stellate ganglia, while in decapods comparable vesicles lie near the brain both on the right and left side of the optic tract.

The origin and impetus for this subject stems from the first piece of research by John Young, which he undertook upon completing his undergraduate studies at Oxford, thus embarking on a career of research on the cephalopod nervous system that by now has extended over more than four decades. It is fitting to quote in translation the opening paragraph of his paper entitled "Sopra un nuovo organo dei cefalopodi" ("On a new organ in cephalopods")—written in Italian, be it noted—which reported briefly his findings at the Stazione Zoologica in Naples and appeared in the Bollettino della Societa Italiana di Biologia Sperimentale ("Bulletin of the Italian Society of Experimental Biology"), August 1929:

> "In the course of research of which I give an account elsewhere, on the phenomenon of nervous degeneration in different cephalopods

> and particularly in *Eledone moschata,* I had occasion to note a small yellow area on the ventral side of the posterior end of the stellate ganglion. In the literature I was able to find only one indication relative to this formation; Bauer, in his well-known work on the cephalopods [1909], records the existence of a pigmented spot (Pigmentfleck), but does not give particulars about it."

After noting further on that the epistellar body is present in several other octopods, albeit colourless in the common *Octopus vulgaris,* the author reports without comment: "However, I have not been able to demonstrate the existence of an analogous body in *Loligo* and *Sepia,* at least in the same position, that is to say, in the proximity of the stellate ganglion". The paper concludes by noting that two cell types make up the epistellar body and that sufficient evidence is lacking to ascribe a definite function to it though the author hints at a glandular function as a possibility.

A full-length paper by John Young appeared in 1936 which gives a detailed account of the histological structure of the epistellar body, resulting from an analysis covering some nine species of octopods. What is remarkable about this paper is that it opens by focussing attention on the absence of the epistellar body in squid and indeed points out for the first time that in the stellate ganglion there exist small nerve cells whose processes fuse to form giant axons. In the case of *Loligo* the nerve cells are localized to a "giant fibre lobe" while in *Sepia* they are dispersed throughout the ganglion. The giant axons ("giant fibres") leave the ganglion via the stellar nerves, one of which, being the largest, located at the hind-end of the ganglion, is now known as the "squid giant axon". Thus, in looking for the epistellar body in the decapods, John Young found instead the giant fibre system. We see, then, the first important result of the research on the epistellar body, namely, the discovery of the giant axon in decapods which made possible the remarkable advances in the biophysical analysis of the ionic currents underlying the action potential in the squid giant axon by Hodgkin, Huxley & Katz in 1949.

The major part of the 1936 paper deals with the structure of the epistellar body in some detail. It will suffice here to note that one of the two cell types present gives rise to a long process extending radially into the "cavity" of the epistellar body; and the hypothesis is put forward that these cells have a neurosecretory function just as, for example, the cells of the adrenal medulla.

The next series of papers by John Young are now classical for in them is established the detailed neuroanatomy of the first-, second-

and third-order neurons that make up the giant fibre system (reviewed by Dr Rainer Martin in this volume, see his chapter). In devoting his full attention to the very important subject of the giant fibres in the decapods no further mention of the epistellar body appears in his research papers. However, his hypothesis that this body has a neurosecretory function was destined some years later to play a role in calling attention again to it. Indeed in 1962 Nishioka, Hagadorn & Bern, being dedicated scholars of neurosecretory organs, became curious about the epistellar body having a presumptive neurosecretory function and proceeded to study its ultrastructure by electron microscopy in the Pacific octopus, *Octopus bimaculatus.* It soon became evident to them that the cells with the long processes described by John Young earlier are in fact processes with microvilli as seen in rhabdomeric photoreceptor cells. Moreover, in some cells the microvilli are juxtaposed to form rhabdomes as seen in the retina of cephalopods. They concluded from their ultrastructural analysis that

> "the epistellar body does not possess the usual ultrastructural attributes of a neurosecretory system; rather, it shows a striking resemblance to photosensitive structures present in invertebrate animals." (Nishioka, Hagadorn *et al.*, 1962: 406).

It may be recalled that in John Young's note of 1929 he comments on his failure to find an epistellar body in decapods but cautions the reader that this body is absent at least near the stellate ganglion. This caveat proved to be insightful because in 1939 Thore reported the presence of vesicles below the optic stalk in *Illex.* And in 1954 Haefelfinger, extending Thore's findings, described in *Illex* two structures above and below the optic stalk which, he noted, had histological features in common with the epistellar body. In their review paper on organs associated with the optic tract in cephalopods Boycott & Young in 1956 again commented on the presence of vesicles like the epistellar body in decapods, giving photomicrographic evidence from *Sepia* and *Loligo* and citing the work of Thore and Haefelfinger. These were called "parolfactory vesicles" by Boycott & Young (1956) because of their location near the olfactory lobe. Thus by 1956 it was established that in decapods vesicles like the epistellar body are not present on or near the stellate ganglion but rather are intimately associated with the central nervous system above and below the optic stalk.

In a second paper by Nishioka, Yasumasu, Packard, Bern & Young in 1966, spectrophotometric and biochemical data were reported on the photopigment contained in the epistellar body from *Eledone moschata* and in the parolfactory vesicles from *Loligo vulgaris*. In both instances the absorption spectrum obtained indicated unambiguously the presence of rhodopsin in the epistellar body and in the parolfactory vesicles. Moreover, in this study electron microscopic data was obtained which clearly showed that the parolfactory vesicles contained cells with microvilli characteristic of rhabdomeric photoreceptors. It was very clear from these studies that the ultrastructural and biochemical data gave strong support for the hypothesis that the vesicles present in octopods and decapods contain photoreceptors. In the following year Messenger (1967) reported histological findings of large rhabdomere-containing cells appearing near the optic tract of *Liocranchia*, a cranchiid squid. In this paper Messenger expressed the view that "in this class of molluscs extra-ocular photoreception is widespread".

Thus, by 1967 it only remained to carry out a direct physiological test of the photoreceptor hypothesis, namely, to record the generator potential associated with the illumination of photoreceptor cells and the resulting train of impulses conveying information to other regions in the nervous system. In the following year electrophysiological evidence of photoreceptors in the epistellar body of *Eledone moschata* was obtained by Mauro & Baumann (1968), and subsequently in the parolfactory vesicles of *Todarodes sagittatus* by Mauro & Sten-Knudsen (1972).

Recently several new species of deep-sea squid have been reported that have vesicles of unusual size near the central nervous system containing cells with the features of rhabdomeric photoreceptors (R. E. Young, 1972, 1973).

In concluding this brief historical review it is fitting to note that Richard Young and John Young have proposed jointly that the term "photosensitive vesicles" be adopted in place of "parolfactory vesicles" and "epistellar body" since these organs are in fact photoreceptors; this new nomenclature will be adhered to in this review, as consistently as possible.

EXTRA-OCULAR PHOTORECEPTORS IN OCTOPODS

Review of the micro-anatomy

In octopods the photosensitive vesicle appears as a single bean-shaped structure on each stellate ganglion at its hind end near the largest stellar nerve (Fig. 1). This structure—the epistellar photo-

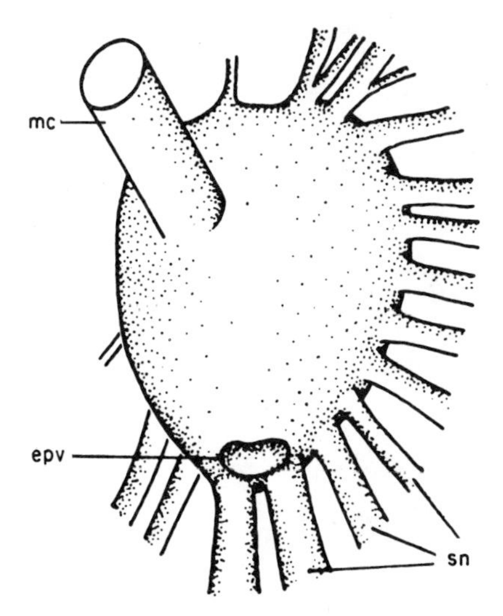

FIG. 1. Diagram of stellate ganglion of *Eledone moschata*. The ovoid epistellar photosensitive vesicle (epv) lies at the posterior end of the stellate ganglion near the origin of several stellar nerves (sn); (mc) mantle connective. (Reprinted with permission from Nishioka, Yasumasu *et al.*, 1966.)

sensitive vesicle (the epistellar body)—lies on the stellate ganglion confined within a common capsule of connective tissue; but it can be found, though rarely, completely separated from the surface of the ganglion. The vesicle consists of two cell types. The first is seen as a large bipolar sensory cell ovoid in shape, about 15 by 30 μm; the cell body extends in a long apical process to form a rhabdomeric structure whereas the opposite extremity extends to form an axon (Fig. 2). Interspersed between the sensory cell bodies are seen cells of the second type, the so-called supporting or epithelial cells, which are smaller in size and stain darkly. The rhabdomeric processes extend more or less radially toward the centre of the vesicle. In poorly fixed preparations these processes are extracted to some degree whereupon the vesicle gives the appearance of having a "lumen". The axonal processes proceed around the periphery converging to form a nerve, which enters the ganglion. In most octopods the vesicles, in a given preparation, are either pink, yellow or orange; in *Octopus vulgaris* they are colourless. It is important to note that there is no dioptric apparatus associated with the vesicle, which is also the case in decapods.

The ultrastructure of the sensory cell was established in all essential details by Nishioka, Hagadorn *et al.* (1962) and Nishioka,

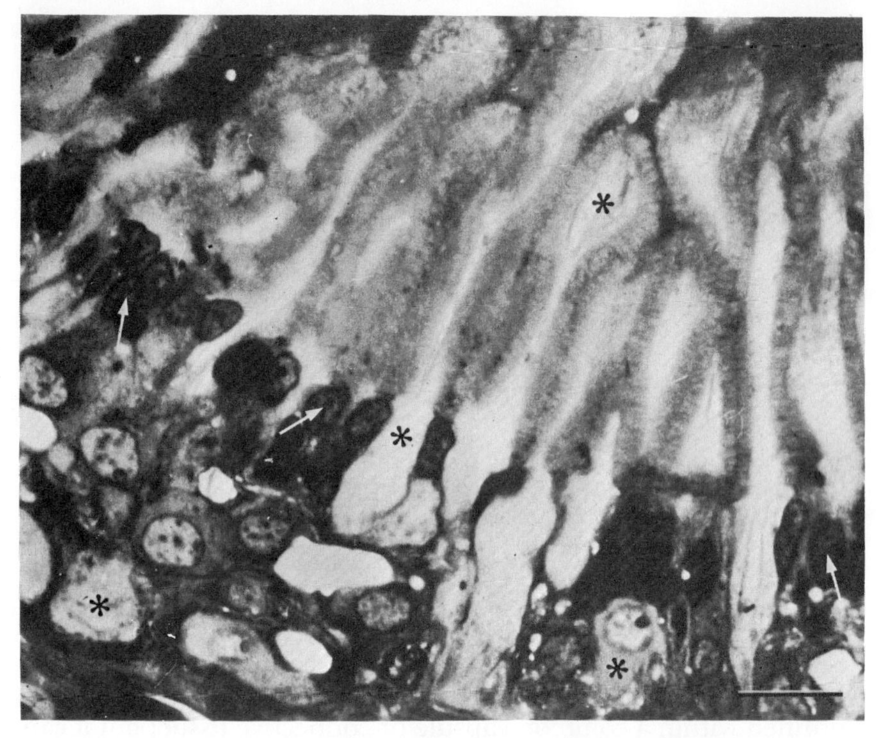

FIG. 2. Light micrograph of the epistellar vesicle of *Eledone moschata*. The vesicle is formed by two types of cells: large photosensitive cells (*) which send elongated processes in the centre of the vesicle, and epithelial or supportive cells with small dark nuclei (→). The bar represents 20 μm (×700).

Yasumasu *et al.* (1966). The electron micrographs obtained in these studies showed clearly the rhabdomeric nature of the long apical process by revealing the typical microvillar extensions of the cell membrane (Fig. 3). This fact alone was convincing evidence that the bipolar cells are photoreceptors of the rhabdomeric type. Recently these results were confirmed in a study which was undertaken primarily to establish the ultrastructure of the nerve associated with a photosensitive vesicle in both octopods and decapods (Perrelet & Mauro, 1972). A new fact which emerged from this study was the existence of a spectrum of fibre diameters in the epistellar nerve. For example, in a nerve which was 50 μm in diameter and contained about 1500 fibres the fibre diameters ranged from 0·3 to 1·5 μm (Fig. 4).

Another new finding is the presence of closely packed tubular profiles in the cytoplasm of the photoreceptor cell body. The

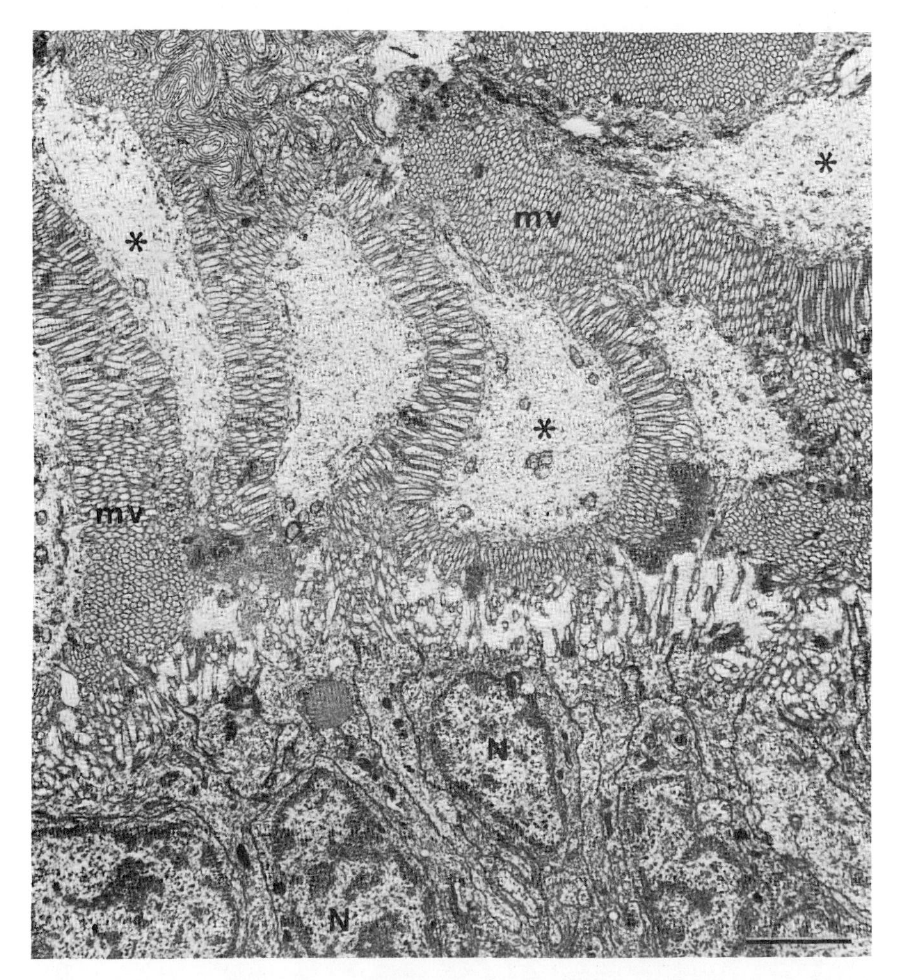

FIG. 3. Electron micrograph of the epistellar vesicle of *Eledone moschata*. The centre of the vesicle, illustrated here, contains closely-packed microvilli (mv) or rhabdomeres which stem from the processes (*) of the photoreceptor cells. Several nuclei (N) of the supporting cells can be seen beyond the rhabdomeric processes. The bar represents 0·5 μm (×28 000).

tubules appear to be limited by unit membranes; they measure about 80 nm in diameter and are arranged in a paracrystalline formation (Fig. 4). Often these tubules lie close to arrays of stacked membranes. Nerve sections close to the cell body in some areas reveal axons with a paracrystalline array of tubules, presumably extending from the cell body. The physiological significance of these tubules remains to be elucidated.

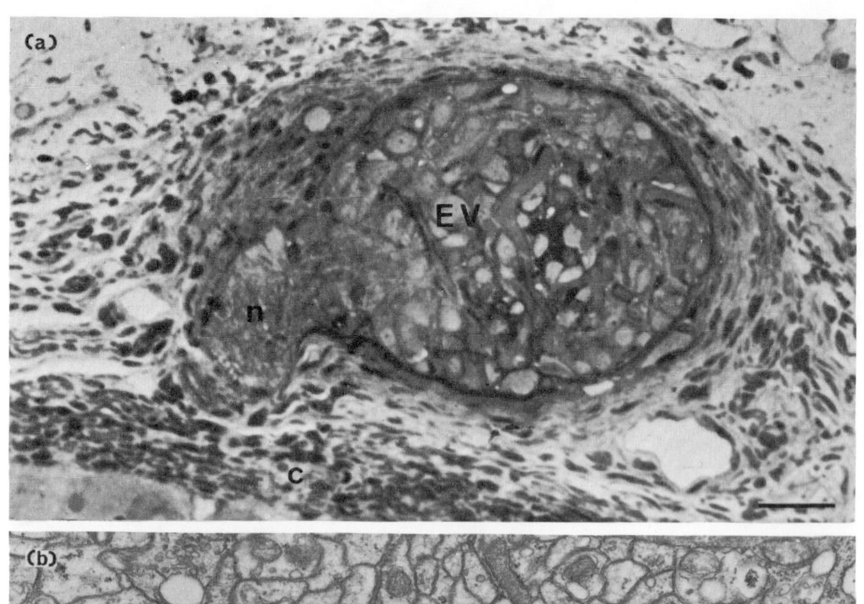
(a)
EV
n
c

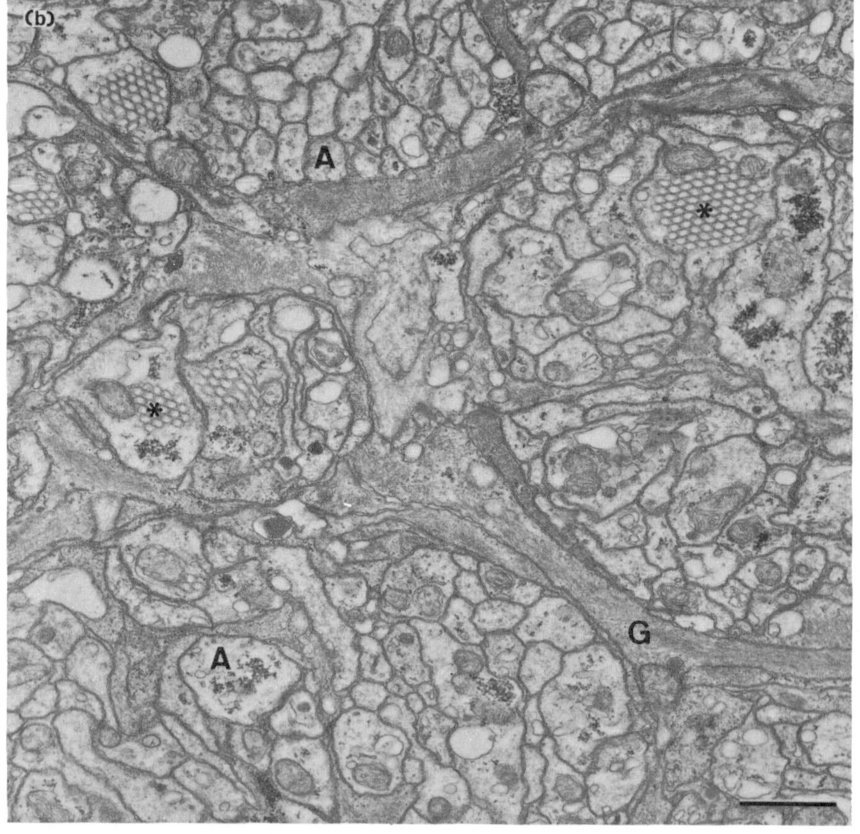
(b)
A
*
*
G
A

Spectrophotometric data

Nishioka, Yasumasu *et al.* (1966) obtained the absorption spectrum of an extract of 30 epistellar photosensitive vesicles from *Eledone moschata.* The extract had an initial λ_{max} of 470 nm; after irradiation at higher pH the λ_{max} shifted to 380 nm. Moreover, the extract of eyes from the same group of animals gave nearly an identical absorption spectrum. These results, they concluded, were consistent with the photopigment rhodopsin. To substantiate further the presence of rhodopsin the antimony chloride method for the determination of vitamin A_1 was used on an extract of vesicles. The spectrophotometric data obtained by this method agreed with that of commercial vitamin A_1.

Electrophysiological evidence of photoreceptors

In a study carried out at the Stazione Zoologica in the summer of 1968, Mauro & Baumann (1968) succeeded in recording electrical responses from the epistellar photosensitive vesicle in *Eledone moschata.* The recordings were made with conventional micropipette electrodes. In the course of a given experiment, the responses were predominantly extracellular (Fig. 5). However in a small number of cells it was possible to record intracellularly. Successful penetration of the cell membrane usually gave resting potentials of approximately 40 mV which in some cells could be maintained for as long as 10 min. In Fig. 6 recordings are shown of the intracellular response obtained with light flashes at three different intensities, increasing from top to bottom. The characteristic components of the generator potential associated with invertebrate photoreceptors are clearly evident. The transient component arises with the onset of the light flash and subsides to a plateau, the steady state component, which is maintained according to the duration of the light stimulus. Small fluctuations or "quantum bumps" can be seen in the steady state response, which is another feature of the electrical response of invertebrate photoreceptors to light, especially at low light intensities.

FIG. 4. (a) Light micrograph of the epistellar vesicle of *Eledone moschata.* The vesicle (EV) was serially sectioned and this particular section involves the point of emergence of the epistellar nerve (n) from the vesicle. Note the dense connective capsule (c) of the stellate ganglion. The bar represents 40 μm (×250). (b) Electron micrograph of the epistellar nerve. Axons (A) of different sizes can be seen, some of them containing the typical paracrystalline arrays of tubules (*) extending from the photosensitive cell body. Axons are arranged in discrete bundles separated by glial cell (G) profiles. The bar represents 1 μm (×13 000).

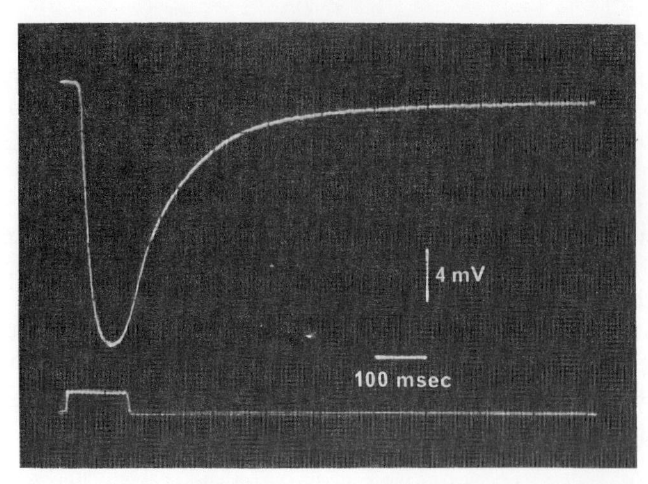

FIG. 5. Extracellular recording of the generator potential from an epistellar vesicle of *Eledone moschata* with a micropipette electrode. The light flash is recorded in the lower tracing.

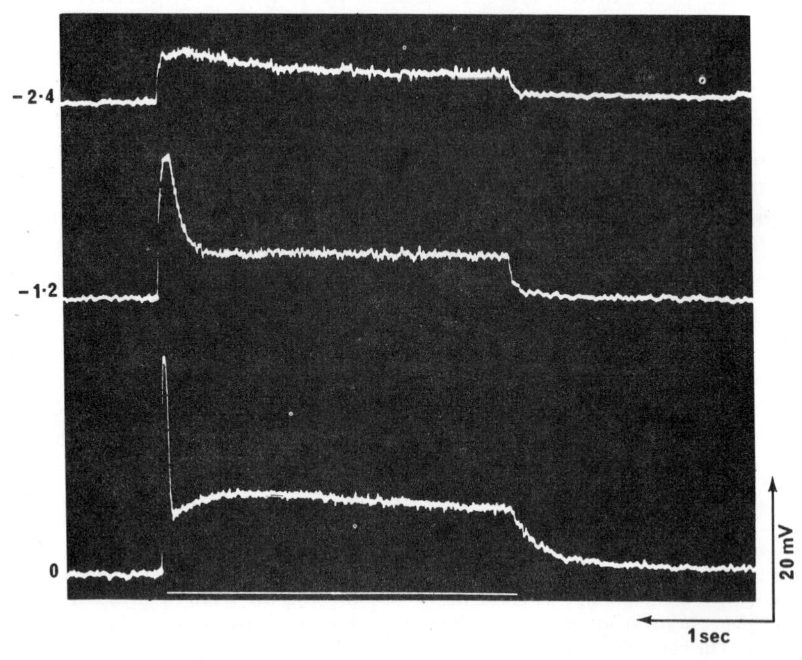

FIG. 6. Generator potential recorded intracellularly from a cell in the epistellar vesicle with light of increasing intensity lasting 2·4 sec. Relative light intensity, expressed in log units, is indicated at the left of each response. Light flash indicated by white line. (Reprinted with permission from Mauro & Baumann, 1968.)

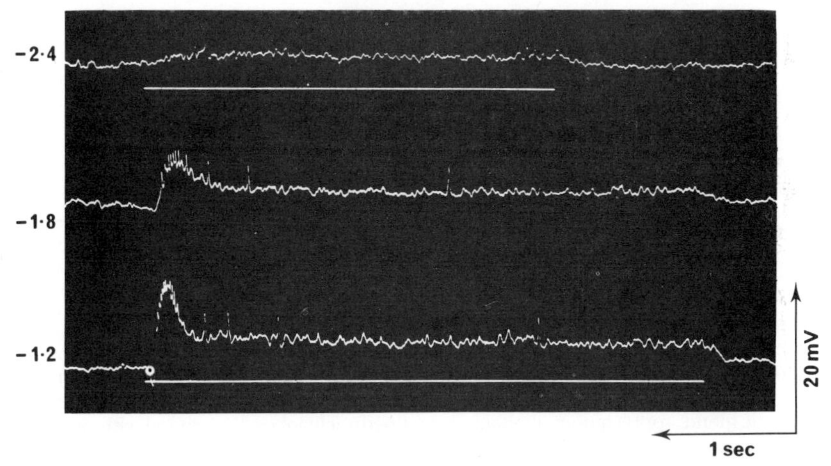

FIG. 7. Responses to light with repetitive firing. Stimuli lasted 4·3 sec for the weak light intensity of −2·4 log units and 5·8 sec for the two stronger intensities of −1·8 and of −1·2. As indicated by the negative potential at the beginning of the response to the light intensity of −1·2 log units, the penetration of the cell with the microelectrode was probably not complete. (Reprinted with permission from Mauro & Baumann, 1968.)

The absence in these records of repetitive firing in response to steady state illumination can be accounted for as arising from damage to the trigger zone, located between the cell body and the axon, upon penetration of the micropipette electrode. Indeed, on rare occasions, cells were found that displayed a train of impulses during the steady state component of the generator potential (Fig. 7). Repeated attempts to record impulses with metallic micro-electrodes from either the ganglion or the mantle connective, for signs of information being transmitted to the central nervous system, have thus far been unsuccessful. It is relevant to note here that in the squid, *Todarodes sagittatus*, light-evoked action potentials have been recorded from the nerve from the ventral photosensitive vesicles (ventral parolfactory vesicles) (Mauro & Sten-Knudsen, 1972). Thus, since the micro-anatomical evidence has established that the vesicles in the decapods and octopods are analogous structures, there is no reason to doubt that also in the epistellar photosensitive vesicles the photoreceptor cells give rise to propagated nerve impulses.

EXTRA-OCULAR PHOTORECEPTORS IN SQUID

Review of the micro-anatomy

As shown in Fig. 8, in *Loligo vulgaris* (and *Sepia*) photosensitive vesicles lie on the optic tract near the peduncle lobe mostly on the

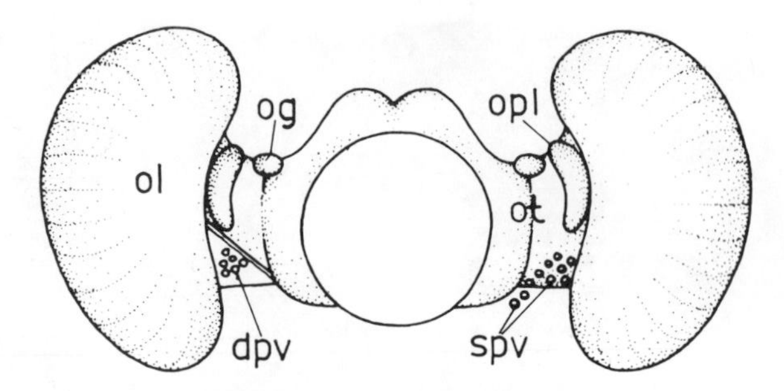

FIG. 8. Diagram of posterior view of brain of *Loligo vulgaris*. Most of the superficial photosensitive vesicles (spv) lie on the optic tract (ot) but a few also are attached to cartilage. Part of the optic tract has been cut away on the left side of figure to show the position of the deep photosensitive vesicles (dpv) which lie embedded in optic tract tissue; (ol) optic lobe; (og) optic gland; (opl) combined olfactory and peduncle lobes. (Reprinted with permission from Nishioka, Yasumasu *et al.*, 1966.)

ventral surface; others are imbedded within the tract. The vesicles in squid are usually smaller than in octopods, e.g., about one-fifth of the size of those found in *Eledone*, in animals of comparable size. The histological and ultrastructural studies carried out by Nishioka, Yasumasu *et al.* (1966) and more recently by Perrelet & Mauro (1972) have shown that vesicles in squid have features in common with those in octopods.

In the midwater squids *Todarodes* and *Illex* the photosensitive vesicles appear with a remarkable modification in that two distinct sets of vesicles are present, a dorsal and ventral set. The dorsal photosensitive vesicles lie near the base of the vertical lobe; the associated nerve fibres are short and enter the optic tract below. The ventral photosensitive vesicles on the other hand are more remote from the optic tract in that they give rise to a long nerve bundle which terminates in the peduncle lobe (Fig. 9). A striking feature of the vesicles in the ventral set is that they form a grape-like structure consisting of many vesicles (Fig. 10).

In the deep-sea squid *Bathyteuthis* only a ventral pair of organs is present. According to R. E. Young (1972) "each organ is very large; relative to the size of this squid, it is the largest that I have observed in any cephalopod". In *Galiteuthis*, another deep-sea species, each ventral organ is located close to the optic tract embracing part of the peduncle lobe (R. E. Young, 1972). This close association of the vesicles with the optic tract is also seen in *Liocranchia*, a cranchiid squid (Messenger, 1967).

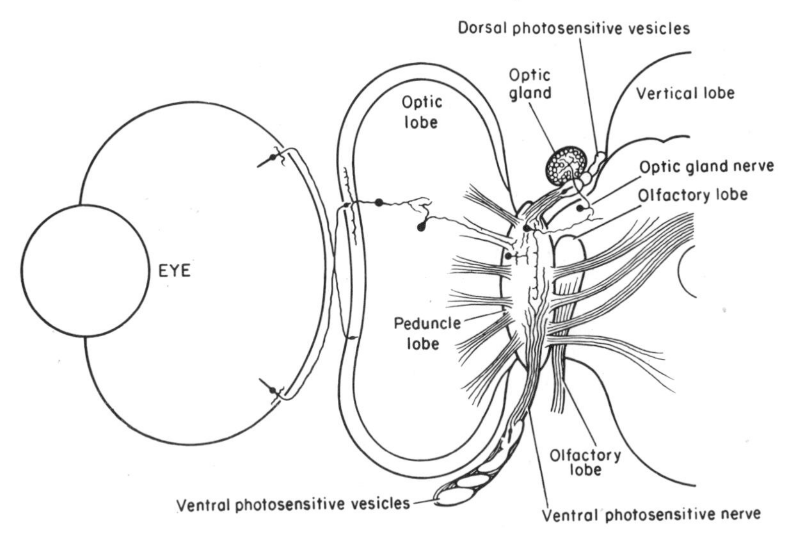

FIG. 9. Diagram showing position and connections of the ventral photosensitive vesicles of the squid *Todarodes*. (Reprinted with permission from Baumann *et al.*, 1970.)

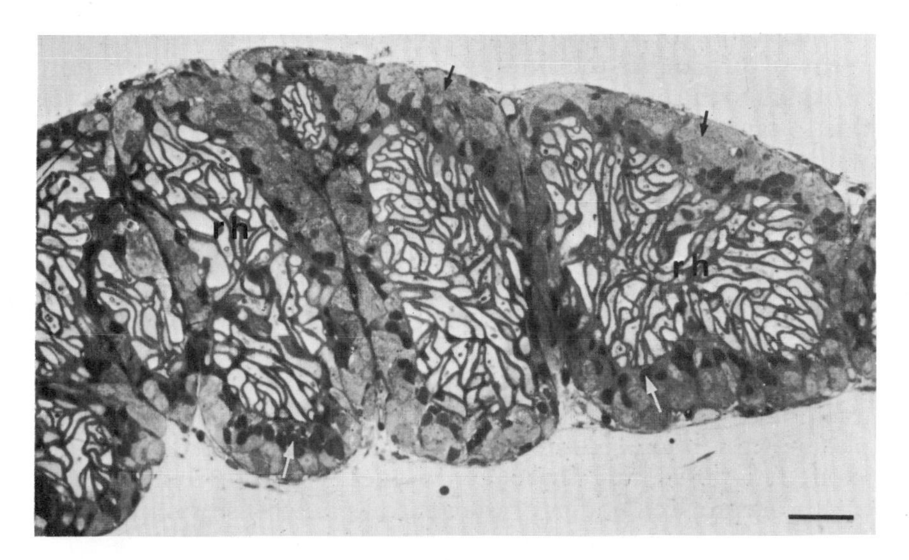

FIG. 10. Light micrograph of the ventral photosensitive vesicles of *Todarodes sagittatus*. Each vesicle is formed by a wall showing clear and dark nuclei and a centre with a rhabdomeric pattern (rh). Clear nuclei (black arrows) belong to the photosensitive, rhabdomeric cells, while dark ones (white arrows) are from the supporting cells. Individual vesicles aggregate so as to form a grape-like structure (×220).

One of the most remarkable variations reported thus far in the extra-ocular system of vesicles is seen in *Abraliopsis*. In one species three sets of vesicles are present and in another species there are four (R. E. Young, 1973). This author notes that a "window" is provided by a sparsity of chromatophores on the dorsal surface of the integument overlying the dorsal set of vesicles. Indeed on the ventral surface the absence of chromatophores provides a virtually transparent window for the ventral set of vesicles. However, while such modification may occur in the integument of some cephalopods it should not be inferred that presence of chromatophores renders the surface opaque to light.* Moreover in the case of octopods the rhythmic parting of the mantle wall near the funnel, during breathing movements, permits the entry of both sea-water to the mantle cavity and light to the epistellar photosensitive vesicles since they lie on the stellate ganglia and thus "look" into the cavity.

As in the epistellar nerve in octopods, the nerve bundle from the ventral set of vesicles shows a spectrum of fibre sizes ranging from 0·8 to 0·2 μm in diameter. In this case the small fibres appear in small bundles as islets interspersed among large fibres (Fig. 11). Careful study of sections close to the sensory cell bodies show that the large fibres originate from the latter and therefore are afferent. What is puzzling is the finding that sections of the nerve more proximal to the brain show mostly large fibres. This feature may be due to tapering of some of the large fibres to smaller diameters, leading one to suspect that such fibres are efferent. However, no evidence has been found of synaptic vesicles or other synaptic apparatus in the proximity of the cell body to support the hypothesis of efferent fibres. Clearly more ultrastructural analysis is necessary to resolve the significance of the small fibres. That small fibres are present is supported by electrophysiological evidence, which will be discussed presently.

In commenting on the ultrastructure of the epistellar vesicle it was noted that paracrystalline tubules were present in the cytoplasm of the photoreceptor cell body and even extended into the axon. It is intriguing that such tubules were not found in the photoreceptor cell bodies of the ventral photosensitive vesicles.

* It is relevant here to note that even in the case of mammals penetration of light through the skull has been established by implanting photo-electric cells in different regions of the brain and measuring the light transmitted under daylight conditions (van Brunt, Shepherd, Wall, Ganong & Clegg, 1964).

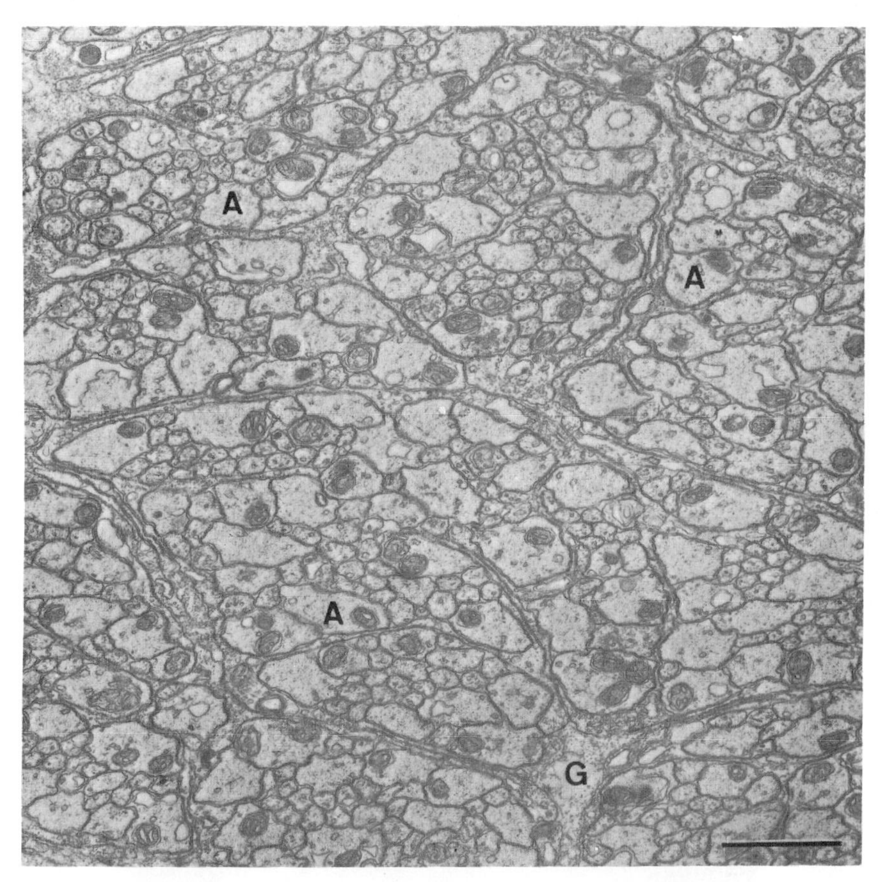

FIG. 11. Electron micrograph of the nerve to the ventral set of vesicles of *Todarodes sagittatus*. The nerve issued from the vesicles contains axons of varying diameters. The larger profiles (A) could be related to the photosensitive cells in the vesicle. Glial cells (G) can be seen in between axons. The bar represents 2 μm (×8000).

Spectrophotometric data

Spectrophotometric measurements carried out by Nishioka, Yasumasu *et al.* (1966) on an extract of photosensitive vesicles obtained from *Loligo vulgaris* yielded an initial λ_{max} of 493 nm and a final λ_{max} of 380 nm upon irradiation at a higher pH. These data were virtually identical with the absorption spectra obtained on an extract of eyes from the same animals. Moreover, the presence of vitamin A_1 was confirmed by the antimony chloride method. Nishioka, Yasumasu *et al.* (1966) concluded that these data pointed to rhodopsin as the photopigment, as in the octopods.

Spectral sensitivity

We attempted to establish the spectral sensitivity of the photoreceptors in the ventral photosensitive vesicles of *Todarodes sagittatus* by recording extracellularly the generator potential excited by six different wavelengths of light. Six interference filters calibrated at 586, 542, 503, 453 and 437 nm were used in conjunction with a xenon light source. In two separate experiments the maximum response occurred at 503 nm.

Electrophysiological evidence

We have recorded light-evoked electrical responses extracellularly from the vesicles of *Sepia, Loligo* and *Todarodes* both by suction and micropipette electrodes. An example of an extracellular recording of the generator potential by a micropipette electrode from a dorsal photosensitive vesicle in *Todarodes* is shown in Figs 12 and 13. On the other hand, in attempting to record intracellularly, our success has been rather meagre in comparison with the results obtained in octopods (e.g. Figs 6 and 7) in that the responses lasted for a few seconds owing to poor impalement. While we failed to obtain intracellular recordings comparable to those in octopods, we did succeed in recording light-evoked action potentials from the nerve from the ventral set of vesicles in a specimen of *Todarodes*,

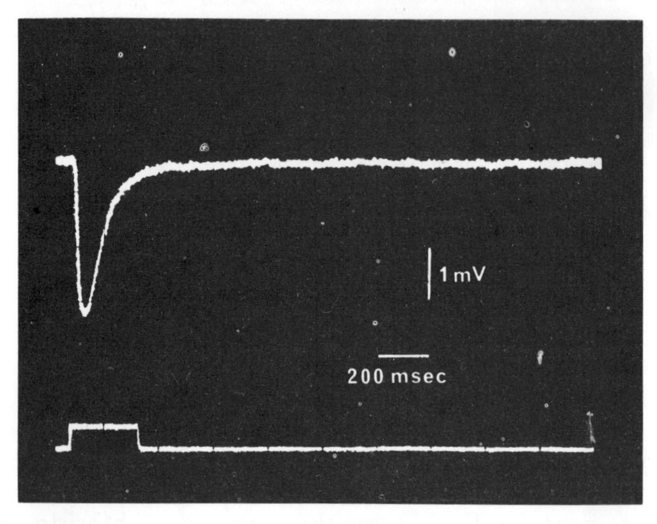

FIG. 12. External micropipette recording of the generator potential from the dorsal vesicle of *Todarodes sagittatus*. Light flash is registered in lower tracing. Intensity is supermaximal.

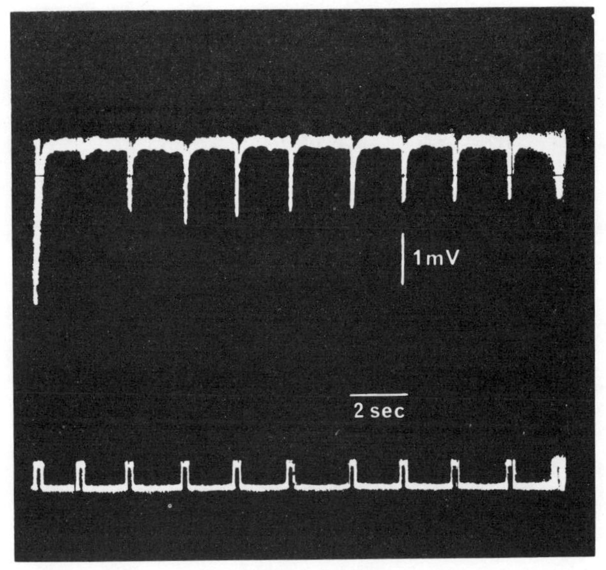

FIG. 13. Recording from same cell showing "suppression" and subsequent recovery of the generator potential to repetitive light flashes. Preparation was dark adapted.

as shown in Fig. 14. A comparable experiment failed in the octopod, *Eledone moschata*, because the epistellar nerve, as commented earlier, is virtually inaccessible. The vesicle end of the preparation was pinned to the base of the chamber and a small nerve bundle dissected from the main bundle and lifted in air by a wick electrode, the second electrode being in the bath. The monophasic compound action potential (upper tracing) resulted from a synchronous excitation of many fibres upon the onset of the light pulse (lower tracing). Asynchronous firing of nerve impulses was recorded earlier on the same preparation as shown in Fig. 15. This is the clearest evidence we obtained that the vesicles contain functional photoreceptors, in that light-evoked impulses are in fact propagated to the central nervous system. These results are consistent with the recordings reported by MacNichol & Love (1961) showing nerve impulses from the retinal nerves and the optic lobe of *Loligo pealeii* upon illumination of the eye. It is relevant to note that in molluscs the axons from the photoreceptor cells are electrically excitable whereas in arthropods—thus far reported*—the

* Very recently Belmonte & Stensaas (1975) have reported that light evokes repetitive spikes in the axons of the photoreceptor cells of the scorpion eye.

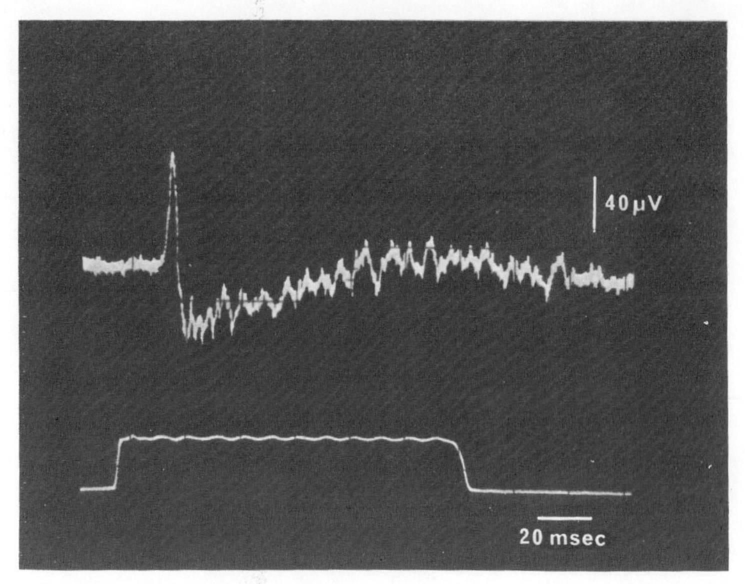

FIG. 14. Light-evoked electrical response of small bundle dissected from the nerve from the ventral photosensitive vesicles. Initial diphasic recording is due to synchronous excitation of nerve fibres with onset of generator potential ("transient component" of generator potential). Asynchronous firing follows in response to steady state component of generator potential. Light pulse in lower tracing was recorded by photo-electrical cell. (Reprinted with permission from Mauro & Sten-Knudsen, 1972.)

axons are electrically inexcitable. This property of arthropod photoreceptors is seen very clearly in the extra-ocular photoreceptor cells in *Limulus polyphemus*, for example, in the ventral photoreceptors which are located along the so-called lateral olfactory nerves (Millecchia & Mauro, 1969). Only a photoreceptor cell very close to the central nervous system is able to excite ganglionic pathways by adequate electronic spread of the generator potential (Snodderly, 1969).

In the light of the small fibre system which was discovered by an electron microscopic study, as mentioned previously, it should be noted that electrical stimulation of the ventral nerve bundle also reveals the presence of a small fibre population in the compound action potential. The velocity of propagation of this slow group is about 0·1 m/sec. The main group—consisting of the large fibres—propagates at about 1·5 m/sec. It is interesting that stimulating the nerve at the peduncle-end gives a slow fibre response as far as the vesicle-end of the nerve. This implies that tapering of some fibres as they approach the distal end of the nerve does occur, as noted earlier. However, while the electrophysiological evidence supports

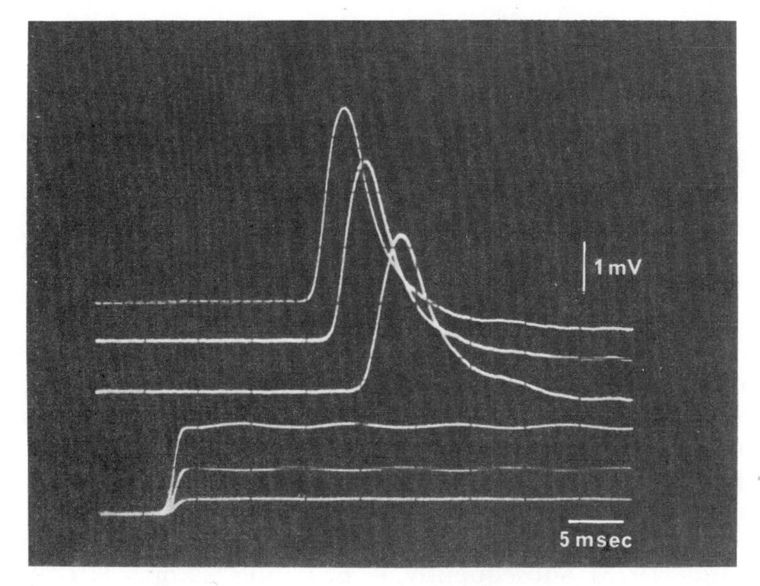

FIG. 15. Monophasic recording obtained when distal segment of nerve bundle was allowed to dry and thus became inexcitable. Three upper recordings given by the lower, middle and upper tracing show response to low, medium and intense light pulse, respectively. Baseline displaced upward manually between successive recordings. Corresponding light stimulus registered in lower set of recordings.

the ultrastructural evidence for small fibres in the nerve, it in no way resolves the physiological role played by such fibres.

Recently, Sperling, Lisman & Godfrey (1973) have briefly reported light-evoked responses obtained by intracellular recordings from the ventral photosensitive vesicles of *Loligo pealeii*.

DISCUSSION

We have reviewed the anatomical, biochemical and electrophysiological data that clearly establish the presence of extra-ocular photoreceptors in cephalopods, appearing as vesicles—photosensitive vesicles—near the optic tract in decapods and on the stellate ganglia in octopods. It is logical now to ask: what physiological purpose is served by these photosensitive vesicles that lie within the body cavity? To dismiss such organs by simply stating that they are vestigial and therefore serve no function is unsatisfying intellectually, if not meaningless. The absence of two crucial features as seen in the eye, namely, an oriented array of radially extending rhabdomeric processes and a lens apparatus, clearly eliminates a visual function for the extra-ocular photoreceptors; rather it appears that they serve as light-sensing organs informing the nervous system of light levels in the animal's environment. The electrophysiological evidence in *Todarodes*, which shows that the ventral nerve conducts light-evoked impulses from the ventral set of photosensitive vesicles to the peduncle lobe (J. Z. Young, see his chapter, this volume) and, it may be inferred, to other parts of the central nervous system, gives strong support for the hypothesis that the extra-ocular photoreceptors in cephalopods—at least in decapods—have a significant role as light detectors in controlling physiological function.

This function might be the regulation of cyclical activity, either circadian or seasonal, pertaining, for example, to spawning and reproductive behaviour (Baumann *et al.*, 1970). And recently it has been suggested that in some deep-sea squid, e.g. *Abraliopsis*, the presence of both elaborate photophores and photosensitive vesicles may imply a control of the luminescence of the photophores by a feedback system activated by the extra-ocular photoreceptors (R. E. Young, 1973). Subsequently, R. E. Young & Roper (1976) have reported the results of experiments with live specimens of *Abraliopsis* which they have interpreted as supporting the feedback hypothesis.

Undoubtedly the elucidation of the role of the extra-ocular system of photoreceptors in various cephalopods will require the combined efforts of a variety of investigators, e.g. electrophysiologists, neuroanatomists, biochemists and behaviourists, as in the case of the photoreceptors located in the epiphyseal stalk (the pineal body) of some of the lower vertebrates such as salamanders (Kelley, 1962; Adler, 1976), frogs, lizards and fishes (Dodt & Jacobson, 1963; Dodt, 1973).

Acknowledgements

I am deeply grateful to my friends, Dr Alain Perrelet, Dr Fritz Baumann and Dr Ove Sten-Knudsen for their collaboration in different phases of the work reviewed here. It is a pleasure to express our thanks to Professor John Young for his enthusiastic encouragement and help in pursuing these investigations.

References

Adler, K. (1976). Extraocular photoreception in amphibians. *Photochem. Photobiol.* **23**: 275–298.

Bauer, V. (1909). Einführung in die Physiologie der Cephalopoden. Mit besonderer Berüsksichtigung der im Mittelmeer haüfigen Formen. *Mitt. zool. Stat. Neapel* **19**: 149–268.

Baumann, F., Mauro, A., Millecchia, R., Nightingale, S. & Young, J. Z. (1970). The extra-ocular light receptors of the squids *Todarodes* and *Illex*. *Brain Res.* **21**: 275–279.

Belmonte, C. & Stensaas, L. J. (1975). Repetitive spikes in photoreceptor axons of the scorpion eye. Invertebrate eye structure and tetrodotoxin. *J. gen. Physiol.* **66**: 649–655.

Boycott, B. B. & Young, J. Z. (1956). The subpedunculate body and nerve and other organs associated with the optic tract of cephalopods. In *Bertil Hanström: Zoological papers in honour of his sixty-fifth birthday*: 76–105. Wingstrand, K. G. (ed.). Lund: Zool. Inst.

Dodt, E. (1973). The parietal eye (pineal and parietal organs) of lower vertebrates. In *Handbook of sensory physiology* **7**: 113–140. Jung, R. (ed.). Berlin: Springer Verlag.

Dodt, E. & Jacobson, M. (1963). Photosensitivity of a localized region of the frog diencephalon. *J. Neurophysiol.* **26**: 752–758.

Haefelfinger, H. R. (1954). Inkretorische Drüsencomplexe in Gehirn decapoder Cephalopoden. *Rev. suisse Zool.* **61**: 153–162.

Hodgkin, A. L., Huxley, A. F. & Katz, B. (1949). Ionic currents underlying activity in the giant axon of the squid. *Archs Sci. Physiol.* **8**: 129–150.

Kelley, D. E. (1962). Pineal organs: photoreception, secretion and development. *Am. Scient.* **50**: 597–625.

MacNichol, E. F. Jr. & Love, W. E. (1961). Impulse discharges from the retinal nerve and optic ganglion of the squid. In *The visual system: neurophysiology and psychophysics*. Jung, R. & Kornhuber, H. (eds.). Berlin: Springer Verlag.

Mauro, A. & Baumann, F. (1968). Electrophysiological evidence of photoreceptors in the epistellar body of *Eledone moschata*. *Nature, Lond.* **220**: 1332–1334.

Mauro, A. & Sten-Knudsen, O. (1972). Light-evoked impulses from extra-ocular photoreceptors in the squid *Todarodes*. *Nature, Lond.* **237**: 342–343.

Messenger, J. B. (1967). Parolfactory vesicles as photoreceptors in a deep-sea squid. *Nature, Lond.* **213**: 836–838.

Millecchia, R. & Mauro, A. (1969). The ventral photoreceptor cells of *Limulus*. II. The basic photoresponse. *J. gen. Physiol.* **54**: 310–330.

Nishioka, R. S., Hagadorn, I. R. & Bern, H. A. (1962). Ultrastructure of the epistellar body of the octopus. *Z. Zellforsch. mikrosk. Anat.* **57**: 406–421.

Nishioka, R. S., Yasumasu, I., Packard, A., Bern, H. A. & Young, J. Z. (1966). Nature of vesicles associated with the nervous system of cephalopods. *Z. Zellforsch. mikrosk. Anat.* **75**: 301–316.

Perrelet, A. & Mauro, A. (1972). Ultrastructure of the nerves associated with the epistellar body of the octopod *Eledone moschata* and the parolfactory vesicles of the squid *Todarodes sagittatus*. *Brain Res.* **37**: 161–171.

Snodderly, M. (1969). *Processing of visual inputs by the ancient brain of* Limulus. Ph.D. Thesis: The Rockefeller University.

Sperling, L., Lisman, J. E. & Godfrey, A. (1973). Light-evoked responses from the ventral parolfactory vesicles of *Loligo pealei*. *Biol. Bull. mar. biol. Ass. Woods Hole* **145**: 456.

Thore, S. (1939). Über ein neues Organ bei den Decapoden Cephalopden. *J. fysiogr. Sällsk. Lund Förh.* **9**: 105–111.

van Brunt, E. E., Shepherd, M. D., Wall, R. J., Ganong, W. F. & Clegg, M. T. (1964). Penetration of light into the brain of mammals. *Ann. N. Y. Acad. Sci.* **117**: 217–224.

Young, J. Z. (1929). Sopra un nuovo organo dei cefalopodi. *Boll. Soc. ital. Biol. sper.* **4**: 1022–1024.

Young, J. Z. (1936). The giant nerve fibres and epistellar body of cephalopods. *Q. Jl microsc. Sci.* **78**: 367–386.

Young, R. E. (1972). Function of extra-ocular photoreceptors in bathypelagic cephalopods. *Deep-sea Res.* **19**: 651–660.

Young, R. E. (1973). Information feedback from photoreceptors and ventral countershading in mid-water squid. *Pacif. Sci.* **27**: 1–7.

Young, R. E. & Roper, C. F. E. (1976). Bioluminescent countershading in midwater animals: evidence from living squid. *Science, Wash.* **191**: 1046–1048.

Symp. zool. Soc. London (1977) No. 38, 309–324.

STRUCTURE AND FUNCTION OF THE ANGULAR ACCELERATION RECEPTOR SYSTEMS IN THE STATOCYSTS OF CEPHALOPODS

B. U. BUDELMANN

Zoological Institute, University of Regensburg, Regensburg, W. Germany

SYNOPSIS

The information on the structure and function of the angular acceleration receptor systems (crista/cupula systems) of the statocysts of octopod and decapod cephalopods is reviewed. Particular reference is made to the ultrastructure of the hair cells, their morphological polarization, and the cupula attached to them.

In cephalopods the cristae are divided into sections, nine in octopods and four in decapods. They are arranged in the three mutually perpendicular planes, transverse, longitudinal and vertical. The cristae are composed of hair cells (larger and smaller ones), supporting cells, and large neurons below. The larger hair cells have generally been considered to be primary receptor cells. Recent electron micrographs of serial sections, however, do not support this hypothesis.

The hair cells of the cristae are oriented in line with the long axis of the crista ridge. Their morphological polarization is defined by the orientation of the basal foot structure and the internal $9 \times 2 + 2$ filament content of their kinocilia. The direction of polarization has been determined for each row of larger and smaller hair cells for all the crista sections of *Octopus*, *Sepia* and *Loligo*.

A cupula is attached to each crista section. It is a sail-like structure composed of a matrix of medium electron density and fibrous strands. It is in contact only with the distal ends of the kinocilia of the hair cells.

The morphological results are compared with the physiological findings.

INTRODUCTION

Since the thorough light microscopical study of Young (1960) there has been renewed interest in the structure and function of cephalopod statocysts. Cephalopod statocysts are the best-developed equilibrium organs found among invertebrates. Their structure is elaborate, with systems for the detection of linear acceleration (gravity) as well as angular acceleration. They provide the organism directly with information regarding its attitude in space. Together with data from other proprioceptive, visual and tactile systems this information enables the animal to control its motor activities relative to gravity.

The structure and function of cephalopod statocysts have recently been reviewed in great detail (Vinnikov, Gasenko, Titova *et al.*, 1971; Budelmann, 1975, 1976). This paper is concerned

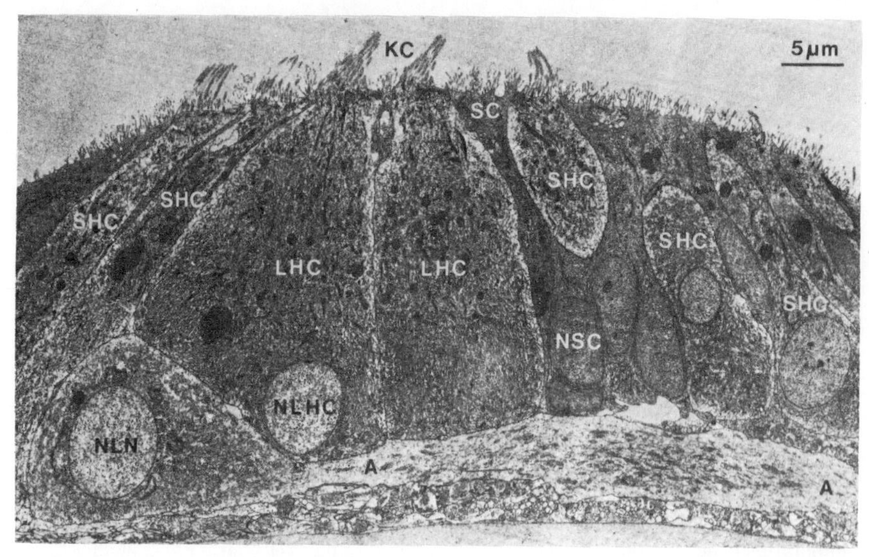

FIG. 1. Cross-section through the crista of the left statocyst of *Octopus vulgaris* (section 3, crista transversalis). The cupula is removed. Dorsal side to the right. A: axon of large neurone; KC: kinocilia; LHC: large hair cell; NLHC: nucleus of large hair cell; NLN: nucleus of large neurone; NSC: nucleus of supporting cell; SHC: small hair cell; SC: supporting cell.

only with the angular acceleration receptor systems of the statocysts. In addition to summarizing the current state of knowledge it presents new data regarding the fine structure of the sensory elements and the cupula attached to them. It also describes the complete polarization pattern of each system in *Octopus*, *Sepia* and *Loligo*.

MORPHOLOGY OF THE ANGULAR ACCELERATION RECEPTOR SYSTEMS

In octopods and decapods the angular acceleration receptor systems are a part of the statocysts. Their epithelium is composed of sensory (hair) cells and supporting cells; large neurones lie underneath (Fig. 1).

These elements in the main form the crista ridge, which winds over three planes almost perpendicular to one another: transverse, longitudinal and vertical. From each hair cell a large number of kinocilia protrude into the fluid-filled cavity. Attached to the kinocilia is a sail-like cupula.

Structure of the crista

The hair cells of the crista bear numerous kinocilia and microvilli at their distal end, and no stereocilia. Normally 100–130 kinocilia project from each single hair cell. They are arranged to form an elongated ciliary group with each kinocilium normally about 6 μm long and 0·24 μm in diameter (Fig. 2A). The ciliary groups are aligned in rows along the length of the crista (Barber, 1966b; Budelmann, Barber & West, 1973).

Each hair cell is morphologically polarized. This polarization is defined by the orientation of the basal foot structure (Fig. 2B) and the internal $9 \times 2 + 2$ filaments contained in each kinocilium. All kinocilia of a single cell are polarized in the same direction, namely at right angles to the long axis of the ciliary group (Fig. 2C), and therefore at right angles to the course of the crista ridge (Barber, 1966a). Normally each group of kinocilia forms an angle of 45 to 60° with the cell surface. The direction of polarization is always away from this acute angle.

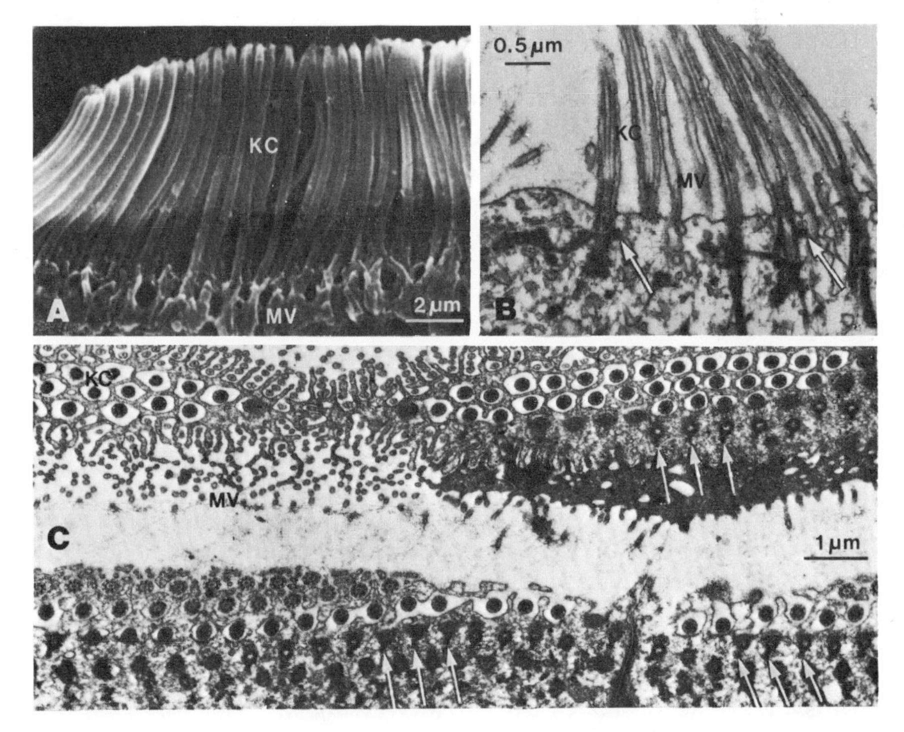

FIG. 2. Fine structure of the kinocilia of crista hair cells. (A) Group of kinocilia of a hair cell of the crista transversalis posterior of the left statocyst of *Sepia officinalis*. (B and C) Orientation of basal feet. (B) Transverse section through apical region of a single crista hair cell of the statocyst of *Octopus vulgaris*. (C) Oblique tangential section through apical region of four crista hair cells of the statocyst of a newly hatched *Sepia officinalis*; arrows point to basal feet. KC: kinocilia; MV: microvilli.

The hair cells are generally considered to be primary receptor cells with an axon running from their bases (Hamlyn-Harris, 1903; Young, 1960; Barber, 1966b; Vinnikov, Gasenko, Bronstein *et al.*, 1967). Unequivocal proof for this opinion is offered nowhere, and Young (1960) showed that nerve cells with dendrites are also present below the hair cells. Kolmer (1926) on the other hand suggests that the hair cells in the crista (and in the macula of the gravity receptor system) of *Sepiola* just might be secondary receptors; Klein (1932) confirmed this view for the larger hair cells of the crista of *Sepia*. In recent electron micrographs of complete serial sections of two large hair cells of the *Octopus* crista no axon has been found leaving the cells. In addition, afferent synapses have been found at the base of the cells as well as efferent ones. The afferent synaptic vesicles lie inside the hair cells. This arrangement clearly indicates that these large hair cells are secondary receptors.

Afferent synapses at the base of *Octopus* hair cells were described by Barber (1966b). It is still unclear, however, whether all hair cells are secondary receptors; some of the smaller hair cells may be primary (Klein, 1932). So far in none of the electron micrographs of cross-sections of the cristae of *Octopus*, *Sepia* and *Loligo* has a large hair cell been found with an axon projecting from it to make it a primary receptor. But the serial sections in progress should clarify whether all crista hair cells (the larger and smaller ones) are secondary receptor cells or not.

Octopods

In octopods, such as *Octopus vulgaris*, the crista ridge runs in three main planes around the statocyst sac. According to Young (1960) it also has three main parts: two horizontally oriented, the crista transversalis and the crist longitudinalis, and one vertically, the crista verticalis. Each of these three parts is further subdivided into three sections, thus giving a total of nine sections (Fig. 3A). The division of the crista ridge into only three main parts, however, does not correspond exactly with its topographical orientation. The crista runs half-way around the inside of the sac and then turns sharply upwards. Thus, to follow its course more precisely, sections 1 and 2 (of the c. transversalis) are anterior and run almost transversely, section 3 (of the c. transversalis) and section 4 (of the c. longitudinalis) continue laterally in a general longitudinal direction, sections 5 and 6 (of the c. longitudinalis) are posterior and again run almost transversely, and sections 7, 8 and 9 are also posterior but run almost vertically (cf. Fig. 4, and arrangements in decapods).

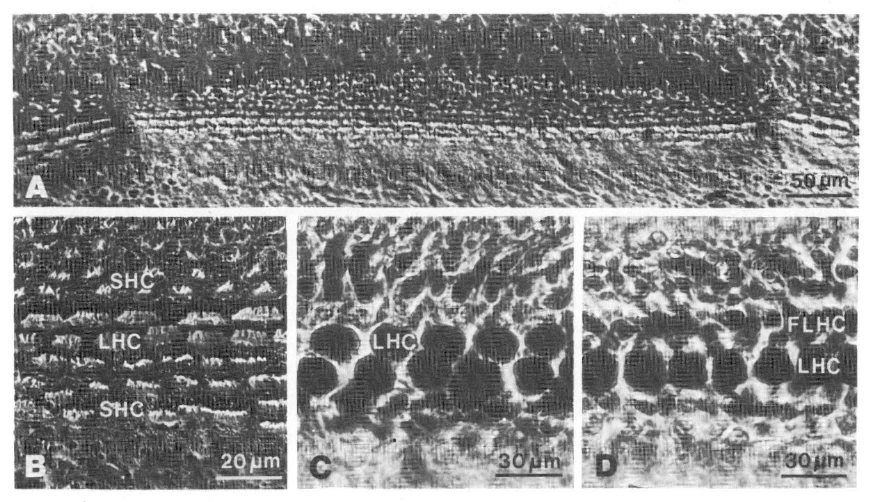

FIG. 3. Crista of the statocyst of *Octopus vulgaris.* (A) Surface view of a single row section (section 4, crista longitudinalis) of the left statocyst. The cupula is removed. Dorsal is top of page. At both sides of the crista section portions of double row sections can be seen. (From Budelmann *et al.*, 1973; published with permission of ASP Biological and Medical Press, Elsevier division.) (B) Surface view of a part of a double row section (section 5, crista longitudinalis) of the left statocyst. The cupula is removed. Dorsal is top of page. (C and D) Part of a double (C) and a single (D) row section of the right statocyst. (The hair cells are stained by immersion of the cristae into $CoCl_2$ solution and subsequent treatment with ammonium sulphide.) FLHC: fairly large hair cell; LHC: large hair cell; SHC: small hair cell.

Each of the nine crista sections is characterized by either a single or a double row of large hair cells running along the middle of the crista ridge (Fig. 3B–D). Double-rowed and single-rowed sections alternate, with a double-rowed section at the beginning and the end (Young, 1960; fig. 4).

The average length of a crista section is 0·43 mm, with about 35 large hair cells per large hair cell row. Rows of smaller hair cells have also been described lying dorsally and ventrally or laterally and medially to the rows of larger hair cells (Fig. 3A and B). Because the crista runs horizontally from the front to the rear of the statocyst sac and then turns vertically up, the structures that are dorsal to the crista's middle ridge while the crista is horizontal become lateral when it rises vertically, and those that are ventral become medial. A regular row of fairly large cells (upper inner hair cells; Young, 1960) lies dorsal to the large cells of the single row of sections 2, 4 and 6, and lateral to that of section 8. Dorsal to the row of fairly large cells are up to five additional rows of smaller hair cells (upper outer hair cells; Young, 1960); these cells run parallel to the crista ridge, but do not form continuous rows. Fairly large hair cells are lacking in the double row sections 1, 3, 5, 7 and 9. But up to five rows of smaller hair cells (upper outer hair cells; Young, 1960) dorsal or lateral to the double rows of large hair cells persist (in sections 1, 3 and 5, dorsal; in sections 7 and 9, lateral). One to three other rows of fairly regularly arranged smaller hair cells (lower inner and lower outer hair cells; Young, 1960) lie ventrally or medially to the single and double rows of large hair cells in all sections (sections 1 to 6, ventral; sections 7 to 9, medial) (Budelmann *et al.*, 1973).

Both the larger and smaller hair cells are morphologically polarized. Their direction of polarization was determined in each row. As Fig. 4 illustrates, both possible directions of polarization

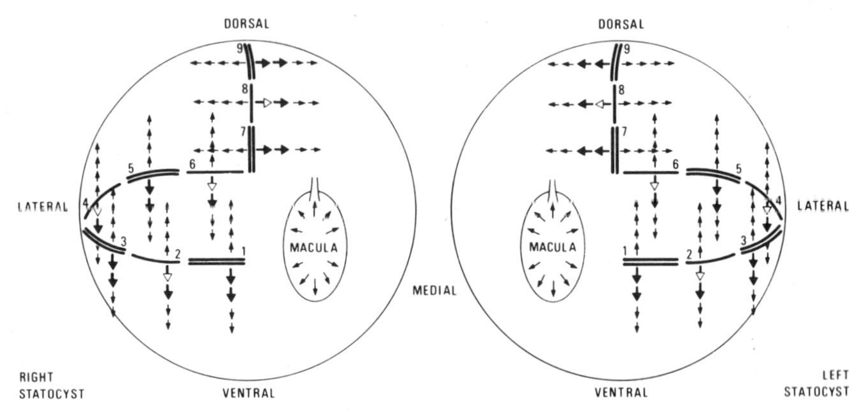

FIG. 4. Pattern of morphological polarization of the hair cells in the different crista sections of the statocysts of the octopod *Octopus vulgaris* (frontal view). Sections 1 to 6 run in the horizontal plane, sections 7 to 9 in the vertical. Each arrow stands for a row of hair cells and indicates the direction in which the basal feet of all their kinocilia point. Large dark arrows refer to large hair cells (double row sections 1, 3, 5, 7 and 9; single row sections 2, 4, 6 and 8); light arrows refer to fairly large hair cells (sections 2, 4, 6 and 8); small arrows refer to small hair cells.

are present in all nine crista sections. (This was described by Barber, 1968, but for unspecified sections.) The large and fairly large hair cells, together with the smaller hair cells ventral or medial to them, are all polarized ventrally (sections 1 to 6) or medially (sections 7 to 9). The dorsal or lateral smaller hair cells are polarized either dorsally (sections 1 to 6) or laterally (sections 7 to 9).

According to Young (1960), each of the three crista parts is connected with the CNS by a separate nerve: the crista transversalis by the anterior crista nerve, the crista longitudinalis by the middle crista nerve, and the crista verticalis by the posterior crista nerve (the latter one also contains axons coming from cells in the wall of the posterior sac). Recent morphological studies, however, have shown that axons coming from cells of section 3 (the longitudinally oriented section of the crista transversalis) join the middle crista nerve. The number of fibres whose diameter ranges from 2 to 22 μm was determined light microscopically for the different crista nerves (Young, 1965). However, counts in progress using the electron microscope show that their number will run much higher, since there are many fibres considerably smaller than 2 μm. A large number of the fibres are efferent.

The three crista nerves run straight upwards and then pass through the cartilage separately. The anterior crista nerve joins the macula nerve; these two together and the other two crista nerves enter the CNS at the level of the lateral pedal lobe where they branch out (see Young, 1971; Hobbs & Young, 1973, for further details).

Decapods

In decapods, such as *Sepia officinalis* and *Loligo vulgaris*, the crista ridge again runs in the three main planes along the statocyst wall. It is divided into four sections: three oriented horizontally, the crista transversalis anterior (CTA), the crista longitudinalis (CL) and the crista transversalis posterior (CTP); and one oriented vertically, the crista verticalis (CV). In *Sepia* the beginning and end of each section of the crista is roofed over by a cartilaginous anticrista lobe leaving only the middle portions exposed (Budelmann *et el.*, 1973).

In each crista section large hair cells are arranged in four continuous rows in the middle of the crista ridge. The two ventral rows of the CTA, CL and CTP and the two medial rows of the CV consist of larger hair cells than the other two. Smaller hair cells are also present, dorsal to the four hair cell rows of the CTA, CL and CTP, and lateral to those of the CV. They are arranged in fairly regular rows parallel to the crista ridge. Individual cells with considerably longer cilia (15–20 μm) are scattered from 30 to 40 μm ventrally to the hair cell rows of the CTA, CL and CTP, and medially with respect to those of the CV (Budelmann *et al.*, 1973; Fig. 5).

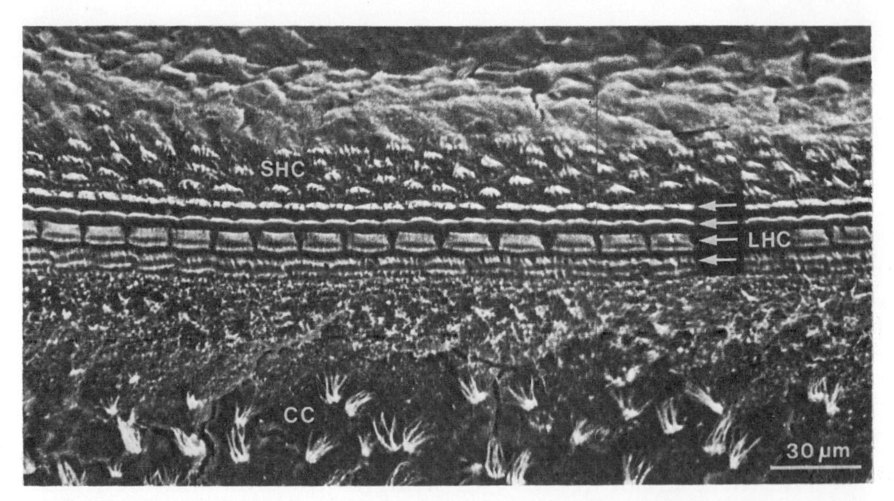

FIG. 5. Part of the crista longitudinalis of the left statocyst of *Sepia officinalis*. The cupula is removed. Dorsal is top of page. (From Budelmann *et al.*, 1973; published with permission of ASP Biological and Medical Press, Elsevier division.) CC: ciliated cells; LHC: large hair cells (4 rows, as indicated by the arrows); SHC: small hair cells.

In *Sepia* the average length of the crista sections is 1·60 mm, each with about 117 large hair cells per row. In *Loligo* the average section length is 1·19 mm and the average number of large hair cells per row is 100 (Budelmann *et al.*, 1973).

As in octopods, both the larger and the smaller hair cells are morphologically polarized. In *Sepia* and *Loligo* the direction of polarization is shown in Fig. 6. In each of the four crista sections both possible polarization directions are present. In those sections oriented horizontally (CTA, CL and CTP) the three ventral rows of larger hair cells are polarized in the ventral direction, and in the sections oriented vertically (CV) they are polarized in the medial direction. In each section the fourth row of larger hair cells together with the smaller hair cells is polarized in the direction opposite to the other three rows of larger hair cells. An identical pattern of polarization is already present in the crista hair cells of the imperfectly developed statocysts of newly hatched *Sepia*.

In decapods little is known about the exact course of the crista nerves. Only two nerves have been described, one apparently containing fibres coming from the crista transversalis anterior (in addition to fibres from the macula neglecta posterior, i.e. the sensory epithelium of one of the three gravity receptor systems), and the other apparently containing fibres coming from the other three crista sections. The two nerves pass through the cartilage

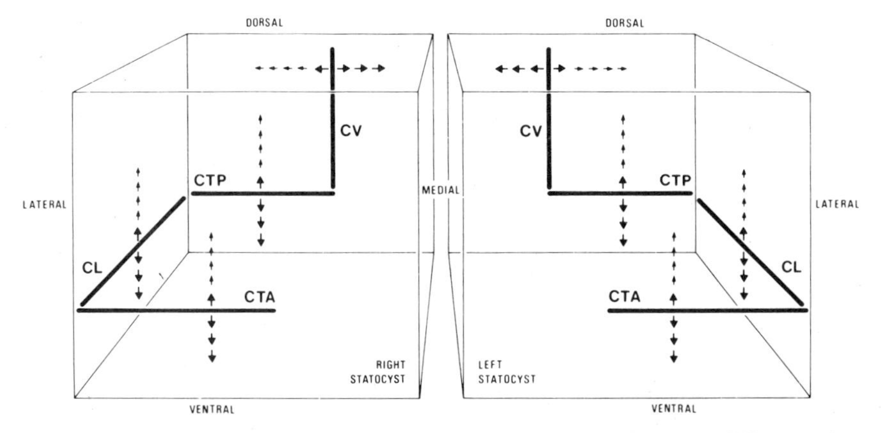

FIG. 6. Pattern of morphological polarization of the hair cells in the different crista sections of the statocysts of the decapods *Sepia officinalis* and *Loligo vulgaris* (frontal view). Sections CTA, CL and CTP run in the horizontal plane, section CV in the vertical. Each arrow stands for a row of hair cells and indicates the direction in which the basal feet of all their kinocilia point. Large arrows refer to rows of large hair cells, small arrows to those of small hair cells. CL: crista longitudinalis; CTA: crista transversalis anterior; CTP: crista transversalis posterior; CV: crista verticalis.

separately and enter the CNS at the level of the pedal lobe (Hamlyn–Harris, 1903; Klein, 1932; Thore, 1939; for further details see also Hobbs & Young, 1973).

Structure of the cupula

In octopods and decapods a cupula is attached to each section of the crista (Young, 1960; Budelmann *et al.*, 1973). It is a sail-like structure protruding freely into the fluid-filled cyst cavity (Fig. 7A and B). The cupula is composed of a matrix of medium electron density and fibrous strands. These strands, up to 50 nm in diameter, are tube-like in structure and often show a dense core (Fig. 7F).

In *Octopus* the cupula is about 0·4 mm long and at least 0·2 mm wide (Fig. 7A); its maximum height (up to 0·4 mm) lies about half-way along the crista section to which it is attached. In *Sepia* the cupula is about 1·6 mm long, its height also increases towards the middle; there the cupula becomes voluminous, at least 0·16 mm thick and 0·7 mm high (Fig. 7B and C).

Each cupula lies above the hair cell rows of its crista section. The cupula contacts the kinocilia of the hair cells at their distal ends only (Fig. 7D and E). Fibrous strands are sometimes seen attaching the cupula to parts of the crista outside the hair cell rows (Fig. 7G) (Budelmann *et al.*, 1973).

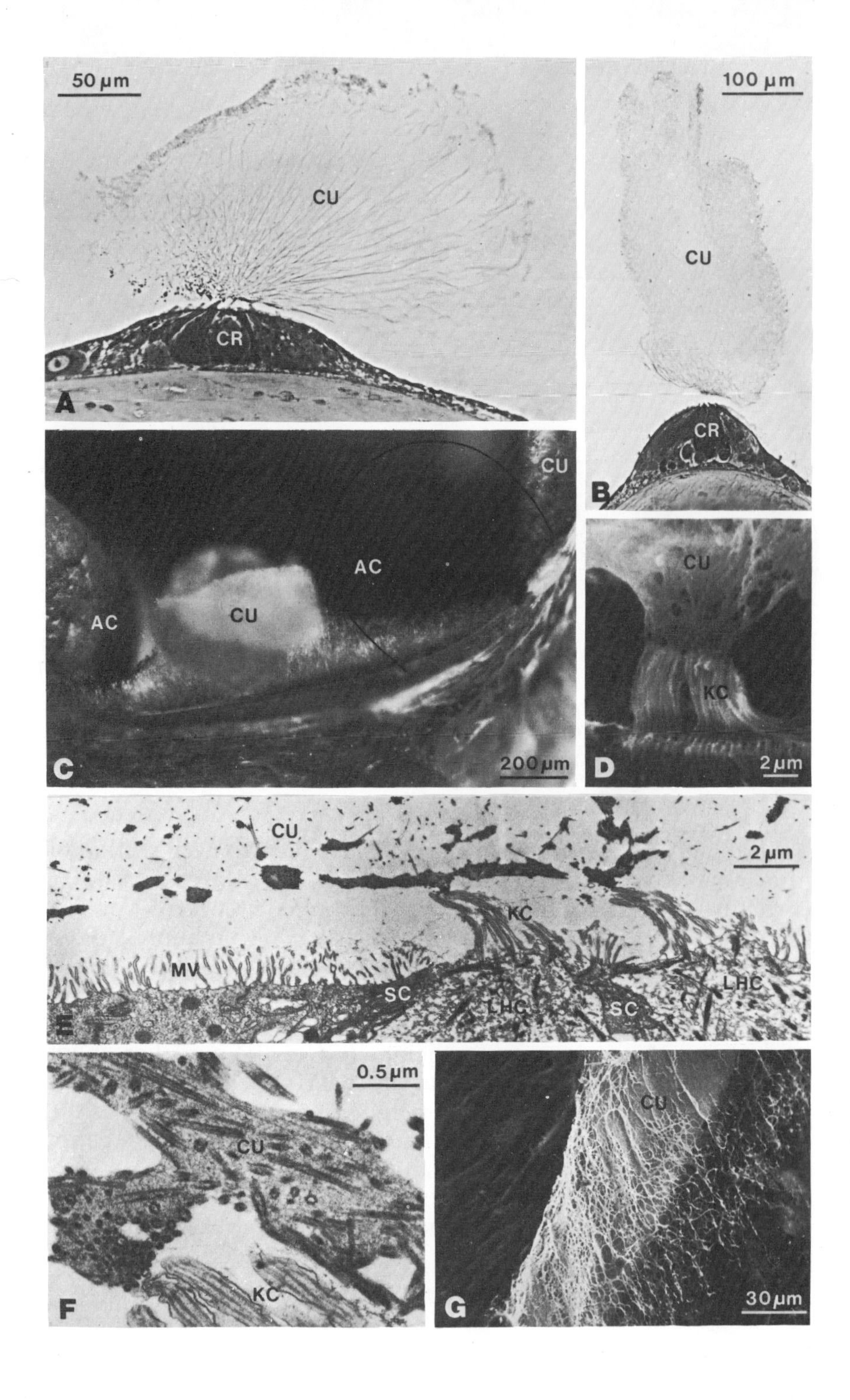
50 µm
CU
CR
A
100 µm
CU
CR
B
CU
AC
CU
AC
C
200 µm
CU
KC
D
2 µm
CU
2 µm
KC
MV
SC
LHC
SC
LHC
E
0.5 µm
CU
KC
F
CU
G
30 µm

PHYSIOLOGY OF THE ANGULAR ACCELERATION RECEPTOR SYSTEMS

General

According to the morphological organization of these receptor systems any angular acceleration (either actively caused by the animal or passive) will result in a movement of the endolymph fluid relative to the sensory (crista) epithelium. The cupula follows this fluid movement to a certain extent and thereby causes the kinocilia to shear in one direction or the other, depending on the direction of angular acceleration. Shear forces on the kinocilia result in a change in the receptor potential of the hair cells. By analogy with the results from the vertebrate vestibular and lateral line receptors (for references see Flock, 1971) it has been suggested (Barber, 1968) that in cephalopods a passive deflection (shear) of the kinocilia of the hair cell in the direction of the basal foot structure (i.e. the direction of the outer filaments 5 and 6) causes a maximal excitation (depolarization) of the hair cell, correlated with a maximal increase of the firing rate of the nerve fibre originating at this cell, and a maximal inhibition (hyperpolarization), correlated with a maximal decrease of the firing rate when shear occurs in the opposite direction. The information regarding the angular acceleration is encoded quantitatively in the sensory output.

Although the cartilaginous anticrista lobes (one in octopods, two to 13 in decapods; Ishikawa, 1924) influence the movement of the endolymph to a certain extent, nevertheless the very topographical arrangement of the crista/cupula systems implies that the transverse cristae systems should function best for detection of angular acceleration during rotation around any transverse axes of the animal, the longitudinal cristae systems for those around any

---

FIG. 7. Structure of cupula. (A) Cross-section (4 μm) through the crista (section 1 of the crista transversalis) of the left statocyst of *Octopus vulgaris.* Dorsal side is to the right. (B) Cross-section (4 μm) through the crista (crista transversalis posterior) of the left statocyst of *Sepia officinalis.* (The cupula is slightly detached from the kinocilia.) Dorsal side is to the left. (C) Cupula of the crista (crista transversalis posterior) of the left statocyst of *Sepia officinalis,* as viewed from below with a stereo-microscope. Part of the cupula of the crista longitudinalis is seen to the right. The anticrista lobe which formed a roof over the division between the two crista sections has been removed. Position indicated by black line. (D) Contact of the cupula to the distal ends of the kinocilia (crista verticalis of the right statocyst of *Sepia officinalis*). (E) Cross-section through part of the apex of a crista ridge (section 1 of the crista transversalis of the left statocyst of *Octopus vulgaris*). Dorsal side is to the left. (F) Section through the distal ends of kinocilia in contact with the cupula (section 1, crista transversalis of the left statocyst of *Octopus vulgaris*). (G) Part of the cupula overlying a crista section (crista transversalis posterior of *Sepia officinalis*) (G: from Budelmann *et al.,* 1973; published with permission of ASP Biological and Medical Press, Elsevier division). AC: anticrista lobe; CR: crista ridge; CU: cupula; KC: kinocilia; LHC: large hair cell; MV: microvilli; SC: supporting cell.

longitudinal axes, and the vertical cristae systems for those around any vertical axes (Young, 1960; Budelmann *et al.*, 1973). This hypothesis has been confirmed electrophysiologically as well as in behaviour experiments.

Electrophysiological experiments

Electrophysiological recordings have so far been carried out only in octopods. Extracellular recordings from afferent neurones of the different crista sections have shown that each crista/cupula system is most sensitive to angular acceleration and deceleration during rotation around its appropriate axis, as was predicted by its topography. Angular acceleration and deceleration stimuli during rotation at 20° per sec elicited a clear change in the discharge frequency, rotation at 4·8° per sec elicited only a minimal response and rotation at 2·67° per sec was below threshold (Wolff & Budelmann, in prep. b).

The discharge patterns of the angular acceleration receptors are comparable to those in vertebrates: rotation to the one side elicits increased activity during acceleration and inhibition during deceleration (Fig. 8, upper trace). Rotation to the opposite side

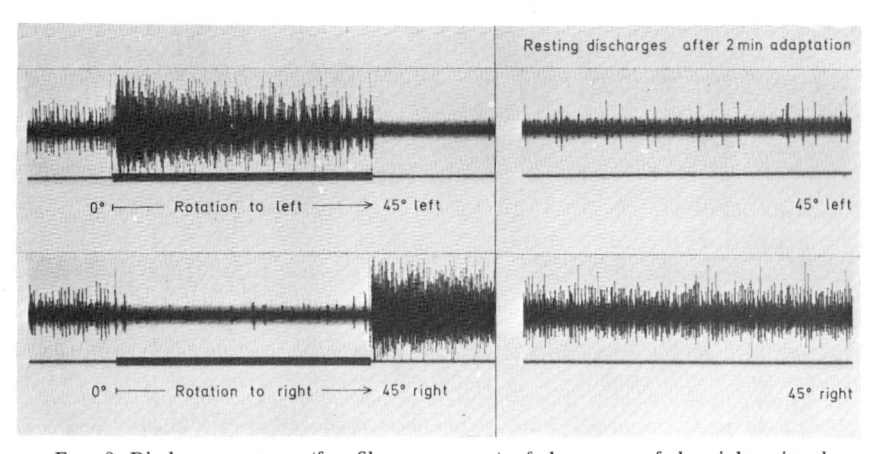

FIG. 8. Discharge pattern (few-fibre-responses) of the nerve of the right crista longitudinalis (horizontal angular acceleration receptor system) of *Octopus vulgaris* after intracranial section of all right statocyst nerves. Horizontal broad traces indicate the positional changes of 45° with an angular acceleration stimulus at the beginning and a deceleration stimulus at the end of each constant speed rotation of 3·33 rpm. Each horizontal axis rotation was around the animal's longitudinal axis and was carried out after a 2-min adaptation to the initial resting position (start position). To the right: resting discharges after a 2-min adaptation to the final resting position (stop position). Upper trace: rotation to the left from 0° (normal position) to 45° left. Lower trace: rotation to right from 0° (normal position) to 45° right. (From Budelmann & Wolff, 1973; published with permission of Springer Verlag.)

elicits the reversed discharge patterns: inhibition during acceleration and increased activity during deceleration (Fig. 8, lower trace). These results clearly demonstrate a single direction of physiological polarization within the receptor system (Wolff & Budelmann, in prep. b).

Such a physiological polarization has been found in all crista systems. In the horizontal sections 1 to 6 a ventral deflection of the cupula causes excitation and dorsal deflection inhibition. Similarly in the vertical sections 7 to 9 a medial cupula deflection causes excitation and lateral cupula deflection inhibition (Wolff & Budelmann, in prep. b). The pattern of physiological polarization of these crista sections is well correlated with the morphological polarization pattern of the large hair cells of the single and double rows.

Exceptions to the above results were recorded from the anterior transverse sections of the crista (sections 1 and 2). In these sections a few units were excited during ventral cupula deflection but others during dorsal deflection (Wolff & Budelmann, in prep. b; cf. also behavioural experiments). This would imply that some units in these sections with morphological polarization opposite to that of the larger hair cells are also active as angular acceleration receptors. Further experiments should clarify this point.

In addition to the sensitivity to angular acceleration and deceleration stimuli, high sensitivity to vibration has been found in all crista systems of *Octopus* (Maturana & Sperling, 1963; Wolff & Budelmann, in prep. a).

Behavioural experiments

Dijkgraaf (1959, 1961) examined the role of the vertically oriented crista sections (crista verticalis) on compensatory reactions in *Octopus* (head and eye movements, including nystagmus and after-nystagmus). Post-rotatory reflexes in response to angular deceleration stimuli could still be observed in bilaterally blinded animals with all statocyst nerves cut, except for the one innervating the crista verticalis (n. crista posterior). Most animals reacted equally well to rotations in either direction; a few, however, showed slightly stronger reactions when the cupula was deflected medially. These reflexes were abolished after additional transection of the n. crista posterior.

In *Sepia* too post-rotatory nystagmus has been related to the function of the crista/cupula system (Dijkgraaf, 1963). In a more extensive study Messenger (1970) confirmed these observations for blinded animals also. In addition he established that post-rotatory nystagmus is almost completely suppressed by visual cues, and is abolished by bilateral destruction of the statocysts. In *Sepia* good reactions were normally found at rotation speeds of more than 14·7° per sec, at 5·9° per sec the reactions were usually much smaller (Collewijn, 1970). These behavioural results correspond well with the electrophysiological findings in *Octopus*.

Gravity sensitivity

The crista/cupula system is primarily a receptor system for the detection of angular acceleration. Its structural organization, however, does not necessarily exclude an additional function as a gravity receptor system, under the supposition that the cupula has a specific gravity different from the surrounding fluid.

An unequivocal response of the crista/cupula system to linear acceleration (gravity) was found electrophysiologically in *Octopus* during rotation around a horizontal axis (Budelmann & Wolff, 1973). Extracellular recordings showed different responses to identical angular acceleration (deceleration) depending on the position of the preparation relative to gravity at the onset of acceleration (deceleration). In addition different resting discharge rates were elicited as a function of angular position with respect to gravity (cf. Fig. 8). Furthermore, during constant velocity rotation the impulse frequency was found to oscillate in phase with rotation.

This unambiguous sensitivity of the crista/cupula system to the linear acceleration of gravity need not necessarily affect their fundamental role as a receptor system for the detection of angular acceleration. Only behavioural experiments, however, can clarify whether this property actually influences behavioural responses for which the gravity receptor systems should be adequate.

Acknowledgements

The author thanks Dr R. Loftus for improvements in the English text and Miss G. Thies for skilful help in transmission electron microscopy. The work was supported in part by grant Wo 160/3 of the Deutsche Forschungsgemeinschaft given to Dr H. G. Wolff.

References

Barber, V. C. (1966a). The morphological polarization of kinocilia in the *Octopus* statocyst. *J. Anat., Lond.* **100**: 685–686.

Barber, V. C. (1966b). The fine structure of the statocyst of *Octopus vulgaris*. *Z. Zellforsch. mikrosk. Anat.* **70**: 91–107.

Barber, V. C. (1968). The structure of mollusc statocysts, with particular reference to cephalopods. *Symp. zool. Soc. Lond.* No. 23: 37–62.

Budelmann, B.-U. (1975). Gravity receptor function in cephalopods, with particular reference to *Sepia officinalis*. *Fortschr. Zool.* **23**: 84–96.

Budelmann, B.-U. (1976). Equilibrium receptor systems in molluscs. In *Structure and function of proprioceptors in the invertebrates*. Mill, P. J. (ed.). London: Chapman & Hall.

Budelmann, B.-U., Barber, V. C. & West, S. (1973). Scanning electron microscopical studies of the arrangements and numbers of hair cells in the statocysts of *Octopus vulgaris, Sepia officinalis* and *Loligo vulgaris*. *Brain Res.* **56**: 25–41.

Budelmann, B.-U. & Wolff, H. G. (1973). Gravity response from angular acceleration receptors in *Octopus vulgaris*. *J. comp. Physiol.* **85**: 283–290.

Collewijn, H. (1970). Oculomotor reactions in the cuttlefish, *Sepia officinalis*. *J. exp. Biol.* **52**: 369–384.

Dijkgraaf, S. (1959). Kompensatorische Kopf bewegung bei Aktivdrehung eines Tintenfisches. *Naturwissenschaften* **46**: 611.

Dijkgraaf, S. (1961). The statocyst of *Octopus vulgaris* as a rotation receptor. *Publ. Staz. zool. Napoli* **32**: 64–87.

Dijkgraaf, S. (1963). Nystagmus and related phenomena in *Sepia officinalis*. *Experientia* **19**: 29–30.

Flock, Å. (1971). Sensory transduction in hair cells. In *Handbook of sensory physiology* **1**: *Principles of receptor physiology*: 396–441. Loewenstein, W. R. (ed.). Berlin, Heidelberg, New York: Springer Verlag.

Hamlyn-Harris, R. (1903). Die Statocysten der Cephalopoden. *Zool. Jb.* (Anat.) **18**: 327–358.

Hobbs, M. J. & Young, J. Z. (1973). A cephalopod cerebellum. *Brain Res.* **55**: 424–430.

Ishikawa, M. (1924). On the phylogenetic position of the cephalopod genera of Japan based on the structure of statocysts. *J. Coll. agr. Tokyo Imp. Univ.* **7**: 165–210.

Klein, K. (1932). Die Nervenendigungen in der Statocyste von *Sepia*. *Z. Zellforsch. mikrosk. Anat.* **14**: 481–516.

Kolmer, W. (1926). Statoreceptoren. In *Handbuch der normalen und pathologischen Physiologie* **11**: 767–790. Bethe, A., v. Bergmann, G., Embden, G., Ellinger, A. (eds). Berlin: Springer Verlag.

Maturana, H. M. & Sperling, S. (1963). Unidirectional response to angular acceleration recorded from the middle cristal nerve in the statocyst of *Octopus vulgaris*. *Nature, Lond.* **197**: 815–816.

Messenger, J. B. (1970). Optomotor responses and nystagmus in intact, blinded and statocystless cuttlefish (*Sepia officinalis* L.) *J. exp. Biol.* **53**: 789–796.

Thore, S. (1939). Beiträge zur Kenntnis der vergleichenden Anatomie des zentralen Nervensystems der dibranchiaten Cephalopoden. *Publ. Staz. zool. Napoli* **17**: 313–506.

Vinnikov, Ya. A., Gasenko, O. G., Bronstein, A. A., Tsirulis, T. P., Ivanov, V. P. & Pyatkina, G. A. (1967). Structural, cytochemical and functional organization of statocysts of Cephalopoda. In *Symposium on neurobiology of invertebrates*: 29–48. Salánki, J. (ed.). New York: Plenum Press.

Vinnikov, Ya. A., Gasenko, O. G., Titova, L. K., Bronstein, A. A., Tsirulis, T. P., Pevznier, R. A., Govardovskii, N. A., Gribakin, F. G., Aronova, M. Z. & Tchekhonadskii, N. A. (1971). [The balance receptor. The evolution of its structural, cytochemical and functional organization.] In [*Problems in Space Biology*], **12**. Tchernigovskii, V. N. (ed.). Leningrad: NAUKA [in Russian].
Wolff, H. G. & Budelmann, B.-U. (in prep. a). *Electrical responses of the* Octopus *statocyst receptors to vibration.*
Wolff, H. G. & Budelmann, B.-U. (in prep. b). *Properties of angular acceleration receptors in* Octopus vulgaris.
Young, J. Z. (1960). The statocysts of *Octopus vulgaris*. *Proc. R. Soc.* (B) **152**: 3–29.
Young, J. Z. (1965). The diameters of the fibres of the peripheral nerves of *Octopus*. *Proc. R. Soc.* (B) **162**: 47–79.
Young, J. Z. (1971). *The anatomy of the nervous system of* Octopus vulgaris. Oxford: Clarendon Press.

Note Added in Proof

Recent osmium-fixed material of the angular acceleration receptor system of the statocyst of *Octopus vulgaris* revealed an obvious difference in form and size of the cupulae of the various crista sections: a smaller cupula is attached to the sections 1, 3, 5, 7 and 9 (i.e. the sections with the double row of large hair cells) and a larger cupula is attached to the sections 2, 4, 6 and 8 (i.e. the sections with the single row of large hair cells).

Symp. zool. Soc. Lond. (1977) No. 38, 325–346.

COMPARATIVE BIOCHEMISTRY OF THE CENTRAL NERVOUS SYSTEM

J. J. BARLOW

Department of Neurochemistry, Institute of Neurology, The National Hospital, London, England

SYNOPSIS

Cephalopod central nervous tissue contains enzymes for the synthesis and degradation of biologically active amines likely to be responsible for cholinergic and adrenergic synaptic transmission. These amines include acetylcholine, dopamine, noradrenalin and 5-hydroxytryptamine, and the available quantitative evidence for regional and species specific variations in their concentrations in the cephalopod brain is summarized. The distribution of cholinesterases determined histochemically in some octopod and decapod brains is compared. The cholinesterase of *Sepia officinalis* and *Loligo vulgaris* is localized in the neuropil as is the acetylcholinesterase of *Octopus vulgaris*. Histochemical evidence for the localization of monoamines in specific neuronal pathways supports the view that these amines could function as transmitter substances. Possible explanations for the unusually high concentrations that occur in certain regions of the cephalopod brain are discussed.

INTRODUCTION

In a recent review of chemical transmission in invertebrates Gerschenfeld (1973) emphasized that there are very few examples where the chemical responsible for mediating impulse transmission at a particular synapse is known with any degree of certainty. The best established cases are acetylcholine at the vertebrate nerve–muscle junction, noradrenalin at some peripheral adrenergic synapses and γ-aminobutyric acid at the crustacean inhibitory nerve–muscle junction. The requisite electrophysiological evidence is more difficult to obtain from the central nervous system. Here we have to rely on a general consensus of more indirect morphological, biochemical and pharmacological evidence to suggest likely transmitter candidates. This paper summarizes present information on the nature and regional distribution of putative transmitters in the central nervous systems of cephalopods. There are essentially three reasons for our interest in the biochemistry of transmitters in cephalopods as they may (i) provide models for studying mechanisms underlying chemical transmission, (ii) provide insight into the ways in which complex nervous systems have evolved, and (iii) form a basis for relating behavioural experiments to anatomical studies.

DISTRIBUTION OF CHOLINERGIC COMPOUNDS IN CEPHALOPOD BRAIN

Quantitative analyses in different cephalopods

The three biochemical components required for cholinergic synaptic transmission—choline acetylase which synthesizes the transmitter acetylcholine, and acetylcholinesterase which hydrolyses it—occur in extremely high concentrations in cephalopod central nervous tissue. In 1951 Feldberg, Harris & Lin examined *Sepia officinalis* for evidence of alternating cholinergic and non-cholinergic pathways. The activity of the enzyme choline acetylase was high in the optic lobe. Florey (1963) pointed out that the acetylcholine concentrations in *Sepia officinalis* and *Octopus dofleini* optic lobes were similar to those found in insect brain, and considerably higher than those in mammalian tissue. Enormous amounts of acetylcholinesterase were found in *Loligo pealeii* optic lobe (Nachmansohn & Meyerhof, 1941) and in *Octopus dofleini* (Loe & Florey, 1966). These concentrations were thought to be comparable only with those occurring in the highly specialized electric organs of fish. Results such as these supported to some extent speculation that acetylcholine might have a general role in cephalopod nerve metabolism rather than a specific function in synaptic transmission. Subcellular fractionation of octopod optic lobe, however, showed that acetylcholine was concentrated in synaptosomal fractions which contained the nerve endings (Florey & Winesdorfer, 1968; Jones, 1970). The optic lobe of *Octopus dofleini* contained up to 350 μg acetylcholine/g wet weight, and synaptosomes isolated from this tissue had 100 times the amount of acetylcholine found in comparable fractions from mammalian brain (Florey & Winesdorfer, 1968). The distribution of acetylcholine and acetylcholinesterase showed almost parallel variations in the nervous system of *Octopus dofleini* (Loe & Florey, 1966). Further evidence suggesting that the acetylcholinesterase may be intimately involved in synaptic transmission derives from subcellular fractionation of *Loligo pealeii* optic lobe (Welsch & Dettbarn, 1972). The enzyme was recovered in a synaptosomal fraction which also contained the most acetylcholine. Several groups of workers have utilized these interesting neurochemical properties of cephalopod brain in studies of cholinergic mechanisms (Prince, 1967; Heilbronn, Hause & Lundgren, 1971; Dowdall & Whittaker, 1973; Dowdall & Simon, 1973). We wished to have more precise information on the localization of cholinesterases in cephalopod brain in order to take advantage of the detailed anatomical studies made by Professor J. Z. Young.

Histochemistry of cholinesterase in octopods

The histochemical method of Shute & Lewis (1960) has been used to determine the localization of cholinesterase in *Octopus vulgaris* brain (Barlow, in prep. a). By varying the nature of the substrate, the pH of the medium or the inhibitor used, butyrylcholinesterase (BuChE = "non-specific" cholinesterase) can readily be distinguished from acetylcholinesterase (AChE = "specific" cholinesterase). In the optic lobe of *Pteroctopus tetracirrhus* BuChE is confined to the optic nerve bundles and outer granule cell layer (Fig. 1).

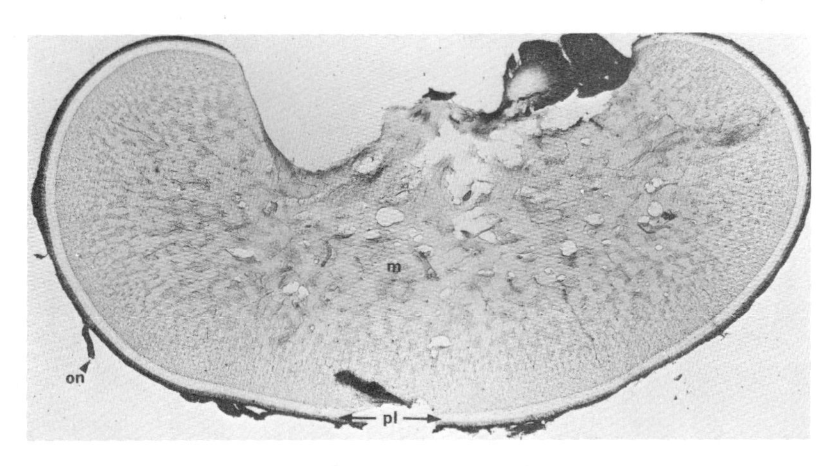

FIG. 1. *Pteroctopus tetracirrhus* optic lobe butyrylcholinesterase. The tissue was fixed by immersion in formaldehyde solution, cryostat sectioned (80 μm), and incubated with butyrylthiocholine medium (pH 5·6) according to the method of Shute & Lewis (1960). Reaction occurs in the optic nerve bundles (on) and is absent from the plexiform layer (pl). Weak reaction in the medulla (m) is in blood vessels. ×13.

AChE occurs throughout the neuropil but is absent from the cell perikarya of both cortex and medulla (Fig. 12). *Octopus vulgaris* shows particularly well the variation in intensity of AChE reaction in bands within the plexiform layer (Fig. 8). This striking difference in the distribution of the two cholinesterases had been shown previously to occur in the optic lobes of *Octopus vulgaris* and *Eledone moschata* (Drukker & Schadé, 1964, 1967). The apparent excess of AChE activity over BuChE is in agreement with the direct enzymic assay of Loe & Florey (1966) in *Octopus dofleini* optic lobe.

The circumoesophageal ganglia reacted for BuChE are shown in sagittal section (Fig. 2). Enzyme is present in the connective

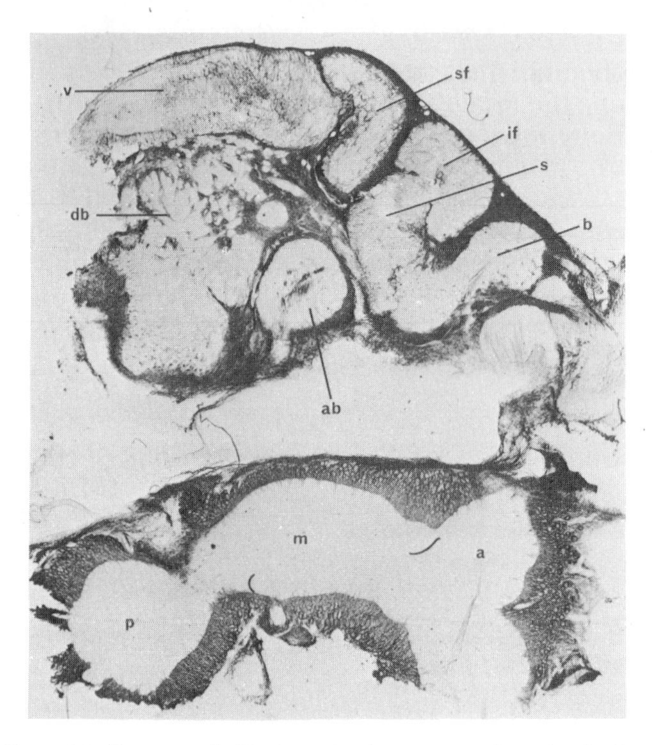

FIG. 2. Butyrylcholinesterase in *Octopus vulgaris* supraoesophageal and suboesophageal ganglia. Formaldehyde fixed, freezing microtome section (100 μm) incubated with butyrylthiocholine (pH 5·6). Reaction is concentrated in the cortical layers of the ganglia. a: Anterior; m: median and p: posterior suboesophageal lobes. In the supraoesophageal ganglion ab: anterior basal; b: superior buccal; db: dorsal basal; if: inferior frontal; s: subfrontal; sf: superior frontal and v: vertical lobes. ×11.

tissue envelope and surrounds the cell perikarya in the cortical areas of each lobe, and is almost entirely absent from the neuropil, except where it is associated with blood vessels. The distribution of AChE (Fig. 3) is essentially the reverse image of that shown for BuChE, the reaction being almost entirely confined to the neuropil. There is clearly considerable variation, however, in AChE activity in different brain regions. There is little activity in the subfrontal and vertical lobes, whereas reaction in the superior median frontal, subvertical and basal lobes is intense. Higher magnification reveals that some of the neuronal cell perikarya,

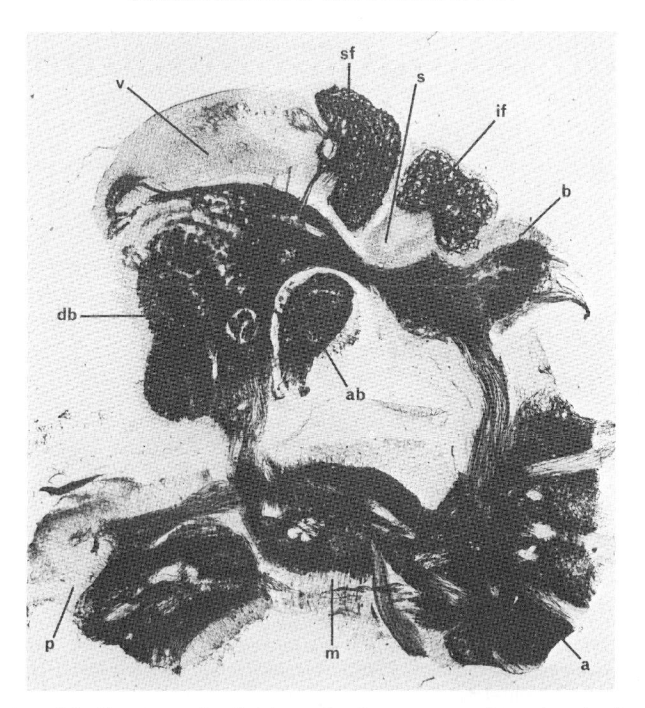

FIG. 3. Acetylcholinesterase in *Octopus vulgaris* supraoesophageal and suboesophageal ganglia. Section prepared as for that in Fig. 2 but incubated with acetylthiocholine medium (pH 4·8). Most intense reaction occurs in the neuropil although it is noticeably decreased in the vertical (v) and subfrontal (s) lobes. Symbols and magnification as for Fig. 2.

particularly larger cells in the suboesophageal ganglion, react positively for AChE.

The validity of this histochemical localization of AChE in *Octopus vulgaris* oesophageal ganglia is supported by direct titrimetric assay of AChE in homogenates of dissected brain areas (Fig. 4). AChE activity, expressed as the cholinesterase coefficient, is significantly higher in the superior frontal ($P<0{\cdot}001$) and subvertical plus basal lobes ($P<0{\cdot}001$) than in the vertical lobe. The value for the suboesophageal ganglia is lower than that for the superior frontal ($P<0{\cdot}05$) and subvertical plus basal lobes ($P<0{\cdot}05$), and higher than the vertical lobe ($P<0{\cdot}01$). These estimates of cholinesterase coefficients are higher than those reported for *Octopus dofleini* circumoesophageal ganglia (Loe & Florey, 1966) and lower than those in optic lobe of *Loligo pealeii* (Nachmansohn & Meyerhof, 1941) or *Octopus vulgaris* (Barlow, 1971).

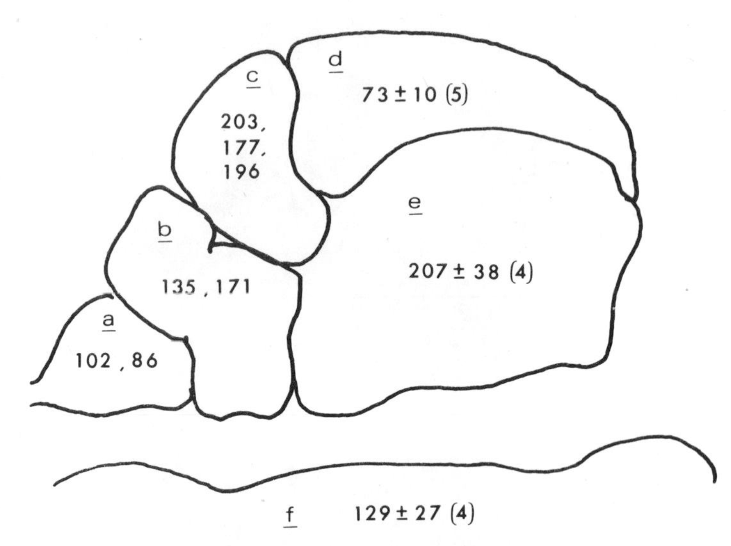

FIG. 4. Titrimetric assay of cholinesterase activity in areas dissected from the brain of *Octopus vulgaris* by the method described previously (Barlow, 1971). Values of the cholinesterase coefficient (mg acetylcholine chloride hydrolysed/hr 100 mg^{-1} wet weight) are individual determinations, or, when more than three, are ± one s.d. with the number of determinations in parentheses. Area a: superior buccal; b: median inferior frontal + lateral inferior frontal + subfrontal + posterior buccal; c: median superior frontal + lateral superior frontal; d: vertical; e: subvertical + anterior, median and dorsal basal; f: suboesophageal ganglion.

Histochemistry of cholinesterase in decapods

In decapod brains the histochemical localization of BuChE appears to be identical to that of AChE determined by a variety of methods (Drukker & Schadé, 1967; Turpaev *et al.*, 1968; Chichery & Chichery, 1974; Barlow, in prep. b). The kinetic analysis of *Ommastrephes sloanei-pacificus* optic lobe cholinesterase by Turpaev *et al.* (1968), using selective enzyme inhibitors, suggested that only a single enzyme was present. Cholinesterase from squids has a greater affinity for acetylcholine than butyrylcholine (Turpaev *et al.*, 1968; Welsch & Dettbarn, 1972) but the substrate specificity is unusually broad.

The distribution of cholinesterase in the circumoesophageal ganglia of *Sepia officinalis* (Fig. 5) and *Loligo vulgaris* is similar to that of acetylcholinesterase in *Octopus vulgaris*, being concentrated in the neuropil. The relative amounts of reaction seen in different brain areas are also basically similar to those in *Octopus vulgaris*. Decapod cholinesterase is low in the vertical lobe and high in the subvertical and basal lobes. Within the superior frontal and anterior basal lobes the enzyme is discretely localized in a way that reflects the complex organization of the neuropil in these regions (Fig. 6).

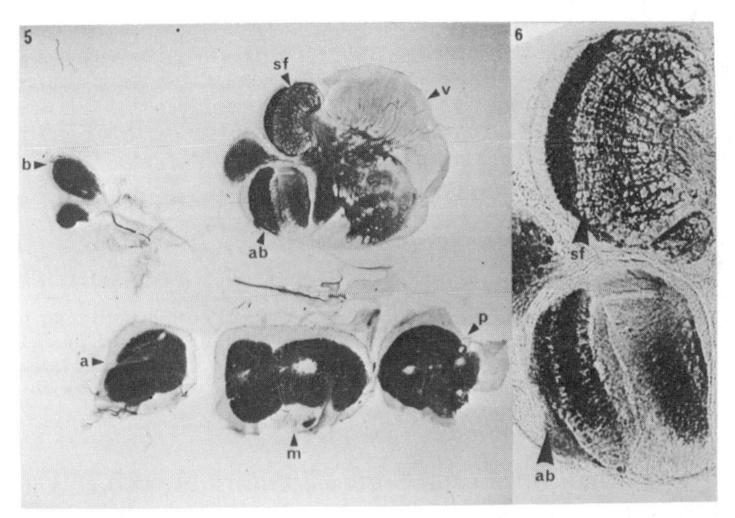

FIG. 5. Cholinesterase in *Sepia officinalis* supraoesophageal and suboesophageal ganglia. Formaldehyde fixed, freezing microtome section (100 μm) incubated with acetylthiocholine medium (pH 4·8). Reaction is confined to the neuropil. Symbols as for Fig. 2 except b: superior and inferior buccal lobes. Anterior to the left. ×6.

FIG. 6. Higher magnification (×18·5) of cholinesterase in the superior frontal (sf) and anterior basal (ab) lobes of *Sepia officinalis* supraoesophageal ganglion.

Cholinesterase in the optic lobe is also restricted to the neuropil, except for the reaction in the spider cells of the peripheral outer granule cell layer. Reaction is not uniform throughout the neuropil, however, as can be seen in the plexiform layer where the neuropil is highly ordered (Fig. 7). This section from the optic lobe of *Loligo vulgaris* was over-reacted in order to show sites of lower enzyme activity, since the method is normally selective for regions of high activity (Shute & Lewis, 1960). Reaction is intense in the first tangential and second radial layers, and light reaction occurs in some tangential and radial fibres lying deeper in the plexiform layer. In *Illex coindetii* AChE and BuChE reaction was the same and the distribution was indistinguishable from that found for *Loligo vulgaris* optic lobe (Figs 9 and 10). This distribution is very different from that occurring in *Octopus vulgaris* (Fig. 8) and *Pteroctopus tetracirrhus* (Fig. 12) where there are additional bands of reaction in the first radial and fourth tangential layers. The outer granule cell

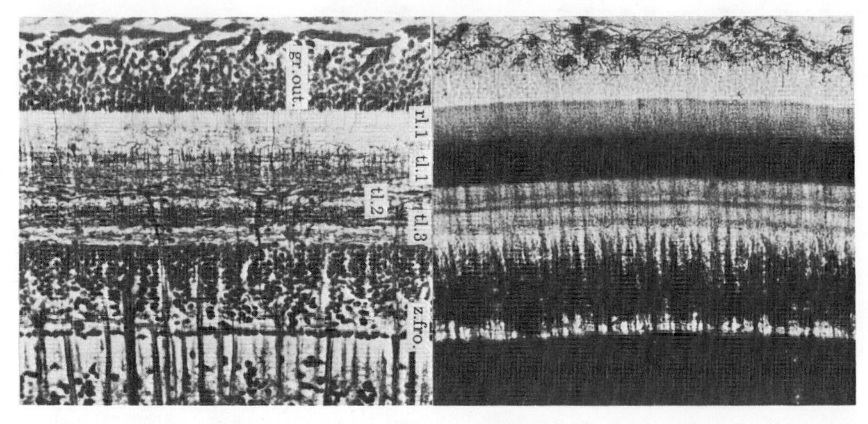

FIG. 7. *Loligo vulgaris* optic lobe, plexiform zone. Cajal silver stain (left) compared with cholinesterase (right). Cell layers (gr.out: outer granule cells; z.fro: frontier zone), lie either side of the plexiform zone (rl.1: 1st radial layer; tl.1–tl.3: 1st–3rd tangential layers). Cajal preparation reproduced from Young (1974) with permission of the Royal Society. ×225.

layer is somewhat thinner in *Sepia officinalis* but both BuChE and AChE reaction are again most pronounced in the first tangential and second radial layers (Fig. 11). *Sepia elegans*, like *Sepia officinalis*, shows no cholinesterase reaction in the optic nerves or first radial layer, but there is slightly greater reaction in one of the deeper plexiform zones (Fig. 14). In the plexiform layer of *Rossia macrosoma* there is reaction in the first radial layer, and also in the third and fourth radial and fourth tangential layers, close to the outer and inner granule cell layers respectively (Fig. 13). The presence of reaction in these extra bands resembles more closely the staining in octopods (Figs 8 and 12) than that in the other decapods (Figs 9, 11 and 14). Apart from these minor species variations in enzyme distribution, all these decapods have in common the parallel localization of BuChE and AChE which is quite unlike the situation in octopods.

Both octopod acetylcholinesterase and decapod cholinesterase, therefore, although present at very high levels in the central nervous system, show well defined anatomical localization. While this does not necessarily tell us which pathways are likely to be cholinergic, the correlation of enzyme activity with finer anatomical features within a particular brain region is more consistent with a function related to cholinergic transmission than to any generalized role in cell metabolism (Florey, 1963; Turpaev *et al.*, 1968).

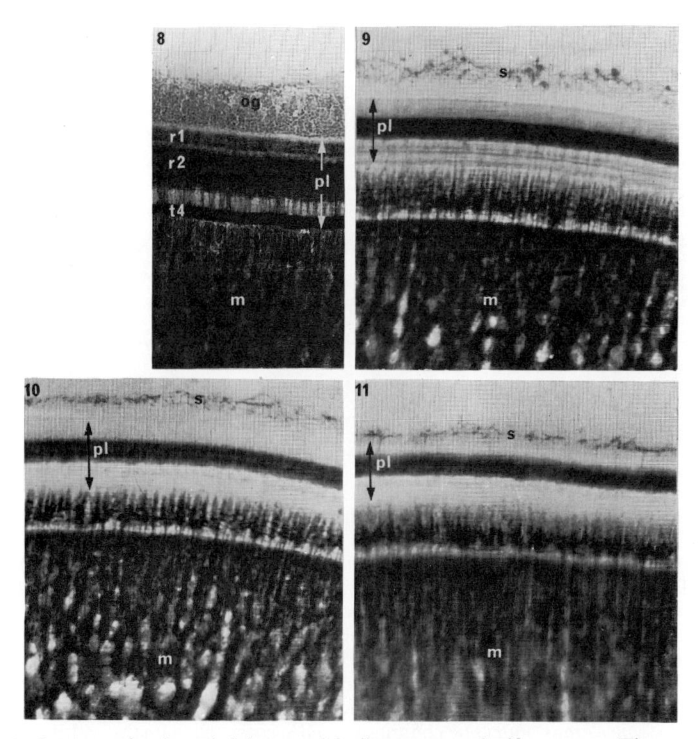

FIG. 8. *Octopus vulgaris* optic lobe acetylcholinesterase, plexiform zone. Tissue prepared as described for Fig. 3. Reaction is absent from the outer granule cell layer (og) and optic nerves, and present in the plexiform zone (pl) and medulla (m). Within the plexiform zone AChE occurs in the first radial layer (r1), second radial and first tangential layers (r2) and fourth tangential layer (t4). ×128.

FIG. 9. *Loligo vulgaris* optic lobe cholinesterase, plexiform zone (pl), medulla (m) and spider cells (s) in the outer granule cell layer. Formaldehyde fixed freezing microtome section (100 μm) incubated with acetylthiocholine medium (pH 4·8) for 2 hr. ×120.

FIG. 10. *Loligo vulgaris* optic lobe cholinesterase. Section from same optic lobe as Fig. 9 but incubated with butyrylthiocholine medium (pH 4·8) for 2 hr. ×120.

FIG. 11. *Sepia officinalis* optic lobe cholinesterase, plexiform zone. Section prepared as Fig. 9. Reaction is most intense in the first tangential and second radial layers. ×120.

Some evidence for a specific cholinergic pathway

Lam, Wiesel & Kaneko (1974) have described the synthesis of acetylcholine from isotopically-labelled choline in the isolated retina of four species of cephalopods: *Octopus joubini*, *Octopus vulgaris*, *Lolliguncula brevis* and *Loligo pealeii*. In both octopods the activity of choline acetyltransferase in the optic nerves was similar to that in the retina. The optic nerves contain the axons from the retinal photoreceptors (Alexandrowicz, 1927; Young, 1962a,b) which form large synaptic bags, densely packed with small clear

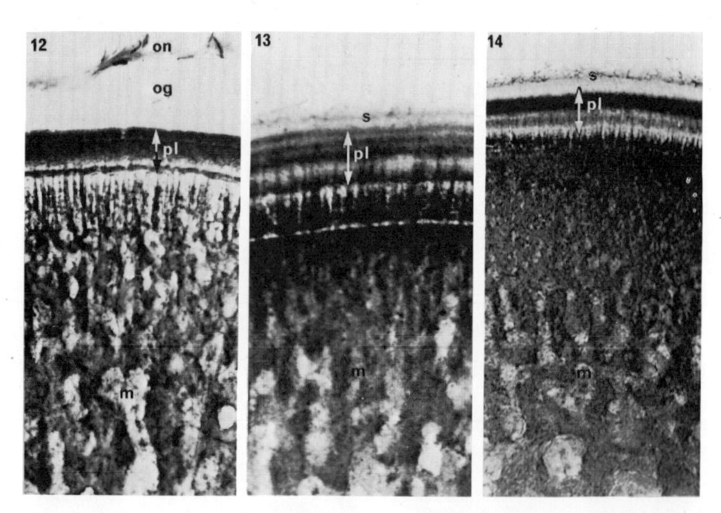

FIG. 12. *Pteroctopus tetracirrhus* optic lobe acetylcholinesterase. Plexiform zone (pl), medulla (m), outer granule cell layer (og) and optic nerves (on). Arrows delineate the width of plexiform zone which shows reaction in the innermost and outermost layers. Section prepared as described in Fig. 1 but incubated with acetylthiocholine medium (pH 4·8). ×60.

FIG. 13. *Rossia macrosoma* optic lobe cholinesterase. Arrows delineate the width of plexiform zone (pl) which shows reaction in the innermost and outermost layers (compare Fig. 8). Besides the reaction in tangential and radial fibres in the plexiform zone, there is reaction in the neuropil of the medulla (m) and spider cells (s) of the outer granule cell layer. Acetylthiocholine or butyrylthiocholine media gave the same enzyme distribution. ×60.

FIG. 14. *Sepia elegans* optic lobe cholinesterase. Arrows delineate the width of plexiform zone (pl). Reaction is most intense in the first tangential and second radial layers and there is additional reaction deeper in the zone (compare Fig. 11). Acetylthiocholine or butyrylthiocholine gave the same enzyme distribution. ×33·5.

synaptic vesicles, immediately on entering the first radial layer of the plexiform zone in the optic lobe (Dilly, Gray & Young, 1963). The visual cells also send out axon collaterals in the retinal neuropil which form presynaptic knobs containing numerous small clear vesicles (Tonosaki, 1965). Lam *et al.* (1974) suggested that acetylcholine synthesis occurred in the cephalopod photoreceptors, and that it could function as the neurotransmitter in these cells. We have attempted to establish whether choline metabolism *in vivo* is influenced by light stimulation of the photoreceptors (Barlow & Martin, in prep.). After injecting radioactive choline into the eye fluid of *Octopus vulgaris* the animals were returned to sea-water tanks kept either under daylight conditions or in a photographic dark room. The rate of transport of labelled choline compounds from the eye to the optic lobe was of the order expected for rapid axoplasmic transport of small molecules. The ratio of the radioactivity recovered in acetylcholine/choline was higher in the retina and optic lobe than in the optic nerves. The greatest turnover of acetylcholine therefore occurred in the tissues which contain nerve terminals. The ratio of radioactivity in acetylcholine/choline in the retinas of animals kept in daylight (1·25±0·36) was significantly ($P<0{\cdot}01$) higher than in those kept in the dark (0·61±0·15), suggesting that acetylcholine turnover is increased by photoreceptor stimulation. While this is consistent with acetylcholine being a transmitter in the photoreceptors it is also likely that the activity of efferent fibres is increased by light stimulation. If the octopod photoreceptors are indeed cholinergic we might expect to find acetylcholinesterase localized where their axons synapse in the first radial layer of the plexiform zone. In *Octopus vulgaris* acetylcholinesterase is found evenly distributed throughout the first radial layer (Drukker & Schadé, 1964; Barlow, in prep. a) but not specifically in association with the optic nerves. The latter do, however, stain for butyrylcholinesterase and it will require electron microscopical evidence to determine whether or not either enzyme is localized at the photoreceptor terminals. The absence of cholinesterase from the first radial layer of the decapods *Loligo vulgaris* and *Sepia officinalis* (Figs 9, 10 and 11) argues against their photoreceptors being cholinergic even though acetylcholine is synthesized in the decapod retina (Lam *et al.*, 1974).

DISTRIBUTION OF AMINERGIC COMPOUNDS IN CEPHALOPOD BRAIN

Quantitative analyses in different cephalopods

A considerable number of potential transmitter substances have now been identified in the central nervous tissue of a variety of cephalopods (Table I). Much less is known regarding the enzymes required for their synthesis and degradation, with the exception of acetylcholine. Both dopamine and noradrenalin can be formed from tyrosine or dihydroxyphenylalanine in *Octopus vulgaris* (Juorio & Barlow, 1973), but the enzymic steps have not been identified. Dopamine is metabolized by the action of a monoamine oxidase and no evidence for the presence of catechol-*O*-methyl transferase has yet been obtained. Monoamine oxidase is present in the central nervous system of *Octopus vulgaris* (Blaschko & Hawkins, 1952), *Loligo forbesii* and *Sepia officinalis* (Blaschko & Himms, 1954). Cephalopod monoamine oxidase may have a particularly broad substrate specificity; the enzyme from *Eledone cirrosa* oxidizes *N*-methyl histamine as well as histamine (Boadle, 1969).

TABLE I

Biogenic amines which have been found in cephalopod central nervous tissue

Amine	Cephalopod	References
Acetylcholine	*Octopus vulgaris*	Bacq & Mazza (1935); Jones (1970)
	Octopus dofleini	Loe & Florey (1966); Florey & Winesdorfer (1968)
	Loligo pealeii	Heilbronn *et al.* (1971); Welsch & Dettbarn (1972); Dowdall & Whittaker (1973); Dowdall & Simon (1973)
	Sepia officinalis	Florey & Florey (1954)
Dopamine	*Eledone cirrosa*	Cottrell (1967); Juorio (1971)
	Eledone moschata	Juorio & Killick (1972a); Juorio & Barlow (1974)
	Octopus vulgaris	Juorio (1971); Juorio & Killick (1972a,b); Barlow *et al.* (1974); Juorio & Barlow (1973, 1974); Juorio & Molinoff (1974)
	Octopus macropus	Juorio & Killick (1972a); Juorio & Barlow (1974)
	Octopus dofleini	Juorio & Philips (1975)
	Loligo vulgaris	Juorio & Killick (1972a); Juorio & Barlow (1974)
	Sepia officinalis	Juorio (1971); Juorio & Killick (1972a); Juorio & Barlow (1974)
Noradrenalin	*Eledone cirrosa*	Cottrell (1967); Juorio (1971)
	Eledone moschata	Juorio & Killick (1972a); Juorio & Barlow (1974)
	Octopus vulgaris	Juorio (1971); Juorio & Killick (1972a,b); Barlow *et al.* (1974); Juorio & Barlow (1973, 1974); Juorio & Molinoff (1974)
	Octopus macropus	Juorio & Killick (1972a); Juorio & Barlow (1974)
	Loligo vulgaris	Juorio & Killick (1972a); Juorio & Barlow (1974)
	Sepia officinalis	Juorio (1971); Juorio & Killick (1972a); Juorio & Barlow (1974)

5-Hydroxytryptamine	*Eledone cirrosa*	Juorio (1971)
	Eledone moschata	Bertaccini (1961); Juorio & Killick (1972a)
	Octopus vulgaris	Welsh & Moorhead (1960); Juorio & Killick (1972a,b); Juorio & Molinoff (1974)
	Octopus briareus	Welsh & Moorhead (1960)
	Octopus macropus	Juorio & Killick (1972a)
	Sepia officinalis	Juorio & Killick (1972a)
	Loligo vulgaris	Juorio & Killick (1972a)
	Loligo pealeii	Welsh & Moorhead (1960)
	Dosidicus gigas	Roseghini & Ramorino (1970)
Tyramine	*Octopus dofleini*	Juorio & Philips (1975)
Octopamine	*Eledone moschata*	Juorio & Molinoff (1974)
	Octopus vulgaris	Juorio & Molinoff (1974)
	Loligo vulgaris	Juorio & Molinoff (1974)
	Sepia officinalis	Juorio & Molinoff (1974)
Histamine	*Eledone moschata*	Bertaccini (1961)
	Octopus vulgaris	Lorenz *et al.* (1973)
	Dosidicus gigas	Roseghini & Ramorino (1970)
N-Acetyl histamine	*Dosidicus gigas*	Roseghini & Ramorino (1970)
γ-Aminobutyric acid	*Eledone cirrosa*	Cory & Rose (1969)

Preliminary experiments using the histochemical method of Glenner, Burtner & Brown (1957) suggested that monoamine oxidase is distributed throughout the brain of *Octopus vulgaris* and *Sepia officinalis* (Barlow, unpublished observations).

Early workers found that the posterior salivary glands of octopods were a rich source of pharmacologically active amines (Bacq, 1935; Erspamer & Boretti, 1951; von Euler, 1952). With the advent of sensitive spectrofluorimetric methods a similar range of compounds have been identified in the brain (see Table I), and, since small amounts of tissue can be assayed quantitatively for dopamine, noradrenalin and 5-hydroxytryptamine, Juorio and co-workers have produced precise data on their regional distribution in several cephalopods.

Juorio (1971) showed that wide variations in the levels of these amines occurred within a particular cephalopod nervous system. For example, *Eledone cirrosa* has little 5-hydroxytryptamine ($<0{\cdot}19\ \mu g/g$) in the inferior frontal lobe, but 30 to 40 times this concentration is present in the optic lobe and inferior buccal lobe. In both *Eledone cirrosa* and *Octopus vulgaris* dopamine and noradrenalin are relatively low in the median suboesophageal ganglion (range $1{\cdot}34$–$1{\cdot}84\ \mu g/g$) and high in the superior buccal lobe (range $10{\cdot}0$–$16{\cdot}7\ \mu g/g$).

Besides many similarities there are pronounced species specific variations in catecholamine levels in functionally and morphologically similar brain regions. The noradrenalin to dopamine ratio shows relatively little variation between optic lobes from different cephalopods, whereas in the vertical lobe there are major species differences (Fig. 15). The reason for comparing the ratios of amine concentrations is that if they occur only in the neuropil, for example, differences in the proportion of neuropil to cell cortex in an area, such as the vertical lobe, will not influence the comparison between cephalopods. It is somewhat surprising that *Eledone cirrosa* appears to be closer to the decapods *Loligo vulgaris* and *Sepia officinalis* judged by these criteria, than to the other octopods.

A comparative study of the various buccal ganglia in octopods and decapods (Juorio & Barlow, 1976), has revealed even more striking biochemical differences in anatomically and functionally related areas. While the level of dopamine is greater than noradrenalin in the posterior buccal ganglia, the reverse is the case in the inferior buccal ganglia (Table II). In decapods the latter contain quite remarkable concentrations of noradrenalin.

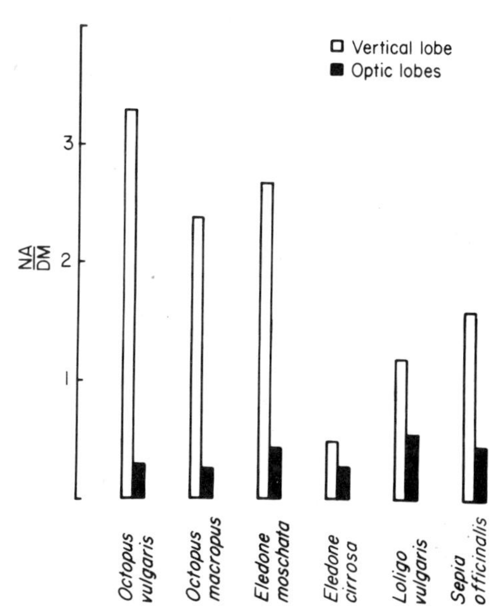

FIG. 15. Ratio of noradrenalin (NA) to dopamine (DM) in the vertical and optic lobes of some cephalopods. Reproduced from Juorio & Barlow (1974) with permission.

TABLE II

Concentration of dopamine (DM) and noradrenalin (NA) in some cephalopod buccal ganglia

	DM(μg/g)	NA (μg/g)
Inferior buccal ganglia		
Loligo vulgaris	7·9	307·0
Sepia officinalis	6·2	44·3
Octopus vulgaris	5·8	7·9
Eledone cirrosa	1·4	3·8
Superior buccal ganglia		
Loligo vulgaris	8·5	5·8
Sepia officinalis	11·2	12·6
Octopus vulgaris	10·0	16·7
Eledone cirrosa	10·3	12·2
Posterior buccal ganglia		
Loligo vulgaris	7·3	2·7
Sepia officinalis	4·3	1·5
Octopus vulgaris	7·1	2·6
Eledone cirrosa	4·1	2·1

From Juorio & Barlow (1976).

Histochemical localization of monoamines

Application of the Falk–Hillarp technique for the histochemical localization of monoamines after reaction with formaldehyde vapour may be hampered by the high salt concentration in cephalopod tissue. Nevertheless, excellent micrographs have been obtained by Matus (1973) for the brain of *Octopus vulgaris*. The fluorescent amine conjugates were clearly localized in certain neuronal perikarya, axonal varicosities and what are most likely synaptic terminals in the neuropil. The regional distribution observed by Matus correlated very closely with the quantitative chemical determinations made by Juorio (1971). In the optic lobe of *Octopus vulgaris*, where the catecholamine levels are high, amine fluorescence was in the neuropil of the medulla and in discrete bands in the plexiform layer (Fig. 16). The first tangential and second radial layers showed the most intense reaction, as was also the case when cephalopod optic lobe was stained for acetylcholinesterase (Figs 7 and 8). This may well represent a major site where cholinergic and adrenergic fibres interact. Whereas the first radial layer does not

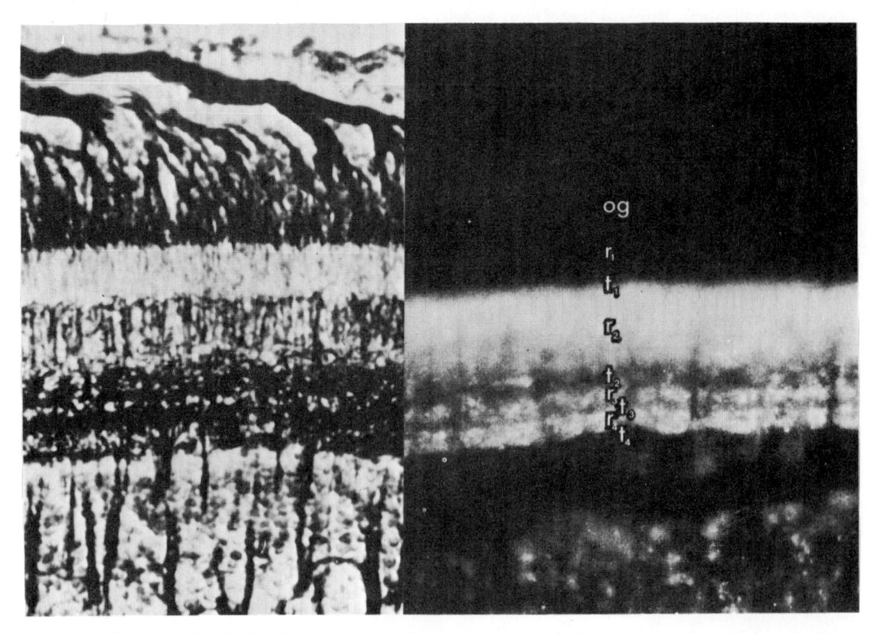

FIG. 16. Formaldehyde-fluorescence histochemistry (right) compared with Cajal silver stained preparation (left) of *Octopus vulgaris* optic lobe, plexiform zone. og: Outer granule cell layer; t_1–t_4: tangential layers of plexiform zone; r_1–r_4: radial layers of plexiform zone. ×265. Reproduced with permission from Matus (1973).

show fluorescence it does react for acetylcholinesterase. The third radial or tangential layers appear to lack both monoamines and acetylcholinesterase. Similar parallels and differences between these two biochemical parameters can be drawn from the supraoesophageal ganglia.

Some evidence for specific adrenergic pathways

It has been suggested that some of the monoamines present in the vertical lobe of *Octopus vulgaris* could be localized in the afferent fibres from the superior median frontal lobe (Juorio, 1971), since these contain dense-core synaptic vesicles (Gray & Young, 1964; Gray, 1970). Barlow & Martin (1971) found the highest proportion of nerve terminals containing dense-core and dense-content vesicles in the superior buccal lobe which contains the highest monoamine concentration in the supraoesophageal ganglion (Juorio, 1971). Histochemical studies of the vertical lobe have indicated that the monoamines are localized in fibres from the superior median frontal lobe (Barlow, 1971; Matus, 1973), rather than in the amacrine cells which contain large agranular vesicles (Gray, 1970; Barlow & Martin, 1971). Cutting the superior median frontal–vertical lobe tract causes degeneration of the afferent terminals which synapse on the amacrine trunks (Gray, 1970). The decrease in dopamine and noradrenalin levels in the vertical lobe which follows this operation (Juorio & Barlow, 1974) supports the histochemical evidence that at least some of the superior median frontal–vertical lobe fibres are catecholaminergic. The evidence, however, is still largely circumstantial and open to alternative interpretations.

Noradrenalin occurs at the highest concentration in the superior buccal lobe of the supraoesophageal ganglion in *Octopus vulgaris* (Juorio, 1971). There are many neurons showing monoamine fluorescence in this lobe and fluorescent fibres have been identified in the efferent posterior salivary gland nerves (Matus, 1973). A ligature applied to the posterior salivary gland nerves results in an accumulation of monoamine fluorescence on the proximal side (Martin & Barlow, 1972). The amine which accumulates to the greatest extent at such a ligature has been identified as noradrenalin (Barlow, Juorio & Martin, 1974). There are reasonable grounds, therefore, for thinking that noradrenergic efferent fibres arise in the superior buccal lobe of the brain, and that these include the large cells at the rear of the lobe which give rise to the posterior salivary gland nerves (Young, 1965).

DISCUSSION

There are clearly considerable regional variations in the levels of acetylcholine, acetylcholinesterase, catecholamines and 5-hydroxytryptamine in cephalopod central nervous systems. The results of histochemical studies support the biochemical evidence which together suggest that the relative proportion of cholinergic, adrenergic or serotoninergic fibres could vary greatly within a given nervous system and between similar regions in different cephalopods. While there are specific brain regions where the levels of some of these potential transmitter substances are very low, or undetectable, in general the levels are high compared with mammalian brain. The concentrations of acetylcholine in *Octopus dofleini* optic lobe (350 μg/g) and of noradrenalin in *Loligo vulgaris* inferior buccal lobe (307 μg/g), although remarkable, are not unique among the Mollusca; *Mercenaria mercenaria* ganglia contain 261 μg dopamine/g (Sweeney, 1963) and 40 μg 5-hydroxytryptamine/g wet weight (Welsh & Moorhead, 1960). These tissues should prove to be particularly useful preparations for biochemical and pharmacological studies, although it is not clear why they should possess such high concentrations of transmitters. The original proposal of Florey (Florey, 1963: Loe & Florey, 1966; Florey & Winesdorfer, 1968) that the endings in octopod optic lobes are predominantly cholinergic is supported by histochemical evidence showing intense AChE activity in the neuropil. While the question as to whether the optic nerve endings are likely to be cholinergic or non-cholinergic has not yet been resolved, there are many unidentified monoaminergic fibres present. Therefore, even if the high acetylcholine level is explained by the presence of a majority of cholinergic fibres, the high dopamine level (15 μg/g) in *Octopus vulgaris* (Juorio, 1971) must be related to some other factor. We can speculate briefly what this might be.

Assuming that higher concentrations of catecholamines really are stored in cephalopod adrenergic nerves than in their mammalian counterparts, the greater storage capacity might be required if the rate of transmitter replenishment was slower. This would be dependent on the rate of synthesis, the proportion synthesized in the terminals or cell body, the rate of axonal transport, and the presence or absence of a re-uptake mechanism. Alternatively, greater amounts of transmitter may have to be released either because cephalopod receptors have not evolved the same degree of sensitivity, or because transmitter release is more closely allied to a basic neurosecretory function. The high noradrenalin content of the posterior salivary gland nerves in *Octopus vulgaris* which terminate on glandular cells, may be an example of the latter phenomenon. These points have yet to be investigated experimentally and may provide further insight into the evolution of chemical neurotransmission.

Cephalopod central nervous systems lack the advantage of having large, readily accessible and identifiable nerve cells found in other molluscs such as *Aplysia* or *Helix*. They do, however, contain discrete brain regions with relatively homogeneous populations of certain types of neurons, for example, the amacrine cells of the vertical lobe. This region is morphologically and biochemically distinct from other brain lobes. In several different cephalopods it contains relatively low levels of cholinesterase and monoamines. The considerable importance of the vertical lobe in cephalopod learning and behaviour justifies a search for alternative transmitter candidates here.

Acknowledgements

The author wishes to thank Dr A. V. Juorio and the Editor of the *Journal of Comparative and General Pharmacology* for permission to reproduce Fig. 15; Dr A. I. Matus and the Editor of *Tissue and Cell* for permission to reproduce Fig. 16; Professor J. Z. Young and the Editor of the *Philosophical Transactions of the Royal Society of London* for permission to reproduce part of Fig. 7. Mr J. Mills provided expert photographic assistance and The Worshipful Company of Pewterers gave financial support.

References

Alexandrowicz, J. S. (1927). Contribution à l'étude des muscles, des nerfs et du mécanisme de l'accommodation de l'oeil des céphalopodes. *Archs Zool. exp. gén.* **66**: 71–134.

Bacq, Z.-M. (1935). Recherches sur la physiologie et la pharmacologie du système autonome. XVII Les esters de la choline dans les extraits de tissu des Invertébrés. *Archs Int. Physiol.* **42**: 24–42.

Bacq, Z.-M. & Mazza, F. (1935). Identification d'acetylcholine extraite des cellules ganglionnaires d'*Octopus*. *C.r. Séanc. Soc. Biol.* **120**: 246–247.

Barlow, J. J. (1971). The distribution of acetylcholinesterase and catecholamines in the vertical lobe of *Octopus vulgaris*. *Brain Res.* **35**: 304–307.

Barlow, J. J. (in prep. a). *The distribution of cholinesterase in the brain of* Octopus vulgaris.

Barlow, J. J. (in prep. b). *Cholinesterase distribution in the brain of* Loligo vulgaris *and* Sepia officinalis.

Barlow, J. J., Juorio, A. V. & Martin, R. (1974). Monoamine transport in the *Octopus* posterior salivary gland nerves. *J. comp. Physiol.* **89**: 105–122.

Barlow, J. J. & Martin, R. (1971). Structural identification and distribution of synaptic profiles in the *Octopus* brain using the zinc-iodide-osmium method. *Brain Res.* **25:** 241–253.

Barlow, J. J. & Martin, R. (in prep.). *Transport of choline compounds in* Octopus *optic nerves.*

Bertaccini, G. (1961). Discussion to H. H. Adam. "Histamine in the central nervous system". In *Regional neurochemistry*: 305. Kety, S. S. & Elkes, J. (eds). Oxford: Pergamon Press.

Blaschko, H. & Hawkins, J. (1952). Observation on amine oxidase in cephalopods. *J. Physiol., Lond.* **118**: 88–93.

Blaschko, H. & Himms, J. M. (1954). Enzymic oxidation of amines in decapods. *J. exp. Biol.* **31**: 1–7.

Boadle, M. C. (1969). Observations on a histaminase of invertebrate origin: a contribution to the study of cephalopod amine oxidases. *Comp. Biochem. Physiol.* **30**: 611–620.

Chichery, M. P. & Chichery, R. (1974). Histochemical study of the localization of the cholinesterases in the central nervous system of *Sepia officinalis. Cell Tiss. Res.* **148**: 551–560.

Cory, H. T. & Rose, S. P. R. (1969). Glucose and amino acid metabolism in octopus optic and vertical lobes *in vitro. J. Neurochem.* **16**: 979–988.

Cottrell, G. A. (1967). Occurrence of dopamine and noradrenaline in the nervous tissue of some invertebrate species. *Br. J. Pharmac. Chemother.* **29**: 63–69.

Dilly, P. N., Gray, E. G. & Young, J. Z. (1963). Electron microscopy of optic nerves and optic lobes of *Octopus* and *Eledone*. *Proc. R. Soc.* (B) **158**: 446–456.

Dowdall, M. J. & Simon, E. J. (1973). Comparative studies on synaptosomes: Uptake of (N-Me-3H) choline by synaptosomes from squid optic lobes. *J. Neurochem.* **21**: 969–982.

Dowdall, M. J. & Whittaker, V. P. (1973). Comparative studies in synaptosome formation: The preparation of synaptosomes from the head ganglion of the squid, *Loligo pealii. J. Neurochem.* **20**: 921–935.

Drukker, J. & Schadé, J. P. (1964). Neurobiological studies on cephalopods. III Histochemistry of 24 enzymes in the optic system. *Netherl. J. Sea Res.* **2**: 155–182.

Drukker, J. & Schadé, J. P. (1967). Neurobiological studies on cephalopods. VII Histochemistry and electrophoretic properties of some esterases in the optic lobe of cephalopods. *Pubbl. Staz. zool. Napoli* **35**: 374–401.

Erspamer, V. & Boretti, G. (1951). Identification and characterization by paper chromatography of enteramine, octopamine, tyramine, histamine and allied substances in extracts of posterior salivary glands of Octopoda and in tissue extracts of vertebrates and invertebrates. *Archs int. Pharmacol. Thér.* **88**: 296–332.

Euler, U. S. von (1952). Presence of catecholamines in visceral organs of fish and invertebrates. *Acta physiol. scand.* **28**: 297–308.

Feldberg, W., Harris, G. W. & Lin, R. C. Y. (1951). Observations on the presence of cholinergic and non-cholinergic neurones in the central nervous system. *J. Physiol., Lond.* **112**: 400–404.

Florey, E. (1963). Acetylcholine in invertebrate nervous systems. *Can. J. Biochem. Physiol.* **41**: 2619–2626.

Florey, E. & Florey, E. (1954). Ueber die moegliche Bedeutung von Enteramin (5-oxy-tryptamin) als nervoeser Aktionssubstanz bei Cephalopoden und Dekapoden Crustaceen. *Z. Naturf.* **9B**: 53–68.
Florey, E. & Winesdorfer, J. (1968). Cholinergic nerve endings in *Octopus* brain. *J. Neurochem.* **15**: 169–177.
Gerschenfeld, H. M. (1973). Chemical transmission in invertebrate central nervous systems and neuromuscular junctions. *Physiol. Revs.* **53**: 1–119.
Glenner, G. G., Burtner, H. J. & Brown, G. W. (1957). Histochemical demonstration of monoamine oxidase activity by tetrazolium salts. *J. Histochem. Cytochem.* **5**: 591–600.
Gray, E. G. (1970). The fine structure of the vertical lobe of octopus brain. *Phil. Trans. R. Soc.* (B) **258**: 379–395.
Gray, E. G. & Young, J. Z. (1964). Electron microscopy of synaptic structure of octopus brain. *J. Cell Biol.* **21**: 87–103.
Heilbronn, E., Hause, S. & Lundgren, G. (1971). Chemical identification of acetylcholine in squid-head ganglion. *Brain Res.* **33**: 431–437.
Jones, D. G. (1970). The isolation of synaptic vesicles from *Octopus* brain. *Brain Res.* **17**: 181–193.
Juorio, A. V. (1971). Catecholamines and 5-hydroxytryptamine in nervous tissue of cephalopods. *J. Physiol., Lond.* **216**: 213–226.
Juorio, A. V. & Barlow, J. J. (1973). Formation of catecholamines and acid metabolites by *Octopus* brain. *Experientia* **29**: 943–944.
Juorio, A. V. & Barlow, J. J. (1974). Catecholamine levels in the vertical lobes of *Octopus vulgaris* and other cephalopods and the effect of experimental degeneration. *Comp. gen. Pharmac.* **5**: 281–284.
Juorio, A. V. & Barlow, J. J. (1976). High noradrenaline content of a squid buccal ganglion. *Brain Res.* **104**: 379–383.
Juorio, A. V. & Killick, S. W. (1972a). Monoamines and their metabolism in some molluscs. *Comp. gen. Pharmac.* **3**: 283–295.
Juorio, A. V. & Killick, S. W. (1972b). The effect of drugs on the synthesis and storage of monoamines in nervous tissues of molluscs. *Int. J. Neurosci.* **4**: 195–202.
Juorio, A. V. & Molinoff, P. B. (1974). The normal occurrence of octopamine in neural tissues of the *Octopus* and other cephalopods. *J. Neurochem.* **22**: 271–280.
Juorio, A. V. & Philips, S. R. (1975). Tyramines in *Octopus* nerves. *Brain Res.* **83**: 180–184.
Lam, D. K., Wiesel, T. N. & Kaneko, A. (1974). Neurotransmitter synthesis in cephalopod retina. *Brain Res.* **82**: 365–368.
Loe, P. R. & Florey, E. (1966). The distribution of acetylcholine and cholinesterase in the nervous system and innervated organs of *Octopus dofleini*. *Comp. Biochem. Physiol.* **17**: 509–522.
Lorenz, W., Matejka, E., Schmal, A., Seidel, W., Reimann, H. J., Uhlig, R. & Mann, G. (1973). A phylogenetic study on the occurrence and distribution of the histamine in the gastrointestinal tract and other tissues of man and various animals. *Comp. gen. Pharmac.* **4**: 229–250.
Martin, R. & Barlow, J. J. (1972). Localisation of monoamines in nerves of the posterior salivary gland and salivary centre in the brain of *Octopus*. *Z. Zellforsch. mikrosk. Anat.* **125**: 16–30.
Matus, A. I. (1973). Histochemical localization of biogenic monoamines in the cephalic ganglia of *Octopus vulgaris*. *Tissue Cell* **5**: 591–601.

Nachmansohn, D. & Meyerhof, B. (1941). Relation between electrical changes during nerve activity and concentration of choline esterase. *J. Neurophysiol.* **4**: 348–361.

Prince, A. K. (1967). Properties of choline acetyltransferase from squid ganglia. *Proc. nat. Acad. Sci. U.S.A.* **57**: 1117–1122.

Roseghini, M. & Ramorino, L. M. (1970). 5-hydroxytryptamine, histamine and N-acetyl histamine in the nervous system of *Dosidicus gigas*. *J. Neurochem.* **17**: 489–492.

Shute, C. C. D. & Lewis, P. R. (1960). The salivary centre in the rat. *J. Anat., Lond.* **94**: 59–73.

Sweeney, D. (1963). Dopamine: Its occurrence in molluscan ganglia. *Science, N.Y.* **139**: 1051.

Tonosaki, A. (1965). The fine structure of the retinal plexus in *Octopus vulgaris*. *Z. Zellforsch. mikrosk. Anat.* **67**: 521–532.

Turpaev, T. M., Abashkina, L. I., Brestkin, A. P., Brick, I. L., Grigorjeva, G. M., Pevzner, D. L., Rozengart, V. J., Rozengart, E. V. & Sakharov, D. A. (1968). Cholinesterase of squid optic ganglia. *Europ. J. Biochem.* **6**: 55–59.

Welsch, F. & Dettbarn, W. D. (1972). The subcellular distribution of acetylcholine, cholinesterases and choline acetyltransferase in optic lobes of the squid *Loligo pealeii*. *Brain Res.* **39**: 467–482.

Welsh, J. H. & Moorhead, M. (1960). The quantitative distribution of 5-hydroxytryptamine in the invertebrates, especially in their nervous systems. *J. Neurochem.* **6**: 146–169.

Young, J. Z. (1962a). The retina of cephalopods and its degeneration after optic nerve section. *Phil. Trans. R. Soc.* (B) **245**: 1–18.

Young, J. Z. (1962b). The optic lobes of *Octopus vulgaris*. *Phil. Trans. R. Soc.* (B) **245**: 19–58.

Young, J. Z. (1965). The buccal nervous system of *Octopus*. *Phil. Trans. R. Soc.* (B) **249**: 27–44.

Young, J. Z. (1974). The central nervous system of *Loligo*. I. The optic lobe. *Phil. Trans. R. Soc.* (B) **267**: 263–302.

Symp. zool. Soc. Lond. (1977) No. 38, 347–376.

PREY-CAPTURE AND LEARNING IN THE CUTTLEFISH, *SEPIA*

J. B. MESSENGER

Department of Zoology, Sheffield University, Sheffield, England

SYNOPSIS

This paper examines prey-catching behaviour in one particular cephalopod to show how it presents a number of features of general interest in that they can be exploited in subsequent experiments on learning.

The tentacle-attack that cuttlefish make on fast-moving prey comprises a sequence of three phases, the first two of which are controlled by visual feedback. The final *strike*, when the tentacles are very quickly ejected at the prey, is under "open-loop" control, but it is very accurate, even in an experimental situation that minimizes visual cues, and evidence is presented that it depends on a binocular depth-perception mechanism.

Because the attack on prey is visually induced, cuttlefish presented with prawns in a glass tube go through all the motions of the attack. Yet they soon learn to stop striking, and it is shown that this must be at least partially the result of their receiving aversive or negative reinforcement ("pain") when the tentacular clubs are forcibly thrown against the glass. It is also shown that the later phases of the attack wane faster than the earlier ones, suggesting that learning to suppress one phase of a motor sequence may be contingent upon having learned to suppress the subsequent phase. And because the final motor act of this sequence need only be initiated for learning to commence, the hypothesis is advanced that efferent copy may be important in establishing memories. Evidence is also presented that there are separate short-term and long-term memory stores with parallel entry.

Finally we examine the development of learning ability with age and correlate improvements in performance with the development of the vertical lobe system.

INTRODUCTION

As far as we know, all coleoid cephalopods are carnivores and many of them are active predators that capture relatively large prey such as fish, crustaceans and other cephalopods. In this paper we shall examine in some detail the prey-catching behaviour of one particular cephalopod: the common cuttlefish, *Sepia officinalis* L.

Sepia is a very successful genus of shallow inshore waters (see Voss, his chapter, this volume). A highly specialized predator, with elongate fins for delicately controlled movements (Boycott, 1958), it is neutrally buoyant (Denton, 1974) and, perhaps as a result of this, its large, well developed eyes have what are probably areas of acute vision. It has an elaborate skin that permits superb cryptic colouration (Holmes, 1940; Boycott, 1958) and it also has specialized feeding appendages. As in other decapods the eight short *arms*, serving a variety of functions, are supplemented by two elongate *tentacles* whose sole function is prey capture: the long shafts are naked and only on the expanded, terminal club are there prehensile suckers, some of which are very large (Nixon & Dilly, their chapter, this volume). When not in use these tentacles are withdrawn into pockets below the eye (Fig. 1).

Cuttlefish have been observed to feed on crabs, prawns, *Squilla*, fish and, of course, smaller cuttlefish. Slow-moving prey, such as crabs, are generally (though not exclusively: Boycott, 1958) seized with the arms, but swifter creatures, such as teleosts (with their Mauthner cells) or prawns (with their giant fibres) are captured by the tentacles in the characteristic attack that is described here. Most of the present experiments utilized prawns (*Leander* sp.) or, for very young cuttlefish, *Mysis*.

It is worth recalling that we know very little about prey-capture in other cephalopods and that field or laboratory data are urgently needed. While it seems clear that the loliginids and ommastrephids, for example, behave like *Sepia*, it seems unlikely that the cranchiids, or genera like *Mastigoteuthis*, stalk and capture large prey in quite the same way. Among the octopods, too, there are striking differences in mode of life and feeding behaviour; not all are agile pouncers like *Octopus vulgaris*, whose attacks are generally monocular, incidentally. For example *Argonauta*, a slow-moving pelagic form, seems to brush small planktonic animals into its mouth with its webbed arms (Young, 1960).

It is not always possible to relate these differences to differences in brain structure (but see J. Z. Young, his chapter, this volume). The centres controlling the visual attack of *Sepia* are the optic, peduncle and basal lobes (Boycott, 1961; Messenger, 1967) and these are well developed in all coleoids, emphasizing that they have other functions, too. Equally it seems that the vertical-superior frontal lobe system, with its large input from the optic lobes, is well developed in most coleoids, reminding us that they probably all seek their prey and avoid their predators by sight and that learning about things seen must play an important part in their life.

Indeed our interest in the attack lies in this very fact. Though it is in itself a fascinating piece of behaviour, what is much more important is that we can use it as a system to consider some aspects of the learning process itself. So famous has Professor Young's work on *Octopus* become that it may surprise some readers to learn that his very first experiments on learning in cephalopods utilized *Sepia* (Sanders & Young, 1940). Cuttlefish are difficult to keep and handle, however, and for this reason Young soon turned his attention to the octopus, which, for most purposes, is a much more convenient experimental animal than *Sepia*, especially because of its ability to withstand surgical insult. Yet *Octopus* also has its disadvantages, especially in visual learning experiments: the large span of its long arms, with their apparently random movements, and the "fluid", ill-defined nature of the attack combine to make it very difficult sometimes to establish whether or not an animal has made an attack in a discrimination situation. The cuttlefish, on the other hand, behaves almost like an arthropod when it attacks a prawn (cf. *Mantis*, Mittelstaedt, 1957; Maldonado, Levin & Barros Pita, 1967). There is a series of precise separate motor acts, in a relatively strict sequence (but see Neill & Cullen, 1974), terminating in an abrupt, all-or-none act: these events, especially the last, can provide a quantitative measure of the behaviour especially as it changes when the animal is placed in a learning situation.

THE ATTACK WITH THE TENTACLES

The three phases of the attack

When a prawn is placed in the visual field of a hungry cuttlefish so that it can be viewed monocularly (Fig. 2) the cuttlefish makes a series of motor acts (Messenger, 1968) that can be summarized as follows.

There are eye movements (first the ipsilateral then the contralateral eye turning to fixate the prawn), head movements and body movements (the latter being brought about by funnel and fin activity). These have the effect of aligning the cuttlefish so that it faces the prawn directly. Meanwhile there may be colour changes over the entire body and erection of the first, and perhaps the second, pair of arms (Fig. 3). These seem to act as distractors or lures and remain erect until the end of the next phase of the attack, when they may be moved from side to side. As a result of these actions, which together constitute the first phase of the attack, or *attention*, the prawn will now be fixated binocularly (Fig. 3); it will lie directly ahead of the cuttlefish and its image will presumably lie on the posterior "fovea", the specialized area in the retina on the equatorial strip (Young, 1963; and Muntz, his chapter, this volume).

It is simple to demonstrate that this phase of the attack is visually initiated and maintained. It never occurs in darkness or dim red light; it does occur if the prawn is visible behind glass or even in a separate tank; and Boletzky (1972) has described

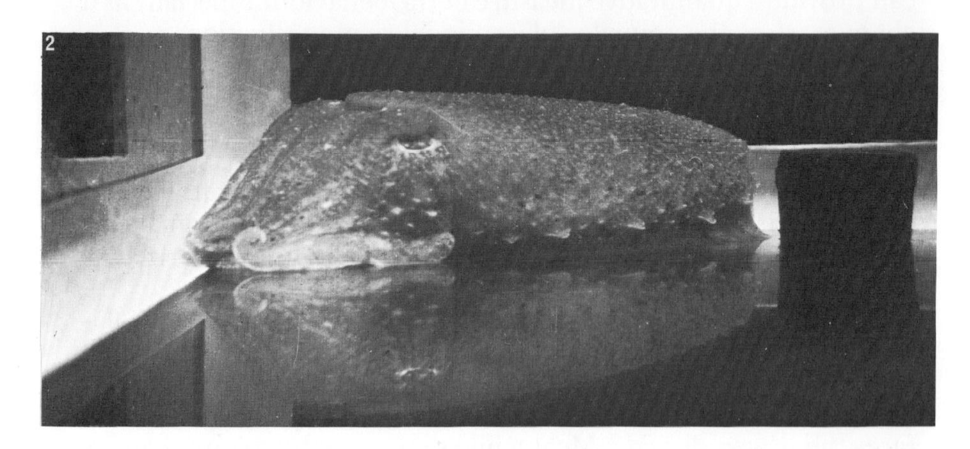

cuttlefish fixating crabs suspended in air above the water of their tank. Moreover, if the prawn moves the cuttlefish turns too, so as to keep it directly ahead and fixated binocularly (Fig. 4). If the prawn is on string, the cuttlefish can actually be induced to turn on its own axis through 360° to maintain fixation.

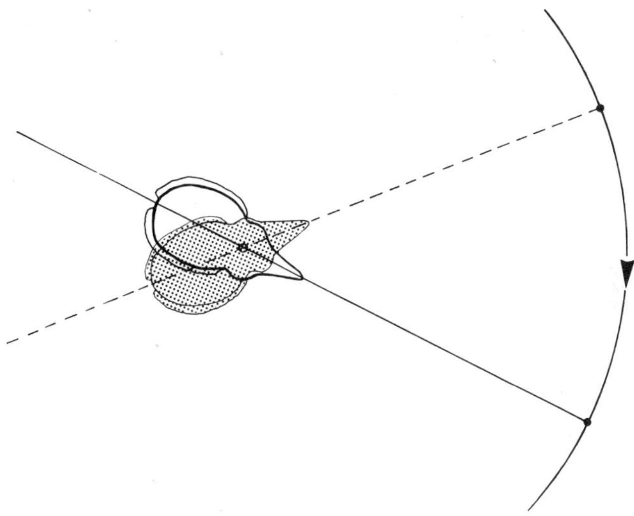

FIG. 4. Visual feedback operates during attention: movement of the prawn will be compensated for by movement of the head or the whole animal, here shown yawing about a vertical axis to keep the prawn fixated.

The next phase of the attack, *positioning*, is usually an approach. The cuttlefish swims straight towards the prawn that lies directly ahead of it. Sometimes, however, if the prawn is too close the cuttlefish may back off by gentle swimming. As a result of such a manoeuvre, whichever it is, the cuttlefish will come to lie at the *attacking distance* (Fig. 5) from the prawn where it pauses for up to 10 sec or more but generally for 2 or 3 sec. The attacking distance is a function of mantle length as would be expected (Messenger,

---

FIG. 1. A cuttlefish showing how the retracted tentacle is folded up on itself inside a sac, here cut open.

FIG. 2. Prawn's-eye view of a cuttlefish gazing at it monocularly, prior to the onset of attention.

FIG. 3. Prawn's-eye view of a cuttlefish fixating it binocularly at the end of positioning. Note the raised "lure" arms with their dark stripe. N.B. the mantle length of all these animals is about 120 mm.

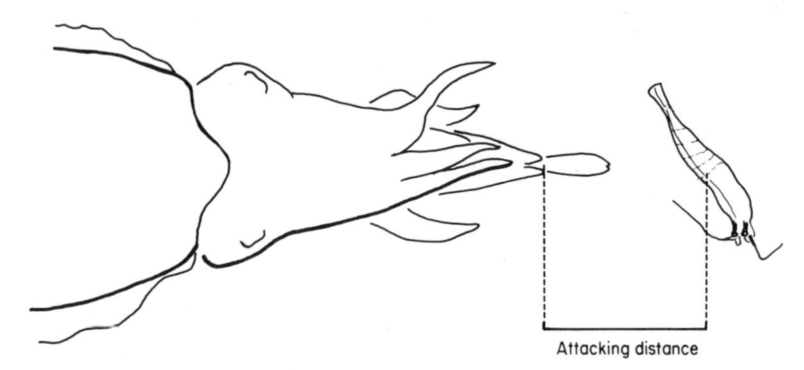

FIG. 5. The attacking distance of a particular individual on one occasion (40 mm). The tips of the tentacles can be seen protruding from between the arms; at this stage they are still under voluntary control and can be withdrawn.

unpublished observations) but it is not absolutely constant for any one individual and measurements have shown that small differences in attacking distance can be correlated with changes in the degree of ocular convergence (Messenger, 1968).

If, during the pause, the prawn moves to one side, the cuttlefish re-aligns itself by gently yawing with its fins and funnel; if the prey moves further away the cuttlefish follows it to maintain the attacking distance. Visual feedback is evidently still being employed at this stage to control the attack, as we can demonstrate simply by turning off the light: positioning will be abandoned after a few seconds. There is no need to suppose olfactory cues play any part in prey capture by this species.

It must be emphasized that if the prawn is presented behind glass the cuttlefish will position itself in the usual way. However, unless it can gain its attacking distance it will never enter the next phase of the attack: tentacle ejection. This fact is made use of in one series of experiments discussed below.

The final phase of the attack is initiated after this pause: this is the tentacle ejection or *strike*. The whole animal lunges forward slightly and the tentacles, which may have been visible previously just protruding from between the arms (Fig. 5), are thrown out so quickly that high-speed cinematography is necessary to resolve the details (Messenger, 1973c; Messenger & Lucey, in preparation). At first each tentacle shaft emerges quite straight as it travels towards the prey separate from its fellow (Fig. 6); as the prey is approached the tentacle clubs open outwards slightly presenting more sucker surfaces to the side of the prawn. The tentacles strike the prawn so strongly that they invariably carry it further away from the body: at this juncture the shafts begin to buckle (Fig. 7) and it is not until the end of full excursion that they once more straighten out as they are pulled back towards the arms and mouth. The arms meanwhile have opened and spread out ready to deal with the prawn (Fig. 8), which they manipulate so that it can be bitten in the mid-dorsal abdomen and injected with a poison that kills it within about 5 sec (Messenger, 1968).

The strike is very fast: at 25°C the tentacles can reach the prawn in less than 15 msec which is rather faster than the strike of the *Mantis* (e.g. 42 msec at 29°C, Maldonado *et al.*, 1967). Subsequently, events proceed more slowly: full excursion may be completed by about 100 msec but final withdrawal of the tentacles with their load is not completed until about 300 msec. It is worth emphasizing that the tentacles are always shot out straight ahead, and that their trajectory is flat.

The high speed of the strike precludes its control by visual feedback and, as in other fast motor acts in a variety of organisms, it seems that it is under "open-loop" control; that is, the strike is programmed on the assumption that the prawn will not move. If it does, the strike will fail. Indeed the prawn's best strategy is to wait until the tentacles have been shot towards it and only then flick its telson; and laboratory observations suggest that most unsuccessful strikes are the result of a "last-minute" move by the prawn. Another way of demonstrating that the control is open-loop is to turn off the light as the tentacles are being shot out. Not only will the strike continue but it will invariably be successful, suggesting that the cuttlefish has estimated the exact position of the prawn prior to extending its tentacles.

The accuracy of the attack

The tentacle-attack is remarkable not only for its speed but for its accuracy. At the Naples laboratory, in grey rectangular tanks, about 91% of prawns are seized first time (that is without error) and the percentage of encounters with prawns that are successful with one error or less is 98% (Messenger, 1968). Admittedly the experimental situation favours the cuttlefish because of the bare plastic substrate and absence of hiding places, but nonetheless it is clear that cuttlefish are able to locate their prey very accurately.

To explore this ability further we put healthy *Sepia* singly into large spherical tanks (60 cm in diameter) that were a matt grey inside and, as far as possible, contained no visual cues. Even with

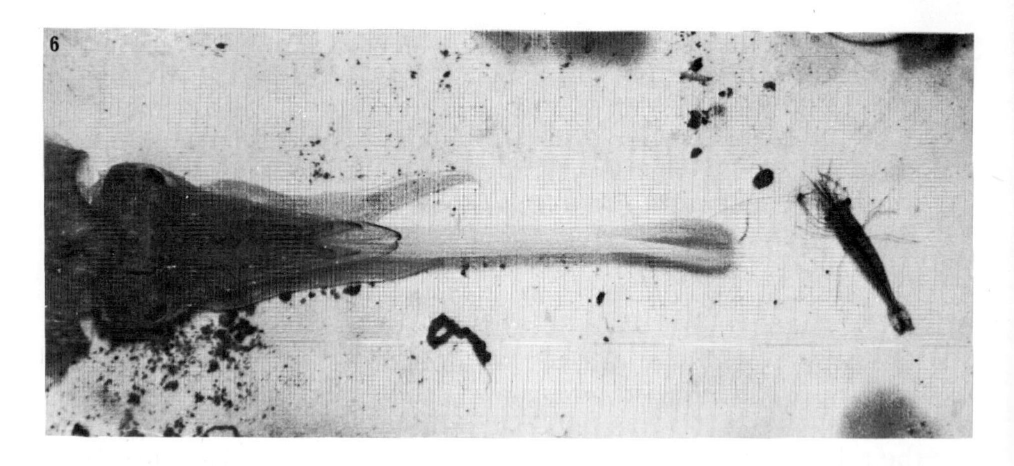
6

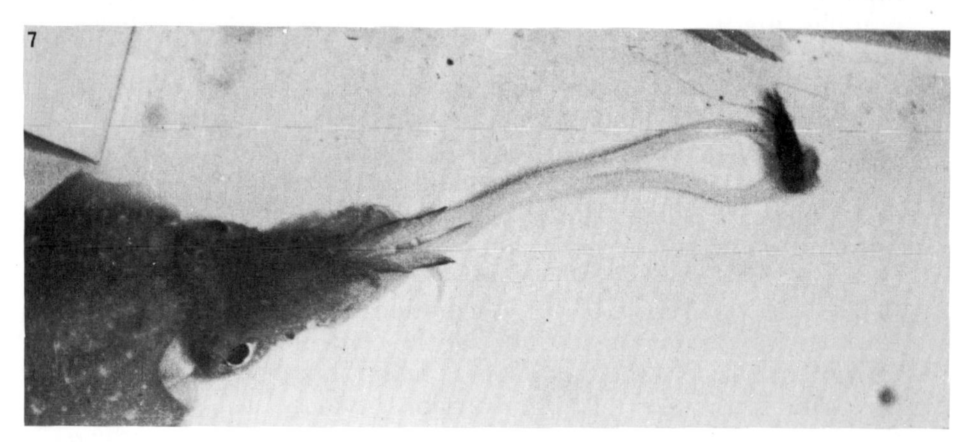
7

8

such a uniform background cuttlefish still captured prawns very efficiently: 82% of encounters terminated successfully without error (Messenger, in prep.).

Whatever the situation, if the strike does fail the cuttlefish must withdraw its tentacles and re-enter the attack program at a suitable point, perhaps in the positioning, perhaps in the attention phase.

Evidence for binocular depth perception

We have already stressed that *Sepia*, unlike *Octopus*, attacks its prey after binocular fixation and this clearly implies that it must have a binocular system for depth estimation. Although there is no direct physiological evidence for this (cf. Barlow, Blakemore & Pettigrew, 1967) we have at least three kinds of behavioural evidence to support this contention.

First, as can be seen in Fig. 9, interference with one eye lowers the accuracy of the strike: unilaterally blinded *Sepia* (for method

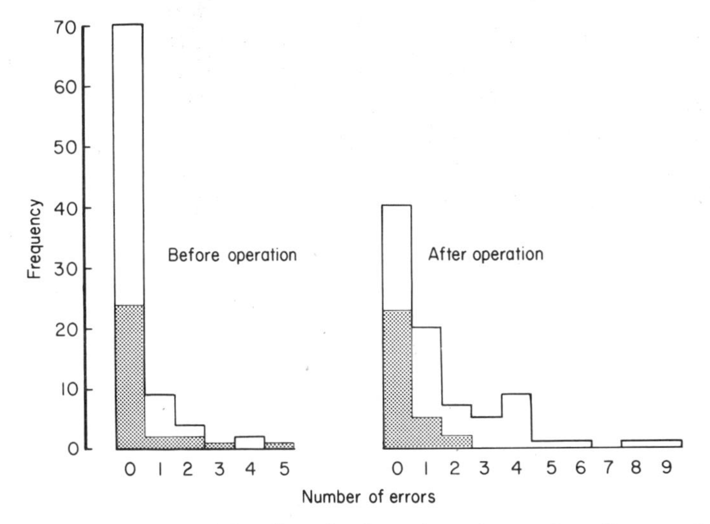

FIG. 9. Strike-accuracy and the effect of unilateral eye closure. A small group of control animals (stippled; N = 5) performs as well *after* operation as before but operated animals (N = 17) make significantly fewer error-free strikes afterwards ($t = 5{\cdot}99$; d.f. = 32; $P < 0{\cdot}001$).

FIGS 6, 7 and 8. The tentacle ejection seen from above. In Fig. 6 the tentacles are being shot out towards the prawn; in Fig. 7 they have seized it with the suckers on the tentacular club, pushing it further away, and the shafts are beginning to buckle; in Fig. 8 the arms are opening to receive the prawn as it is drawn back towards the mouth. N.B. prawn length about 50 mm; exposures 1/1000 sec.

see Messenger, 1970) tested in the grey spheres make significantly more errors on the first strike after operation than they do before the operation. This is also true of cuttlefish tested in ordinary rectangular tanks, though it is interesting that their performance appears to be better than that of operated animals in the sphere, presumably because of the extra visual cues available to them.

Second, size-constancy seems to operate. It is possible to train cuttlefish to discriminate between squares of different sizes by either the successive or the simultaneous presentation method. For example, when a pair of squares (4×4 cm and 2×2 cm) is presented to the cuttlefish at the same distance (60 cm) and attacks on the large square lead to a prawn reward, the animals very quickly come to restrict their attacks to the large square. In one such experiment, where we had trained a group of *Sepia* on this discrimination, we then began to present the small square at half the distance (30 cm) while continuing to present the large square at 60 cm. It can be seen from Fig. 10 that in these circumstances the animals not merely continue to discriminate: their score actually improves. Although it has proved impossible to perform the critical experiment, using monocularly-blinded *Sepia*, these results suggest very strongly that *Sepia* has a depth-estimating system, presumably binocular. There is evidence from experiments with young cuttlefish, isolated after hatching from their eggs and tested during the first few days of life, that supports this idea. Such

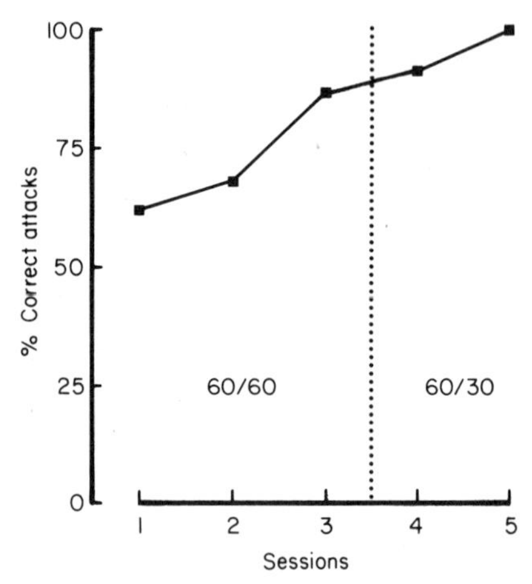

FIG. 10. Size constancy. Cuttlefish rapidly learn to discriminate between a large and a small white square presented successively at a distance of 60 cm. When the smaller square is presented at 30 cm they continue to improve their score (N = 8).

juveniles feed on *Mysis*, and Wells has shown (1958) that their binocular attack, which is in every way like the adult's, is quite as accurate even on the first few times it is made. Furthermore, we have been able to demonstrate that juvenile cuttlefish never strike at *Mysis* presented at greater than the attacking distance (Fig. 11). The young cuttlefish merely "swims into" the glass (see below) and this is true of even the first occasion in its life that it is shown a prawn (Messenger, unpublished observations). The inference is clear: even newly-hatched cuttlefish possess a mechanism for estimating distance: in fact, the distance at which it is reasonable to start ejecting their tentacles.

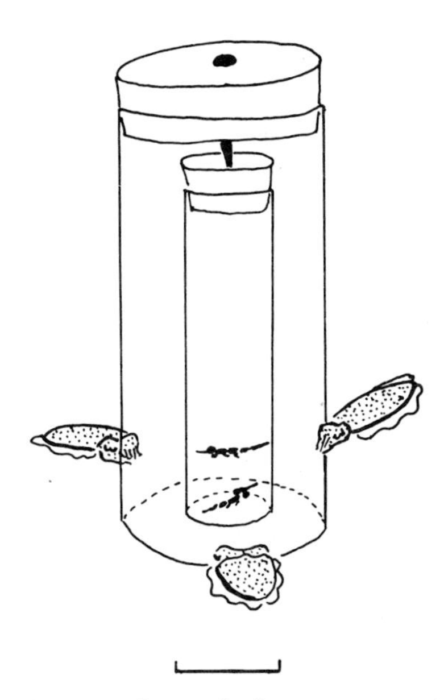

FIG. 11. Depth perception operating on the first encounter with prey. By placing *Mysis* inside a double tube filled with sea-water we ensure that a young cuttlefish can never gain its attacking distance: in these circumstances it swims continually against the glass and never ejects its tentacles. The scale represents 10 mm.

Third, it is possible to drastically reduce the accuracy of attack by splitting the supraoesophageal brain in the sagittal midline plane, so that the optic commisure and the basal lobes are divided (Messenger, unpublished observations). This operation has only been performed on four animals (two-months-old) but the results were unequivocal.

To summarize, *Sepia* captures a prawn after binocular fixation and after manoeuvring so that the prawn lies directly ahead of its arms. The attack is visually initiated and in its earlier phases visually controlled; the final strike is a rapid "open-loop" motor act. It is extremely accurate even where background cues are limited and there is evidence that this is the result of binocular depth perception. Any error in estimation may, perhaps, be compensated for by the habit of "overshooting" with the tentacles, which, it will be recalled, actually carry the prawn further away once they have seized it (p. 353 above). But that *Sepia* has evolved a fast and very effective method of capturing animals capable of sudden rapid movements seems beyond dispute.

LEARNING NOT TO ATTACK

A simple experiment

The foregoing account has emphasized that the attack by *Sepia* is visually guided; as one might expect therefore, and as Sanders & Young (1940) showed long ago, prawns behind glass are equally effective in eliciting attacks from cuttlefish. The basic experiment utilized here consists of placing a large "Perspex" tube containing two prawns (about 50 mm long) in a tank with a hungry cuttlefish. When this is done all three phases of the attack appear with the same components and in the same sequence: the only difference now is that when the tentacles are shot out at the prawns they strike the wall of the tube (Fig. 12) with considerable force and the animal gains no food. The tentacles then have to be drawn back into their pockets before they can be ejected a second time.

What makes the attack so attractive for our purpose is that the strike is a discrete, all-or-none event; there can be no doubt that it has occurred, so that the number of strikes in a given period can provide a useful quantitative measure of the level of attacking and enable us to follow changes in behaviour as a result of experience.

Cuttlefish shown prawns in a tube do not continue to make abortive attacks on them for long. As can be seen from Fig. 13, the number of strikes wanes rapidly in a series of repeated presentations. There is evidence that during continuous presentation the strike rate falls off even more rapidly; indeed over two-thirds of all the strikes made in a 20-minute presentation period are made in the first three minutes (Messenger, 1973a). Experiments suggest that this diminishing strike rate is not the result of motor fatigue or sensory adaptation but is a manifestation of *learning*; for example,

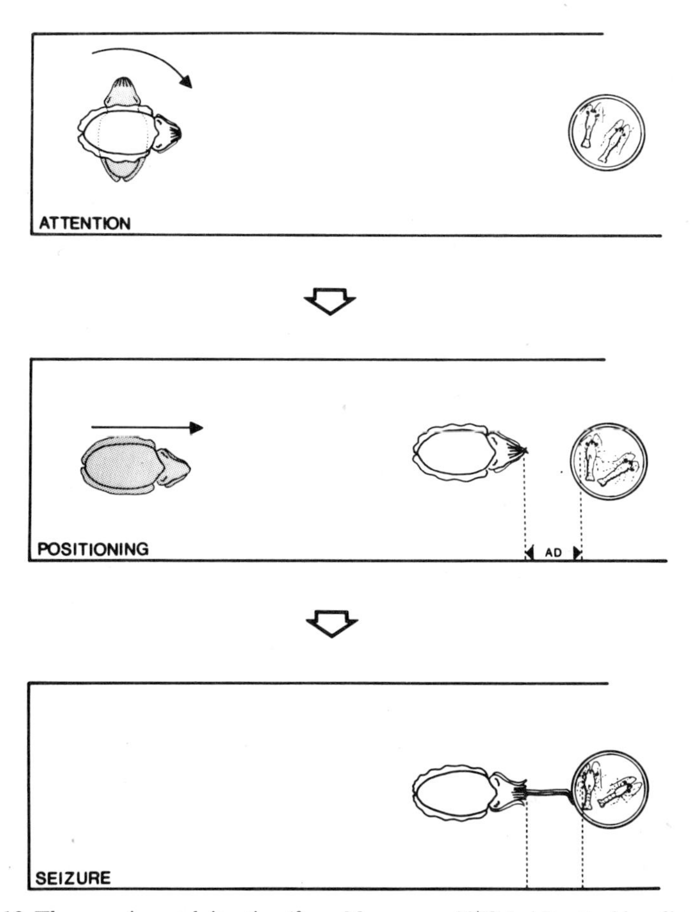

FIG. 12. The experimental situation (from Messenger, 1973a). AD: attacking distance. In bottom diagram read "STRIKE" for "SEIZURE".

the response is specific to prawns and the attack will re-appear instantly if, say, a crab in a tube is introduced into the tank.

The effect of reinforcement on the rate of learning

Such a stimulus-specific waning of the attack response could result from the lack of positive reinforcement (food reward) or the presence of negative reinforcement ("pain") or perhaps both in combination; it is difficult to separate these factors. Certainly increasing the positive reinforcement, by occasionally giving food on a wire directly to the arms for striking at prawns in a tube, diminishes the rate of waning and lack of food reward must presumably contribute to the waning process. Experimentally,

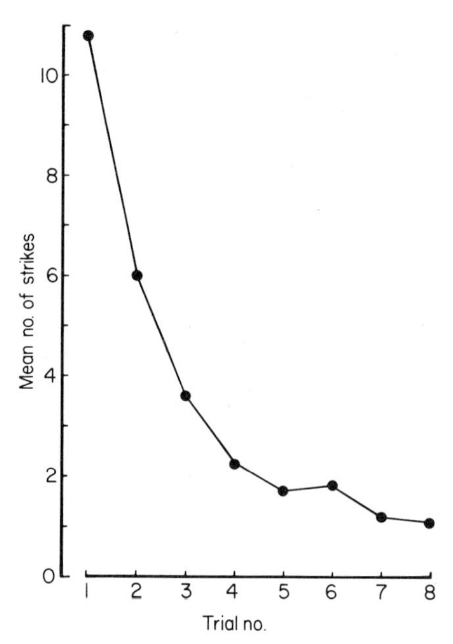

FIG. 13. Learning during repeated presentation (eight 3-min trials; data from Messenger, 1973a). N = 40.

however, it is easier to manipulate the levels of *negative* reinforcement and in this way we have been able to show that the level of pain associated with a strike at inaccessible prawns is undoubtedly a factor influencing subsequent behaviour. As might be expected, the relative rate of waning is faster as the pain level is increased (Fig. 14) (Mackintosh, 1974).

To increase pain, we gave the cuttlefish a small (10 V) electric shock every time the tentacles struck the tube. This was so effective at lowering striking that, as can be seen from Fig. 15, the total number of strikes in the first five minutes was less than half that of control animals.

It is presumably via the tentacles that pain signals reach the CNS in this situation and Graziadei (1964) has shown that the suckers contain large numbers of mechanoreceptors. Furthermore Messenger (unpublished observations) has shown that after severing the tentacular nerve, degeneration granules are found in areas of the brain associated with learning (see p. 374). To decrease pain input, therefore, we resorted to the simple expedient of cutting off the tentacles at the base, which previous experiment had shown to be quite feasible. Animals that have undergone such an operation

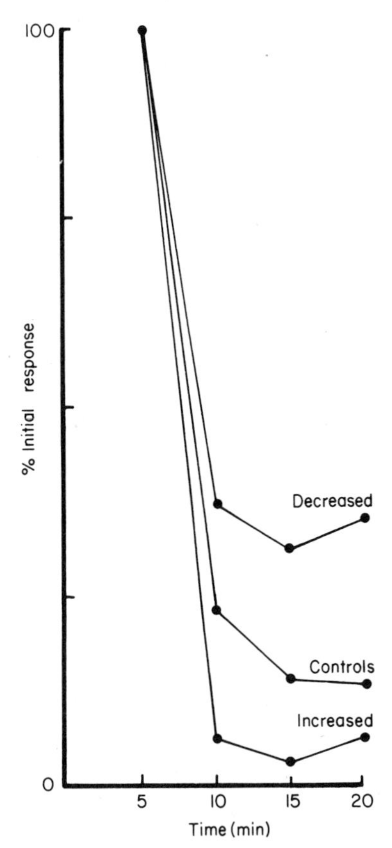

FIG. 14. The effect of increasing or decreasing negative reinforcement (from data in Messenger, 1973a).

recover instantly; they move well, change colour, copulate, capture crabs with their arms and generally survive very well indeed. Their only abnormality is that they pursue prawns all over the tank in the usual way, despite the fact that such attacks are unproductive. The motor sequence now culminates in a short, unmistakeable, forward lunge, which we term a *pseudo-strike*. What is important about this from the viewpoint of the present experiment is that in a pseudo-strike no part of the animal ever touches the tube, so that repeated attacks on prawns in a tube cannot deliver signals of mechanical shock and/or "pain" to the CNS. In these circumstances the level of attack increases threefold (Fig. 15). In fact, the no-tentacle preparation, after 15 minutes exposure to prawns in a tube, still attacks at the level of control animals during the first five minutes. It seems

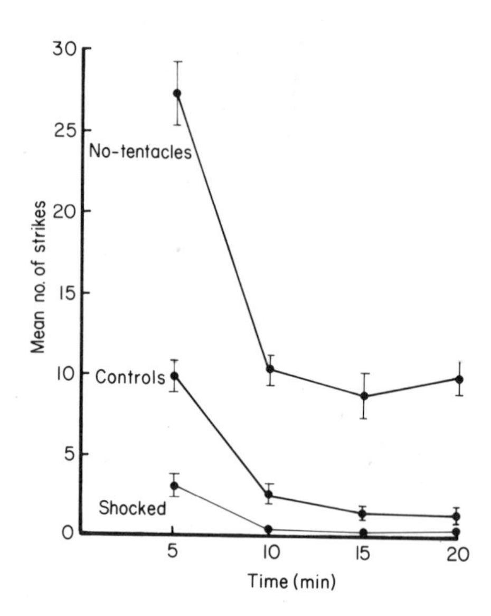

FIG. 15. The elevated level of attack in cuttlefish without tentacles. Strikes are totalled for each 5-min period of a continuous 20-min presentation (from Messenger, 1973a). N = 30 for control, N = 10 for each other group.

safe to infer that removing the pain input has a dramatic effect on the waning of the attack and that what we are observing in these experiments is what Thorpe (1963) styles *associative learning*.

The enhanced level of attacks by cuttlefish without tentacles is remarkable behaviour. Apart from demonstrating that the operation itself has had no deleterious effects, such behaviour implies that the animal cannot "see" that it has no tentacles. Since the cuttlefish can see the prawn, it must be able to see its tentacle-tip when this reaches the prawn, as it normally does so quickly and so accurately. Yet this information is not apparently integrated with any proprioceptive information that might be available about the tentacle; and the open-loop control system for striking proceeds not only on the assumption that the prawn will not move but also that the tentacles are present and fully functional. In fact this is further evidence that *there is no visual feedback system controlling tentacle behaviour*. Indeed the pause made by a no-tentacle animal, after an abortive attack, before it darkens all over, re-positions itself and pseudo-strikes again, suggests that it is the failure of food signals to reach the CNS from receptors round the mouth and lips (Graziadei, 1960) that initiates the next pseudo-strike.

This finding, curious though it may seem by vertebrate standards, is in full agreement with other data from cephalopods suggesting that this class of mollusc has been under no selective pressure to evolve a mechanism in the CNS to correlate visual with tactile or proprioceptive information (see, for example, Wells, 1963).

Sequential learning and efferent copy

So far we have only considered the strike, the final motor act of the attack; and we have seen that it wanes dramatically over a period of 20 minutes or so. But what of the earlier phases of the attack? Do these persist unaltered or are there changes here too?

We recorded the time spent in positioning and in attention in one group of animals and compared it with the actual number of strikes made during a 20-minute period. It emerged (Fig. 16) that the level of all three phases of the attack wanes in the experimental situation but it does so at very different rates. Striking wanes much more, and more quickly, than positioning, which falls to a lower level more quickly than attention. This reminds us that in prey capture, as in any complex piece of behaviour, each motor event must appear in the correct order. Here this involves populations of neurons in the optic lobes, controlling others in the anterior- and median-basal lobes, and in the interbasal lobes that in turn control the motor system via cells in the three suboesophageal areas (Boycott, 1961). And the suggestion is made that if this, or any

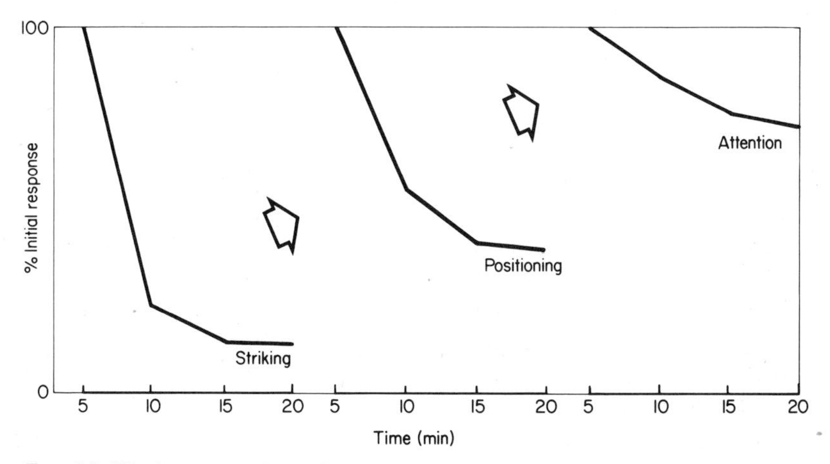

FIG. 16. Waning rates of the three phases of the attack over a 20-min period as a percentage of the initial response. The arrows emphasize the sequential, retrograde effect.

other, motor sequence has to be modified because of changing circumstances in the environment then it can only do so retrogradely by beginning at the final motor act and working back. That is, until striking has faded positioning will have to be maintained and similarly until positioning has been abandoned attention must persist.

In a sense this might seem obvious, yet it serves to focus our attention on the sequential nature of the events constituting the attack and the hypothesis is at least open to experiment. Knowing that cuttlefish will not eject their tentacles until they are within the attacking distance (p.352 above), we can ask: what would happen if we interposed a transparent screen across the tank to prevent access to the tube of prawns (Fig. 17)?

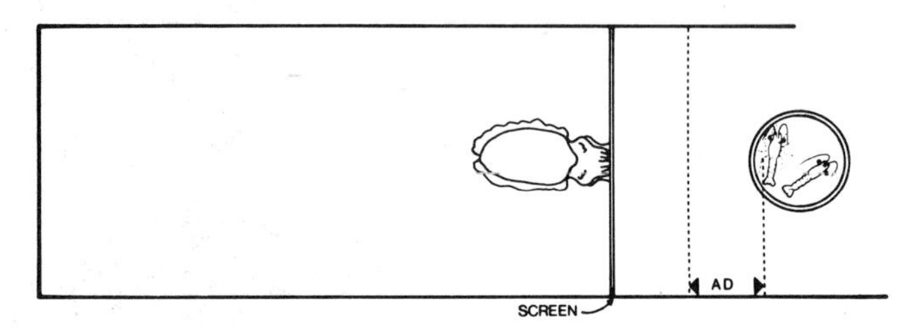

FIG. 17. A transparent screen is used to prevent the cuttlefish from completing its positioning program (from Messenger, 1973a).

In this situation a cuttlefish goes through all the movements of attention in the usual way and then enters the positioning phase, swimming down the tank towards the tube of prawns. When it encounters the screen, however, it continues to swim into it, rubbing the glass with its arms like a small boy pushing his nose against a sweet-shop window. As it does so the rate of fin-beat increases and waves of colour-changes pass over its body; often the animals will turn nearly black all over. What is remarkable, however, is the way this behaviour persists. In one series of repeated presentations, each trial lasting three minutes, this movement never even momentarily ceased throughout six trials; and attention was also maintained. In other words, preventing the animals from completing the positioning sequence maintains that sequence and the previous one.

This experiment can also be used to show that until a cuttlefish starts to strike at prawns in a tube it cannot learn to stop. In Fig. 18 we see the results of an experiment where three groups of cuttlefish

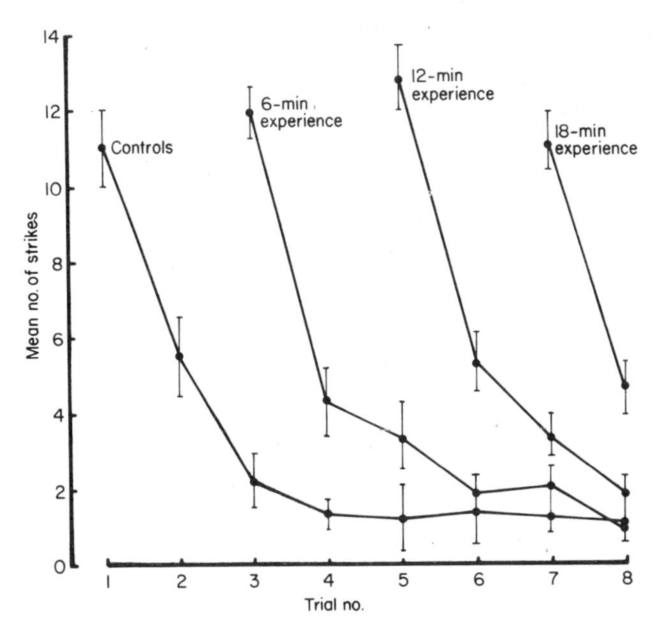

FIG. 18. Visual experience behind the screen has no effect on the level of strikes on the first (or second) trial with access (data from Messenger, 1973a). N = 10 for controls; N = 20 for each other group.

were shown prawns in a tube behind a screen for either the first two trials (6 min), the first four trials (12 min) or the first six trials (18 min) of a series of eight. Subsequently they were allowed access to the tube. The control group was allowed access at all eight trials. Two facts emerge from this experiment. First, there is clearly no significant difference in strike-level on the first trial with access, despite the very different lengths of visual experience that each group has had of prawns in a tube. Second, the rate of learning is always about the same (cf. the slopes between the first and second trials with access), suggesting that the experimental treatment has had no deleterious effect on the learning process. Once allowed to start striking they will learn to stop at the usual rate.

To point up this result it is useful to compare the performance of such "screened off" cuttlefish with that of no-tentacle animals allowed access (Fig. 19). For example if we take the group in Fig. 18 that was prevented from gaining access for the longest time, it will be seen that at the commencement of trial 7 it had had 18 minutes visual experience of prawns in a tube. Yet in effect this group show no learning (because we know that these animals would have made about the same number of strikes on the first trial 1 if access had been allowed). The no-tentacle animals, however, appear to have learned very well by this time, judging by the considerable fall in their strike-level (Fig. 19). This brings out very clearly the fact that

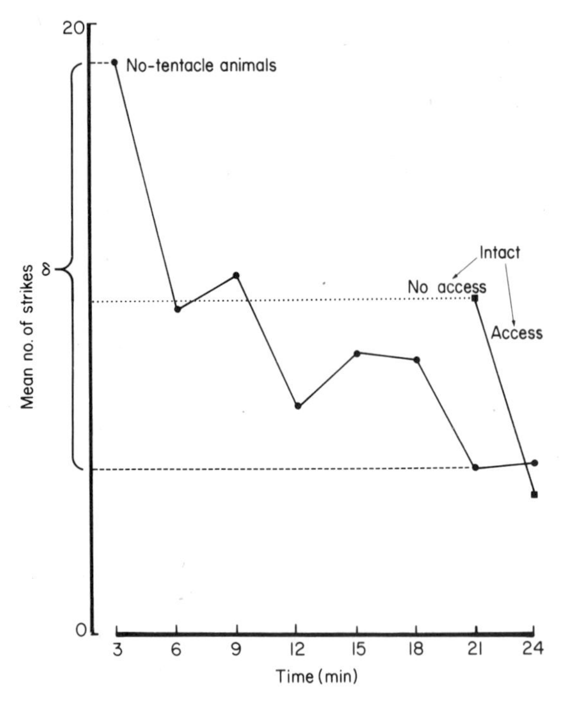

FIG. 19. Learning does not commence until the final motor act is initiated: after 21 min (i.e. 7 trials of 3 min each) the strike level of screened intact cuttlefish has not waned but the pseudo-strike-level of animals without tentacles has fallen considerably (δ) (data from Messenger, 1973a).

a cuttlefish can learn more by pseudo-ejecting phantom tentacles than it can by *not* ejecting real ones. When no-tentacle animals are tested with a screen they, too, never begin to pseudo-strike until they are given access, when they perform at their usual (very high) level (Messenger, 1973a), so that it seems as if the *initiation* of the final motor act rather than its completion is what is necessary for learning to commence. The motor command: "strike!" is made in the optic lobe and is effected via neurons in the interbasal and the anterior pedal lobes (Boycott, 1961). These experiments suggest that efferent copy of the motor command may be all that is necessary to start the learning process in this context. Unfortunately, however, we cannot say at which level in the motor hierarchy the copy is operating, although it seems most likely that it would be in the optic lobe itself.

Are there separate short- and long-term memory stores?

The question of memory span arises in cephalopods as in other animals and there is a variety of evidence (see Messenger, 1973a; Sanders, 1975) that cephalopods too have a labile short-term memory (STM), lasting minutes, and a relatively permanent long-term memory (LTM), lasting at least four months, that is insensitive to ECS or anaesthesia. The much more interesting general problem here is whether STM is "converted" or "consolidated" into LTM, or whether they are separate systems with parallel entry and their own time constants. Although present evidence inclines to the second view this has not been resolved so that it is interesting to note two recent experiments with cephalopods that have produced evidence compatible with this idea. We need only touch on the findings of Sanders & Barlow (1971) with *Octopus* because they are discussed elsewhere in this volume (see Sanders' chapter) but we must briefly examine the evidence in *Sepia* (Messenger, 1971) because it was gained using the experimental situation described here.

In our experiment, over 200 naïve cuttlefish were shown prawns in a tube continuously for 20 min, by which time they had learned to stop striking. The tube was then removed, to be replaced a second time after an interval of between 2 min and 2 days; each animal was naïve when tested and was used once only, at one of 14 different intervals, before being discarded. It was found that the mean number of strikes made by groups of cuttlefish during the *second* presentation was not the same at all the intervals tested, as can be seen in Fig. 20. Since the strike-level is a measure of learning (the *fewer* the strikes the better the memory) this complex curve represents some kind of "forgetting" curve. We should note that the curve is essentially biphasic in appearance with a peak at 22 min and a "dip" at 60 min, which in operational terms indicates a *rise* in retention. At intervals of 90 min and more the response is steady, and it is worth emphasizing that even after 2 days the response only recovers to about 25%. Another way of expressing this is to say that retention never drops much below 75% (see Fig. 22).

Reference to Fig. 20 reveals that cuttlefish that last saw prawns in a tube an hour ago attack less than cuttlefish that saw them only

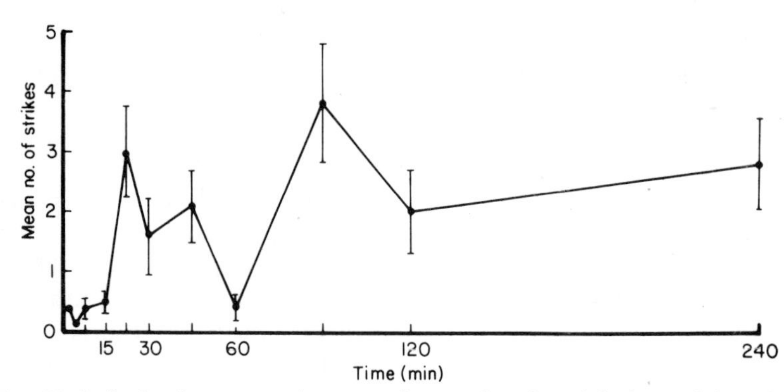

FIG. 20. Strike-level on a second presentation as a function of the interval since the first presentation. Notice the remarkably low strike-level shown by animals (N = 19) tested 1 hr after first presentation (from Messenger, 1973a).

20 minutes ago. This is the paradox of our results: *retention appears to be better at the longer than at the shorter interval.* Because of the large number of animals used in this experiment we are confident that this effect is a genuine result of our experimental manipulation; we can be less sure about how to interpret it, however. The explanation that we favour is that the recovery curve reflects the activity of separate STM and LTM systems. These could have the time courses shown in Fig. 21: STM builds up rapidly but decays fairly quickly (so that by 20 min or so it has gone) and LTM builds up more slowly *under these conditions*, becoming fully effective by 60 min, then diminishing slightly but persisting at about the same level for at least 2 days, probably much more.

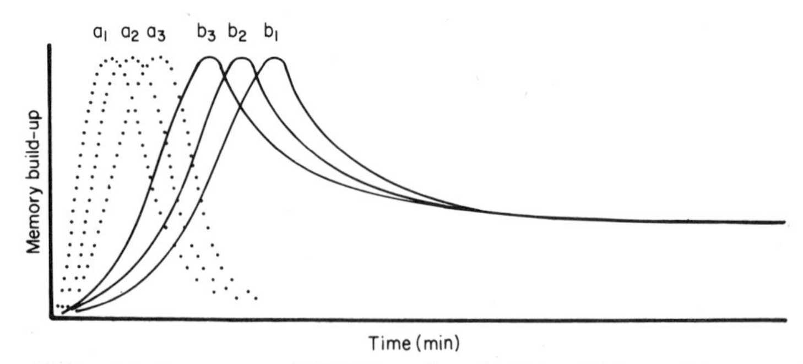

FIG. 21. Possible time course of STM (dotted) and LTM (solid line). If these systems are separate, with parallel entry, then we might expect that under different conditions each would be affected differently. Normally their time courses would overlap considerably (for example, curves a_3, b_3) to conceal their separateness but certain conditions could produce the time characteristics shown by curves a_1 and b_1, which could produce the effect shown in Fig. 20 (modified from Messenger, 1973a).

The most plausible alternative explanation for the appearance of this so-called Kamin-effect (Kamin, 1957) is that it results from the development of fear but the arguments for this, even in mammals (e.g. Steranka & Barrett, 1973) are not always convincing; in cephalopods we know nothing whatever about possible differences in emotional state so that it is fruitless to speculate (but see Sanders, this volume, p. 435).

What seems certain is that by offering just the right amount of negative reinforcement we have artificially separated the effects of STM and LTM: the rather non-adaptive increase in strikes at 20 minutes, when STM has faded before LTM has built up, may be seen as the fortuitous result of a particular combination of circumstances, circumstances that are unlikely to obtain in the wild. For we should note that if we repeat this experiment (in an abbreviated form) using shorter presentation times the inflexion in the recovery curve gradually disappears (Fig. 22). It remains to be tested whether a *longer* presentation would further separate the two components apparent in the 20-minute curve, as we would predict if these are indeed the separate read-outs of STM and LTM.

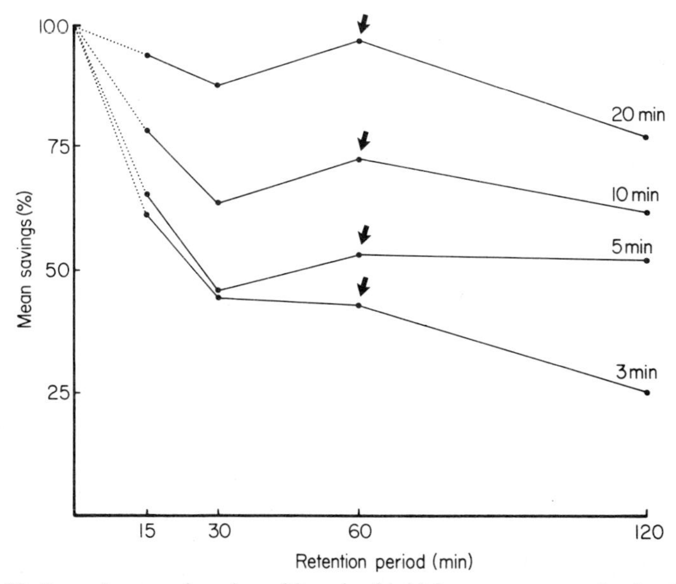

FIG. 22. Retention as a function of length of initial exposure to a stimulus (indicated on right-hand side). None of these curves, which all show retention fading with time, is smooth. The top curve re-plots data appearing in Fig. 20: the arrow marks the improvement in retention that is implicit in the *fall* of strike level in that figure. And in all curves there are signs of an anomaly about 1 hr after the end of the first trial.

Incidentally the data in Fig. 22 provides further evidence that the level of the strike response is sensitive to reinforcement: thus the retention score on a second presentation is always directly related to the amount of negative reinforcement previously received (see also Messenger, 1973a).

LEARNING AND THE VERTICAL LOBE SYSTEM

In their now classic paper of 1940 Sanders & Young established that a cuttlefish could discriminate between a prawn and a prawn next to a white circle, and, more important, they demonstrated that after a lesion that removed most of the closely associated vertical and superior frontal lobes* this ability was lost. This was the first experiment to establish that the cephalopod VSF system (Figs 24, 25 and 26) was associated in some way with the setting up or reading-out of a memory system, although as long ago as 1867 Bert had shown that this area of the brain was electrically "silent". Even more interestingly, Sanders & Young found that once cuttlefish were deprived of the VSF system they could no longer "hunt": that is, they did not pursue their prey once it had left the visual field.

Much later Wells (1958, 1962) carried out some very interesting experiments with newly-hatched cuttlefish, using the prawn-in-a-tube technique employed here. The juveniles hatch from large eggs and look like miniature replicas of the adult animals. Many features of their behaviour were adult-like, too, yet Wells found important differences in situations requiring new and adaptive responses. Thus young cuttlefish presented with *Mysis* in a tube continue to attack them for much longer than the adults would; indeed, until they seem virtually exhausted. And if given even a brief rest interval, the attacks would soon be resumed at full level (Wells, 1958). The lack of plastic, modifiable behaviour appears in other situations, too; for example young *Sepia* are extremely conservative in their choice of prey, and during the first week or so of life they will attack only *Mysis*. Later they begin to attack a variety of objects although by the time they are fully adult they seem to have learned to restrict their attacks to objects that in the past have proved edible (Wells, 1962).

If young cuttlefish do behave so differently from the adults are there any different features in the CNS that could account for this? Wells was quick to draw attention to the small size of the VSF

* Hereafter the VSF system.

system in young cuttlefish; this had already been shown by Mangold (née Wirz, 1954), whose results make it clear that it is only the VSF system that is relatively undeveloped (see Fig. 23 top right corner). The suboesophageal and basal lobes, which are the hierarchically "lower" parts of the CNS involved in motor control

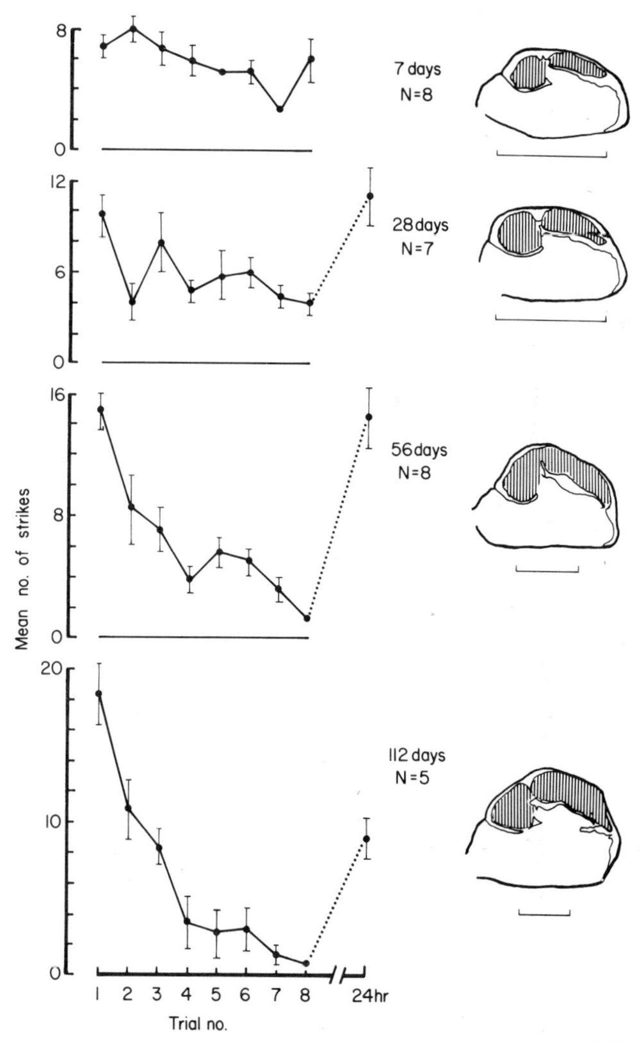

FIG. 23. Improvement in learning with age and the development of the VSF system. Bar indicates s.e. mean. On the right are drawn midline sagittal sections of the brain of a representative individual from each group; the scale bar represents 1 mm and the VSF system is hatched. The small number of animals in each group reflects the difficulties of rearing young *Sepia* (from Messenger, 1973b). Interval between trials was 30 min.

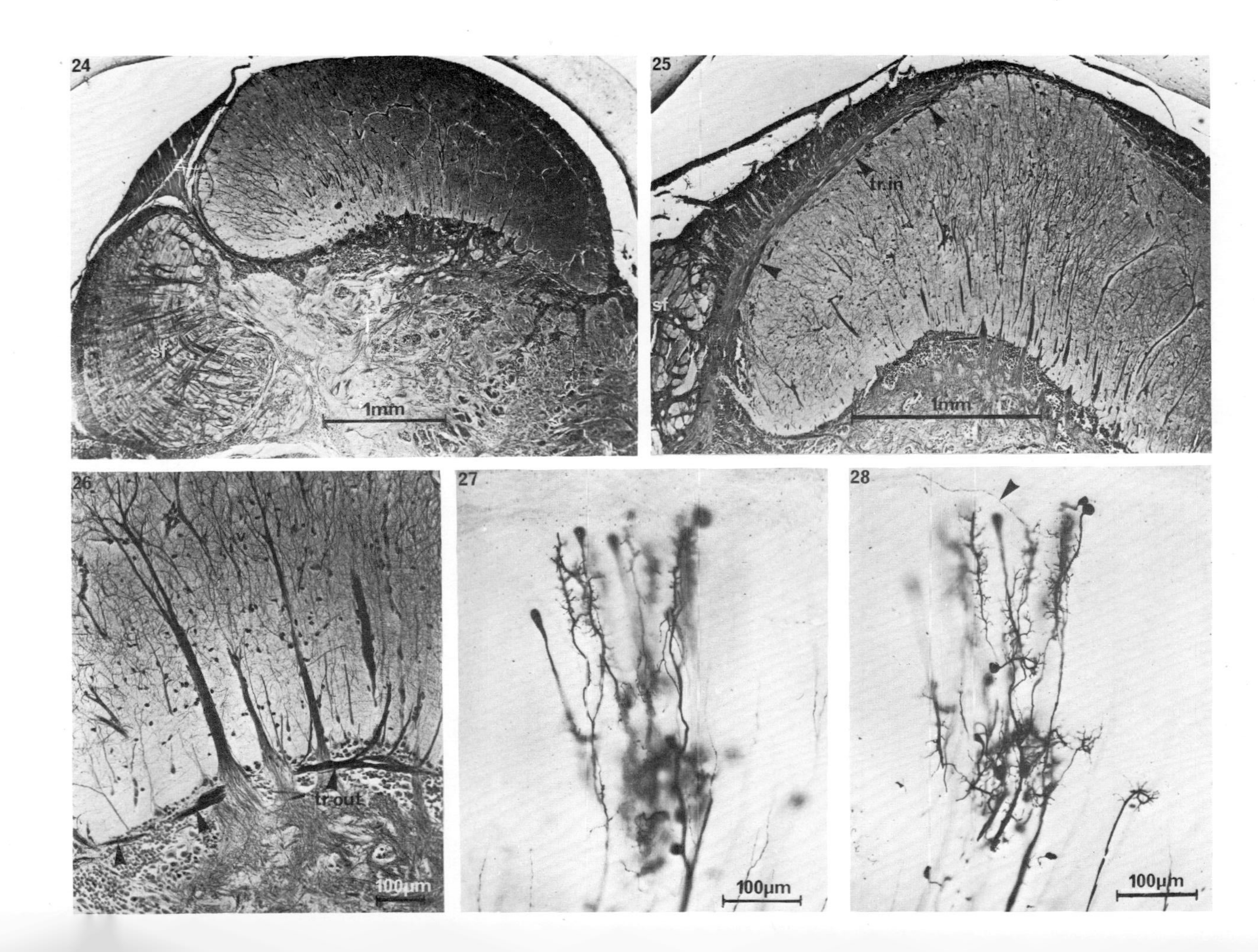
24
sf
1mm
25
tr.in
sf
1mm
26
tr.out
100µm
27
100µm
28
100µm

(Boycott, 1961), are of similar proportion to those of the adult and we can note that *Sepia* swims excellently and captures prey accurately from the moment of hatching (Wells, 1958).

Recently we have had an opportunity to re-examine the behaviour of young *Sepia* and measure learning at different stages of development using the same criteria as in the adult experiments. Cuttlefish hatched and reared in the laboratory (and therefore of known age) were shown prey animals in a tube (*Mysis* for the youngest group, prawns of varying sizes for the others), for a series of eight repeated trials of three minutes each. After an interval of 24 hours they were shown the tube of prawns once again. Four different groups were tested, once only, at ages from one week to four months.

The results of these experiments (Messenger, 1973b) are shown in Fig. 23, which needs little further amplification. Cuttlefish show virtually no evidence that they can learn at seven days. By a month there are signs of learning, but although the subjects made fewer attacks on the eighth than on the first trial their performance had deteriorated to the initial level by the next day. Two-month-old animals showed an unmistakeable, steep learning curve, although their retention performance, after 24 hours, was again very poor. Four-month-old animals not merely learn well (note how the slope of the four curves is progressively steeper) but they exhibit retention. When tested after an interval of 24 hours they made significantly fewer strikes than on the first trial ($t = 2{\cdot}79$; d.f. $= 8$; $P < 0{\cdot}05$). Taken by themselves these results are suggestive but we have also been able to show, in a very crude way, that this change in learning ability over the first four months of life is paralleled by a dramatic increase (Fig. 23) in the relative size of the VSF system, which by the time the animal is showing good learning and retention has attained proportions similar to those of the adult (cf. Fig. 24). These experiments, then, take advantage of an anomaly in development to support what Young found by surgical interference: if the VSF system is absent or small in size, learning performance is impaired.

Finally it is worth noting that we have examined the behaviour of immature cuttlefish to prey passing out of sight. Although there

FIGS 24–28. Fig. 24, the vertical (v) and superior frontal (sf) lobes in adult *Sepia*; midline sagittal section. Fig. 25, lateral section showing the superior frontal to vertical tract (tr. in). Fig. 26, fibres leaving the vertical lobe to form the vertical to superior frontal tract (tr. out). (Figs 24–26 Cajal silver.) Fig. 27, single cells in the vertical lobe, with an incoming superior frontal fibre visible in Fig. 28: Golgi–Kopsch.

is insufficient data to be certain, our findings appear to support Sanders & Young's original observations. If a prawn disappears from the visual field of a newly-hatched cuttlefish, that cuttlefish immediately abandons attention and the attack ceases. If it is already in the positioning phase it stops or backs off. This behaviour persists for the first month or two but gradually some degree of hunting begins to appear and there is no doubt that four-month-old animals follow further than younger ones: however, they too never embark on a chase proper.

It may seem curious to find a hunting animal that will not follow its prey once it is out of sight but presumably there is an adaptive advantage in staying put, rather than risking possible danger for a dubious food reward. That hunting behaviour is absent when the VSF system is poorly developed is most suggestive, however; indeed all these experiments point to the fact that fully functional vertical and superior frontal lobes are necessary for an effective STM. It is interesting that we have recently discovered that some fibres in the tentacular nerve reach as high up in the supra-oesophageal lobes as the superior frontal lobes. Whether these fibres bring "pain" or "taste" input to the system remains to be established, although we would suspect the former. Undoubtedly, however, receptors in the tentacles channel some of their information into a brain area that is *not* concerned with motor control.

We do not yet know what it is about the adult VSF system that facilitates learning in these various contexts. If the vertical lobe is the memory store it could be that a minimal number of cells is necessary for learning, as in the subfrontal lobe of *Octopus* in touch learning (Wells & Young, 1965), and that this number is not approached until about the fourth month of life; but we are only beginning to collect data about cell numbers in the growing or adult *Sepia* brain. In any case it seems more likely to be changes in connections that are crucial in the establishment of memories: for example, are serial synapses formed where the superior frontal fibres contact the cells of the vertical lobe (see Figs 25, 27 and 28; cf. *Octopus*, Gray, 1970)? Of course we are still not certain that the vertical lobe is acting as a memory store: it could be a read-in or read-out device associated with a store elsewhere, presumably in the optic lobes. It may be that changes in the optic lobe with age are even more important in the emergence of the plastic, adaptive behaviour that characterizes the adult. At the gross microscopical level it is clear that the nature of the cell islands in the optic lobe is very different in newly-hatched cuttlefish. These are some of the problems now being examined, using, as an experimental model, the simple visual-attack learning system described here.

Acknowledgements

The experiments described here were carried out over a number of years at the Stazione Zoologica, Napoli, Italy and also at the Laboratoire Arago, Banyuls-sur-Mer, France; it is a pleasure to thank the staff of those institutes for their kindness and help. For financial support thanks are due to the Science Research Council (B/SR/5287; B/RG/7281.1), the University of Sheffield Research Fund (888, 971, 82) and the Royal Society. Permission to reproduce Figs 12, 15, 17 and 20 has been given by *Animal Behaviour*, and Fig. 23 by *Brain Research*. I would also like to thank Dr S. v. Boletzky for rearing young cuttlefish for me and Dr J. J. Barlow for innumerable helpful discussions.

References

Barlow, H. B., Blakemore, C. & Pettigrew, J. D. (1967). The neural mechanism of binocular depth discrimination. *J. Physiol., Lond.* **193**: 327–342.

Bert, P. (1867). Mémoire sur le physiologie de la seiche (*Sepia officinalis* L.). *Mém. Soc. Sci. Phys. nat. Bordeaux* **5**: 114–138.

Boletzky, S. v. (1972). A note on aerial prey-capture by *Sepia officinalis* (Mollusca, Cephalopoda). *Vie Milieu* **23**: 133–140.

Boycott, B. B. (1958). The cuttlefish—*Sepia. New Biol.* **25**: 98–118.

Boycott, B. B. (1961). The functional organisation of the brain of the cuttlefish *Sepia officinalis. Proc. R. Soc.* (B.) **153**: 503–534.

Denton, E. J. (1974). On buoyancy and the lives of modern and fossil cephalopods. *Proc. R. Soc.* (B.) **185**: 273–299.

Gray, E. G. (1970). The fine structure of the vertical lobe of *Octopus* brain. *Phil. Trans. R. Soc.* (B) **258**: 379–395.

Graziadei, P. (1960). Prima dati sul corredo nervoso del labbro orale di *Sepia officinalis. Atti Accad. naz. Lincei rc.* **24**: 398–400.

Graziadei, P. (1964). Receptors in the sucker of the cuttlefish. *Nature, Lond.* **203**: 384–386.

Holmes, W. (1940). The colour changes and colour patterns of *Sepia officinalis. Proc. zool. Soc. Lond.* **110**: 17–36.

Kamin, L. J. (1957). The retention of an incompletely learned avoidance response. *J. comp. physiol. Psychol.* **50**: 457–460.

Mackintosh, N. J. (1974). *The psychology of animal learning.* London and New York: Academic Press.

Maldonado, H., Levin, L. & Barros Pita, J. C. (1967). Hit distance and the predatory strike of the praying mantis. *Z. vergl. Physiol.* **56**: 237–257.

Messenger, J. B. (1967). The peduncle lobe: a visuo-motor centre in *Octopus. Proc. R. Soc.* (B.) **167**: 225–251.

Messenger, J. B. (1968). The visual attack of the cuttlefish, *Sepia officinalis. Anim. Behav.* **16**: 342–357.

Messenger, J. B. (1970). Optomotor responses and nystagmus in intact, blinded and statocystless cuttlefish (*Sepia officinalis* L.). *J. exp. Biol.* **53**: 789–796.
Messenger, J. B. (1971). Two-stage recovery of a response in *Sepia. Nature, Lond.* **232**: 202–203.
Messenger, J. B. (1973a). Learning in the cuttlefish, *Sepia. Anim. Behav.* **21**: 801–826.
Messenger, J. B. (1973b). Learning performance and brain structure: a study in development. *Brain Res.* **58**: 519–523.
Messenger, J. B. (1973c). The strike of *Sepia.* 16 mm film. *Edinburgh University Audio-Visual Services.*
Messenger, J. B. (in prep.). *The visual attack of the cuttlefish: evidence for binocular distance estimation.*
Messenger, J. B. & Lucey, E. C. A. (in prep.). *The tentacle ejection of the cuttlefish.*
Mittelstaedt, H. (1957). Prey capture in mantids. In *Recent advances in invertebrate physiology. A symposium*: 51–71. Sheer, B. T. (ed.). Eugene: University of Oregon Publications.
Neill, S. R. St. J. & Cullen, J. M. (1974). Experiments on whether schooling by their prey affects the hunting behaviour of cephalopods and fish predators. *J. Zool., Lond.* **172**: 549–569.
Sanders, F. K. & Young, J. Z. (1940). Learning and the functions of the higher nervous centres of *Sepia. J. Neurophysiol.* **3**: 501–526.
Sanders, G. D. (1975). The cephalopods. In *Invertebrate learning* **3**: 1–101. Corning, W. C., Dyal, J. A. & Willows, A. O. D. (eds). New York: Plenum.
Sanders, G. D. & Barlow, J. J. (1971). Variations in retention performances during long-term memory formation. *Nature, Lond.* **232**: 203–204.
Steranka, L. R. & Barrett, R. J. (1973). Kamin effect in rats: differential retention or differential acquisition of an active-avoidance response? *J. comp. physiol. Psychol.* **85**: 324–330.
Thorpe, W. H. (1963). *Learning and instinct in animals.* London: Methuen.
Wells, M. J. (1958). Factors affecting reactions to *Mysis* by newly-hatched *Sepia. Behaviour* **13**: 96–111.
Wells, M. J. (1962). Early learning in *Sepia. Symp. zool. Soc. Lond.* No. 8: 149–169.
Wells, M. J. (1963). The orientation of *Octopus. Ergeb. Biol.* **26**: 40–54.
Wells, M. J. & Young, J. Z. (1965). Split-brain preparations and touch learning in the octopus. *J. exp. Biol.* **43**: 565–579.
Wirz, K. (1954). Études quantitatives sur le système nerveux des Céphalopodes. *C.r. hebd. Séanc. Acad. Sci., Paris* **238**: 1353–1355.
Young, J. Z. (1960). Observations on *Argonauta* and especially its method of feeding. *Proc. zool. Soc. Lond.* **133**: 471–479.
Young, J. Z. (1963). Light- and dark-adaptation in the eyes of some cephalopods. *Prod. zool. Soc. Lond.* **140**: 255–277.

Symp. zool. Soc. Lond. (1977) No. 38, 377–434.

BRAIN, BEHAVIOUR AND EVOLUTION OF CEPHALOPODS

J. Z. YOUNG

Department of Anatomy, University College, London
and
Wellcome Institute for the History of Medicine, London, England

SYNOPSIS

Features of the feeding and swimming mechanisms and the brains, eyes and statocysts of various cephalopods are compared. Several have special systems for feeding on small planktonic crustaceans. Phylogenetically primitive features are discussed especially in *Vampyroteuthis*, cirroteuthids and *Spirula*.

INTRODUCTION

Over many years now we have been making sections of the brains of various cephalopods. Through the kindness of our friends we have been provided with well preserved material of many species that are difficult to obtain. I should like to thank especially Drs Malcolm Clarke, Peter Herring, Katharina Mangold, Clyde Roper, Gilbert Voss, and Richard Young for their help in providing this material. The brains have mostly been prepared by Miss Stephens using our modification of Cajal's method (Stephens, 1971). The sections include the arms and buccal mass, and for many species the whole viscera as well. The collection is fully catalogued and is available for use by anyone. There must be many features of the internal anatomy that could profitably be described from this material. The method provides an excellent general view of most tissues, in addition to staining the nerve fibres. It has the advantage that the stain does not fade, and some of our best sections are now 45 years old.

With this material we propose to produce a book describing the brains and receptors of as many cephalopods as possible. *The anatomy of the nervous system of* Octopus vulgaris (Young, 1971) will serve as a basis, but much of this we now realize to be incomplete, especially, for example, our analysis of the basal lobes. Before publishing the book we shall give a full account of the nervous system of *Loligo*, of which the first three parts have been published (Young, 1974, 1976, 1977). By referring to these full descriptions of *Octopus* and *Loligo* we hope to be able to give a concise account of the variations found in other genera in relation to particular habits.

It is perhaps even yet not realized what an enormous variety the cephalopods exhibit, inhabiting every part of the ocean, its surface, midwaters and depths, its shores and sea bottoms. Each habitat requires different behaviour. We therefore have here a great range of natural experiments to study. By examining the details of the organization of the brain appropriate to each habitat we can draw conclusions about the significance of the patterns of connectivity. The method has its risks, but it is in some ways closer to nature than the use of experiments in the laboratory, say by excision and/or electrical stimulation. Clearly the two approaches are complementary and we have tried to use them both.

At the same time the brain provides much evidence about evolution. It is also not widely realized that of the living cephalopods several can be regarded as relics, retaining features of earlier levels of organization. Not only *Nautilus*, but also *Vampyroteuthis*, *Spirula* and the cirroteuthids all show such "primitive" features. In the present paper I shall deal especially with these forms, trying to show how the method can work to illuminate both behaviour and phylogeny. I shall also use these examples to contrast the features of the brain found in the deep sea, in midwaters, and at the surface, both in octopods and decapods.

CHARACTERS OF OCTOPODS AND DECAPODS

It may be convenient here to summarize the special features of the brain in the two major groups of coleoids. In decapods the lobes of the brain are widely separated, there is no elaborate inferior frontal system, the ventral magnocellular lobe is large and there is often a system of giant fibres. Conversely in typical octopods the brachial and superior buccal lobes are joined to the brain and there is a large inferior frontal system concerned with touch discrimination and memory (Wells & Young, 1972). There is no ventral magnocellular lobe or commissure and no giant nerve fibres. The brachial lobes are joined by a suprabrachial commissure, which is absent in decapods. Many of these characters are related to the nektonic life of decapods compared with the benthic one of octopods. We shall find that they are variously modified in the members of each group that depart from the typical mode of life. *Vampyroteuthis* carries characters of *both* groups, which is its special interest.

VAMPYROTEUTHIDAE

Vampyroteuthis infernalis

Introduction

Since *Nautilus* has been dealt with in an earlier paper (Young, 1965) we can begin here with the most primitive of the coleoids. *Vampyroteuthis infernalis* provides the best starting point, since it shows in the nervous system, and in the rest of the body, a strange mixture of characters of decapods and octopods. Surveys of the species and discussion of its relationships are given by Robson (1932) and Pickford (1940). The nervous system has never been fully described but a good figure is given by R. E. Young (1967).

The animal is undoubtedly specialized for bathypelagic life, seldom or never rising above 900 m, and usually taken far from the bottom (Clarke & Lu, 1974, 1975). The gelatinous tissues strongly suggest neutral buoyancy, and a large posterior coelom may indicate a head-down floating position. The mantle muscles and retractors are weak, but there are large fins (Fig. 1). The fin nerve arises from a lobe in the posterior suboesophageal mass very similar to that of decapods. It runs with the pallial nerve and passes the stellate ganglion. The web is extensive and the arms have long filamentous ends. Each arm has a single row of suckers which, with the cirri, must have great tactile powers, aided presumably by the enigmatic retractile filaments discussed below. But the eyes are also large. There is a clear space all round the lens, which presumably gives it an "aphakic" character (see Munk & Frederiksen, 1974).

There is a beak and radula, but their muscles contain many spaces and cannot be strong. There is a salivary papilla but only small posterior salivary glands and duct. On the other hand there are large and active subradular glands, opening by several pores on the surface of the salivary papilla (Fig. 2a,b). The subradular ganglia lie far apart and appear to be related to these subradular glands rather than to muscles of the papilla as in *Octopus*. Also there are very large anterior salivary glands, opening at the front of large lateral buccal palps (Fig. 3). All this suggests the production of a mass of secretion, perhaps concerned with entangling small prey rather than chewing off pieces. In fact in one of our specimens the large crop is filled with mixed plankton. As this includes diatoms as well as copepods it can hardly have been taken in the deep sea. However, it may be that *Vampyroteuthis* feeds by collecting small organisms.

A curious feature is that the chromatophores are not provided with muscles and there are no chromatophore lobes in the brain.

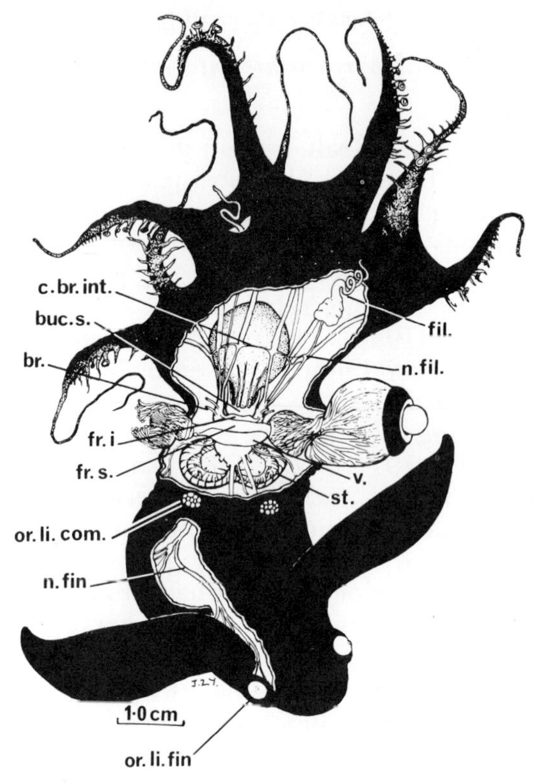

FIG. 1. Drawing of dissection of *Vampyroteuthis*. The light organs have been added from other illustrations. (Key to abbreviations at end of chapter.)

Presumably there is little or no colour change and this may possibly be a primitive feature, preserved in the deep, as may perhaps also be the absence of an ink sac.

The filaments are arms

One of the most difficult of all the questions is whether *Vampyroteuthis* has eight arms or ten. This might seem a simple question of counting, but the nature of the pair of filaments between the first and second arms has produced much speculation. R. E. Young (1967) concluded that they were neither tentacles nor arms and he mentions my suggestion that they might be homologous with the pre-optic tentacles of *Nautilus*. I now think that this is wrong. We have strong evidence that they are indeed the modified second pair of arms of a decapod. Before joining the brain the arm nerves of all

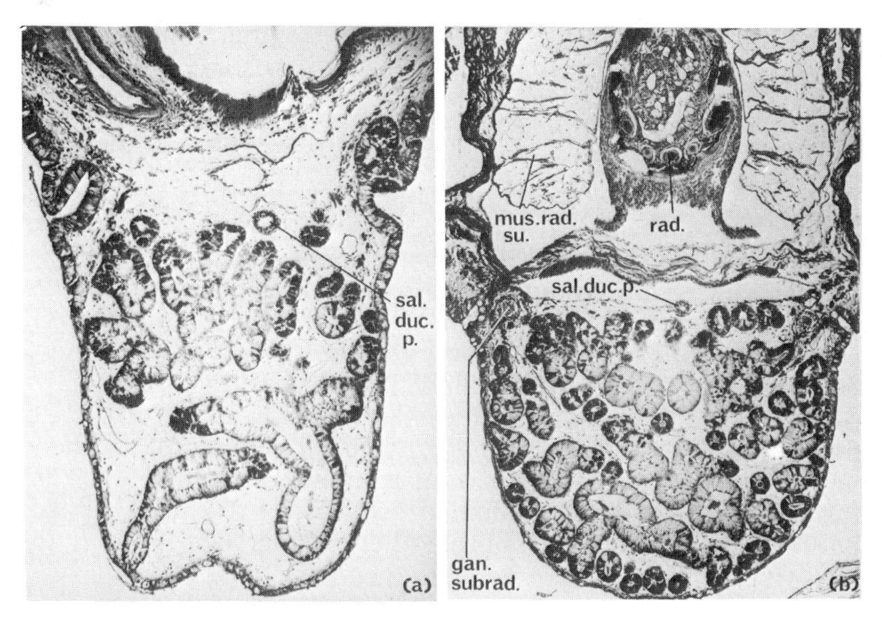

FIG. 2. (a) *Vampyroteuthis*. Transverse section of salivary papilla, showing massive glands. Field width 1·7 mm. (b) Transverse section further back than (a), showing radula and subradular ganglia, widely separated. Note the vacuolated radular support muscles. Field width 2·1 mm. (Key to abbreviations at end of chapter.)

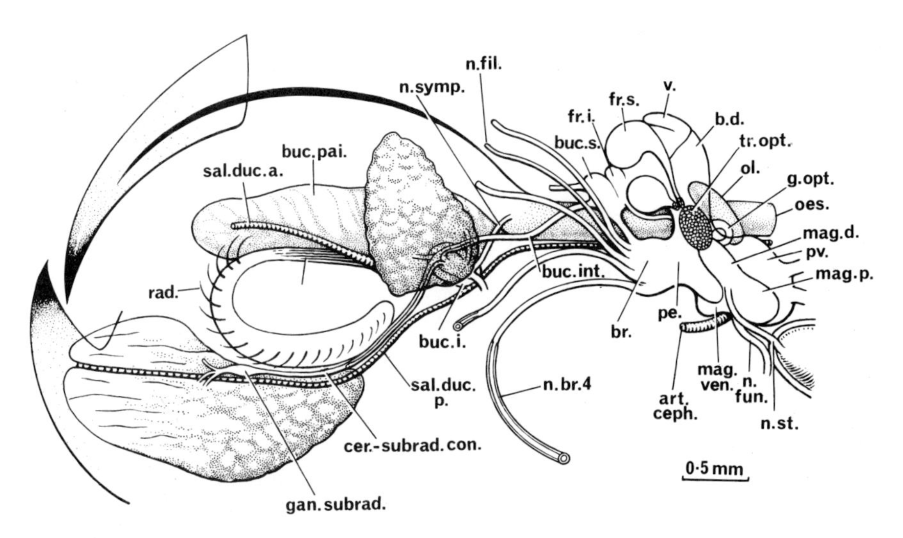

FIG. 3. Reconstruction from serial sections of the buccal mass and brain of *Vampyroteuthis*. (Key to abbreviations at end of chapter.)

cephalopods exchange fibres through a circumoral commissure. In *Vampyroteuthis* the nerves of the filaments contribute to this commissure (Fig. 1). This really establishes that these structures are arms. The further course of the nerves centrally shows other similarities to arm nerves, as well as some differences from them. The other arm nerves each become surrounded by nerve cells before they join each other (Fig. 4). This incidentally is a primitive feature and tells us something about the prebrachial lobes. The filament nerves have no such swelling and remain independent as they pass the front of the brachial lobe. As R. E. Young showed, each nerve contains two bundles, and the more medial of these enters the back of the brachial lobe. He believed that these fibres passed to the middle suboesophageal mass, though he provided no figure to show them and I have not been able to follow them beyond the brachial lobe. But a more important point is that this medial bundle of lightly-staining fibres closely resembles similar bundles coming from each of the other arm nerves.

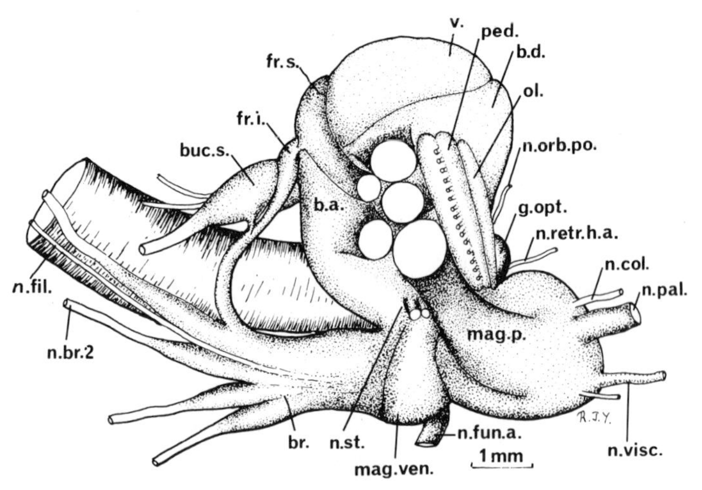

FIG. 4. Drawing of lateral view of dissected brain of *Vampyroteuthis*.

The larger, lateral bundle of the filament nerve, containing rather large, darker-staining fibres, forms the main part of the brachio-magnocellular connective, as R. E. Young describes (though he repudiates the name). The filament nerves do not provide the *whole* of this connective however. Small bundles of fibres from the other arm nerves leave the brachio-palliovisceral connective to join it. The joint bundle then runs on backwards to enter the magnocellular lobe.

The magnocellular lobe

There is therefore little doubt that *Vampyroteuthis* is in the literal sense a decapod, and this agrees dramatically with the presence of a ventral magnocellular lobe (Figs 4 and 5). This is a band of fibres and cells, visible externally and having precisely the relations and connections of the similar lobe and commissure in decapods (Young, 1939). Such an arrangement cannot surely have been separately evolved twice, so the octopods represent a stage either before it appeared or after it had been lost. We shall incline to the second view, and thus *Vampyroteuthis* is an intermediate stage.

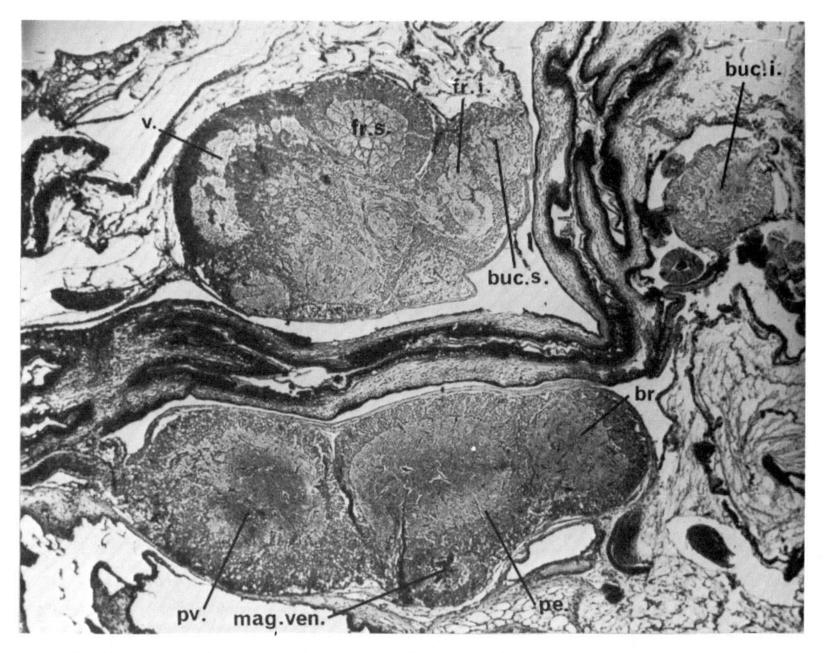

FIG. 5. *Vampyroteuthis*. Sagittal section of the central nervous system to show the ventral magnocellular lobe. Field width 3·6 mm. (Key to abbreviations at end of chapter.)

As R. E. Young says, the filaments are extruded to great lengths and perhaps serve especially to detect predators. The magnocellular lobe is probably concerned mainly with escape reactions. This function of the arms has been developed in these deep-sea forms and we shall find a somewhat similar adaptation in the Chiroteuthidae (p. 388).

Paradoxically, although the magnocellular lobe is one of the most interesting features of the brain, *Vampyroteuthis* has no giant fibre system. There are no especially large cells in the region of entry of the static nerves, where they occur in most decapods. Nor has any sign of giant fibres been seen in the stellate ganglion or stellar nerves. The animal is said to show an escape reaction (R. E. Young, 1967) but the mantle muscles are weak. It remains an interesting question whether *Vampyroteuthis* has lost the giant fibres or never had them.

Compact form of the brain

In spite of the fact that it has essentially ten arms and a magnocellular lobe, the arrangement of the brain is that of an octopod, with the ganglia strongly concentrated (Figs 4 and 5). The superior buccal lobe is attached to the rest of the brain and separated by long connectives from the inferior buccal lobes below the buccal mass (Fig. 3). This is a condition found in all octopods, but neither in *Nautilus* nor any decapod. We must conclude from this and what has been said of the arms and magnocellular lobes, that *Vampyroteuthis* represents a stage in the evolution of octopods from a decapodan condition.

Absence of suprabrachial commissure

This commissure in front of the brain is present only in octopods and is absent from *Vampyroteuthis*. There are at most a very few transverse fibres running through the base of the superior buccal lobe. The presence of this commissure in octopods suggests that the brachial lobes and hence the arms are cephalic in nature, rather than pedal, as is usually supposed (see Young, 1971). If these fibres are an important functional feature, for the much developed arms of bottom-living octopods, why do they cross *above* the oesophagus? If the answer is to be found in phylogeny why are they not present in *Vampyroteuthis*? The most obvious solution is that they are indeed an ancient feature (in octopods) but have been lost as a result of abandoning the benthic habit. This would agree with their absence in the bathypelagic *Japetella* (p. 397).

There is probably a suprapedal commissure in *Vampyroteuthis* as in other coleoids. This connection above the oesophagus suggests that much of the middle suboesophageal mass is also of cerebral nature, for example the lateral pedal lobe controlling the eye muscles which surely cannot be part of the foot! Possibly only the posterior pedal lobe, innervating the funnel, is truly "pedal" (Young, 1976).

The inferior frontal system

The supraoesophageal lobes show a clear octopod character (Fig. 5). The inferior frontal system is better developed than in any decapod, but not as highly differentiated as in *Octopus*. The superior buccal lobes are expanded behind to make posterior buccal lobes, with a dorsal part representing the median inferior frontal lobes. This contains transverse bundles but they are much less differentiated than in typical octopods. Posteriorly there is tissue that may represent a poorly differentiated subfrontal lobe (Fig. 6). This all shows a stage of incipient development of an apparatus for more elaborate processing of tactile information than is present in decapods. Considering the value of such information in the darkness of the deep sea, it is hardly likely that this is a retrogression from an earlier condition. Indeed it suggests the possibility that the octopod characteristics may have first developed in deep water.

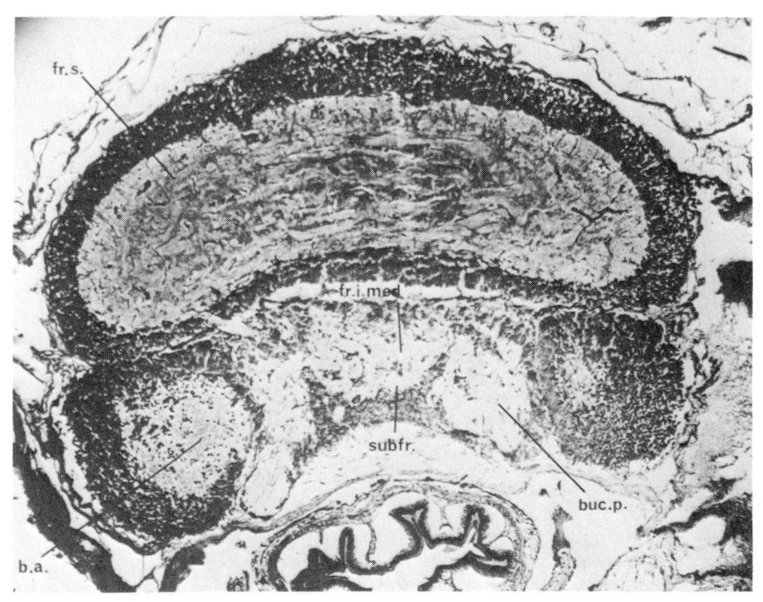

FIG. 6. *Vampyroteuthis*. Transverse section of the front of the brain to show the superior frontal and very small inferior frontal, subfrontal and posterior buccal lobes. Field width 2·2 mm. (Key to abbreviations at end of chapter.)

The vertical lobe system

The remaining higher supraoesophageal centres are well developed in *Vampyroteuthis* and again show features of octopods but with some hints of decapod ones. The median and lateral superior frontal lobes are not sharply differentiated, but there are abundant transverse bundles (Fig. 6). The vertical lobe is represented by islands of neuropil surrounded by masses of small cells (Fig. 7). This is unlike the decapodan plan with a wide neuropil and narrow cell layer (Fig. 35a,b). Yet in some views the neuropil is wider and less interrupted by cells than one expects in an octopod. There is no clear division into five lobules, but it could well be imagined how this might develop. It would be most interesting to discover the significance of the difference in plan of the vertical lobe in the two main groups. It seems probable that each has developed from some simpler ancestral condition. A clue may be that the decapods hunt mainly by sight, the octopods by touch as well as sight.

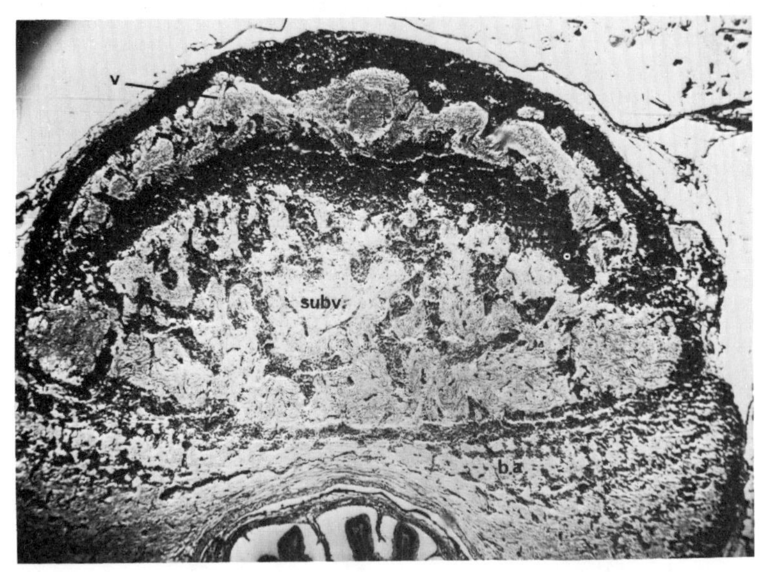

FIG. 7. *Vampyroteuthis.* Transverse section of the vertical and subvertical lobes. Field width 2·2 mm. (Key to abbreviations at end of chapter.)

The basal lobes

These cerebellum-like centres are well-developed in *Vampyroteuthis.* The peduncle lobes are orientated vertically (as in most decapods, except *Sepia*). The anterior basal lobe is large and is divided into two parts (Young, 1977). The median basal lobe is of moderate size but there is no lateral basal lobe (the chromatophore centre).

The statocyst

If one was designing a structure intermediate between octopods and decapods one could hardly invent anything better than the vampyromorph statocyst (Fig. 8). It has inner and outer sacs with endolymph and perilymph, as in octopods, but several anticristae as in decapods (Barber, 1968; Stephens & Young, 1975, 1976). The crista is divided into four parts and each part is not further subdivided, both of these are decapod characters. There are three maculae, as in decapods but with some signs of difference from the latter. These features have of course functional as well as phylogenetic significance. The separation of the sacs and especially the large posterior sac may be related to the measurement of changes of depth, so important for a wholly deep-sea life. This might again indicate the origin of octopods in deep waters. Similar large sacs are found in other deep-water cephalopods (p. 400).

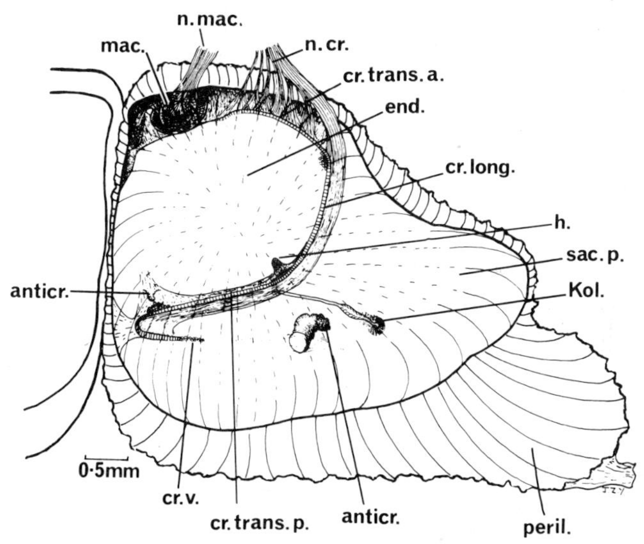

FIG. 8. *Vampyroteuthis*. Drawing of statocyst, dissected from body and seen from above. (Key to abbreviations at end of chapter.)

The photosensitive vesicles

As discussed by R. E. Young (see his chapter, this volume), the photosensitive vesicles are providing increasingly interesting suggestions of correlation with habitat, although we do not yet understand their function. They were first reported as the epistellar bodies of octopods, erroneously thought to be glands (Young, 1929). Bern and his colleagues have shown that they in fact contain photoreceptors (see Mauro, his chapter, this volume). In decapods, where there are no epistellar bodies, there are vesicles of similar structure and physiology around the optic stalk, called, from their position, parolfactory vesicles. The nerve from these latter passes to the peduncle lobe and perhaps connects with the nearby centre that controls the optic gland and hence reproduction (see Wells & Wells, their chapter, this volume). These photoreceptors have of course no lens and must serve only to measure light intensity, perhaps usually in relation to diurnal or seasonal changes of depth.

One of the oddest of all the characters of *Vampyroteuthis* is that its photosensitive vesicles are placed literally half-way between the epistellar bodies of octopods and the parolfactory vesicles of decapods. Richard Young describes them as lying "at the anterior end of the mantle cavity" (1972). They are small but deeply pigmented (orange). Yet because of the dense black pigmentation of the animal it is difficult to know what light they may detect. Young suggests it is light emitted by bioluminescence of prey *within* the mantle cavity, which might be a hazard since the mantle is not quite fully pigmented. The nerve is connected with the posterior funnel nerve and might activate movements of expulsion.

This interesting suggestion prompts a further speculation that the epistellar body of octopods may have a similar function. In some preliminary experiments with octopuses, with their optic nerves cut and kept in the dark, we regularly found that sudden illumination produced an active expiratory movement. Further analyses of this kind would be interesting.

Summary

In nearly all the characters of its brain, *Vampyroteuthis* is a typical coleoid, showing few signs of affinity to *Nautilus*. It represents a stage long after the separation of the coleoid and nautiloid stocks but only shortly after octopods began to diverge from decapods. It tells us something about the ancestors of both groups, especially about those of the octopods.

OCTOPODS

Benthopelagic octopods

Cirroteuthidae

The Cirrata include deep-sea octopods that retain some primitive features. We have sections of only one unidentified cirroteuthid, but it is well preserved and shows a most remarkable combination of features. The brain is greatly shortened, the centres being closer together than in perhaps any other cephalopod (Fig. 9). The superior buccal lobe is closely joined to the brain and in this the

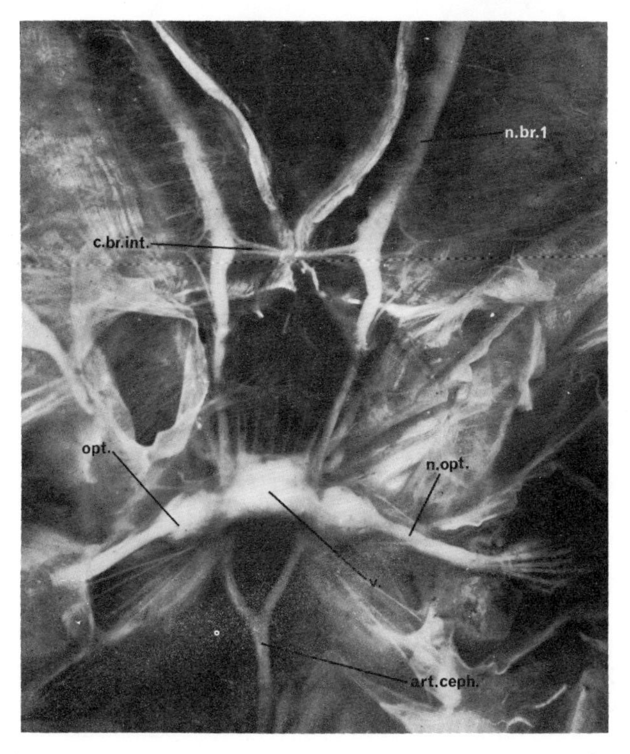

FIG. 9. Cirroteuthid. Photograph of central nervous system from above. (Key to abbreviations at end of chapter.)

cirroteuthid is clearly octopodan. The one remaining eye of our specimen is much distorted but shows quite clearly that the retina is in two parts—a peripherel part of shorter rhabdomes and a central part, where they reach a length of 300 μm (Fig. 10). The optic nerves are very long and most curiously do not show the chiasma inverting the retinal information, which is found in every other cephalopod investigated so far— except *Nautilus* (Young, 1965). It hardly seems possible that this complex feature could have been developed and then lost again and I tentatively conclude that this cirroteuthid represents a stage before it was acquired.

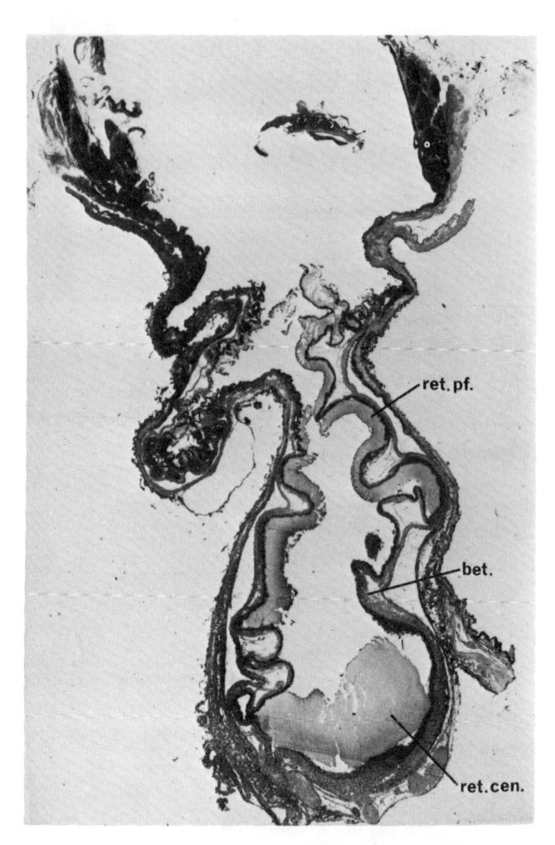

FIG. 10. Cirroteuthid. Transverse section of eye. The lens has been removed and the bulb has collapsed. The retina has a peripheral part (ret.pf.) with short rhabdomes and a central part (ret.cen.) with long ones. In the region between (bet.) there are no rhabdomes at all. Field width 0·5 mm.

The optic lobes are relatively small and the structure of their outer layers is simple and octopodan in that it lacks the characteristic inner plexiform layer seen in decapods.

There is a moderately large suprabrachial commissure (Fig. 11). This is of course a characteristic of octopods (p. 384), but here it lies not in front of the cerebro-brachial connective as is usual, but behind it.

Each arm of a cirroteuthid has a vacuolated outer part, presumably containing a liquid of lower density than sea-water to give buoyancy (Fig. 12). The suckers are powerfully muscular and there are rings of muscles for moving the arm. The cirri have large nerves and are presumably sensory. The intrabrachial nerve cords

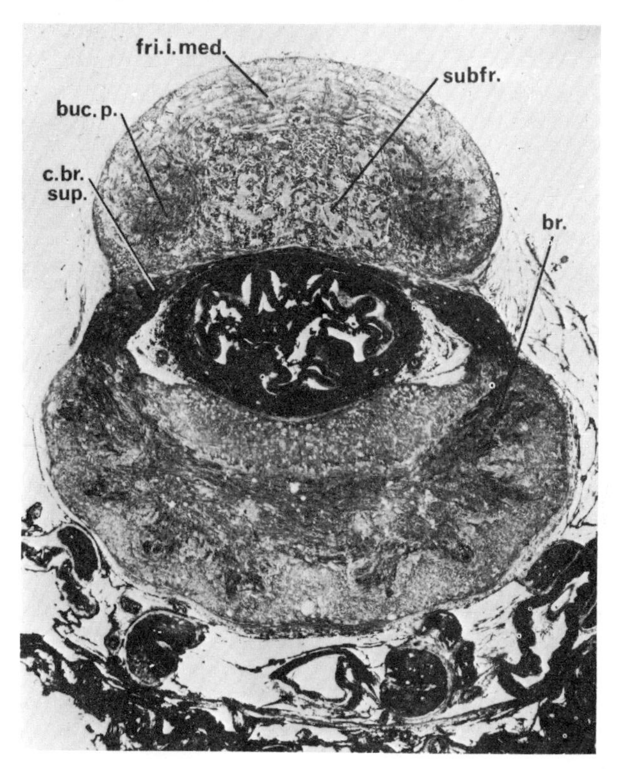

FIG. 11. Cirroteuthid. Transverse section at front of brain, showing suprabrachial commissure and inferior frontal system. Field width 4 mm. (Key to abbreviations at end of chapter.)

are large and ganglionated until just before they enter the brain. At this point they are joined by a circumoral ring (Fig. 9).

The tactile system dominates the supraoesophageal lobes. Fibres stream in at the sides through what could be compared with the lateral inferior frontal lobe of *Octopus*, and onwards to a tissue that is clearly comparable to the median inferior frontal lobe, with its bundles of fibres (Fig. 11). From this fibres proceed down through a maze of lobules with many small cells, which is clearly subfrontal tissue, more abundant even than in *Octopus*, but less compact. This tissue is in turn in connection with a backward extension from the superior buccal lobe, which corresponds to the octopod posterior buccal lobe and sends bundles of fibres, probably efferent, to the brachial lobes.

We have here therefore a very specialized tissue with characteristics that in *Octopus* have been shown to be concerned with

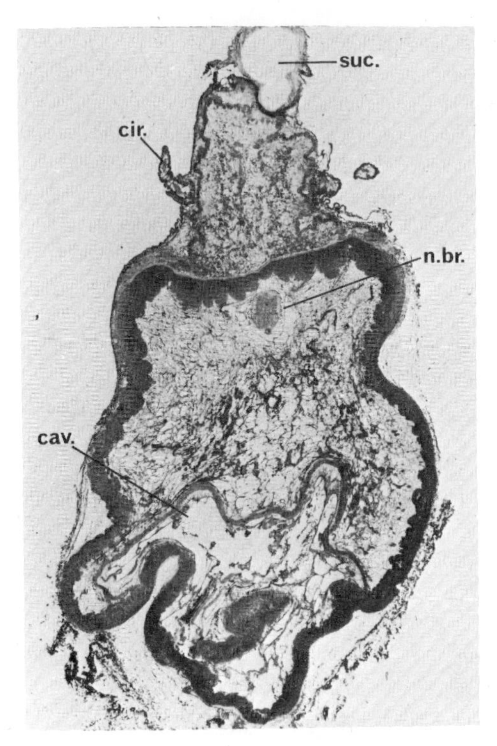

FIG. 12. Cirroteuthid. Transverse section of arm showing cirri, and spaces in the inner part (cav.). Field width 6 mm. (Key to abbreviations at end of chapter.)

tactile learning (Wells & Young, 1972). From the excellent photographs of Roper & Brundage (1972) we know that these animals are what has been called benthopelagic (Marshall & Bourne, 1964), floating just above the bottom. It seems very probable that with the information from their cirri they are able to learn which pieces of material provide food. They have no radula and probably swallow the food whole. From the few reports they eat various deep-living crustaceans and perhaps polychaetes (Scott, 1910; Robson, 1930). The bottom fauna is now known to be very varied and abundant and a tactile memory system is likely to be most useful.

In contrast, the visual learning system of cirroteuthids is reduced. The median superior frontal is a very small lobule, much smaller than the median inferior frontal lobe, from which it is separated only by a thin wall of cells. This median superior frontal lobule appears to receive no input from the optic lobes. It receives bundles from the posterior buccal region and probably serves to carry some information from the arms to the vertical lobes.

The latter are quite different from any cephalopod yet investigated (Figs 13 & 14). They consist of a number of islands of neuropil surrounded by small cells. These are broadly continuous below with further islands and rather larger cells, constituting a lobe corresponding to the subvertical lobe of octopods. This is of course continuous in front with the posterior buccal lobe. It seems likely that the whole of this vertical lobe system functions as an adjunct of the tactile learning system. Experimental work on *Octopus* has shown a definite influence of the vertical lobe in the storing of tactile memories. The small-celled islands of the cirroteuthid are not grouped into definite lobes, but it could well be imagined how they might develop into the condition in *Octopus*.

Further back the subvertical region becomes connected with the optic lobes by fibre bundles, probably passing in both directions (Fig. 14). This tissue merges above the small optic commissure with a region that corresponds to the median basal lobe but is structurally similar to the subvertical lobe. From this region large bundles pass to the pedal, palliovisceral and magnocellular lobes, no doubt controlling locomotion. It might perhaps be that this back part of the vertical lobe system is concerned with learned avoidance reactions, but this is speculative.

The whole of the upper part of the vertical lobe system is, throughout, similar in appearance. Although parts corresponding to those named can be quite clearly recognized they are, as it were, little differentiated from each other structurally. This may be a result of a common concern with tactile functions, but it might also be a primitive feature. There is a certain similarity in the appearance of these islands to those of the cerebral commissure of *Nautilus*.

The magnocellular lobe is not very clearly differentiated from the posterior pedal and palliovisceral lobes. There is no ventral magnocellular commissure beneath the pedal lobe. It was therefore most unexpected to find that there are some elements of a definite giant fibre system. There are large fibres in the anterior part of the magnocellular lobe. They could not be traced to particular cells, but they lie close to the entry of the static nerve, which is the situation of the giant cells of decapods. Only occasional traces of them were seen in the palliovisceral lobe and pallial nerves. However in both stellate ganglia undoubted giant fibres

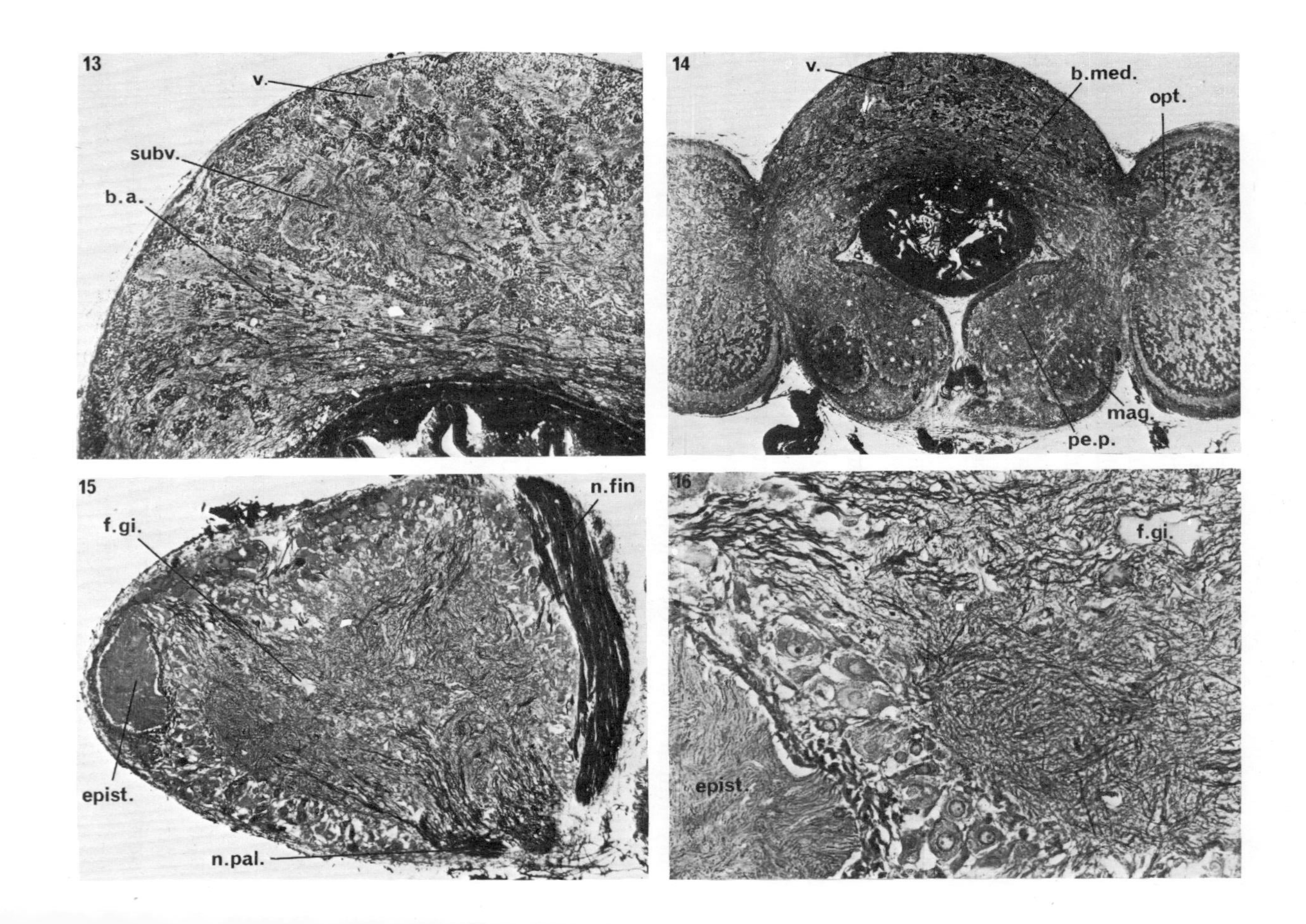
13
v.
subv.
b.a.
14
v.
b.med.
opt.
mag.
pe.p.
15
n.fin
f.gi.
epist.
n.pal.
16
f.gi.
epist.

were seen, one or two running to each stellar nerve (Figs 15 and 16). Their presence alongside the epistellar body finally disproves one of my earliest hypotheses (Young, 1936). More important, it establishes that cirroteuthids are relics of a stage when octopods and decapods had not yet diverged.

The fin nerve passes by the stellate ganglion exactly as in decapods. A distinct lobule on the dorsal hind end of the palliovisceral lobe sends large fibres to the pallial nerve and is no doubt the fin lobe. It is in exactly the same position as the fin lobe of *Vampyroteuthis* and of decapods. There is no reason to doubt that the fin is a genetically homologous system in all coleoids.

The basal lobe system is quite well developed. There is a large peduncle lobe, placed vertically, which is perhaps the primitive orientation as in most decapods, not horizontally as in most octopods and *Sepia*. The anterior basal lobe is quite large but not very clearly differentiated into the typical anterior and posterior parts (Fig. 13) (Young, 1977). These large basal lobes evidently have important actions other than control of the eye movements. The oculomotor centre of the lateral pedal lobe is small. There is no special posterior lateral pedal lobe, a region probably concerned with attack by the tentacles (Boycott, 1961, and see *Pterygioteuthis* p. 415).

There is an optic gland at the back of the optic tract having large cells and a rich blood supply. It receives a nerve from a lobule next to the peduncle lobe, which is probably "olfactory". A very interesting point is that there are no photosensitive vesicles in this region. One suggestion for their function in decapods, where they lie here, is in connection with reproduction. This cirroteuthid has no parolfactory vesicles but has a typical epistellar body, at the back of the stellate ganglion (Fig. 15), which is of course the classic position in which it was first discovered by its yellow colour in *Eledone* (Young, 1929). In this deep-sea species it is larger than has been seen in any other animal and is provided with a central mass of greatly elongated, irregular rhabdomes (Fig. 16). By this feature alone the

FIG. 13. Cirroteuthid. Transverse section of vertical, subvertical and anterior basal lobes. Field width 5·5 mm.

FIG. 14. Cirroteuthid. Transverse section of central nervous system and optic lobes. Field width 5·3 mm.

FIG. 15. Cirroteuthid. Sagittal section of stellate ganglion, showing epistellar body, giant fibres and fin nerve. Field width 1·8 mm.

FIG. 16. As Fig. 15, showing the giant fibre and irregular rhabdomes of the epistellar body. Field width 0·57 mm. (Key to abbreviations at end of chapter.)

cirroteuthids must be decisively associated with the Octopoda, in spite of their giant fibres.

By contrast, the statocyst bears a mixture of characters. It shows the octopodan condition of inner and outer sacs, separated by perilymph (Fig. 17). But the cavity is interrupted by several anticristae, as in decapods. Unfortunately in our specimen the statocyst is so distorted that no detailed functional interpretation can be made. But the anticristae suggest that cirroteuthids need to measure angular acceleration through quite rapid turns (see Budelmann, his chapter, this volume), as indeed the photographs by Roper & Brundage (1972) suggest.

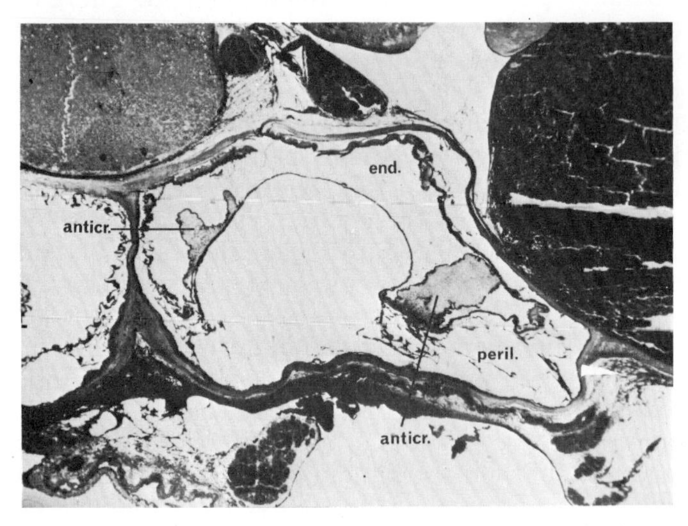

FIG. 17. Cirroteuthid. Transverse section of statocyst showing inner and outer sacs, and several anticristae. Field width 5 mm. (Key to abbreviations at end of chapter.)

In summary we have in cirroteuthids a remarkable mosaic of characters, some primitive, some specialized for their way of life, some octopodan, others decapodan. Can they be a specialized offshoot from a primitive benthic octopod that still retains some characteristics of the common coleoid stock? The general simplicity of the supraoesophageal lobes suggests this, as does the absence of optic chiasmas. But the reduction of the whole optic system is surely secondary. The condition of the statocyst, similar in some ways to *Vampyroteuthis* and decapods, suggests that cirroteuthids are a relic of this early stage of octopod divergence, as indeed do the cirri themselves. But the most striking features of all are the presence side-by-side of giant fibres and an epistellar body. This indeed raises more questions than it solves, but emphasizes the wide separation between the cirroteuthids and the other coleoids.

Bathypelagic octopods

Bolitaenidae, ***Japetella***

We can now examine animals that, although pelagic, are closer to the typical epibenthic octopods and are included with them in the same suborder (Incirrata) but in a distinct tribe (Heteroglossa) and family Bolitaenidae. They depart considerably from the Octopodidae and although the differences are obviously due largely to divergent habits they may include retention of primitive features. The bolitaenids have abandoned the bottom for a bathypelagic life. They have been caught at a wide range of depths from 100 to 1500 m. There is some evidence that they descend as they grow older but it is uncertain if they have a diel migration (Clarke, 1969; Clarke & Lu, 1975).

Japetella has a gelatinous, transparent, rounded body, with short arms (Fig. 18). Analysis of the body fluids has shown that it is deficient in sulphates, sufficient to bring this watery animal about halfway to neutral buoyancy (Denton & Gilpin-Brown, 1973). The tissues of the arms are occupied by a series of chambers presumably containing this lighter fluid (Fig. 19). The animal perhaps lives in a

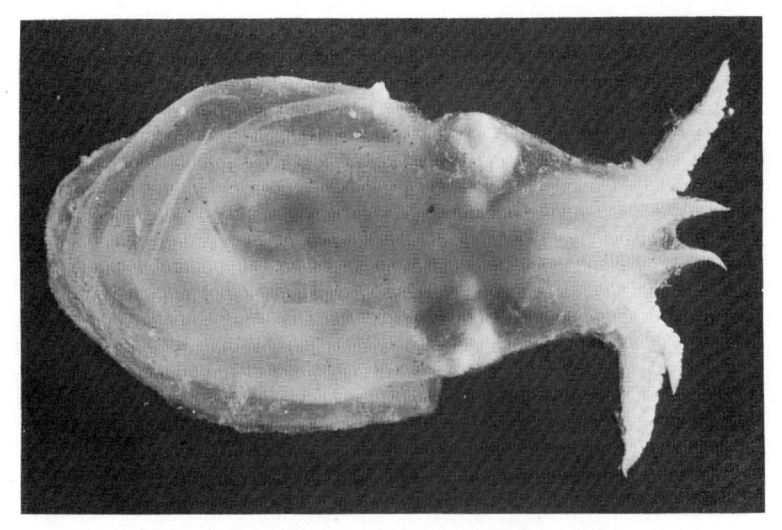

FIG. 18. *Japetella diaphana*. Dorsal mantle length 26 mm; total length 52 mm.

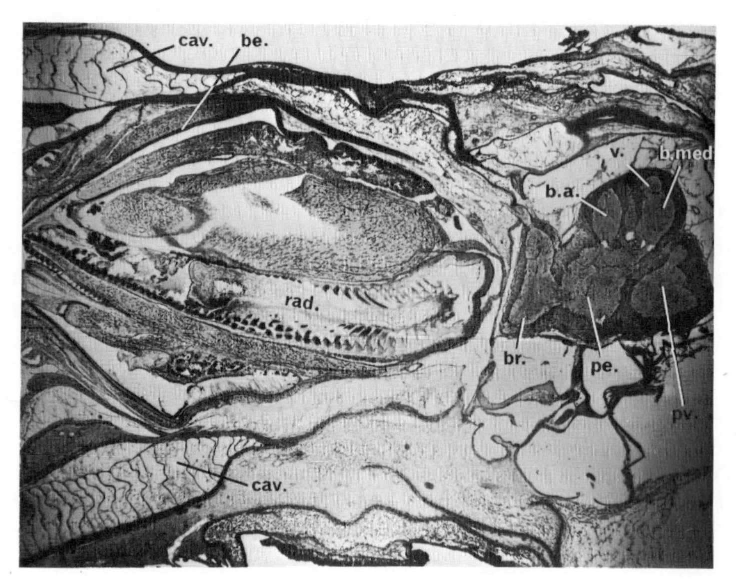

FIG. 19. *Japetella*. Sagittal section showing chambers in the arms, thin, beak, radula and shortened central nervous system. Field width 5 mm. (Key to abbreviations at end of chapter.)

vertical position with the arms upwards, but there are also large fluid cavities in the main body mass.

The brain is typically octopodan in many respects, but the tactile centres are much reduced. The parts are very highly concentrated (Figs 19 and 20). There is a large superior buccal lobe, with paired posterior buccal lobes behind it, quite unlike any decapod. The posterior buccal lobes are joined dorsally at the back representing the median inferior frontal lobe and there are a very few small cells in the position of the subfrontal lobe (Fig. 21). The whole arrangement suggests a secondary reduction from that seen in *Octopus*. The suprabrachial commissure is completely absent, again suggesting that this is functionally related to life on the bottom. The vertical lobe system is quite well developed, but much less so than in epibenthic octopods. There are only three vertical lobules (Fig. 22).

The eyes show a sharp division of the retina into two parts at the front end (Fig. 23). Here a sharp ridge separates a smaller dorsal from a larger ventral chamber of the eye. At the very front the two sets of rhabdomes are completely separated. Further back the ridge becomes less sharp and is covered by rhabdomes. The back-half of the retina is undivided. Corresponding to this division

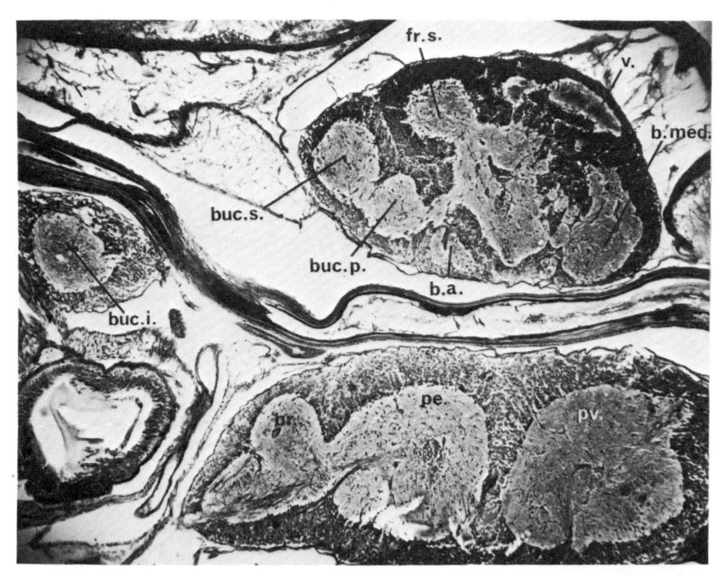

FIG. 20. *Japetella*. Sagittal section of central nervous system. Field width 2·2 mm. (Key to abbreviations at end of chapter.)

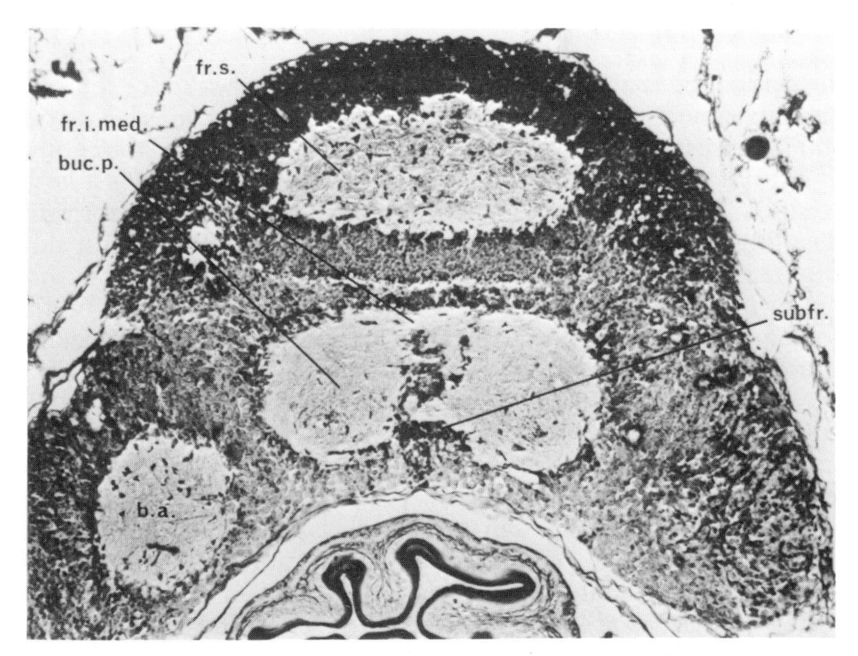

FIG. 21. *Japetella*. Transverse section to show the very reduced inferior frontal system. Field width 0·75 mm. (Key to abbreviations at end of chapter.)

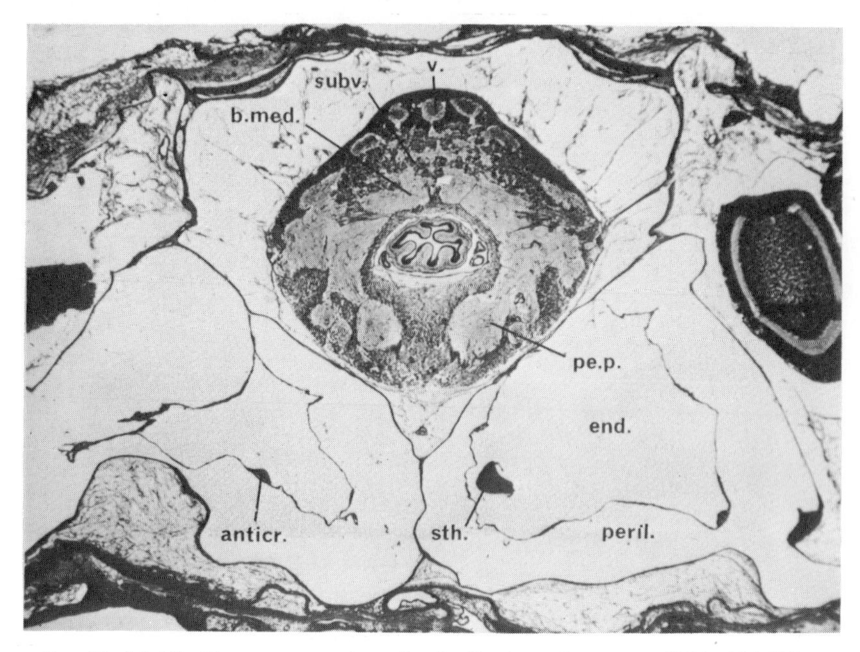

FIG. 22. *Japetella*. Transverse section of back of brain and statocyst. Field width 5·8 mm. (Key to abbreviations at end of chapter.)

the optic lobe is also sharply divided, showing two quite separate parts in some aspects (Fig. 24). At the back it is undivided. The two anterior parts of the optic lobe send separate bundles to the central brain, but their courses have not been followed. There is a large epistellar body of typical octopod form. The statocyst is enormous (Fig. 22) and presumably functions as a dynamic receptor for slow angular accelerations. Its details are not known.

With these many curious features the mode of life of the bolitaenids remains uncertain. They are undoubtedly bathypelagic and possibly feed on plankton. No records of gut contents are known to me. The beaks are thin (as noted also by Robson, 1932). However, the radula is exceptionally large with powerful muscles and expanded at the back (Fig. 19). The sharp recurved lateral teeth seem to be quickly worn away. In contrast to *Vampyroteuthis*, *Spirula* or *Calliteuthis* the salivary papilla is small and not provided with special glands. The submandibular and anterior salivary glands are quite small. There is said to be a large crop and an even larger liver (Robson, 1932). Some of these characteristics suggest a diet of plankton rather than larger prey, but they are not really very similar to those of *Vampyroteuthis* or *Spirula*.

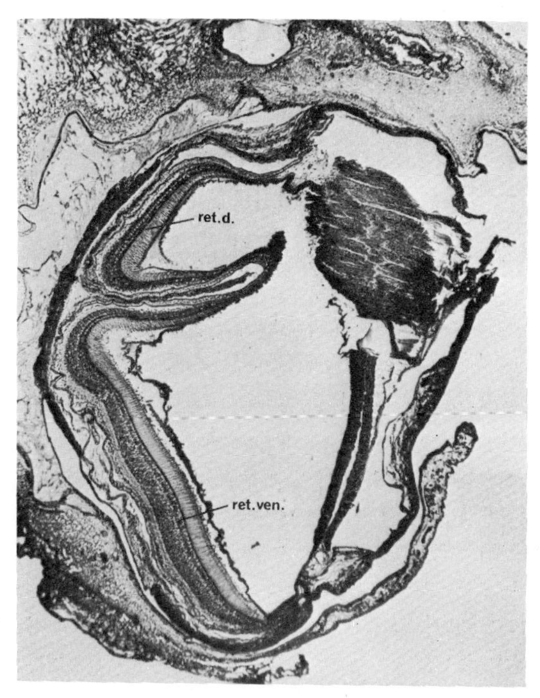

FIG. 23. *Japetella*. Transverse section of front of eye, showing upper and lower sections of the retina, separated by a ridge without rhabdomes. Field width 2·7 mm. (Key to abbreviations at end of chapter.)

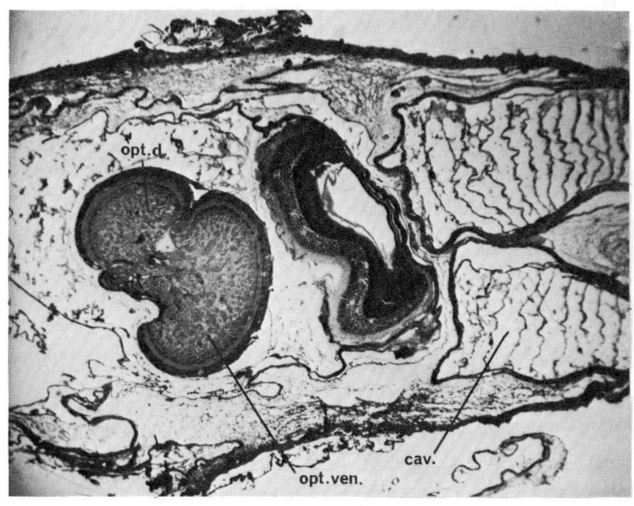

FIG. 24. *Japetella*. Sagittal section to show the optic lobe sharply divided into dorsal and ventral parts. Field width 5·5 mm. (Key to abbreviations at end of chapter.)

Various benthic octopods

Among the more nearly typical benthic octopuses there are many with different ways of life, and distinct optimum depths and temperatures. Some show quite marked divergences related to life in deeper waters, including variations in the central nervous system. For example *Pteroctopus* (=*Scaeurgus tetracirrhus*) is a genus much closer to the littoral *Octopus* than *Japetella*, but yet deviating considerably from the epibenthic form. Its mode of life is not well known but it is recorded as living below 200 m ("epibathyal") by Mangold-Wirz (1963). It has been taken down to 600 m. It is a small creature with short arms united by a web for much of their length (Fig. 25). The brain is characteristically octopodan, but the

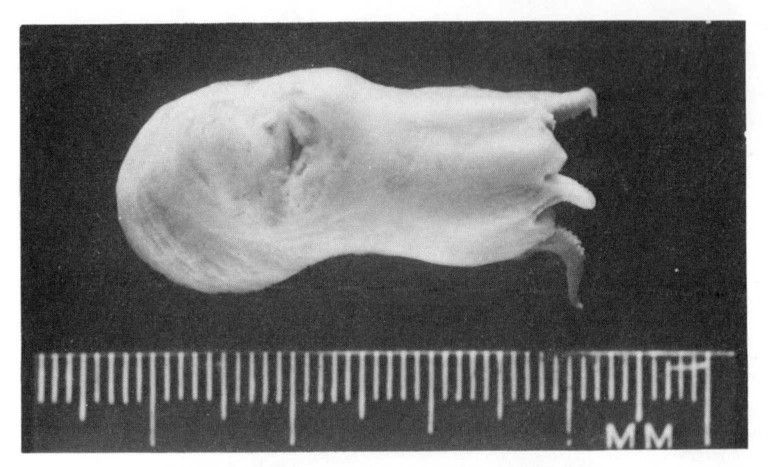

FIG. 25. *Pteroctopus*, seen from the ventral side.

inferior frontal system is small. We can recognize an undoubted subfrontal lobe, and there is a suprabrachial commissure (Fig. 26). There is a large epistellar body, facing towards the viscera (Fig. 27). These characters suggest that it has not abandoned a bottom-living habit, perhaps it lives in mud or sand.

Scaeurgus unicirrhus is still closer to *Octopus* and is recorded from 200–400 m in the Catalan Sea (Mangold-Wirz, 1963), and from the shore to 350 m by Robson (1929). It has a fully developed inferior frontal system.

Benthoctopus is an obvious benthic form, with long arms, called "subabyssal" by Robson (1929, 1932) and taken down to 1200 m. Its brain is hardly distinguishable from that of *Octopus* and the epistellar body is small and embedded in the stellate ganglion. This agrees with the finding of this genus in shallow as well as deep waters.

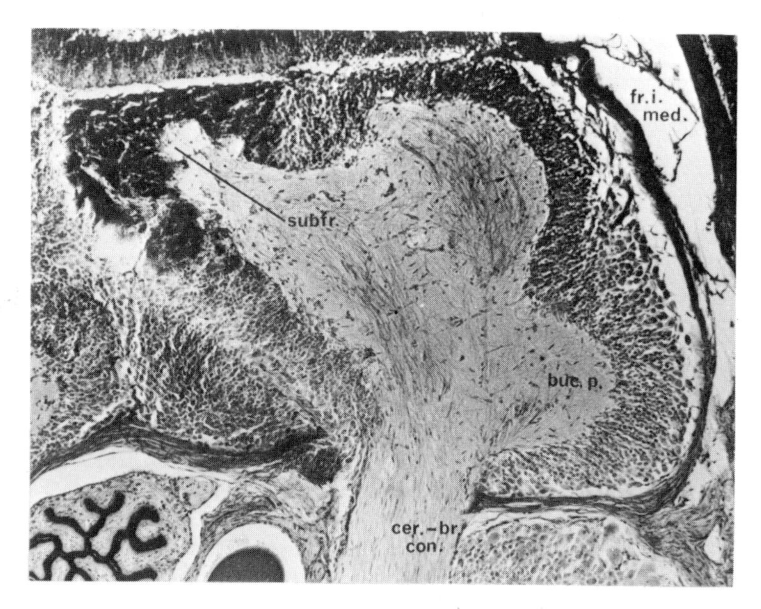

FIG. 26. *Pteroctopus*. Oblique transverse section of the inferior frontal region. The subfrontal is better developed than in *Japetella*, but less so than in *Octopus*. Field width 2·2 mm. (Key to abbreviations at end of chapter.)

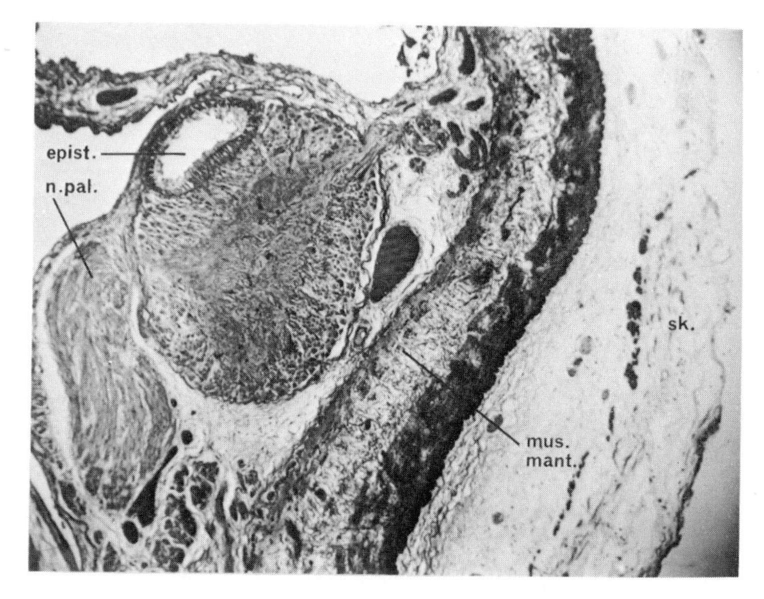

FIG. 27. *Pteroctopus*. Transverse section of stellate ganglion, showing large epistellar body facing inwards. Field width 2·2 mm. (Key to abbreviations at end of chapter.)

Eledone lives in deeper waters than *Octopus*, but is of course benthic. Mangold-Wirz (1963) records it between 35 and 120 m in the Catalan Sea. It would be hard to distinguish the brain from that of *Octopus*. The epistellar body is, however, distinctly larger in *Eledone* and clearly pigmented with orange, which was the characteristic that first led me to notice it. *Octopus vulgaris* is reported in the Catalan Sea only down to 65 m and has a smaller epistellar body than any of those yet described. The epistellar body is also very small in *O. dofleini* from the American Pacific Coast.

Epipelagic octopods

Argonautidae

Finally we find interesting similarities and differences between these epibenthic octopods and those inhabiting the surface of the sea. In *Argonauta, Tremoctopus* and *Ocythoe* we have a striking reduction of the inferior frontal system (Fig. 28). This may at first seem surprising since in these animals the arms and their webs are very large and indeed specialized as a feeding mechanism (Young, 1960). In *Tremoctopus* they form an enormous floating sheet (Voss & Williamson, 1972) with symbiotic nematocysts (Jones, 1963). The

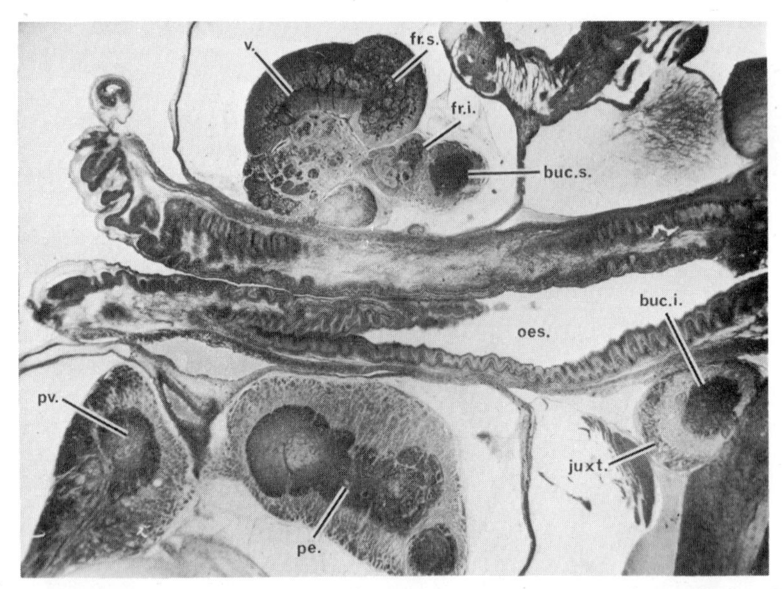

FIG. 28. *Argonauta.* Sagittal section of central nervous system showing small size of the inferior frontal system. Field width 6 mm. Note wide oesophagus; also juxta-ganglionic tissue around inferior buccal ganglion. (Key to abbreviations at end of chapter.)

small size of the inferior frontal system presumably indicates that these animals have little power of chemotactile discrimination and learning. Little is known of their food; perhaps they sweep in small planktonic organisms without discrimination. Their brains have a curious resemblance to *Vampyroteuthis*, but this is only in the supraoesophageal lobes and is presumably due to convergence. They show no sign of a large ventral magnocellular lobe.

Although they differ from benthic octopods in the inferior frontal system they resemble them in having a large statocyst cavity interrupted by only a small anticrista (Young, 1971). They also have a typical small octopodan epistellar body. So far as I can read the evidence, this suggests that these epipelagic forms have arisen secondarily from epibenthic octopods.

THE DECAPODS ARE A HOMOGENEOUS GROUP

In spite of the obvious difficulties it seems to me to be important to retain the decapods as a unit. In the nervous system and other parts there are many striking similarities between sepioids and teuthoids, which strongly suggest that they are a common stock, distinct from that of octopods. These features include:

(1) separation of superior buccal lobes from the brain;
(2) small development of the inferior frontal system;
(3) structure of the vertical lobe, with large neuropil;
(4) presence of ventral magnocellular lobe and giant fibre system;
(5) quadripartite crista;
(6) large anticristae;
(7) photosensitive vesicles in the head.

These are not minor features, but are all of great functional significance and must surely indicate a common ancestry. The best way to keep the two groups together and distinct from the octopods seems to be to call them suborders. Thus we should have:

Subclass Coleoidea
  Order 1 Belemnoidea
  Order 2 Vampyromorpha
  Order 3 Decapoda
    Suborder 1 Sepioidea
    Suborder 2 Teuthoidea
  Order 4 Octopoda

With this system the term Myopsida disappears and this I regard as an advantage. The family Loliginidae is not more different from the other teuthoid families than they are from each other (see below).

In spite of their common features the decapods show a wonderful variety of nervous and other systems in relation to their depth and other conditions of life. We can only survey a few of them, beginning with the enigmatic *Spirula*.

Oceanic sepioid

Spirula spirula

Questions about *Spirula* are of special interest for many reasons. Its unique shell, besides the phylogenetic problem it raises, allows the animal to perform the very marked daily vertical migrations shown by Clarke (1969). Off the Canary Islands they rise from 600 m or below at midday to 250 m or less at midnight. Since they are neutrally buoyant these great movements require little energy (Denton, Gilpin-Brown & Howarth, 1967). They do however suggest the presence of some means of measuring illumination and/or depth.

It is not known for certain how *Spirula* feeds, but in one of our specimens, though the oesophagus was empty, the large muscular gizzard contained small crustaceans undergoing digestion, including copepods and ostracods (Fig. 29). No phytoplankton or non-crustacean organisms were present. The presence of small crustaceans in this part of the stomach was also noted by Kerr (1931). The radula is reduced to a vestige, but there is a large muscular salivary papilla, with quite large posterior salivary glands. The anterior salivary and submandibular glands are absent. How then does *Spirula* take in plankton? The lower border of the salivary papilla is much folded and glandular and may serve to lick in the plankton (Fig. 30), but the large and muscular beak must also be involved (Fig. 31). The tentacles are of course well developed but the arms are all short and their suckers small (Nixon & Dilly, see their chapter, this volume). It seems that the whole apparatus serves to catch individual items of food rather than simply to take in a mass of plankton. The large posterior salivary glands suggest that the prey is poisoned, but they may have some other function.

In its nervous system *Spirula* is a typical decapod, showing none of the features in which *Vampyroteuthis* resembles the octopods. The parts are widely spread, with the superior and inferior buccal

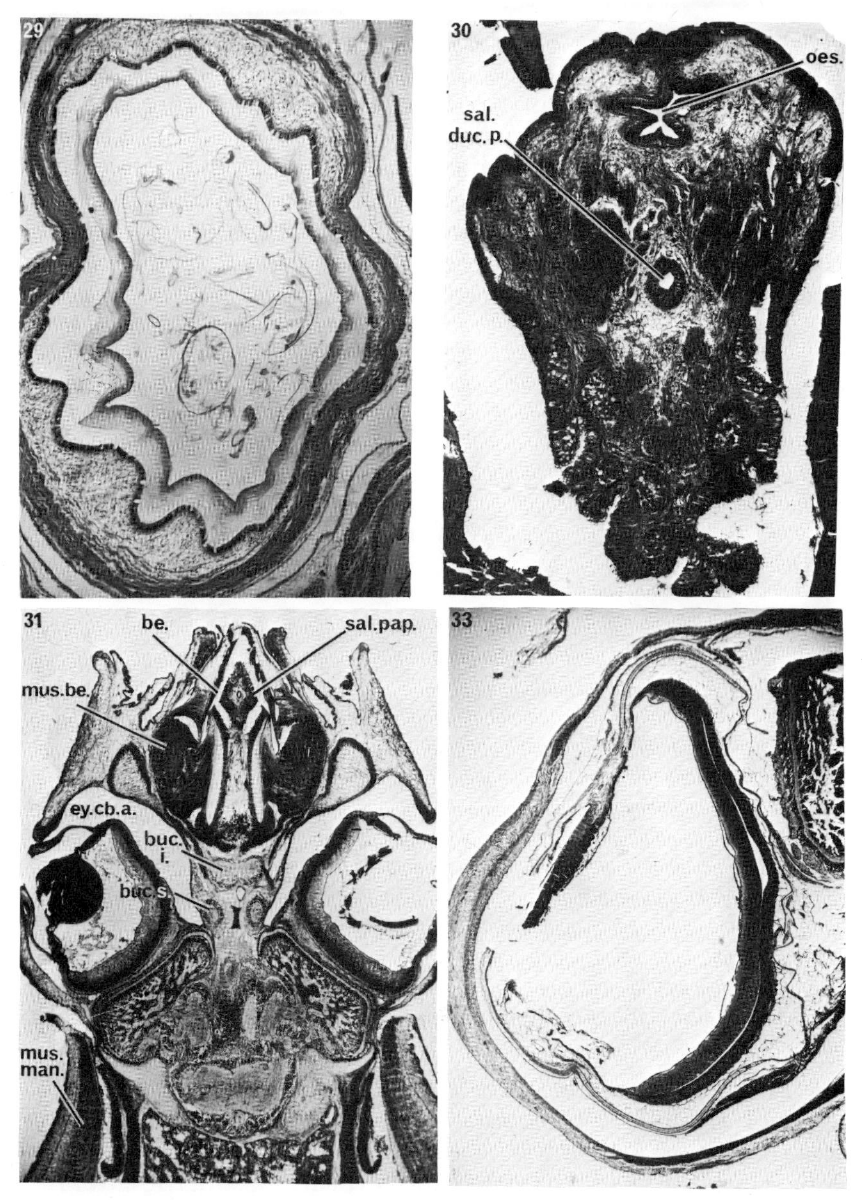

FIG. 29. *Spirula*. Transverse section through gizzard, showing thick muscular wall and partly digested small crustaceans (haematoxylin and eosin). Field width 2·2 mm.

FIG. 30. *Spirula*. Transverse section of salivary papilla, showing glands opening at the surface. Field width 0·9 mm.

FIG. 31. *Spirula*. Horizontal section of a small animal showing buccal mass, eyes and mantle. Note strong muscles of the beak, large, open anterior chamber of the eye and thick mantle muscles. Field width 4 mm.

FIG. 33. *Spirula*. Transverse section at back of eye of large animal, showing rhabdomes longer ventrally than dorsally. Field width 4 mm. (Key to abbreviations at end of chapter.)

lobes far forward, as in *Nautilus* and decapods (Fig. 32). The anterior suboesophageal mass lies widely separated from the middle mass. An interesting point is that in this small brain it can be readily seen that fibres from the brachial nerves reach back past the anterior mass and past the front part of the middle mass to end in the posterior pedal, magnocellular and palliovisceral lobes. There is no separate suprabrachial commissure.

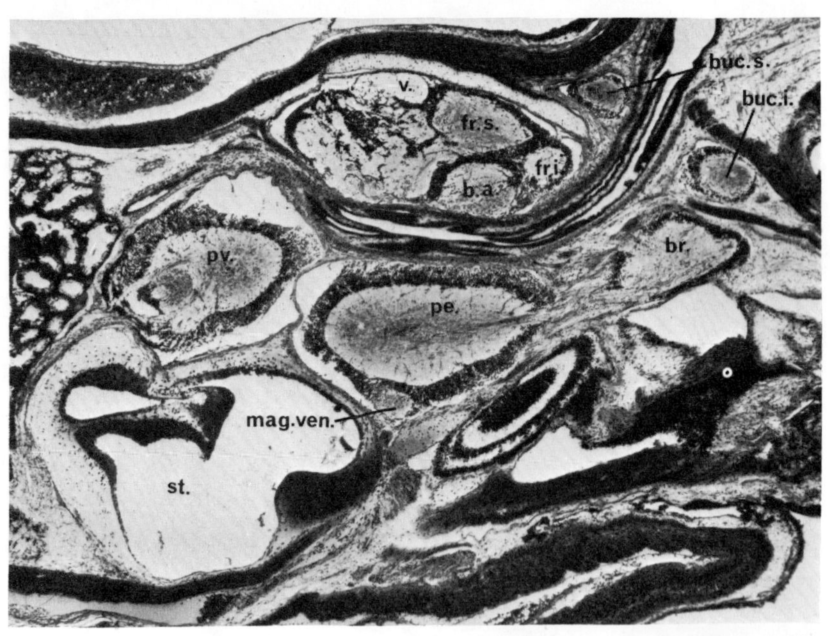

FIG. 32. *Spirula*. Sagittal section of central nervous system of a small animal. Note superior buccal lobe widely separated from the brain, small vertical lobe, large statocyst. Field width 3·6 mm. (Key to abbreviations at end of chapter.)

The inferior frontal lobe is small. There is no sign of median inferior frontal or subfrontal lobes, and presumably no touch-discrimination learning is possible. The median superior frontal lobe is large, indeed larger than the vertical lobe, but the latter has a typical decapodan structure, with a wide neuropil.

The magnocellular lobe is also typical of decapods, with a large ventral portion and commissure below the pedal lobe. There is a typical pair of first-order giant cells. The axons cross in the palliovisceral lobe, probably without fusing. The giant fibre system of the stellate ganglion resembles that of *Sepia*, rather than *Loligo* (see Martin, his chapter, this volume). There are several large fibres in each stellar nerve (up to 20 μm diameter) and they do not arise from a special giant fibre lobe. The mantle muscles are thick and evidently *Spirula* is a strong swimmer with the jet, as evidenced indeed by its vertical migrations.

A feature of the brain is that the cerebellum-like basal lobes are all well-developed. The peduncle lobes are large and orientated vertically, not horizontally as in *Sepia*. Both parts of the anterior basal are large. These centres are especially connected with control of eye-movements, which may be especially important for an animal whose flotation system prevents it from turning the whole body readily.

The eyes are of course very large, with a round pupil. The system is not aphakic, since pigment extends all round to the iris. The rhabdomes are rather long (up to 200 μm). At the back of the eye they are somewhat longer ventrally than dorsally (Fig. 33).

Photosensitive vesicles are present in the tissue of the cortex of the optic lobe, below the optic tract. This is their position in both *Loligo* and *Sepia*. They are much larger than in those genera, but smaller than in many other coleoids. They could be the agents responsible for monitoring the daily vertical migration.

The statocyst is clearly of coleoid structure, but very different from both *Sepia* and the teuthoids (Fig. 34). The macula princeps carries a statolith with a long lateral projection and a short, blunt ventral one. There are large dorsal and ventral maculae neglectae. The whole macular apparatus is set in a distinct pocket at the anterior end of the sac.

The main part of the sac is little obstructed by anticristae, though there are five large hooks at the ends and turns of the crista. There are also four straight anticristae but these are crowded at the hind end. They thus have the effect of producing a transverse chamber in which movement of fluid across the cupula of the vertical crista is more restricted than across the others. This part of the crista is presumably stimulated by angular acceleration about the anatomical longitudinal axis, that is to say in the horizontal earth plane as the animal floats head-downwards. This may be the result of movements by which *Spirula* turns its head to catch food, requiring rapid compensatory eye-movements. This would also agree with the presence of a large anterior basal lobe. The large inertial mass of the fluid in the anterior part of the chamber would allow for great sensitivity of the other parts of the crista to small, slow movements. There is no especially large posterior sac of the statocyst, but hair cells have been seen in various parts of the wall.

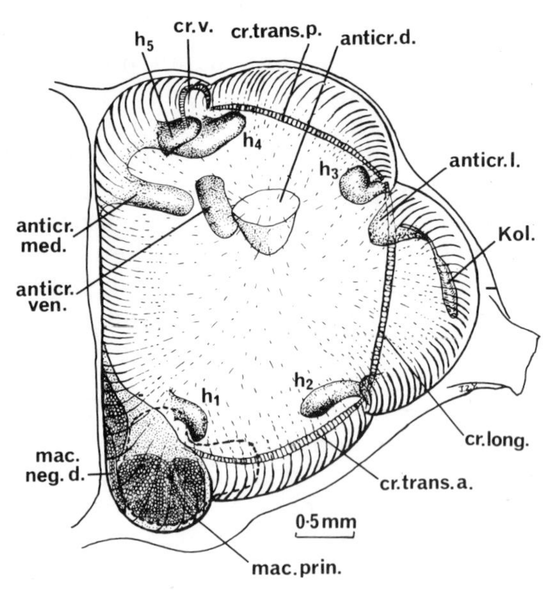

FIG. 34. *Spirula*. Drawing of the statocyst, dissected from the body and shown from dorsal side in the orientation usually adopted in life, with the head downwards. The statolith is shown in outline. h1–5 Are the hamuli and the four anticrista have been named by their anatomical positions. Note that they make an upper chamber limiting the inertial mass moving across the vertical crista. (Key to abbreviations at end of chapter.)

Circalittoral sepioids

The animals of inshore waters are those with which we are most familiar, for obvious reasons, but we must be careful to avoid the easy assumption that they are therefore the most typical of their groups. Both *Sepia* and *Loligo* show very great development of the brains and of other parts of the body. These may be specialized features, reflections of life in varied and well-lit conditions, and not characteristic of all decapods.

Sepia

In *Sepia* we see full development of the brain, eyes and statocyst for an active, visual life in shallow waters (Fig. 35a). The same pattern of the brain is found with surprisingly little variation in the loliginid and ommastrephid squids (Fig. 35b). It is difficult to believe that such an elaborate and detailed system has been evolved in parallel and I suggest that all these animals must be quite closely related. It is not possible here to go into details of this brain plan, which will be described in later papers. It involves full development of the

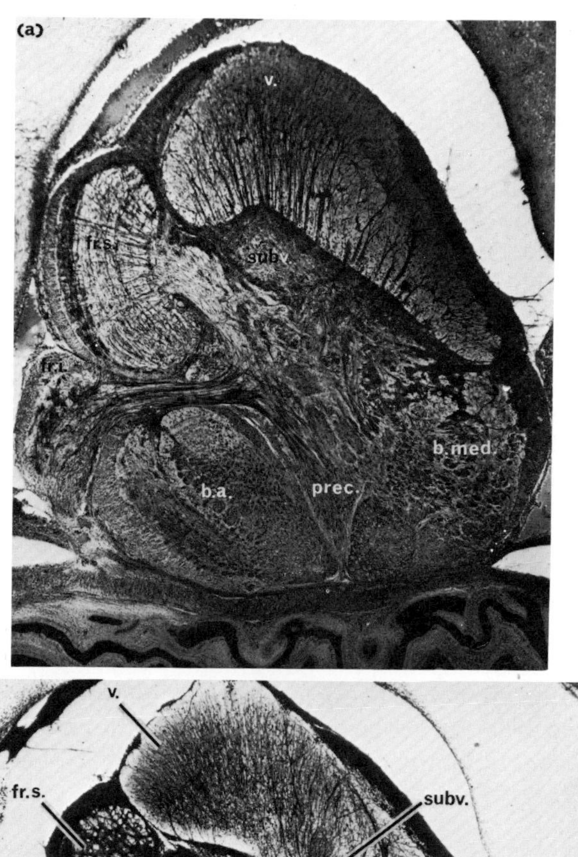

FIG. 35. (a) *Sepia*. Sagittal section of supraoesophageal lobes. Field width 4 mm. (b) *Loligo*. Sagittal section of the supraoesophageal lobes. Note similarity to *Sepia*. Field width 3·3 mm. (Key to abbreviations at end of chapter.)

swimming apparatus with both the jet (using giant fibres) and the fins. Seizure of the prey is by the tentacles, which in *Sepia* are ejectile and used to capture crustaceans, especially prawns, while crabs are usually taken by the arms (Messenger, see his chapter, this volume). The whole visual system is extremely well developed and with it the vertical lobe system. There has been evidence of a visual memory system since 1940 (Sanders & Young, 1940). The tactile system, on the other hand, shows none of the special developments of the inferior frontal system that are seen in octopods (Young, 1963). The basal lobe system is very well developed, no doubt serving for control of movements of the eyes and of the whole animal (Boycott, 1961; Messenger, 1968; Hobbs & Young, 1973). *Sepia* shows an especial development of its colour displays, valuable no doubt against a varied background in a well-lit environment (Holmes, 1940; Boycott, 1953). The statocyst shows very large anticristae, restricting the movement of the contents to narrow channels in certain directions. This is no doubt connected with the rapid movements, especially the rapid turns, of the animal (see Budelmann, this volume; Stephens & Young, 1976).

On the other hand the photosensitive vesicles are little developed in *Sepia*. They are small orange-coloured structures at the back of the optic tract, the parolfactory vesicles. This confirms the suggestion that these vesicles serve to measure light intensity especially in relation to large vertical migrations. These do not occur in *Sepia*, which is not taken much below 100 m (Mangold-Wirz, 1963).

Benthic circalittoral sepioids

Sepiolidae

These are small animals living on the bottom in shallow water, often buried in sand (Mangold-Wirz, 1963). They lack the shell and neutral buoyancy of *Sepia*, but can move rapidly with the jet and fins. We have not yet made a full study of the nervous system but the statocyst shows an interesting contrast to *Sepia* (Fig. 36). There are the usual three maculae, but the statolith on the macula princeps is a simple single mass. The crista has hooks, except at the first turn (where it is lacking also in *Loligo*, though present in *Sepia*). The straight anticristae are reduced to two, one medial and one dorsal. Presumably the large unobstructed space provides high sensitivity for relatively small and slow movements. The question is, why should two quite large peg-like anticristae remain? It is

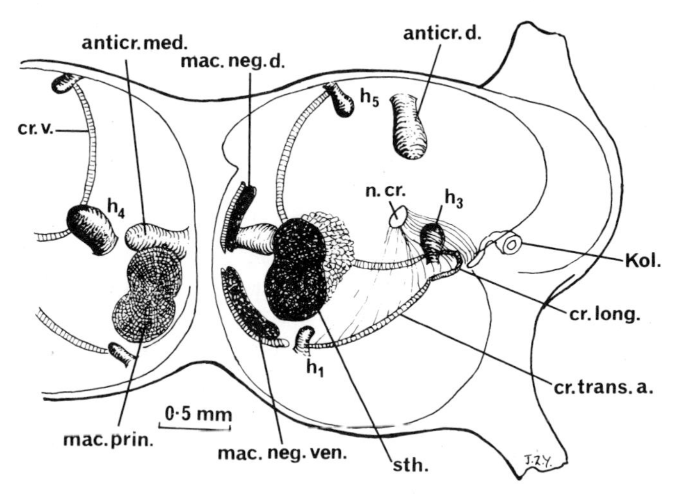

FIG. 36. *Sepiola*. Drawing of statocyst after dissection. Seen from in front. Note that there is no hamulus at the first turn of the crista. There are only two peg-like anticristae and they have been named from their anatomical positions. (Key to abbreviations at end of chapter.)

possible that they carry specific receptor systems operated by hair cells of a plexus that covers them. But no detailed studies have been made of this.

Circalittoral epipelagic squids

Loliginidae

Loligo and *Sepia* are sometimes classified together because of their myopsid condition, and some other features (see Akimushkin, 1963). It may be that this closure of the anterior chamber is an adaptation to life in inshore turbid waters rather than a sign of phyletic affinity. *Octopus* could also be classified as a myopsid! It is true that in the brain and statocyst *Loligo* and *Sepia* are alike in many ways, but this again may be a result of life in somewhat similar conditions and the ommastrephid brains are in any case quite similar. In other respects *Loligo* resembles the ommastrephid squids far more closely than *Sepia*.

Loligo and *Alloteuthis* range from the surface down to 200 m, though *L. forbesi* is reported to live deeper than *L. vulgaris* (Mangold-Wirz, 1963). No full account of the brain of *Loligo* will be given here (Fig. 35b). The giant cell system is perhaps better developed than in any other cephalopods, with fusion of the two first-order giant axons and a special giant fibre lobe in the stellate ganglion (Martin, his chapter, this volume). The supraoesophageal lobes show all the features described already for *Sepia*. There is no experimental evidence of learning in squids, but there can be little doubt that it is possible. After they have been kept for several days at Naples, I have the impression that they dart away more quickly when a net appears than they did at first.

The statocyst shows perhaps even larger anticristae than in *Sepia*, they almost completely divide the cavity in certain planes. Some of the straight anticristae have tapering tips and bend very readily when touched. They are known to be covered with a plexus of hair cells and nerve fibres and may serve as detectors of very rapid acceleration forwards or backwards.

As in *Sepia* the photosensitive vesicles are very small—it would be interesting to compare them in *L. forbesi* and *L. vulgaris*.

Oceanic epipelagic squids

Ommastrephidae

The great variety of oceanic squids gives very good scope for study of variations of the brain and sense organs. We can here only look briefly at a few types. Those living mainly close to the surface show many similarities to *Loligo*. For example *Todarodes* is a large and voracious squid, very widely dispersed and caught in great quantity for food, especially in Japan. It differs from *Loligo* in that it is taken from the surface down to 1000 m and studies of tagged specimens show both diel vertical and some seasonal horizontal migrations (see Clarke, 1966). For us the special interest is that the photosensitive vesicles around the optic tract are much better developed than in *Loligo* (Mauro, his chapter, this volume). This is also true of *illex*, which move inshore each in great quantities from the Newfoundland Great Banks to Labrador. The brains of these species would be hard to distinguish in general from each other or from *Loligo*, though of course there are minor differences.

The giant fibre system of ommastrephids is very well developed and essentially as in *Loligo* [but again with differences (see Martin, his chapter, this volume)]. There is a single giant fibre in each hinder stellar nerve, ranging up to 1 mm in diameter. An interesting difference from *Loligo* is that the giant fibres of ommastrephids do not arise from a special giant fibre lobe. At least in *Stenoteuthis*, of which we have sections, they come from cells of the dorsal side of the ganglion. This has both large and small cells, the ventral side only large ones. The photosensitive vesicles extend above the optic tract as well as below it (Baumann *et al.*, 1970). Their nerves are quite large and join the basal part of the peduncle lobe, close to the origin of the nerve to the optic gland, which controls reproduction (Wells & Wells, their chapter, this volume). We have suggested that the photosensitive vesicles provide a system for measuring light intensity, by-passing the eyes and all the neural equipment for recognition of shapes. The function may then be to monitor daily and perhaps seasonal changes of illumination (see Mauro, his chapter, this volume).

Mesopelagic squids

Enoploteuthidae, ***Pterygioteuthis***

Some of the most interesting squids live largely between 200 and 500 m, probably with marked diel migrations. *Pterygioteuthis* is an example, a very active little animal, without any obvious mechanism for neutral buoyancy. They were found shallower than 200 m at night, but below 400 m in the day (Clarke & Lu, 1974). These animals are especially well provided with light-producing organs (R. E. Young, his chapter, this volume).

Our sections show that these are very muscular animals, without special fluid-filled cavities. As might be expected from their migrations they have very large photosensitive vesicles, dorsal and ventral ones on each side. In the related *Abraliopsis* they are even more complex, and R. E. Young suggests that they may serve to match the down-coming light with that issuing from the animal itself.

In the brain we have all the signs of a well-developed visual system, with large vertical lobes. The particular characteristic of *Pterygioteuthis* is the presence of an extra set of giant fibres controlling the arms (Fig. 37). The first-order giant cells and fibres occupy the usual position. The special set arises from a pair of cells in the lateral wall of the magnocellular lobe. Their trunks run to the midline and there come very close to the first-order axons, probably making synaptic contact. They then turn forwards and make further synapses in the pedal lobe and proceed on forwards to activate a series of second-order anterior giant cells, one for each arm nerve (Fig. 38). We do not know the function of this system. It may be an exaggeration of sets of cells and fibres that are present in all squids. It suggests that *Pterygioteuthis* is able to take especially fast actions with its arms.

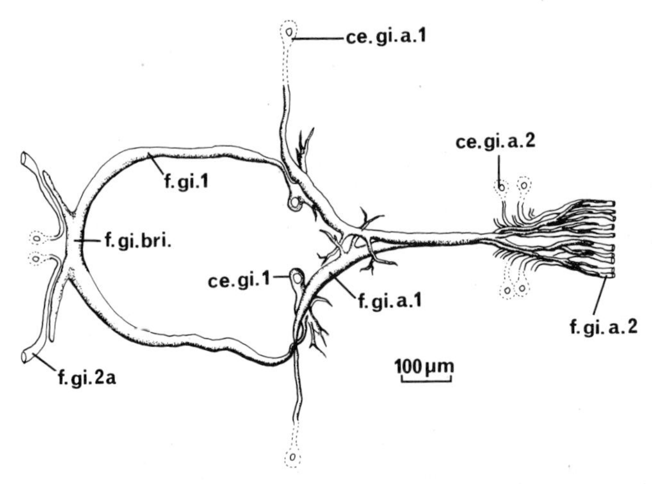

FIG. 37. *Pterygioteuthis*. Drawing of reconstruction of the giant fibre system, showing the anterior giant fibres running to the arms. (Key to abbreviations at end of chapter.)

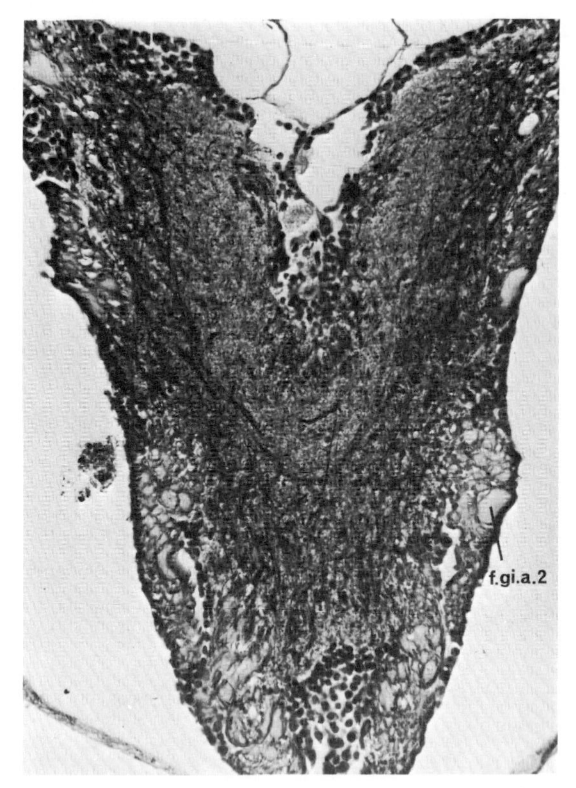

FIG. 38. *Pterygioteuthis*. Transverse section at front of prebrachial lobe, showing roots of brachial nerves, with one giant fibre in each (f.gi.a.2). Field width 0·4 mm.

Architeuthidae, ***Architeuthis***

The giant squids are among the least known of the order, and their depths and habits are problematic. They certainly live at great depths (see Clarke, 1966, and his chapter, this volume). They probably feed on fishes and are themselves the food of sperm whales. The marks often found on the skin of whales (Murray & Hjort, 1912; Roe, 1969) are generally held to be made by *Architeuthis*, so they must be strong fighters, although there is no direct evidence.

Everyone wants to know whether giant squids have *giant* giant fibres. We have no material of the central nervous system but some years ago I was able to dissect the stellate ganglion of an animal washed up at Scarborough in 1933 and sent to the British Museum. The mantle length was 125 cm. The nerves of the mantle muscles are arranged in this genus differently from any other I have seen. Those to the front part of the mantle arise from a relatively small stellate ganglion, in the usual way. The hinder part of the mantle, perhaps more than half of the whole, is supplied from a distinct median nerve, running with the fin nerve and giving off a series of branches to the mantle.

Each of the nerves arising from the ganglion contains one or two large fibres, ranging in diameter from about 80 μm in the more anterior ones to a maximum of 250 μm further back (Fig. 39). The median nerve was poorly preserved but one fibre of about 250 μm could be seen. Two of the more posterior branches contained fibres

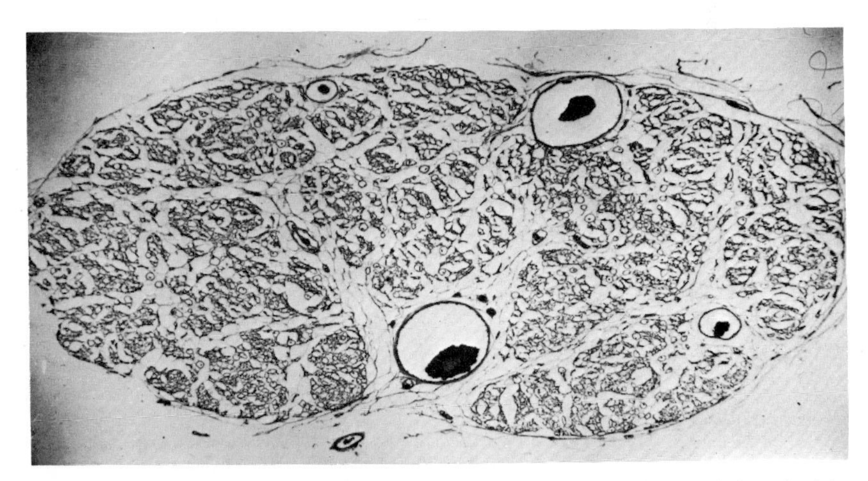

FIG. 39. *Architeuthis*. Hinder stellar nerve, showing giant fibres. The mantle length of the animal was 125 cm. Field width 2·2 mm. In spite of the poor preservation one can estimate that the largest fibre was about 250 μm diameter.

of about 200 μm each. None of the nerves examined contained the exceptionally large fibres reported by Aldrich & Brown (1967). We may conclude that *Architeuthis* is not an especially fast-moving animal. This would agree with evidence that it is neutrally buoyant with a high concentration of ammonium ions in the mantle and arms (Denton, 1974).

Bathypelagic squids

Histioteuthidae

Passing now to other deep-sea squids we find that all have evidence of mechanisms for neutral buoyancy, as shown by the presence of highly vacuolated tissues. Fortunately we now have some evidence about the postures adopted during life, though little about the food or its method of capture. *Gonatus*, *Galiteuthis* and *Histioteuthis* were all photographed from the submersible *Deep Star* at depths of 700–1170 m, in a head down position (Church, 1971). We have some histological material for these, but I prefer now to discuss another histioteuthid, *Histioteuthis* (*Calliteuthis*) *reversa*. This is unusual in the unequal size of its eyes, the left being much larger than the right (Fig. 40). It has been taken at depths that suggest it is bathypelagic, but is also taken near the surface. The tissues are

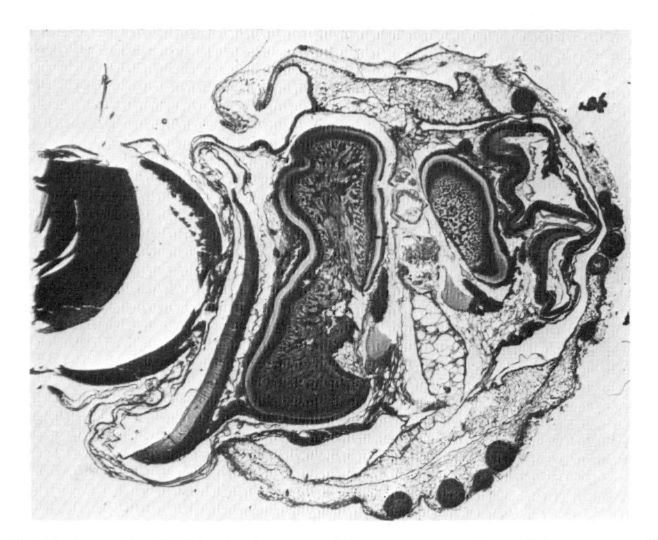

FIG. 40. *Histioteuthis* (*Calliteuthis*) *reversa*. Transverse section of the two eyes. The smaller right eye shows the retina completely divided, while the larger left eye has the retina much thickened ventrally. Note that the plexiform zone of the left optic lobe is much thicker dorsally than ventrally. Field width 8 mm.

extremely thoroughly vacuolated and the animals must surely be neutrally buoyant.

The eyes are indeed highly differentiated (Denton & Warren, 1968; R. E. Young, 1975). The smaller right retina is completely divided at the front into upward—and downward—looking parts. The left retina obviously looks only upwards, since it thins out dorsally (Fig. 40). The right optic lobe is not divided but its ventral part shows an unusual and striking extra development of the layers in the plexiform zone. The vertical lobe of *Histioteuthis reversa* is rather small, as in many of these deep-sea squids (Fig. 44). If its main use is for learning we may suppose that they are rather "stupid" creatures, but this is a most unsatisfactory statement, in more ways than one; no doubt their brains are adequate for the lives that they live.

The gizzard and intestine of one of our specimens of this species was filled with the remains of small crustaceans, including copepods (Fig. 41). This is certainly not "accidental", and the oesophagus is empty. All the remains are crustacean. The jaws are thin and their muscles as well as those of the radula are weak (Fig. 42). It probably cannot handle large prey. But the subradular glands are very large and as in *Vampyroteuthis* they open all over the surface of the salivary papilla (Fig. 43). The buccal membrane and lips are much folded on the inner surface (Fig. 44). The animal perhaps collects these organisms by its arms and by the action of its lips, salivary papilla and labial palps, aided by the radula. The statocyst is large but its detailed organization has not been studied (Fig. 44).

Chiroteuthidae

Mastigoteuthis is an animal living at even greater depths. It has been taken from 500 m down to 1500 m or below and there is no evidence of a diel migration (Clarke & Lu, 1974). The tissues are highly vacuolated and ammonium-rich solutions have been identified both in *Mastigoteuthis* and in the related *Chiroteuthis* (see Denton & Gilpin-Brown, 1973). There is a little evidence about the food of *Mastigoteuthis*; Verrill (1881) found small crustaceans and Rancurel (1971) reported the presence in the stomach of antennae from a crustacean that must have been larger than the predator.

Nixon & Dilly (their chapter, this volume) describe the extraordinary long tentacles covered with minute suckers. From these tentacles and from the arms there proceeds a special nerve, running completely outside the rest of the brain direct to the

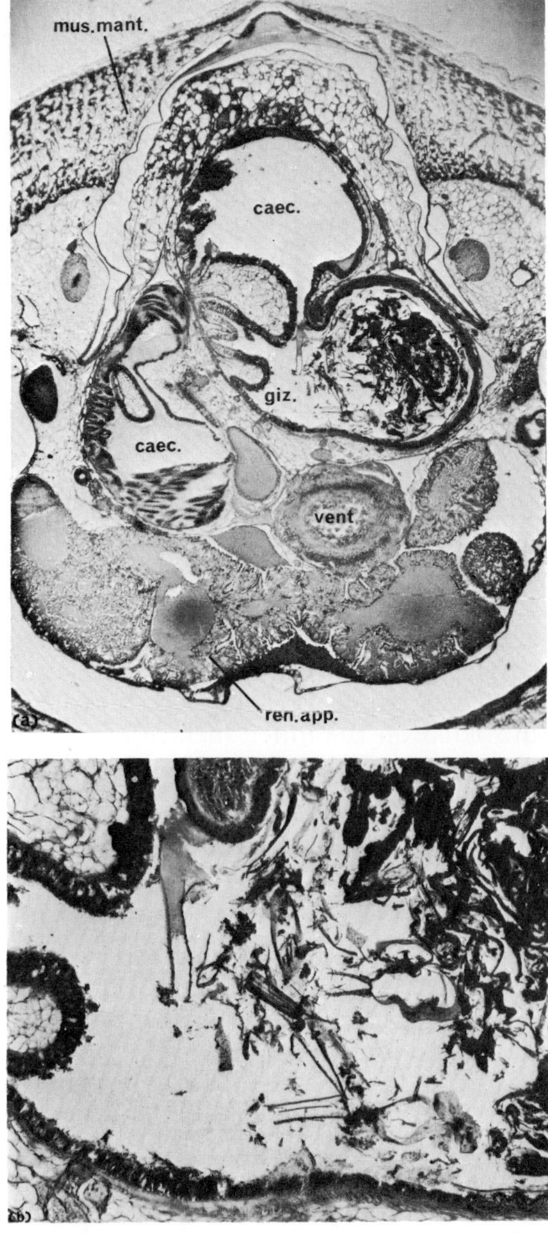

FIG. 41 (a) *Histioteuthis* (*Calliteuthis*) *reversa*. Transverse section showing the gizzard filled with remains of small crustaceans. It communicates with a very large caecum, which has several parts. Note also the vacuolated muscles of the mantle. The renal appendages are remarkably large. Field width 4 mm. (Key to abbreviations at end of chapter.) (b) Enlarged view of some of the contents of the gizzard. Field width 1 mm.

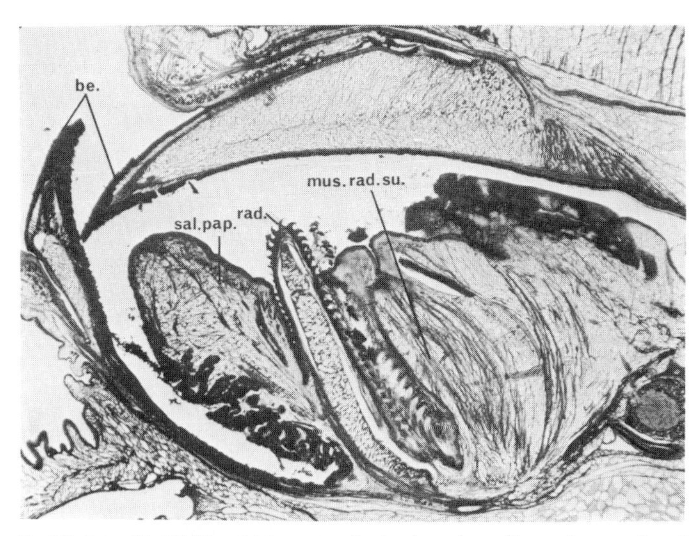

FIG. 42. *Histioteuthis (Calliteuthis) reversa.* Sagittal section of buccal mass, showing rather thin beak, large radula with weak muscles and salivary papilla with glands. Field width 3·5 mm. (Key to abbreviations at end of chapter.)

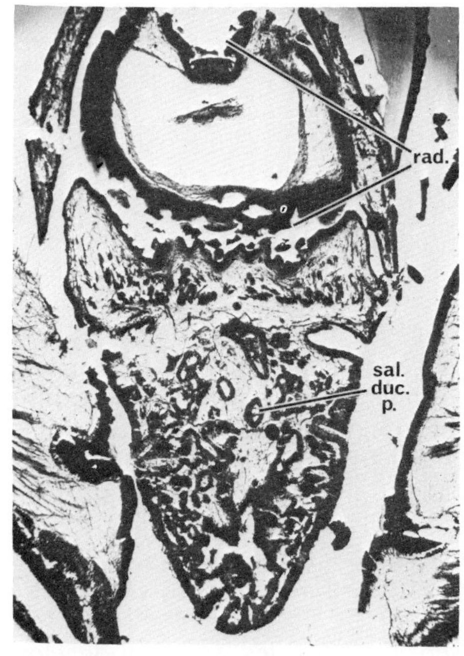

FIG. 43. *Histioteuthis (Calliteuthis) reversa.* Transverse section of salivary papilla, showing its glands and irregular upper surface working against the old teeth at the distal end of the radula. Field width 1·6 mm. (Key to abbreviations at end of chapter.)

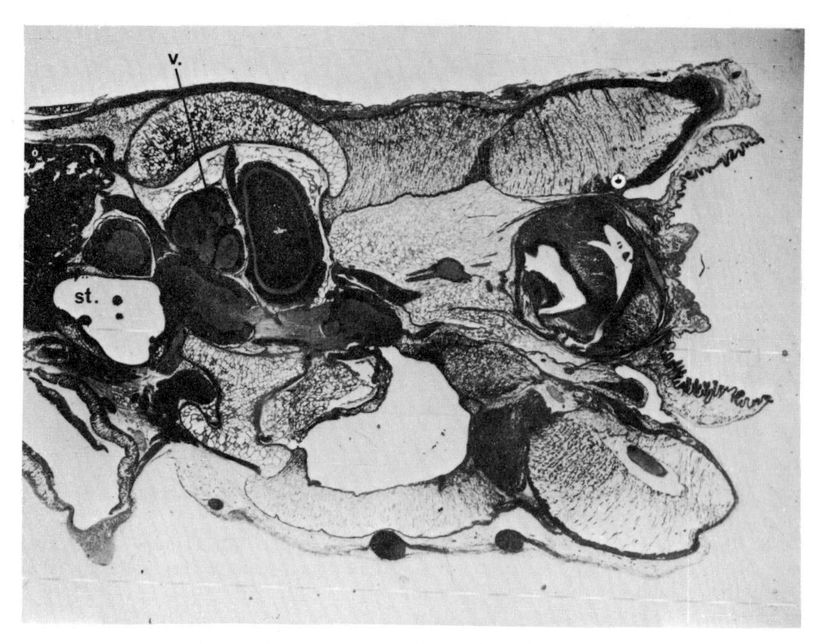

FIG. 44. *Histioteuthis (Calliteuthis) reversa.* Sagittal section of head, showing the greatly vacuolated tissues, oral membrane, small vertical lobe and large statocyst. Field width 8 mm. (Key to abbreviations at end of chapter.)

magnocellular lobe (Fig. 45). This pathway is present in other squids, but never so greatly developed. Moreover, the ventral magnocellular lobe is quite astonishingly large and internally differentiated (Fig. 46). When I first saw it I thought that the sections were upside down and that this was a supraoesophageal lobe! Indeed it is somewhat like the inferior frontal system of octopuses, with many interweaving bundles of fibres. Presumably the need to interchange chemotactile information has led to the development of similar organizations in these different parts of the brain of very different animals. This centre is actually larger than the vertical lobe, which, once again, is rather small in *Mastigoteuthis.*

It is not possible to say how this magnocellular lobe system operates. Paradoxically there are no giant cells or fibres in this creature with an enlarged giant cell lobe. On the other hand the fin lobe is very large and receives connections from the magnocellular lobe.

The photosensitive vesicles are well-developed in *Mastigoteuthis*, forming two sets, an anterior set below the optic stalk and a posterior set at the back of the orbit. Both send nerves to the penduncle lobe (Fig. 45).

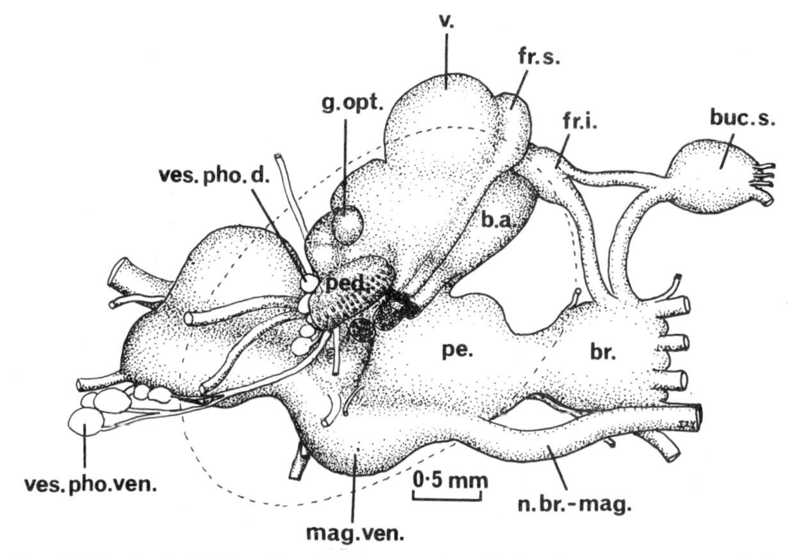

FIG. 45. *Mastigoteuthis.* Drawing of reconstruction of central nervous system and photosensitive vesicles. Note closeness of parts and especially the direct brachio-magnocellular nerve (n.br.-mag.), and the very large ventral magnocellular lobe.

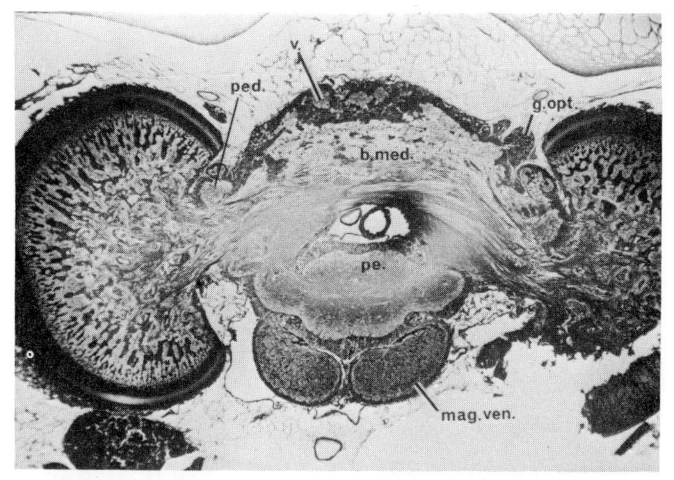

FIG. 46. *Mastigoteuthis.* Transverse section of the brain to show the very large ventral magnocellular lobe. Field width 5·5 mm. (Key to abbreviations at end of chapter.)

Cranchiidae, ***Taonius*** *and* ***Bathothauma***

The cranchiid squids are always considered last, though it is not clear that they really depart more widely than others from whatever should be considered the central type (if any!). They are neutrally buoyant by production of ammonium ions and have a much modified mantle mechanism. The mantle muscles are weak and the fins small. They have been described as drifters rather than swimmers (Clarke, 1966). *Cranchia* and *Taonius* are able to roll themselves up into a ball (R. E. Young, 1972; Dilly, 1972), and so perhaps may others.

These squids begin life with eyes on fantastically long stalks, long tentacles and very short arms (Fig. 47). The buccal mass and mouth are placed far forward on a long head process or "neck". There has been little consideration of the significance for feeding of these curious features (see Young, 1970). There must be some connection between the long tentacles, eyes on stalks and neck. The tentacles recall those of *Spirula* and may be somehow related to plankton feeding. At this stage cranchiids are frequently caught

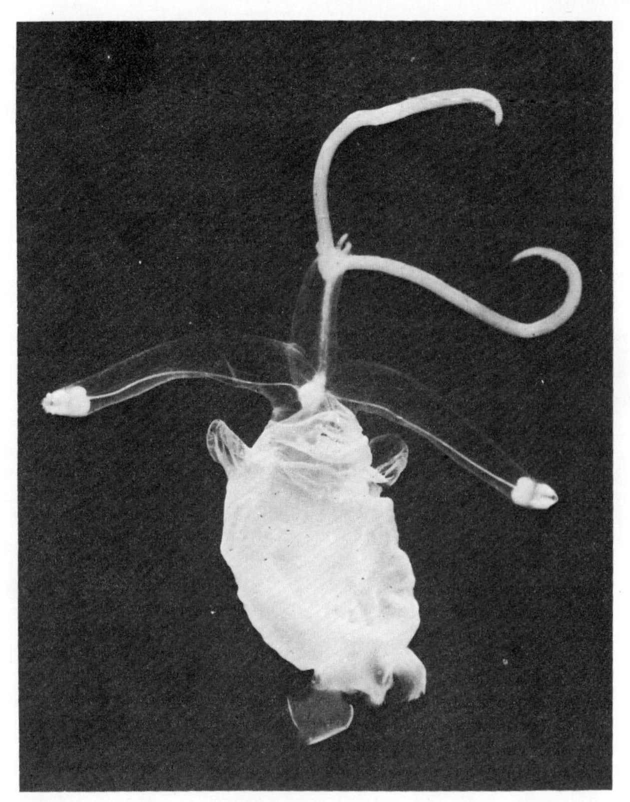

FIG. 47. *Bathothauma*. Larva to show tentacles, short arms, long neck and optic stalks. The optic lobes lie peripherally. Field width 5 mm.

near the surface but also at considerable depths (Clarke & Lu, 1974). It is not clear whether there is a diel migration but there is evidence of a downward ontogenetic one. As the animals grow they lose the eye-stalks, arms grow and the animals become of more usual teuthoid form (Young, 1970; Dilly & Nixon, 1976; Aldred, 1974). They remain neutrally buoyant as adults but these are seldom caught and little is known of them.

The stalked eyes of *Bathothauma* carry a projecting cone of unknown function (Figs 48 and 49) (Young, 1970). It is often called

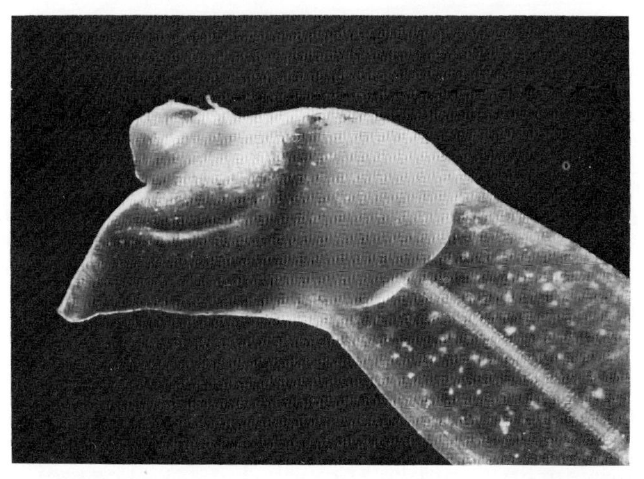

FIG. 48. *Bathothauma*. Eye and optic lobe of larva, showing the pointed "appendage". Field width 7 mm.

a luminous organ, but it differs greatly from the luminous organs of the adult. It probably allows internal reflection and has an open pore at its tip. There is no obvious luminescent tissue, but the organ could allow light to pass either inwards or outwards, perhaps from luminous bacteria. Possibly it serves to direct a light at some forward position where the beam meets the tentacles and/or the mouth. The whole apparatus may be connected with collecting small animals in the dark. Possibly the "appendage" serves, by flashing, to stimulate prey to flash, in the region of the arms or mouth. The food eaten certainly includes crustaceans, remains of which have been found in the stomach (Young, 1970). The retina and optic lobes show two distinctly separate parts (Fig. 49), one presumably for looking upwards or forwards, the other downwards or backwards.

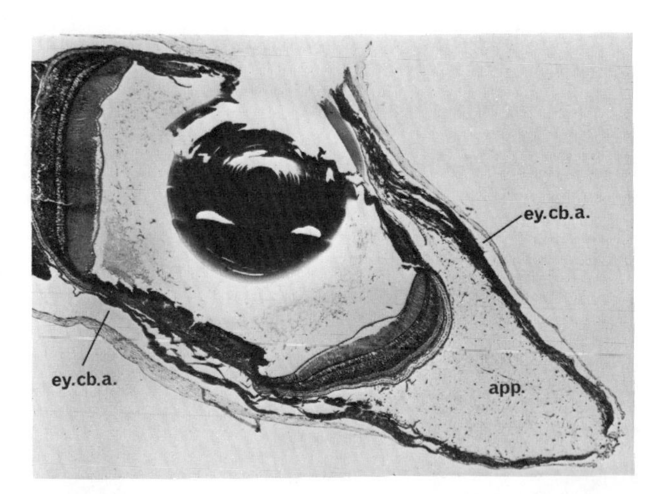

FIG. 49. *Bathothauma*. Section of the eye and its appendage (app.). Note there are two main parts to the retina, widely separated. The appendage is a projection into the expanded anterior chambers (ey.cb.a.). It contains only loose connective tissue with few cells, but its wall perhaps contains reflecting layers with an opening at the tip. There may be luminescent bacteria around the lens. Field width 2 mm.

Rather surprisingly giant cells and fibres are present in cranchiids, though small. The mantle is capable of occasional jet action although there is a long recovery period (Clarke, 1962). On the other hand, the statocyst of *Taonius* has all the characteristics of an animal showing little angular rotation (Fig. 50). The sac is large and there are hamuli on the crista, but no other anticristae to obstruct the cavity (Fig. 51). There is an enormous posterior sac and large Kölliker's canal, once again suggesting some special function for this part (p. 387). The statolith is a tiny structure, possibly connected with the drifting habit.

As in all cranchiids examined (Messenger, pers. comm.) the photosensitive vesicles are very large and are placed in the medial walls of the orbit (Fig. 51).

DISCUSSION

Adaptations to habitat

Even this small selection from the great variety of cephalopods makes it plain that every part of the body becomes modified to suit the particular manner of life of the species. Within both the octopods and decapods we can recognize members suited to various depths. In species living in shallow water we mostly find strong

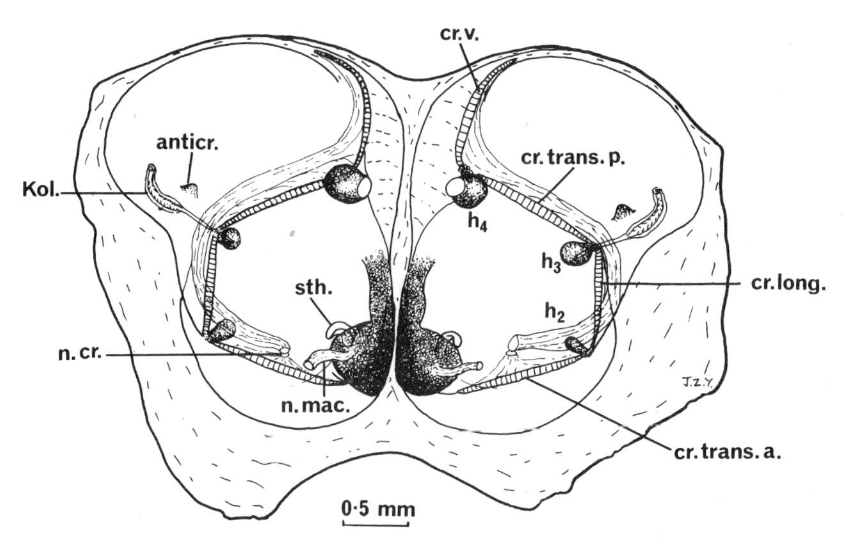

FIG. 50. *Taonius megalops.* Drawing of the statocyst of a juvenile of mantle length 100 mm. Seen from above. The macula is not clearly divided. It carries a very small statolith. There are only three hamuli along the crista and one very small peg-like anticrista. The posterior sac is enormous and there is also a large anterior sac in front of the anterior transverse crista. (Key to abbreviations at end of chapter.)

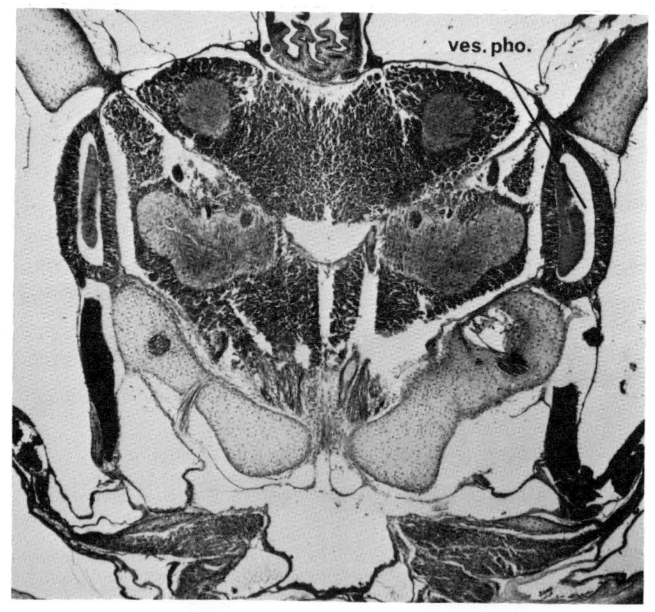

FIG. 51. *Taonius megalops.* Hind end of brain of post-larva of 45 mm mantle length, to show the large photosensitive vesicles (ves.pho.). Field width 3 mm.

swimmers (or "walkers"). Among the decapods they are provided with giant fibre systems. Proceeding to greater depths it becomes increasingly more valuable to float and mechanisms for neutral buoyancy appear in both octopods and decapods, with reduction of the mantle and giant fibre system.

It is less easy to make general statements about food. No doubt each species has its particular feeding habits and mechanisms, but none of them are well understood. We do not know how the radula functions even in *Octopus*. It can bore holes in gastropod shells (Pilson & Taylor, 1961; Arnold & Arnold, 1969) or may pick the meat out of the small legs of crabs (Altman & Nixon, 1970), so it is indeed a versatile organ. Some variations in the buccal mass in pelagic animals suggest feeding on small organisms. These include development of the glands of the salivary papilla and subradular and anterior salivary glands, with reduction of the posterior salivary glands. There are suggestions of this in *Vampyroteuthis, Spirula, Histioteuthis, Japetella* and perhaps the argonautids. The radula may be reduced (as in *Spirula*). Modifications of the brain in connection with this type of feeding include reduction of the inferior frontal system in pelagic octopods, the touch learning system being apparently less useful for feeding on small organisms. Indeed the vertical lobe system is also not very prominent in these animals, but further quantitative studies are needed on this (see Wirz, 1959).

No doubt there are many special developments of the brain connected with particular life patterns. We have met two of them in this account. In *Mastigoteuthis* the extraordinary long tentacles and minute suckers provide information to a greatly developed magnocellular lobe (p. 419). There are signs of a similar plan in *Vampyroteuthis*. It may be related to the collection of small prey. In *Pterygioteuthis*, conversely, there is an extra giant fibre system, with one fibre to each arm (p. 415). This is a very active mesopelagic squid and must take its prey in a special manner. No doubt there are many other such developments in individual genera and families.

The large eyes and long rhabdomes of deep-sea forms have often been described but it is less usually recognized that these eyes often have complicated subdivisions of the retina. We have found this in *Japetella, Cirroteuthis, Histioteuthis* and *Bathothauma* among those described here and probably also in *Vampyroteuthis*. The optic lobes show corresponding subdivisions and presumably the parts of the retina that look in different directions are used in different ways. It would be interesting to pursue this matter, especially in the light of the findings of Muntz (his chapter, this volume).

The eyes and feeding method of young cranchiids is especially mysterious (p. 424). Their stalked eyes, long tentacles and long neck must be used together, presumably to take small prey. The "appendage" of the eye of some species is supposed to be a light organ. If so it must be used in a manner quite different from that of the adults of the same species.

The importance of the photosensitive vesicles is rapidly becoming more evident as Mauro (this volume) and R. E. Young (this volume) emphasize. It is still not clear precisely how they are used, probably in different ways in distinct species. They certainly become increasingly important in animals living at greater depths, at least down to say 1000 m. Thus among octopods the epistellar bodies are smallest in *Octopus*, larger in *Eledone* and *Pteroctopus*, larger still in *Japetella* and largest of all in cirroteuthids. Among decapods they are small in *Loligo* and *Sepia*, moderate in *Spirula*, larger in the ommastrephids and very large in *Pterygioteuthis*, *Mastigoteuthis*, *Histioteuthis* and the cranchiids.

The statocysts also show great variations. In the rapidly moving squids and cuttlefishes the cavity is divided by numerous large peg-like anticristae. These probably serve to increase the sensitivity of the cupulae of the crista to rapid angular accelerations (see Budelmann, this volume). In this respect they function like the semicircular canals of vertebrates and may indeed have been evolved as part of the system for meeting competition with the fishes (see Packard, 1972). Conversely the forms with neutral buoyancy presumably turn only slowly and require a large inertial mass of fluid to give maximum sensitivity to small displacements, as Vinnikov and his colleagues have shown (Govardovskii, 1971). Reduction of the peg-like anticristae is seen in many of the deep-water species here described. It has not proceeded equally far in all and the condition in *Spirula* is especially interesting (p. 409). Some deep-sea forms have a very large posterior sac in the statocyst. It is suggested that this may be somehow concerned in the measurement of depth. Its lining carries an elaborate plexus of hair cells and nerve fibres.

Phylogenetic considerations

With all these special adaptations we can still say something about the evolutionary relationships. The special position of *Vampyroteuthis* is considered on p. 388. We have throughout continued to use the terms decapod and octopod because from the nervous system, statocyst, photosensitive vesicles and many other features, besides the number of arms, it is clear that sepioids and teuthoids are closer to each other than either are to octopods (p. 405). This similarity requires to be somehow recognized, perhaps by including them in a common suborder. I agree that sepioids and teuthoids require to be placed in separate groups, but the myopsid character and perhaps other similarities of *Sepia* and *Loligo* may be mainly due to their circumlittoral habits. In other respects *Loligo* is much closer to the ommastrephid squids than to *Sepia*.

Among sepioids the position of *Spirula* remains puzzling and the nervous system tells us little. It shows a full decapodan condition, including the presence of giant fibres. It is not easy to say how and when *Spirula* diverged from *Sepia*, or when both separated from the squids.

The nervous system of cirrate octopods shows several primitive features and merits further study. The supraoesophageal lobes are very different from all other octopods and may retain primitive features. The finding of giant fibres is quite unexpected. It suggests that cirroteuthids show an early stage of divergence of octopods from decapods, perhaps not very long after the stage represented by *Vampyroteuthis*. The presence of several anticristae in the statocyst agrees with this and the absence of a chiasma of the optic nerve fibres seems an exceedingly primitive feature, found only in *Nautilus*.

In summary, it is clear from this brief survey that the sense organs and nervous system can tell us a great deal about the way of life of various cephalopods. We badly need further information about habits from data obtained at sea. These are of course very difficult to obtain. We are gradually learning how to piece the information together and to correlate it with studies of the structure and organization of the brain. Such knowledge, gained inductively, can be used, with caution, for deductions about the probable habits of species not yet investigated at sea. At least it can give the field observer some suggestions of what to look for when he is fortunate enough to observe these oceanic and deep-sea cephalopods alive.

References

Akimushkin, I. I. (1963). [*Cephalopods of the seas of the USSR.*] Izdatel 'stvo Akademii Nauk SSR, Moskva. [Israel Program for Scientific Translations, Jerusalem 1965.]

Aldred, R. G. (1974). Structure, growth and distribution of the squid *Bathothauma lyromma* Chun. *J. mar. biol. Ass. U.K.* **54**: 995–1006.
Aldrich, F. A. & Brown, E. L. (1967). The giant squid in Newfoundland. *Newfoundld Quart.* **65**: 4–8.
Altman, J. S. & Nixon, M. (1970). Use of the beaks and radula by *Octopus vulgaris* in feeding. *J. Zool., Lond.* **161**: 25–38.
Arnold, J. M. & Arnold, K. O. (1969). Some aspects of hole-boring predation by *Octopus vulgaris*. *Am. Zool.* **9**: 991–996.
Barber, V. C. (1968). The structure of mollusc statocysts, with particular reference to cephalopods. *Symp. zool. Soc., Lond.* No. 23: 37–62.
Baumann, F., Mauro, A., Milecchia, R., Nightingale, S. & Young, J. Z. (1970). The extra-ocular light receptors of the squids *Todarodes* and *Illex*. *Brain Res.* **21**: 275–279.
Boycott, B. B. (1953). The chromatophore system of cephalopods. *Proc. Linn. Soc. Lond.* **164**: 235–240.
Boycott, B. B. (1961). The functional organization of the brain of the cuttlefish *Sepia officinalis*. *Proc. R. Soc.* (B) **153**: 503–534.
Church, R. (1971). "Deepstar" explores the ocean floor. *Nat. geogr.* Mag. **139**: 110–129.
Clarke, M. R. (1962). Respiratory and swimming movements in the cephalopod *Cranchia scabra*. *Nature, Lond.* **196**: 351–352.
Clarke, M. R. (1966). A review of the systematics and ecology of oceanic squids. *Adv. mar. Biol.* **4**: 91–300.
Clarke, M. R. (1969). Cephalopoda collected on the SOND cruise. *J. mar. biol. Ass. U.K.* **49**: 961–976.
Clarke, M. R. & Lu, C. C. (1974). Vertical distribution of cephalopods at 30°N, 23°W in the North Atlantic. *J. mar. biol. Ass. U.K.* **54**: 969–984.
Clarke, M. R. & Lu, C. C. (1975). Vertical distribution of cephalopods at 18°N, 25°W in the North Atlantic. *J. mar. biol. Ass. U.K.* **55**: 165–182.
Denton, E. J. (1974). Croonian lecture, 1973. On buoyancy and the lives of modern and fossil cephalopods. *Proc. R. Soc.* (B) **185**: 273–299.
Denton, E. J. & Gilpin-Brown, J. B. (1973). Flotation mechanisms in modern and fossil cephalopods. *Adv. mar. Biol.* **11**: 197–268.
Denton, E. J., Gilpin-Brown, J. B. & Howarth, J. V. (1967). On the buoyancy of *Spirula spirula*. *J. mar. biol. Ass. U.K.* **47**: 181–191.
Denton, E. J. & Warren, F. J. (1968). Eyes of the Histioteuthidae. *Nature, Lond.* **219**: 400–401.
Dilley, P. N. (1972). *Taonius megalops*, a squid that rolls up into a ball. *Nature, Lond.* **237**: 403–404.
Dilly, P. N. & Nixon, M. (1976). Growth of *Taonius megalops* (Mollusca, Cephalopoda). *J. Zool., Lond.* **179**: 19–83.
Govardovskii, V. I. (1971). Some properties and the dynamics of endolymph fluid in the statocysts of cephalopods. *Zh. evol. Biochem. Physiol.* **7**: 410–416.
Hobbs, M. J. & Young, J. Z. (1973). A cephalopod cerebellum. *Brain Res.* **55**: 424–430.
Holmes, W. (1940). The colour changes and colour patterns of *Sepia officinalis* L. *Proc. zool. Soc. Lond.* (A) **110**: 17–36.
Jones, E. C. (1963). *Tremoctopus violaceus* uses *Physalia* tentacles as weapons. *Science, N.Y.* **139**: 764–766.
Kerr, J. G. (1931). Notes upon the Dana specimens of *Spirula* and upon certain problems of cephalopod morphology. *Dana Rep.* No. 8: 1–34.

Mangold-Wirz, K. (1963). Biologie des céphalopodes benthiques et nectoniques de la Mer Catalane. *Vie Milieu*, (Suppl.) No. 13: 1–285.
Marshall, N. B. & Bourne, D. W. (1964). A photographic survey of benthic fishes in the Red Sea and Gulf of Aden, with observations on their population density, diversity and habits. *Bull. Mus. comp. Zool., Harv.* **132**: 223–244.
Messenger, J. B. (1968). The visual attack of the cuttlefish, *Sepia officinalis*. *Anim. Behav.* **16**: 342–367.
Munk, O. & Frederiksen, R. D. (1974). On the function of aphakic apertures in teleosts. *Vidensk. Meddr dansk naturh. Foren.* **137**: 65–94.
Murray, J. & Hjort, J. (1912). *The depths of the ocean*. London: Macmillan.
Packard, A. (1972). Cephalopods and fish: the limits of convergence. *Biol. Rev.* **47**: 241–307.
Pickford, G. (1940). The *Vampyromorpha*, living-fossil Cephalopoda. *Trans. N.Y. Acad. Sci.* (2) **2**: 169–181.
Pilson, M. E. Q. & Taylor, P. B. (1961). Hole drilling by *Octopus*. *Science, N.Y.* **134**: 1366–1368.
Rancurel, P. (1971). *Mastigoteuthis grimaldi* (Joubin, 1895). Chiroteuthidae peu connu de l'Atlantique tropical (Cephalopoda-Oegopsida). *Cah. ORSTOM* (Océanogr.). **9**: 125–145.
Robson, G. C. (1929 & 1932). *A monograph of the recent Cephalopoda*. I & II. London: Printed by order of the Trustees of the British Museum.
Robson, G. C. (1930). Cephalopoda, I. Octopoda. *Discovery Rep.* **2**: 373–401.
Roe, H. S. J. (1969). The food and feeding habits of the sperm whales (*Physeter catodon* L.) taken off the west coast of Iceland. *J. Cons. perm. int. Explor. Mer* **33**: 93–102.
Roper, C. F. E. & Brundage, W. L. (1972). Cirrate octopods with associated deep-sea organisms: new biological data based on deep benthic photographs (Cephalopoda). *Smithson. Contr. Zool.* No. 121: 1–46.
Sanders, F. K. & Young, J. Z. (1940). Learning and other functions of the higher nervous centres of *Sepia*. *J. Neurophysiol.* **3**: 501–526.
Scott, T. (1910). Notes on crustacea found in the gizzard of a deep-sea cephalopod. *Ann. Mag. nat. Hist.* (8) **5**: 51–54.
Stephens, P. R. (1971). Histological methods. In *The anatomy of the nervous system of* Octopus vulgaris. Young, J. Z. Oxford: Clarendon Press.
Stephens, P. R. & Young, J. Z. (1975). The statocysts of various cephalopods. *J. Physiol., Lond.* **249**: 1*P*.
Verrill, A. E. (1881). Report on the cephalopods, and on some additional species dredged by the U.S. Fish Commission Steamer "Fish-Hawk", during the season of 1880. *Bull. Mus. comp. Zool. Harv.* **8**: 99–116.
Voss, G. L. & Williamson, G. (1972). *Cephalopods of Hong Kong*. Hong Kong: Government Press.
Wells, M. J. & Young, J. Z. (1972). The median inferior frontal lobe and touch learning in the octopus. *J. exp. Biol.* **56**: 381–402.
Wirz, K. (1959). Étude biométrique du systéme nerveux des Céphalopodes. *Bull. Biol. Fr. Belge* **93**: 78–117.
Young, J. Z. (1929). Sopra un nuovo organe dei cefalopodi. *Boll. Soc. ital. Biol. sper.* **4**: 1022–1024.
Young, J. Z. (1936). The giant nerve fibres and epistellar body of cephalopods. *Q. Jl microsc. Sci.* **78**: 367–386.
Young, J. Z. (1939). Fused neurons and synaptic contacts in the giant nerve fibres of cephalopods. *Phil. Trans. R. Soc.* (B) **229**: 465–503.

Young, J. Z. (1960). Observations on *Argonauta* and especially its method of feeding. *Proc. zool. Soc. Lond.* **133**: 471–479.

Young, J. Z. (1963). The number and sizes of nerve cells in *Octopus. Proc. zool. Soc. Lond.* **140**: 229–254.

Young, J. Z. (1965). The central nervous system of *Nautilus. Phil. Trans. R. Soc.* (B) **249**: 1–25.

Young, J. Z. (1970). The stalked eyes of *Bathothauma* (Mollusca, Cephalopoda). *J. Zool., Lond.* **162**: 437–447.

Young, J. Z. (1971). *The anatomy of the nervous system of* Octopus vulgaris. Oxford: Clarendon Press.

Young, J. Z. (1974). The central nervous system (B) *Loligo*. I. The optic lobe. *Phil. Trans. R. Soc.* (B) **267**: 263–300.

Young, J. Z. (1976). The nervous system of *Loligo*. II. Suboesophageal centres. *Phil. Trans. R. Soc.* (B) **274**: 101–167.

Young, J. Z. (1977). The central nervous system of *Loligo*. III. Higher motor centres. The basal supraoesophageal lobes. *Phil. Trans. R. Soc.* (B) **276**: 351–398.

Young, J. Z. & Stephens, P. R. (1976). The statocysts of *Vampyroteuthis infernalis* (Mollusca: Cephalopoda). *J. Zool., Lond.* **180**: 565–588.

Young, R. E. (1967). Homology of retractile filaments of vampire squid. *Science, N.Y.* **156**: 1633–1634.

Young, R. E. (1972). The systematics and areal distribution of pelagic cephalopods from the seas off Southern California. *Smithson. Contr. Zool.* No. 97: 1–159.

Young, R. E. (1975). Function of the dimorphic eyes in the midwater squid. *Histioteuthis dofleini. Pacif. Sci.* **29**: 211–218.

Abbreviations Used in Figures

anticr., anticrista
anticr.d., dorsal anticrista
anticr.l., lateral anticrista
anticr.med., median anticrista
anticr.ven., ventral anticrista
app., appendage of eye of *Bathothauma*
art.ceph., cephalic artery
b.a., anterior basal lobe
b.d., dorsal basal lobe
b.med., median basal lobe
be., beak
bet., region without rhabdomes between peripheral and central retina
br., brachial lobe
buc.i., inferior buccal lobe
buc.int., interbuccal connective
buc.p., posterior buccal lobe
buc.pai., paired buccal lobes
buc.s., superior buccal lobe
c.br.int., interbrachial commissure
c.br.sup., suprabrachial commissure
caec., caecum
cav., cavity for buoyancy
ce.gi.1, giant cell 1
ce.gi.a.1, anterior giant cell 1
ce.gi.a.2, anterior giant cell 2
cer.-br. con., cerebro-brachial connective
cer.-subrad.con., cerebro-subradular connective
cir., cirrus
cr.long., longitudinal crista
cr.trans.a., anterior transverse crista
cr.trans.p., posterior transverse crista
cr.v., vertical crista
end., endolymph
epist., epistellar body
ey.cb.a., anterior chamber of eye
f.gi., giant fibre
f.gi.1, giant fibre 1
f.gi.2a, giant fibre 2a
f.gi.a.1, anterior giant fibre 1
f.gi.a.2, anterior giant fibre 2
f.gi.bri., bridge of giant fibres
fil., filament
fr.i., inferior frontal lobe
fr.i.med., median inferior frontal lobe
fr.s., superior frontal lobe
g.opt., optic gland

gan.subrad., subradular ganglion
giz., gizzard
h_{1-5}, hamuli of crista (1–5)
juxt., juxtaganglionic tissue
Kol., Kölliker's canal
mac., macula
mac.neg.d., dorsal macula neglecta
mac.neg.ven., ventral macula neglecta
mac. prin., macula princeps
mag., magnocellular lobe
mag.d., dorsal magnocellular lobe
mag.p., posterior magnocellular lobe
mag.ven., ventral magnocellular lobe
mus.be., muscles of beak
mus.mant., mantle muscles
mus.rad.su., radula support muscles
n.br. 1–5, brachial nerves
n.br.-mag., brachio-magnocellular nerve
n.col., collar nerve
n.cr., crista nerve
n.fil., filament nerve
n.fin., fin nerve
n.fun., funnel nerve
n.fun.a., anterior funnel nerve
n.mac., macula nerve
n.opt., optic nerves
n.orb.po., postorbital nerve
n.pal., pallial nerve
n.retr.h.a., anterior head retractor nerve
n.st., static nerve
n.symp., sympathetic nerve
n.visc., visceral nerve
oes., oesophagus
ol., olfactory lobe
opt., optic lobe
opt.d., dorsal optic lobe
opt.ven., ventral optic lobe
or.li.com., composite light organ
or.li.fin, fin light organ
pe., pedal lobe
pe.p., posterior pedal lobe
ped., peduncle lobe
peril., perilymph
prec., precommissural lobe
pv., palliovisceral lobe
rad., radula
ren.app., renal appendages
ret.cen., central retina
ret.d., dorsal retina
ret.pf., peripheral retina
ret.ven., ventral retina
sac.p., posterior sac
sal.duc.a., duct of anterior salivary glands
sal.duc.p., posterior salivary duct
sal.pap., salivary papilla
sk., skin
st., statocyst
sth., statolith
subfr., subfrontal lobes
subv., subvertical lobe
suc., sucker
tr.opt., optic tract
v., vertical lobe
vent., ventricle
ves.pho., photosensitive vesicles
ves.pho.d., dorsal photosensitive vesicles
ves.pho.ven., ventral photosensitive vesicles

ACKNOWLEDGEMENTS

It is a very great pleasure to thank those who over many years have helped with this work. I am especially indebted to Miss P. R. Stephens who has prepared many of the sections illustrated here. She has expended very great energy and skill in preparing long series of sections of many species, which have benefited not only our own researches but those of cephalopod workers all over the world. I am also most grateful to Dr M. Nixon for continual help in the preparation of this and other papers. Much of our material has been collected at the Marine Biological Station at Plymouth and the Stazione Zoologica at Naples. We have to thank their Directors and Staff for unfailing support.

Symp. zool. Soc. Lond. (1977) No. 38, 435–445.

MULTIPHASIC RETENTION PERFORMANCE CURVES: FEAR OR MEMORY?

G. D. SANDERS

Psychology Department, City of London Polytechnic, London, England

SYNOPSIS

Typical negatively accelerated monotonic retention curves have been reported for cephalopods, but recently some investigators have found retention performances to be a multiphasic function of time. These curves have been interpreted as showing transitions between temporally different memory processes. Similar experiments with vertebrates have led to a major alternative interpretation in terms of fear. The effects of fear and memory are confounded in retraining measures making this type of data an inadequate basis for choosing between the two alternative explanations. In the present experiment tests with a novel stimulus were employed as an independent measure of fear in an attempt to separate the effects, on retraining performance, of fear and memory.

Octopuses trained not to take a rough sphere and retrained 0·5, 2 or 4 hr later showed a retention performance deficit at 2 hr, but on a series of tests with a novel smooth sphere, which were interspersed between the retraining trials, the performance was of the same order at all three intervals. On the basis of these results it is argued that the performance deficit cannot be accounted for by a change in the level of fear but it may be attributed to a temporary reduction in the strength of the available memory.

INTRODUCTION

Multiphasic retention performance in cephalopods

The typical "forgetting curve", familiar since the turn of the century, is monotonic and negatively accelerated. Such curves have been generated by a wide variety of subjects from man to cephalopods. We should remember, however, that, as conventionally plotted, these "forgetting curves" actually show the strength of a behavioural response which is interpreted as an indication of the amount retained. The curves are, in fact, retention performance curves and, as such, may be open to a number of interpretations.

A typical monotonic retention performance curve was obtained when *Octopus vulgaris* were tested at intervals of five to 120 days after prolonged training on a tactile discrimination task (Sanders, 1970). However, if tests are given at relatively short intervals after a brief period of training, multiphasic retention performance curves may be produced. Thus while studying the habituation of attack in

Sepia officinalis, Messenger (1971) found that recovery of the response was a biphasic function of the time since the last trial. Similarly Sanders & Barlow (1971), who trained *Octopus vulgaris* not to attack a crab by giving a 4-V shock with a criterion of no attack for 1 min, found on retraining at retention intervals of 30 min to 30 hr, a triphasic retention curve with minima at 1 and 8 hr (Fig. 1a, Expt 2). In another experiment (Sanders & Barlow, unpublished data) the use of a 10-V shock with a criterion of 3 min without an attack produced a similar curve (Fig. 1b, Expt 3).

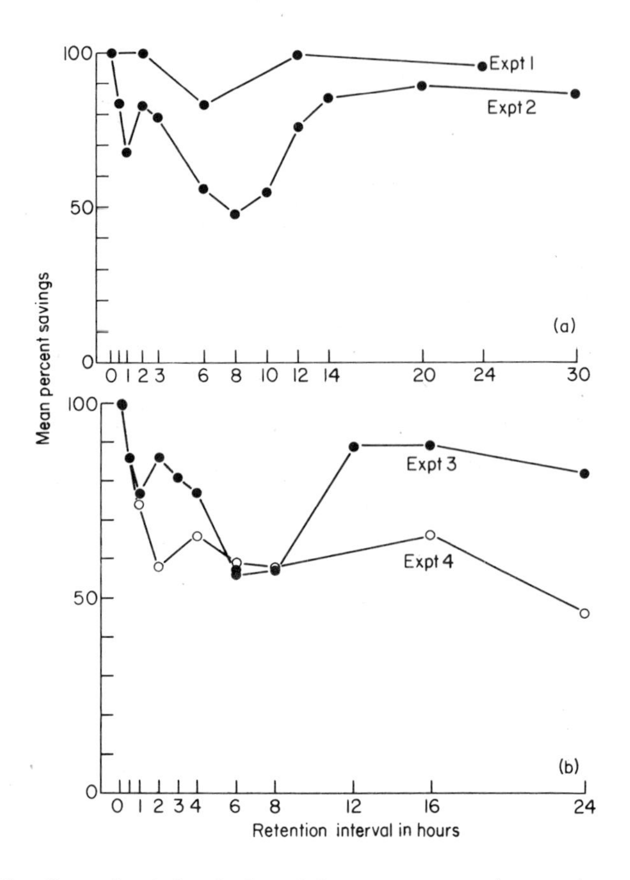

FIG. 1. The effects of variations in the training parameters on the retention performance of octopuses trained not to attack a crab (Sanders & Barlow, 1971; unpublished data). (a) Experiment 1: shock 4 V, criterion 3 min. Experiment 2: shock 4 V, criterion 1 min. (b) Experiment 3: shock 10 V, criterion 3 min. Experiment 4: shock 10 V, criterion 0·5 min.

When the time spent in training was increased by using a 4-V shock with criterion time raised from 1 to 3 min the overall retention performance improved and the minima tended to disappear (Fig. 1a, Expt 1). With a 10-V shock and the criterion time decreased from 3 min to 30 sec the overall retention performance was reduced and the curve again tended to become monotonic (Fig. 1b, Expt 4). Although more training attacks were made with the 4-V than with the 10-V shock it is noticeable that the longer criteria were attained without an increase in the number of attacks (Table I). If we compare the data from the first and second experiments and the third and fourth experiments here, we see that retention performance was improved simply by extending the time spent in the training situation without increasing the number of reinforcements (i.e. shocks) received.

TABLE I

The relationship between the overall retention performance and the training parameters

Overall level of retention performance	Shock (V)	Criterion time (min)	Mean training attacks	Mean duration of training (sec)
Expt 1				
High (Fig. 1a)	4	3	13·8	447
Expt 2				
Medium (Fig. 1a)	4	1	15·6	266
Expt 3				
Medium (Fig. 1b)	10	3	4·1	263
Expt 4				
Low (Fig. 1b)	10	0·5	4·2	80

From Sanders & Barlow, 1971 and unpublished data.

Sanders & Barlow (1971) interpreted the triphasic curve as revealing the state of the memory available for the control of behaviour. According to this hypothesis the initial decline in performance (Fig. 1, Expts 2 and 3) results from the loss of short-term memory (STM) while the peak at two hours occurs as an intermediate memory (IM) becomes available. The loss of IM and the subsequent appearance of long-term memory (LTM) produces the deficit at eight hours. The data summarized in Fig. 1 and Table I suggest that multiphasic retention performance curves may be obtainable only within a limited range of training conditions (see Messenger, his chapter, this volume). With too much training, loss from STM and IM is insufficient to produce the deficits. With too little training there is no significant improvement in performance after the initial loss from STM.

Fear and memory interpretations

Since Kamin (1957) first provided evidence that retention performance may not always be a simple monotonic function of time, many experiments, predominantly with rats, have produced biphasic retention performance curves with the maximum deficit occurring at one to six hours. While some authors have proposed explanations in terms of memory (Klein & Spear, 1969; Bintz, 1970; Spear, Klein & Riley, 1971) the majority have favoured interpretations in terms of fear.

Initially, when active avoidance training was used, the temporary deficit in performance was attributed to a high level of fear producing "freezing behaviour" which is incompatible with the required active avoidance response (Denny & Ditchman, 1962; Brush, Myer & Palmer, 1963; Brush, 1964; Anisman & Waller, 1971; Barrett, Leith & Ray, 1971). However, Pinel & Cooper (1966) reproduced the deficit using passive avoidance training suggesting that poor performance occurred when fear was low. Deficits appearing with both active and passive avoidance tasks may be explained as the result of low motivation produced by a low level of fear. Brush (1971) has suggested that a reduction in the level of fear could be produced by a "parasympathetic over-reaction" causing pituitary inhibition of ACTH release.

When aversive training techniques are used fear must always be considered as a possible influence on performance. Two types of fear may be identified. First, specific fear which becomes attached to the stimuli sampled during training. Second, non-specific fear which is produced simply by exposure to the aversive reinforcement used in training. Specific fear will disappear when the memory of the training is not available. Non-specific fear may be expected to increase to a maximum (Allen & Mitcham, 1970) and then dissipate. It can be assumed that both specific and non-specific fear would be reduced by a "parasympathetic over-reaction".

During retention tests with the original training stimulus the memory of the training experience, if available, would be evoked and the strength of this memory would determine the behavioural response. However, behaviour would also be dependent on the level of motivation produced by the specific fear attached to the original training stimulus. If, because of a "parasympathetic over-reaction" for example, the level of fear was low, motivation would be low and, hence, performance would be poor irrespective of the strength of the available memory. Thus, performance with the original training stimulus confounds fear and memory effects.

Responses to a novel stimulus (i.e. one not experienced during training) would be influenced by the level of non-specific fear only. This non-specific fear would not be affected by the strength of the available memory but could be reduced by a "parasympathetic over-reaction". Therefore, tests with the original training stimulus (retraining trials) and with a novel stimulus (specificity tests) given in parallel should permit the separation of fear and memory effects. The training procedure used by Sanders & Barlow (1971) would not be suitable for this type of experiment because of the difficulty of finding a second stimulus as attractive to the octopus as a crab. Fortunately, a rapid tactile training procedure described by Wells (1959a) proved suitable. The present experiment used a modified version of this procedure.

EXPERIMENTAL PROCEDURE

Octopus vulgaris of both sexes weighing between 150 and 400 g were used for the experiment. The octopuses were blinded by severing the optic nerves under urethane anaesthesia and then fed crabs during a brief recovery period. Pre-training was not started until the octopuses were feeding well. The two stimuli used were both perspex spheres, 2·5 cm in diameter, suspended on nylon line. The surface of one sphere was smooth while the other was roughened by the presence of 14 circumferential grooves. Throughout the experiment training and testing were carefully restricted to one arm because there is a considerable delay before information acquired by one arm is transferred to the others (Wells, 1959b).

Initially, the octopuses were pre-trained for two days to take both the smooth and the rough sphere. Four trials with each sphere were given to the second right and the second left arms making a total of 16 pre-training trials for each octopus. On any trial the appropriate sphere was touched against the suckers of the distal half of the chosen arm, whereupon, one of three things could happen. (1) The octopus passes the sphere under the interbrachial web towards the mouth. The sphere is pulled away, the appropriate reinforcement given and a take recorded. The criterion of a take was that within 20 sec of the sphere touching the suckers it should be passed under the interbrachial web. (2) The octopus does not meet the criterion of a take but holds the sphere for 20 sec. At the end of this time the trial is terminated, the sphere pulled away and a rejection recorded. No reinforcement is given. (3) The octopus pushes the sphere away, or simply releases it, within the 20-sec period. The trial terminates with the release of the sphere, no reinforcement is given and a rejection is recorded. During pre-training, takes of both spheres were reinforced by the presentation of a small (0·5 g) piece of fish.

On the day following pre-training one arm was trained not to take the rough sphere. For approximately half of the octopuses the second right arm was chosen while for the remainder the second left arm was used. The training trials were given at intervals of 3 min and takes were reinforced with a 5-V shock delivered to the mantle through a pair of electrodes mounted on a probe. Training was continued to a criterion of three successive rejections. Octopuses were allotted to one of three groups for retraining after intervals of 0·5, 2 or 4 hr. Retraining was continued for ten trials, which was more than sufficient for all the octopuses to reach criterion for a second time. Ten specificity tests with the smooth sphere were interspersed between the retraining trials, which, like the training trials, were given at intervals of 3 min. For all trials on which a take occurred the length of the trial in seconds was recorded as a measure of response latency.

The significance of performance differences between the three groups was assessed by means of the Wilcoxon Rank Sum Test.

RESULTS

The median response latency per trial is shown in Fig. 2. Pre-training data are given for the arm subsequently used in training. The higher latencies recorded for the smooth sphere early in pre-training occurred because this stimulus was presented before the rough on the first day. On the second day equivalent performances were obtained with the two spheres. Training performance is essentially the same for the three groups with a sudden marked increase in response latency on trials 8-11. The retraining performances of the 0·5- and 4-hr groups are identical and superior to that of the 2-hr group but responses to the smooth sphere are

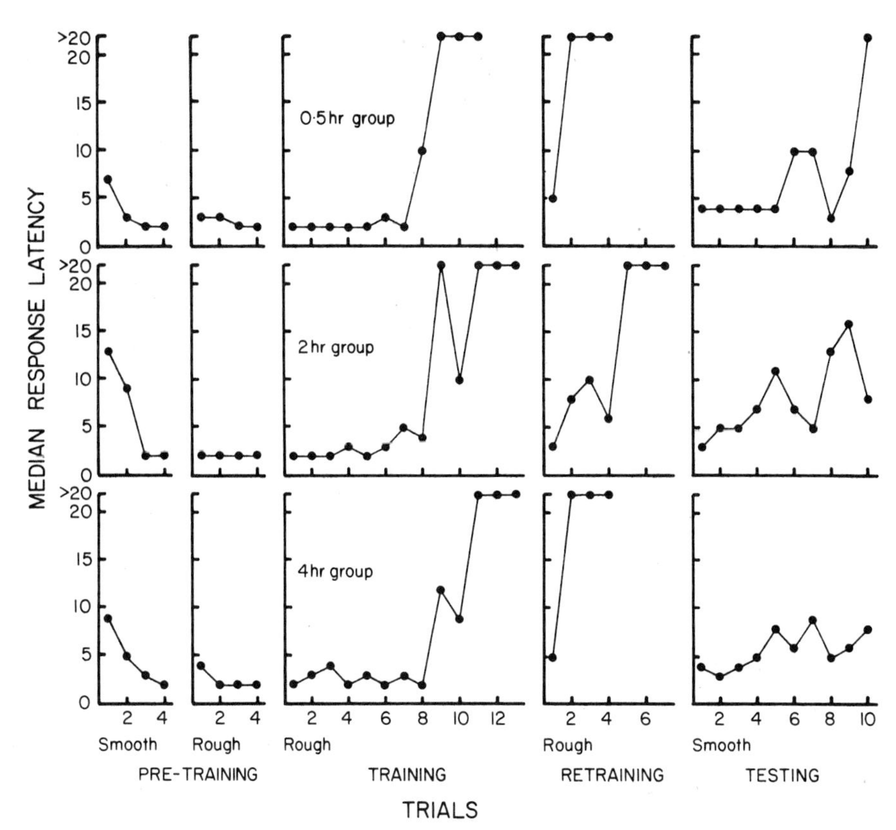

FIG. 2. Data from octopuses trained to reject a rough sphere and retrained after 0·5, 2 or 4 hr. Specificity tests with a smooth sphere were interspersed with the retraining trials. The graphs compare the performances of the three groups in terms of their median response latency per trial during pre-training, training, retraining and testing. Pre-training data is given for the single arm subsequently used throughout the experiment proper.

similar for all three groups. This general picture is confirmed by the analysis illustrated in Figs 3 and 4.

Performance during training in terms of both errors and trials to criterion are almost identical in the three groups (Fig. 3) but on retraining the 2-hr group made more errors and required more trials to reach criterion than the 0·5- and the 4-hr groups ($P < 0{\cdot}02$). The octopuses were, in fact, retrained for ten trials. Having reached criterion, some of the animals made further takes of the rough sphere on post-criterion trials. If retraining performance is measured over all ten trials (Fig. 4) the 2-hr group took the rough sphere more often than either the 0·5- or the 4-hr group ($P < 0{\cdot}02$).

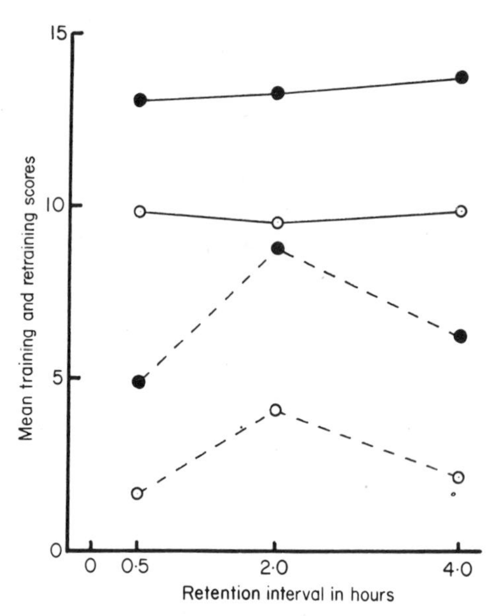

FIG. 3. Training (continuous line) and retraining (broken line) performance measured by trials (filled circles) and errors (open circles) to criterion as a function of retention interval.

Ten test trials with the smooth sphere, which was not used in training, were interspersed between the retraining trials. Performance with this novel stimulus stands in marked contrast to that obtained with the rough sphere during retraining (Fig. 4). All three groups took the smooth sphere more frequently than they took the rough ($P < 0{\cdot}01$). In addition, there were no significant differences between the groups in terms of the number of takes of the smooth made during testing. The smooth sphere was not taken on every trial, however, as it was towards the end of the pre-training: it was, in fact, rejected on approximately 25% of the test trials.

It is possible that the apparent similarity between the groups in their response to the smooth sphere is an artefact produced by the differences in retraining performance. When trials are given at intervals of a few minutes, a shock given on one trial reduces the probability of a take occurring on the next trial (Wells & Young, 1968). The 2-hr group made more errors on retraining and hence received more shocks than the other two groups. It is possible, therefore, that the smooth takes of the 2-hr group were depressed more than those of the 0·5- and 4-hr groups. Such an effect may be more than compensated for by counting as a take each rejection of

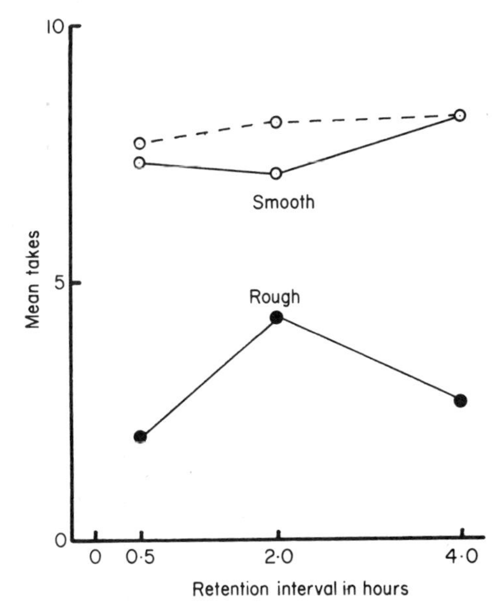

FIG. 4. Retraining with the rough sphere was extended for a total of 10 trials which alternated with 10 specificity tests using the smooth sphere. The continuous lines show the mean number of takes of the rough (filled circles) and smooth (open circles) as a function of the length of the retention interval. The broken line shows the adjusted smooth scores after an allowance has been made for the possible depressive effect of shocks received on retraining trials (see text for a full explanation).

the smooth which occurred immediately after a retraining trial on which a shock was received. Although this computation adds more takes to the 2-hr than to the 0·5- and 4-hr groups (nine as opposed to three and zero respectively) the new mean values thus obtained are more similar than the old (Fig. 4). Furthermore these relative values are not substantially altered by the inclusion of similar data from test trials more distant from the shock.

In summary, although no differences are apparent in the training performances, octopuses retrained after an interval of 2 hr make more errors than others retrained after 0·5 or 4 hr. The response to a novel smooth sphere presented between the retraining trials is lower than that recorded at the end of pre-training but it is of the same order in all three groups.

DISCUSSION AND CONCLUSIONS

As described in the introduction, performance may be worse at intermediate than at shorter and longer retention intervals. Most investigators have interpreted these temporary retention performance deficits as manifestations of changes in the strength of either fear or memory. On retraining with the original stimulus the possible effects of fear and memory are confounded. The present experiment set out to demonstrate that the effect of fear may be separated from that of fear and memory by giving specificity tests with a novel stimulus in parallel with the retraining trials. A rough sphere was used as the original training stimulus and a smooth sphere as the novel stimulus.

The smooth sphere was taken approximately 2·5 times as frequently as the rough sphere during the retraining/specificity testing stage of the experiment (Fig. 4) indicating that learning was largely specific to the original training stimulus. The response to the smooth sphere was, however, some 25% lower than that obtained at the end of pre-training suggesting that non-specific fear depressed the response level. The magnitude of this non-specific fear, as measured by the number of takes of the smooth sphere, was of the same order at 0·5, 2 and 4 hr after training. It is unlikely, therefore, that the poor retention performance at 2 hr, as compared with that recorded at 0·5 and 4 hr, is the result of lower motivation produced by a temporary reduction in the level of fear. These results could be explained by a temporary memory deficit occurring at the intermediate interval. To this extent, these findings are in agreement with the memory interpretation of multiphasic retention performance curves proposed by Sanders & Barlow (1971).

References

Allen, J. D. & Mitcham, J. C. (1970). The time course of fear incubation following single-trial passive-avoidance training. *Psychonom. Sci.* **20**: 169–170.

Anisman, H. & Waller, T. G. (1971). Effects of conflicting response requirements and shock compartment confinement on the Kamin effect in rats. *J. comp. physiol. Psychol.* **77**: 240–244.

Barrett, R. J., Leith, N. J. & Ray, O. S. (1971). Kamin effect in rats: Indice of memory or shock induced inhibition? *J. comp. physiol. Psychol.* **77**: 234–239.

Bintz, J. (1970). Time-dependent memory deficits of adversively motivated behavior. *Learn. Motiv.* **1**: 382–390.

Brush, F. R. (1964). Avoidance learning as a function of time after fear conditioning and unsignalled shock. *Psychon. Sci.* **1**: 405–406.

Brush, F. R. (1971). Retention of aversively motivated behavior. In *Aversive conditioning and learning.* **7**: 401-465. Brush, F. R. (ed.). London and New York: Academic Press.

Brush, F. R., Myer, J. S. & Palmer, M. E. (1963). Effects of kind of prior training and intersession interval upon subsequent avoidance learning. *J. comp. physiol. Psychol.* **56**: 539–545.
Denny, M. R. & Ditchman, R. E. (1962). The locus of the maximal "Kamin Effect" in rats. *J. comp. physiol. Psychol.* **55**: 1069–1070.
Kamin, L. J. (1957). The retention of an incompletely learned avoidance response. *J. comp. physiol. Psychol.* **50**: 457–460.
Klein, S. B. & Spear, N. E. (1969). Influence of age on short-term retention of active avoidance learning in rats. *J. comp. physiol. Psychol.* **69**: 583–589.
Messenger, J. B. (1971). Two-stage recovery of a response in *Sepia. Nature, Lond.* **232**: 202–203.
Pinel, J. P. J. & Cooper, R. M. (1966). Demonstration of the "Kamin Effect" after one-trial avoidance learning. *Psychon. Sci.* **4**: 17–18.
Sanders, G. D. (1970). Long-term memory of a tactile discrimination in *Octopus vulgaris* and the effect of vertical lobe removal. *Brain Res.* **20**: 59–73.
Sanders, G. D. & Barlow, J. J. (1971). Variations in retention performance during long-term memory formation. *Nature, Lond.* **232**: 203–204.
Spear, N. E., Klein, S. B. & Riley, E. P. (1971). The Kamin Effect as "state-dependent learning": memory retrieval failure in the rat. *J. comp. physiol. Psychol.* **74**: 416–425.
Wells, M. J. (1959a). A touch learning centre in *Octopus. J. exp. Biol.* **36**: 590–612.
Wells, M. J. (1959b). Functional evidence for neuronal fields representing the individual arms within the central nervous system of *Octopus. J. exp. Biol.* **36**: 501–511.
Wells, M. J. & Young, J. Z. (1968). Learning with delayed rewards in *Octopus. Z. vergl. Physiol.* **61**: 103–128.

Symp. zool. Soc. Lond. (1977) No. 38, 447–511.

SUCKER SURFACES AND PREY CAPTURE

MARION NIXON* and P. N. DILLY**

Anatomy Department, University College London, London, England

SYNOPSIS

We know much from behavioural studies about the capture of prey by *Sepia officinalis*; this information together with that obtained from an examination of the surface features, with the scanning electron microscope, formed a basis for a study of the surface architecture of suckers of other coleoids with a view to making predictions about their largely unknown feeding habits.

One of the most remarkable features of the "hard" cuticular tissues of the sucker infundibulum is the array of microscopic pores found on the surface, particularly noticeable on the many projecting pegs. In both decapods and octopods the diameter of these pores is between 0·04 μm and 0·10 μm, irrespective of the size of the animal. These openings are, almost certainly, the result of the aggregation of rods arranged in such a way as to form pores on the surface leading to spaces continuing some distance into the interior of the pegs. A theory is developed that these pegs and their pores are involved in the mechanical attachment of the suckers to prey, making intimate contact by means of impact, adhesion, seal and suction.

The growth of the sucker infundibulum takes place by additions at the periphery.

INTRODUCTION

Aristotle recognized that while sepioids, teuthoids and octopods were all swimmers, the octopods walked as well (see D'Arcy Thompson, 1910) but no detailed study was made of the arms and tentacles or their suckers until d'Orbigny (1845) examined the suckers of living cephalopods. Girod (1884) made a histological study of the suckers of *Octopus vulgaris* and *Sepia officinalis*, followed by Niemec (1885) who made a survey of suckers through the animal kingdom. Guérin (1908) gave an excellent historical introduction to his extensive work on the anatomy and histology of the arms, tentacles and suckers of decapods and octopods. Naef (1921, 1923) considered the function and the evolution of the suckers. He recognized two main forms, the stalked, radially or bilaterally symmetrical ones of decapods, and the sessile, radially symmetrical ones of octopods.

During the past 25 years much information about *Octopus vulgaris* has accumulated, especially about its nervous system, learning capacity, chemotactile system, food and feeding habits,

Present addresses: * *The Wellcome Institute for the History of Medicine, London* and ** *Department of Structural Biology, St George's Hospital Medical School, Tooting, London.*

and the receptors in the suckers (see pp. 479–481). We also have experimental analyses of the attack system of *Sepia officinalis*, a study of its nervous system, as well as histological studies of the receptors of the suckers (see p. 452).

In the light of these studies it seemed worthwhile examining the surface of the suckers to see whether they presented features that could be correlated with the known habits of their owners. Examination of the suckers of other cephalopods and comparing the features found with those of the suckers of *Octopus vulgaris* and *Sepia officinalis* might allow us to make some predictions about prey and prey capture in animals so far not observed alive either in the sea or in an aquarium.

GENERAL MORPHOLOGY OF SUCKERS

Suckers are found in all the living coleoids except *Nautilus*, and the three main forms are shown in Figs 1–5. The suckers of sepioids

---

General morphology of suckers

FIG. 1. Photomicrograph of a section cut through a sucker of an oegopsid squid, *Taonius megalops*. The pegs on the infundibulum are easily seen as are the underlying tall, columnar epithelial cells. The inner ring has teeth on its exposed surface, and does not have the same staining properties as the cuticle of the infundibulum. (Cajal.) Abbreviations Figs 1–6: ac., acetabulum; hil., hillock; in.r., inner ring; inf., infundibulum; ped., peduncle; sph., sphincter.

FIG. 2. A sagittal section through a sucker of *Vampyroteuthis infernalis* (the only member of the Vampyromorpha), revealing its relatively simple form in comparison with those of the decapods or the octopods. The sucker has deep radial muscles around the acetabulum. The rather ill-defined peduncle is vacuolated (and see Fig. 36). The suckers have radial symmetry (see Fig. 35). (Cajal.)

FIG. 3. A sagittal section through a small sucker of an octopod, *Octopus vulgaris*. It is very muscular and a sphincter separates the acetabulum from the infundibulum. The cuticular pegs on the infundibular surface are just visible. (Cajal.)

FIG. 4. Scanning electron micrograph of the surface of a sucker from an arm of another oegopsid squid, *Mastigoteuthis* sp. The inner ring has six, long, cone-like teeth (t) that point towards the mouth; those on the remaining part of the ring are wedge-shaped. The cuticular surface of the infundibulum consists of polygonal processes, the majority having a peg projecting from the surface. These processes are bounded by a circle of processes lacking pegs. The size and form of the processes and the pegs vary, those closest to the inner ring being largest although even these have different proportions dependent upon their position on the infundibulum. The pegs close to the pointed teeth of the inner ring are tall and slender while those opposite are much stouter. (Field width 123 μm.)

FIG. 5. The surface of a small sucker of *O. vulgaris*. The rim is a prominent feature with small hillocks superimposed on many of the radially arranged cushions. The infundibulum can be seen below covered with small pegs. (Field width 500 μm.)

FIG. 6. A photomicrograph of a sucker of *O. vulgaris* to show the tall, columnar epithelial cells, with large, dark-staining, elongated nuclei, that secrete the cuticular pegs, each one being formed by the activity of one cell. (Cajal.)

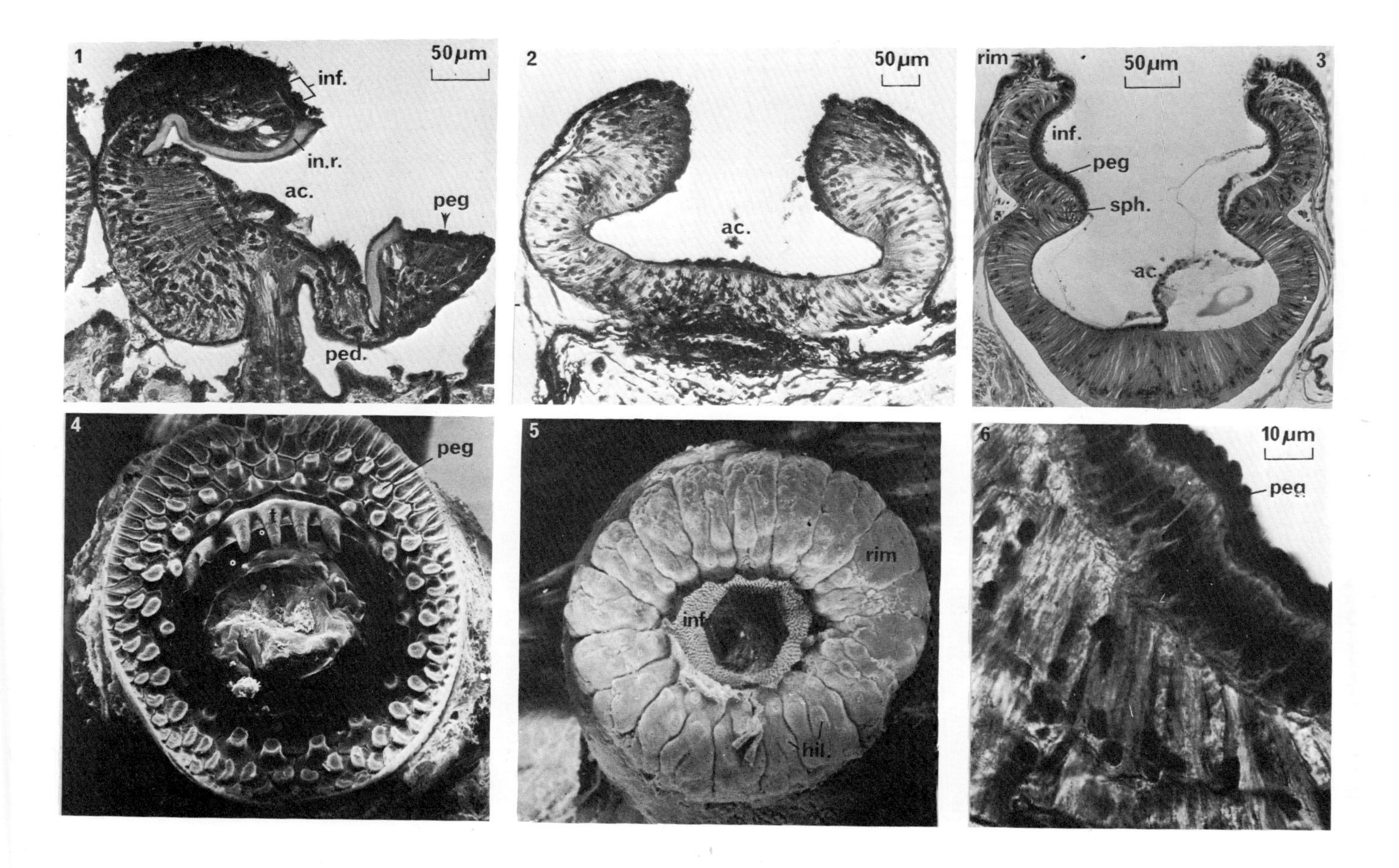
1
50 µm
inf.
in.r.
ac.
peg
ped.
2
50 µm
ac.
rim
50 µm
3
inf.
peg
sph.
ac.
4
peg
5
rim
inf.
hil.
6
10 µm
peg

and teuthoids (decapods) are attached to the arms by a peduncle, pedicle or stalk (Fig. 1). The sucker has a chamber, the acetabulum (ac.) vinegar cup lined by an inner ring (in.r.) which endows it with a firm wall (Fig. 1). This ring although often called chitinous was found not to contain chitin (Rudall, 1955) but it is well organized, secreted presumably by the underlying epithelial cells (Fig. 6). Usually, at the base of the acetabulum, there are powerful radial muscles presumably for lowering the floor of the acetabulum to produce suction. Above the acetabulum, and surrounding the exposed surface of the inner ring, is the infundibulum (inf., funnel). The infundibulum is covered with cuticular, polygonal processes, each with a peg (Figs 1 and 4), secreted by the underlying tall columnar epithelial cells (Fig. 1). The polygonal processes are formed of aggregations of rods, with spaces in between that open at the surface as innumerable minute pores (p. 483; Fig. 49). We suggest that these pores may constitute a mechanism for the adhesion of the sucker to a foreign surface.

The suckers of the deep-sea relic form *Vampyroteuthis* are simple cups, closely attached on the lower surface to the arms (Fig. 2), and perhaps are an indication of an early form of coleoid sucker (p. 473). The acetabulum has a very thin lining and there does not appear to be any distinct sphincter as in octopod suckers. The straight "neck" of the sucker is perhaps analogous to the infundibulum of the decapod and octopod suckers; it has a dense-staining epithelial covering but does not have any of the ornamentation seen in the other suckers.

The suckers of decapods show considerable range of form, the most marked being seen in some families, such as the onychoteuthids and enoploteuthids, where instead of suckers there may be single, strongly curved teeth or claws; this was certainly one of the early forms of attachment mechanisms found amongst cephalopods (see Donovan's chapter, this volume). The variations found may be for capturing different types of prey, the claws for soft-bodied animals and suckers for hard-bodied ones such as crustaceans (Naef, 1921, 1923). The suckers of octopods (Figs 3 and 5), at least superficially, show less variation than do those of the decapods.

We have investigated the suckers of a small number of species, taking examples from each of the four groups of coleoid cephalopods (Table I). For each animal examined we have reviewed the information available about distribution, depth, habits, food, physiology and behaviour, as well as the morphology of the arm, tentacle and suckers.

TABLE I

The animals used in this study, their size, place of capture and the fixatives used

		Mantle length (mm)	Fixative	Area of capture
Sepioidea				
Spirulidae	*Spirula spirula*	31, 32	Formalin	Fuerteventura
Sepiidae	*Sepia officinalis*	8, 35	Bouin's, Formalin	Naples, Plymouth
Teuthoidea				
Myopsida	*Alloteuthis subulata*	50	D404 Mucolex[th]	Plymouth
Oegopsida	*Mastigoteuthis* sp.		Formalin	North Atlantic
	Taonius megalops	4, 34, 46	Formalin	North Atlantic
Vampyromorpha	*Vampyroteuthis infernalis*		70% Alcohol	Mid-Atlantic
Octopoda				
Cirrata	*Cirroteuthis* sp.	68	Formalin	Bay of Biscay
Incirrata	*Octopus vulgaris*	78, 84	Formalin	Banyuls
	Eledone moschata		Formalin	Banyuls
	Octopus salutii	73	Formalin	Banyuls
	Scaeurgus unicirrhus	60	Formalin	Banyuls
	Japetella diaphana	26	Formalin	—
	Argonauta argo	32	Formalin	Naples

ANIMALS AND METHODS

Table I gives details of the fixation fluid, place of capture and mantle length of the animals.

For the scanning electron microscopy the specimens were first washed and then taken through graded ethanol, ethanol and "Freon 113" mixtures, and, finally pure "Freon 113" before they were critical point dried (Boyde, 1974). The specimens were then attached to aluminium rivets, with either transparent adhesive (UHU) or with double-sided sellotape, before being given evaporated electrically conductive coatings of carbon and then gold. The remaining exposed part of the rivet was painted with colloidal silver. The specimens were then examined with a Cambridge Stereoscan S4-10.

The histological material has come largely from the collection of Professor J. Z. Young. The specimens had been fixed in 10%

neutral formalin in sea-water and treated in one of the following ways: (a) Cejal's silver, block-staining method (Stephens, 1971); (b) Holmes' silver stain; (c) haematoxylin and eosin; (d) Azan; (e) Alcian blue.

The actions of suckers were observed after severing the arms from *Eledone moschata* and the tentacles from *Sepia officinalis* under anaesthesia (10% alcohol in sea-water).

SEPIOIDEA

Sepia officinalis

Distribution and depth

This cuttlefish lives in the coastal regions of Europe and South Africa usually in depths of 0–250 m (Mangold-Wirz, 1963; Adam & Rees, 1966).

Habits and food

During the day it lives partially buried in sand or gravel, relying for concealment on chromatophores and iridophores (Holmes, 1940). In the hours of darkness its buoyancy changes and much of this period is spent swimming (Denton & Gilpin-Brown, 1960, 1961).

The attack on fast-moving prey such as prawns or fish is an entirely visual response (Sanders & Young, 1940; Messenger, 1968, 1973, and his chapter, this volume). The cuttlefish first fixates the prawn, then moves so that the prey lies straight ahead, and finally goes forward or backwards until the "attacking distance" is reached (Messenger, 1968). At this point the tentacles are shot out to strike the prey, in less than 30 msec. Crabs may also be taken with the tentacles in the same way as prawns though generally the arms alone are used. Newly hatched *S. officinalis* will attack and capture mysids with the tentacles in the same way as adults (Wells, 1958, 1962; Boletzky, 1974).

The efficiency of the co-ordination of the visual system (Muntz, his chapter, this volume), the central nervous system (Boycott, 1961), and the peripheral nervous system (Graziadei, 1959, 1964a), is matched by that of the suckers of the tentacular club in adhering to a prawn during and after its capture.

Physiology and behaviour

Messenger (1968) tested experimentally-blinded animals with extracts from freshly killed prawns or crabs, but was unable to elicit any response to these substances even when they were pipetted within a few mm of the arms. However, freshly killed prawns were taken when they were presented again but to the oral surface of the arms. The effects of other chemical substances have not been tested in *S. officinalis* as they have in *Octopus vulgaris* (see p. 480).

The arms and tentacles

A brief description of the external features of the arms and tentacles and their suckers was given by Tompsett (1939). Girod (1884) investigated the histology of the arms, tentacles and suckers and found all to have a dense and complicated musculature. More recently the receptors of the suckers were examined by Graziadei (1959, 1964a, 1965). He found multipolar nerve cells at the base of the epithelial cells of the infundibulum, some 300 in a sucker 3 mm in diameter (Graziadei, 1964a). Around the outside of the infundibulum there are about 600 flask-shaped cells, and about 100 tapered cells ending with a tuft of short cilia at their free surface. At the bottom of the acetabulum, the two latter types of receptor cell were found, but in smaller numbers, as well as multipolar cells different in form from those found in the infundibulum. It has been proposed that multipolar cells are probably tension receptors, the flask-shaped cells touch receptors, and the tapered cells chemoreceptors.

The arm suckers

Part of a sucker from an arm of a 12-day-old animal, 8·5 mm dorsal mantle length, is shown in Fig. 7. The inner ring has wide but blunt projections on one half and a few conical ones on the remaining part.

The infundibulum has rings of polygonal processes each with a projecting peg. The size of each peg is almost equal to that of its base, and there is little space between the pegs (Fig. 7). The upper surface of the pegs is slightly concave and somewhat rough. The pegs and the polygonal processes are covered with small pores.

At the periphery of the sucker, processes without pegs are arranged radially. In Fig. 7 three of these flat processes are migrating inwards while the distal region of each appears to be developing into a peg (1, 2 and 3). We have here a process of formation evident also in other cephalopods examined and particularly easy to see in the suckers of young animals (see p. 471). On the outside of the sucker, and slightly raised from the surrounding tissue, are pores with a number of cilia protruding some 0·5–2·5 μm from the surface (arrows, Fig. 7).

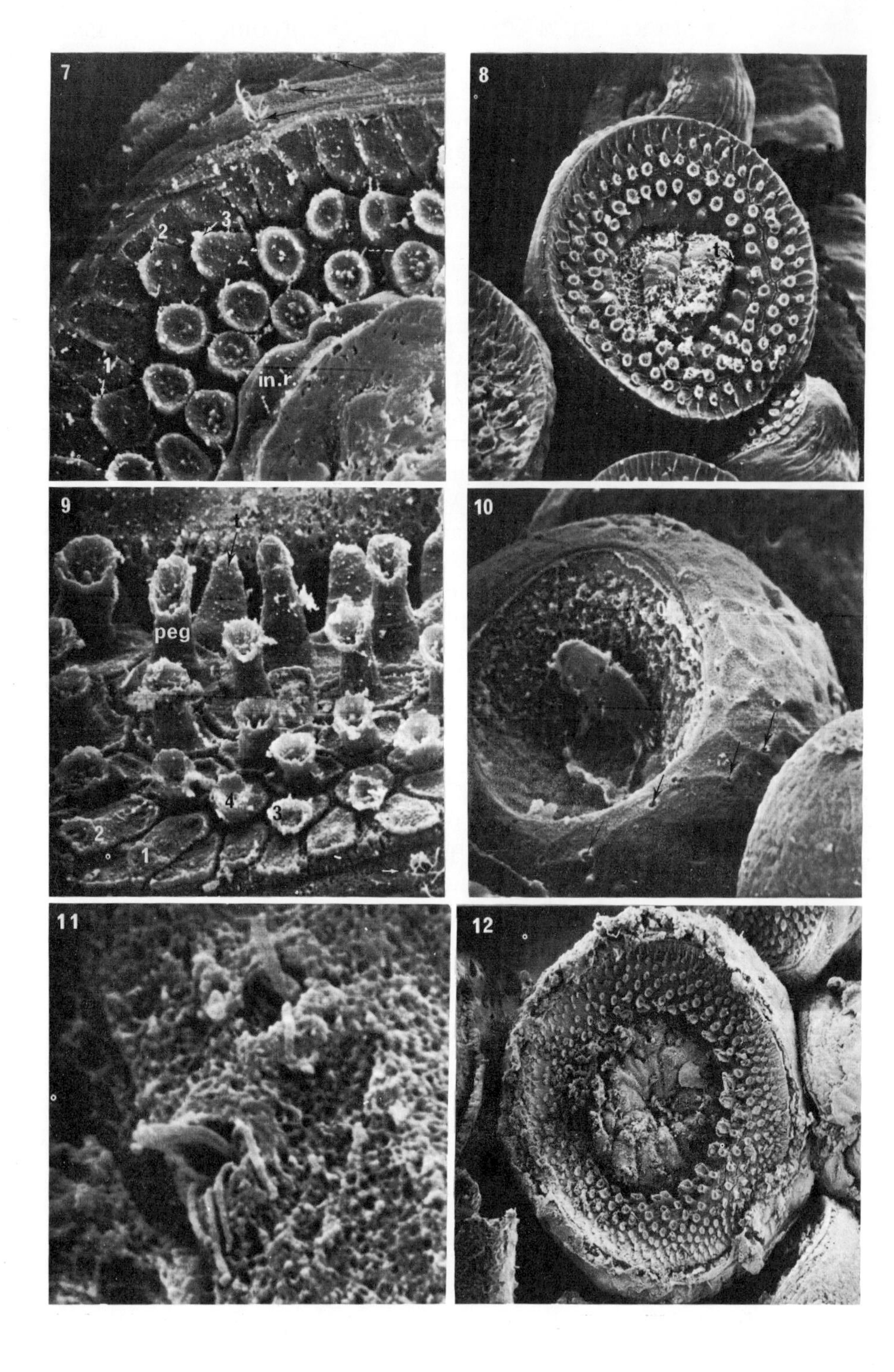
7
2
3
1
in.r.
8
t
9
t
peg
4
3
2
1
10
11
12

The tentacular suckers

The club of the tentacle of the 12-day-old animal was well-developed and had six longitudinal rows of suckers, some of which were larger than the others. The inner ring of the suckers had developing teeth (t) in the form of cones, taller on one side of the ring than the other (Fig. 8). The infundibulum had two or three rings of polygonal processes with pegs. The pegs on the more peripheral processes were shorter than those adjacent to the inner ring (Fig. 9). At the periphery where the processes are flat the inner part of some of these processes widens while the distal part gradually takes on the form of the peg (Fig. 9). A progression through some stages of the development of a peg is indicated by 1, 2, 3 and 4 in Fig. 9.

The suckers on the club were in various stages of development. One sucker in which the infundibular surface was apparent but little differentiated is shown in Fig. 10. The boundaries delimiting the cells can be seen as elevated junctions, and between some of the apical borders of these cells are pores from which cilia protrude (arrowed). Only a few cilia emerge from each of these relatively isolated pores, some ten being visible in Fig. 11. These cilia are

---

Sepia officinalis

FIG. 7. A sucker from an arm of a 12-day-old animal. On the outside of the sucker tufts of cilia can be seen emerging through pores (arrowed). The peripheral ring of processes lacks pegs, but just enclosed by these processes are three others (arrowed 1–3) which are not of the elongate form and whose outer edge is slightly raised. The more central processes have stout pegs with a rough upper surface. The pegs are almost as large as their polygonal bases. The inner ring (in.r.) has blunt teeth. (Field width 43 μm.)

FIG. 8. Some of the suckers on the tentacular club, of the same small animal, whose surfaces show stages in the development of the polygonal processes from the periphery towards the inner ring (t: teeth). (Field width 172 μm.)

FIG. 9. The central sucker in Fig. 8, but at higher magnification, to show the teeth (t) of the inner ring, and the stages in the transformation of the outer flat, peripheral processes into ones with pegs (1–4). Cilia are present around the outside of the sucker (arrowed). (Field width 32 μm.)

FIG. 10. The infundibular surface of another tentacular sucker from the same animal showing little of the complex cuticular formation seen in Figs 7 and 8. A few infundibular pegs can be seen in the lower left corner of the photograph. On the outer surface of the sucker the cells are delimited by their elevated junctions, and at some of their apical borders there are pores (arrowed) from which tufts of cilia emerge. (Field width 172 μm.)

FIG. 11. Two of the pores of the sucker in Fig. 10, seen at higher magnification, from which some three and seven cilia emerge. (Field width 9 μm.)

FIG. 12. A tentacular sucker from a larger animal (35 mm dorsal mantle length) to show the increase in the number of pegs on the infundibulum. There is a gradation in the size of the pegs decreasing from the centre to the periphery. (Field width 363 μm.)

about 20 μm in length, although it is not possible to follow them to the cell surface.

One sucker from a larger animal (35 mm dorsal mantle length) is shown in Fig. 12. There are six rings of polygonal processes in contrast with two rings on the suckers from the small animal.

In an adult animal the suckers of the club are not uniform in size. There are about five notably larger ones on the manus and even the remaining suckers vary in size: those on the manus, apart from the very large ones, are bigger than those on the dactylus and carpus (Roper, Young & Voss, 1969).

The action of the suckers

The suckers of a tentacular club, severed from an anaesthetized animal, were blotted on filter paper and then pressed lightly on to a glass slide so that their action could be observed under a binocular microscope. The suckers would not adhere to thin sheets of gelatine or celloidin, but would stick well onto a finger, a polythene bowl or wood, less well to their own mantle muscle, and only poorly to glass or polystyrene. Unless the surface of the suckers was blotted they would not stick but slid off these last two surfaces.

The rim was found to form a seal as ink, taken from the *Sepia* and pipetted between the suckers, did not pass to the infundibulum. The maintenance of the hold of the sucker on the glass surface depended on the integrity of rim and infundibulum; as a small region became distorted the sucker fell off. The diameter of several of the suckers was measured initially and then again when a considerable pull was being exerted on the peduncle to produce suction. No measurable difference was found in the diameter under the conditions of the tests. This suggests that one function of the infundibulum is to retain its shape and size irrespective of the forces exerted on it by the acetabular muscles, the peduncle and the tentacular muscle. The infundibular surface must in some way make intimate contact with a foreign surface and retain this hold under changing conditions. When the peduncle was pulled light was refracted by the infundibular cuticle, perhaps by distortion of the cuticular pegs in relation to their upper surface which maintained its position on the foreign surface.

When the peduncle was pulled the inner ring was withdrawn leaving only the rim and infundibulum in contact with the glass. The suckers remained attached to the glass, while also being subjected to considerable pulls, for up to about 30 min; this was in marked contrast with *Eledone moschata* (see p. 490).

Discussion

The surface of the suckers of the tentacle must be exceedingly effective in adhering to prey as this presumably sticks initially; the action of the muscles of the sucker to produce suction in the acetabulum being subsequent. As Messenger (1968 and his chapter, this volume) has demonstrated, once an attack has been initiated, the tentacles are ejected and the prawn held by the suckers of the club within 30 msec, to be brought back to the open ring of reflexed arms and to the mouth in 150–300 msec. Not only is this action extremely rapid but the accuracy of the attack is also very high in the laboratory, being 90% on the first presentation of the prawn and 100% at the second. This level of accuracy was reduced somewhat, by the total removal of one tentacle, to 63 and 82% on the first and second presentations respectively (Messenger, 1968). Nevertheless this indicates the extreme efficiency of the surfaces of the suckers in adhering to a prawn, and, in the absence of one tentacle it can only be the suckers which hold the prey. The capture of the prey does not necessitate the apposition of the tentacular clubs. Newly hatched animals are also able to capture a mysid with one tentacular club (Boletzky, 1974). The arms alone can be used in the capture of slow-moving prey such as crabs, which are immobilized by the suckers and paralysed within about 9 sec of capture (Messenger, 1968).

The infundibular surface of the suckers of the arms is very similar to that of the suckers of the tentacles, suggesting that they are used in a similar manner for adhesion to the prey. The teeth on the inner ring of the suckers of both arms and tentacles are small and blunt suggesting they are not much used in the retention of prey.

There may be several stages in the attachment of suckers to prey. The rim is presumably first, and the infundibular surface next, to come into contact with the prey, the pegs perhaps being initially deformed. The fine pores seen at the surface of the pegs may act either by adhesion, or by suction, or perhaps one action is followed by the other. Each pore may contain a glue to help in the action of the suckers when adhering to prey, but this seems unlikely since the suckers would only adhere to glass after being blotted. If the small pores act by suction then a large number of minute chambers would be acting upon the uneven surface of the victim. After the whole of the infundibular surface is attached to the prey and with the rim forming a hermetic seal, then the muscles of the sucker can act to produce suction.

To fill these conditions the infundibular cuticle must have some rather special properties as it has to be sufficiently pliant to follow the shape of the foreign surface, it has to adhere to it during considerable physical changes that must occur when the acetabular muscles act, and it needs a strong connection with the underlying epithelial cells if the whole cuticle is not to be detached readily.

The pores on the outer surface of suckers of the small animal are too well separated to allow the protruding cilia to act in a motile capacity. It seems possible that these may be the cilia from the tapered cells which Graziadei (1964a) suggested were chemoreceptors. The value of chemoreceptors around the suckers of the tentacle may be to provide information about the prey before it reaches the mouth. However, no cilia have been found in similar places on the suckers of larger animals although this may be due to greater dispersion around the sucker or the result of poor fixation.

The scanning electron micrographs have provided some evidence of the manner in which the growth of the infundibulum occurs. The suckers of the small animal showed that there is a transformation of the radially arranged processes without pegs to polygonal processes with pegs. Initially the pegs were short but they appear to increase in height although transmission electron microscopy of the underlying epithelial cells is needed to demonstrate how and where material is added.

Spirula spirula

Distribution and depth

S. spirula lives in tropical and sub-tropical regions of the Atlantic, Pacific and Indian Oceans. It has been caught to depths of 3500 m (Clarke, 1966). Recently the use of closing nets has provided evidence that it lives in groups and exhibits a diel vertical migration (Clarke, 1969).

Habits and food

This pelagic sepioid is neutrally buoyant (Denton, Gilpin-Brown & Howarth, 1967) and usually maintains a vertical position in the water, the tail end being at the top, while at the bottom the arms are kept close together to form a cone (Schmidt, 1922; Bruun, 1943; Colman, 1954). It was known long ago that *Spirula* could ascend and descend (Clausen quoted by Gray, 1845). Colman (1954) said it swims up and down slowly with the fins but can also jet itself violently, tail first.

Schmidt (1922) found the eight arms to be highly sensitive to touch, while the tentacles were less, if at all, responsive. When disturbed it withdraws the head, arms and tentacles into the mantle cavity, thus resembling a rather stout rod (Brunn, 1943). When brought into contact with the side of the aquarium the animal spread its arms and clung to the glass (Schmidt, 1922).

There do not appear to be any records of the animal feeding in captivity but it is capable of inflicting a powerful bite with the sharp black beak (Schmidt, 1922). We have evidence that it feeds upon small pelagic crustaceans as their remains have been found in the stomachs of several specimens (Kerr, 1931; J. Z. Young, see his chapter, this volume). This may indicate prey selection. The food is presumably bitten by the beaks and then taken by the lateral buccal palps towards the oesophagus, as *Spirula* has no radula (Kerr, 1931).

The arms and tentacles

Both the arms and the tentacles have dense and complex muscles. The eight arms are short and tapering with four longitudinal rows of suckers, each attached by a short peduncle (see Naef, 1921, 1923).

The arm suckers

The surface features of a sucker are shown in Fig. 13. The sucker is strengthened by an inner ring, which appears as a delicate iridescent horny hoop (Owen, 1879). It has blunt teeth all fairly similar in size and height, each with a fairly flat surface with pores, like the rest of the surface of the ring. The infundibulum has rings of polygonal processes. Their cuticular surface is entirely covered with pores approximately 0·07 μm in diameter (Fig. 14). Each polygonal process has a projecting peg whose upper surface is somewhat concave and rather rough. The peripheral, radially arranged processes lack pegs.

The tentacular suckers

The suckers are slightly smaller than those on the arms but their surface features are similar (Fig. 15). The inner ring has blunt teeth, those on one half of the ring being taller and more widely separated than the remaining ones. The whole ring and its teeth as well as the pegs of the infundibulum are covered with innumerable small pores.

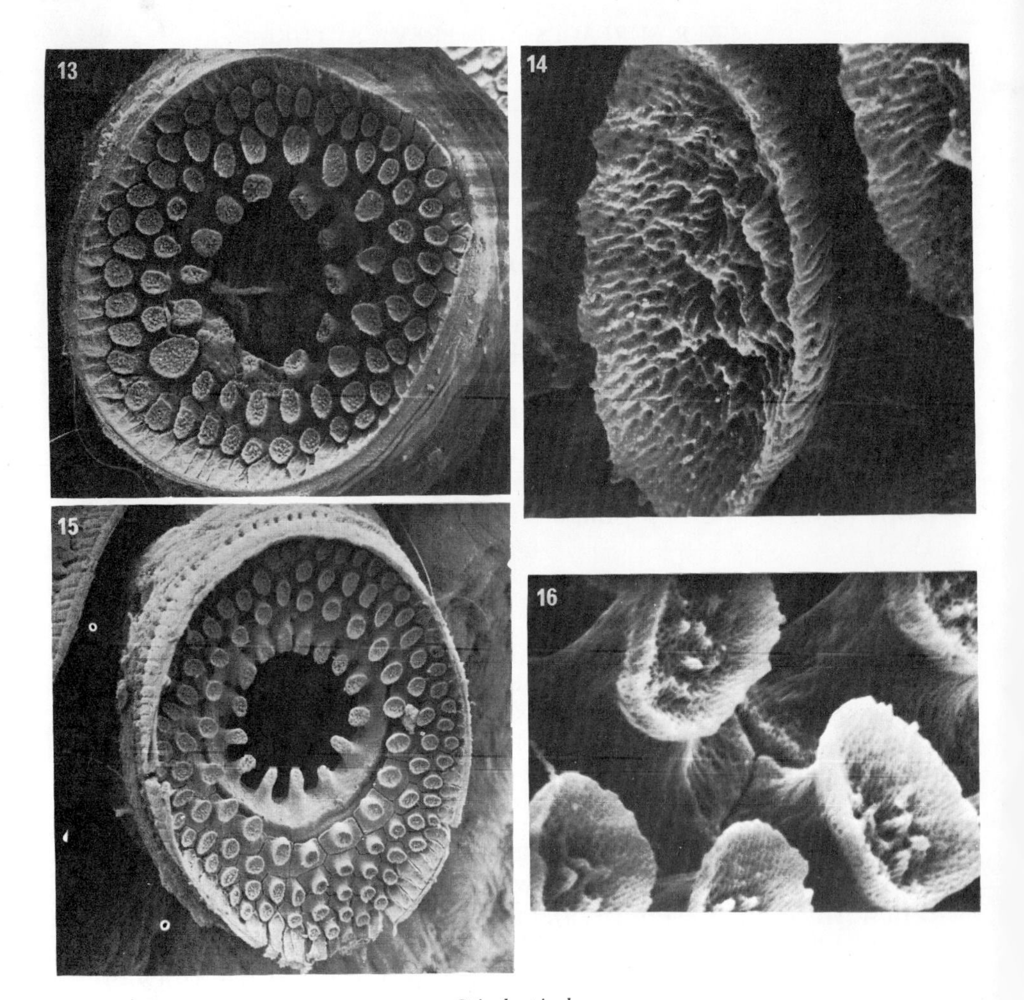

Spirula spirula

FIG. 13. The surface of a sucker from an arm. The inner ring has blunt pegs, exhibiting little differentiation, projecting from its exposed surface. The pegs on the infundibular surface are almost as large as their polygonal bases, although there is some differentiation in the size of the pegs. The surrounding processes lack pegs. (Field width 181 μm.)

FIG. 14. One of the pegs on the infundibulum of an arm sucker showing the pores on the surface; the upper surface of the peg is slightly concave and rough. (Field width 9 μm.)

FIG. 15. This sucker from a tentacle shows close similarity to those of the arms. The inner ring has blunt pegs of varying height on its free surface. The surrounding cuticular, polygonal processes on the infundibulum are largest adjacent to the inner ring. They also show differences in relative proportions in the size of the pegs and their bases as did those on the arm sucker. The processes of this innermost ring are larger than the surrounding ones and they become progressively smaller towards the periphery. (Field width 93 μm.)

FIG. 16. The pegs of the infundibulum at high magnification to show the pores over their surfaces. The roughness of the upper surfaces is similar to that seen on the infundibular pegs on the suckers of the arms in Fig. 14. (Field width 15 μm.)

The polygonal processes of the infundibulum are closely similar to those of the arm sucker. Again the surface is covered with small pores each approximately 0·10 μm (Fig. 16). At the periphery the processes lack pegs, but do have a radial arrangement.

Discussion

We know nothing of how *S. spirula* may take planktonic crustaceans but the complicated muscles of both arms and tentacles suggest that these appendages and their suckers are active in the capture of prey, especially in view of the absence of a radula, almost unique amongst cephalopods.

TEUTHOIDEA

Myopsida, ***Alloteuthis subulata***

Distribution and depth

This squid is found in the Mediterranean (Mangold-Wirz, 1963), the eastern North Atlantic, the English Channel, the North Sea and the Baltic Sea (Muus, 1963). It is a coastal animal common in the Plymouth area although there appear to be no formal records of the depths in which it is captured; a study of this is now in progress (M. R. Clarke & A. D. Mattacola).

Habits and food

It is a nektonic squid with shoaling habits. Much of its time is spent swimming backwards. One of us (M. N.) observed these squid in a long tank and saw them swim with apparent ease close to the bottom suggesting that they are familiar with it under natural conditions.

A. subulata feeds on fish and crustaceans, and is also cannibalistic (Bidder, 1950); the prey is bitten with the beak and rapidly swallowed. Pieces of flesh can be seen passing along the oesophagus while the feeding is still in progress; the time for the whole meal taking 15-20 min (Bidder, 1950).

The arms and tentacles

Each arm has two longitudinal rows of fairly small suckers (Naef, 1921, 1923). The tentacular club is 11 mm long in an animal of 50 mm mantle length (Muus, 1963) and it has four longitudinal rows of suckers, those on the middle two rows of the manus being large and the remaining ones smaller (Naef, 1921, 1923). Both arms and tentacles have dense muscles, some longitudinal, others transverse (Fig. 17).

The arm suckers

The main features of the surface of a sucker are shown in Fig. 18. There are seven wide, compressed, blunt teeth pointing towards the mouth on half of the inner ring (Fig. 19); the remaining half is a semi-crescentic plate with elevated ridges dividing the upper surface into polygons.

The infundibulum has four to five rows of polygonal processes with projecting pegs. The height and diameter of the pegs, and the size of the polygonal processes all increase towards the inner ring (Figs 18–19).

On the outermost ring of polygonal processes the pegs are on the periphery of the process and there is a progressive change to a more central position and finally to that part of the process nearest the ring (Fig. 19). At the edge of the infundibulum the radially arranged processes lack pegs and are of various sizes. Centrally the larger of these processes spread laterally, while their narrower, distal ends appear at an early stage of transformation into a peg (arrowed in Fig. 19). At the extreme edge very small immature processes are replacing those which are already developing pegs.

Alloteuthis subulata

FIG. 17. A transverse section through an arm and tentacle near to the mouth. There is dense and complex musculature surrounding the axial nerve cord in both. (Holmes.)

FIG. 18. The inner ring of a sucker from an arm has wide, blunt teeth on part of the ring while the remainder is a somewhat flattened hemicrescent. The infundibulum has pegs showing some difference in size and shape. (Field width 567 μm.)

FIG. 19. The same sucker as in Fig. 18 to show the transformation of a flat process (arrow) into one with a peg. The blunt teeth of the inner ring show signs presumably due to wear in one plane. (Field width 100 μm.)

FIG. 20. The tentacular suckers contrast with those of the arms in having quite long, slender, cone-shaped teeth around one half of the ring. The infundibulum has polygonal processes only some of which have pegs, the remaining ones being flat but not just restricted to the periphery. (Field width 2564 μm.)

FIG. 21. Part of the same sucker as in Fig. 20 but at higher magnification. The surfaces of the teeth of the inner ring and of the pegs of the infundibulum are in the same plane, about 45° relative to their bases. The polygonal processes develop slender pegs which are quite tall when adjacent to the inner ring. (Field width 964 μm.)

FIG. 22. A photomicrograph of a sucker at high magnification, to show the inner ring (in.r.) and the pegs of the infundibulum, each one above an epithelial cell. At the periphery these cells are both smaller and shorter. (Holmes.)

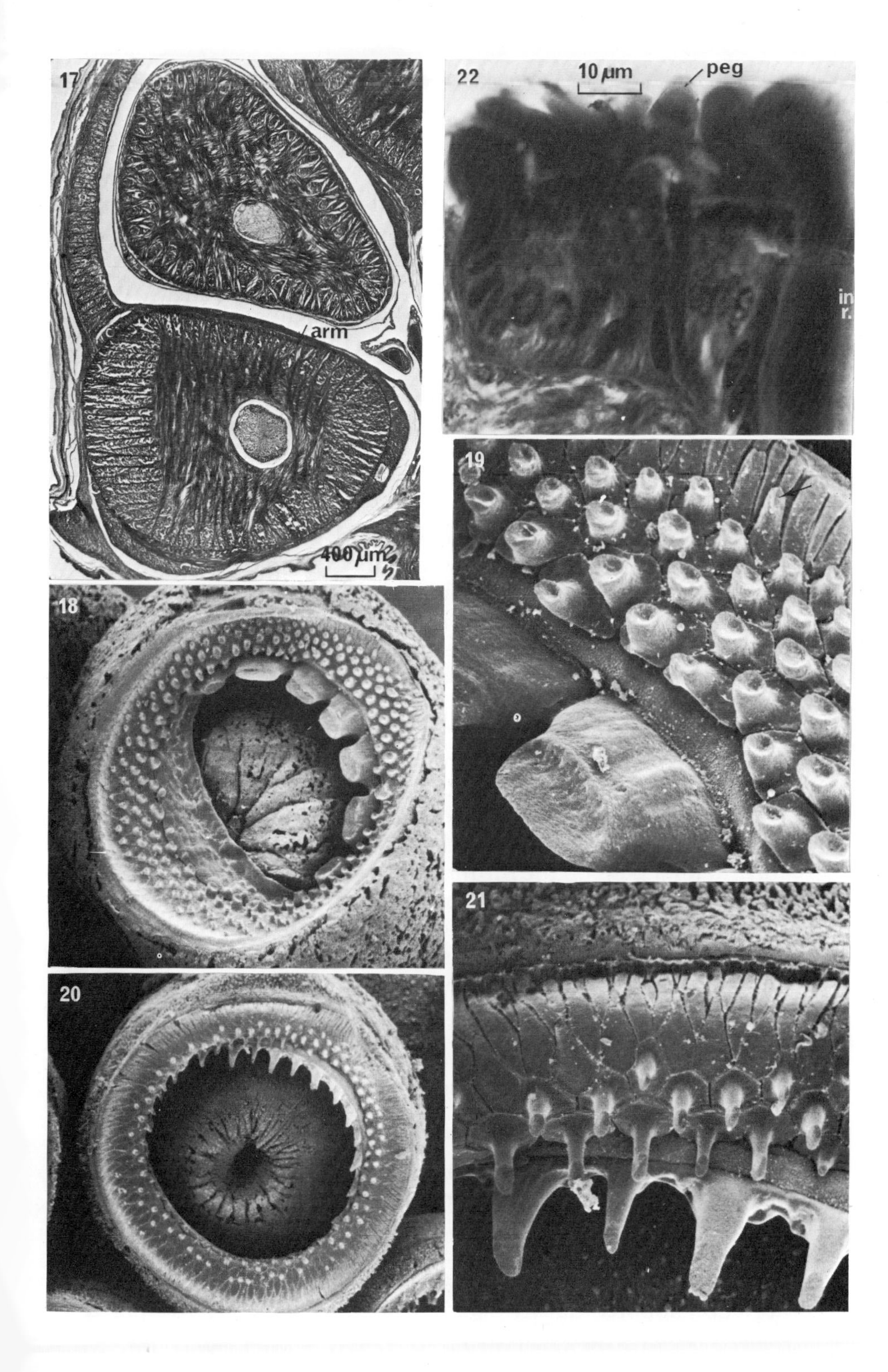

17
arm
400 µm
22
10 µm
peg
in
r.
19
18
21
20

The tentacular suckers

The large suckers of the manus (Roper, Young & Voss, 1969) are shown in Fig. 20. There are a number of fairly sharp, conical teeth, pointing towards the mouth, around some two-thirds of the inner ring, the remaining part being relatively smooth. The infundibulum has polygonal processes with relatively tall pegs, those nearest the inner ring being tallest. The change in the size of the processes and their pegs follows that seen in the arm sucker above. The transformation from a flat process to one with a peg does not take place as near the periphery (Fig. 21) as in the arm suckers (Fig. 19). The pegs of the tentacular sucker are tall, slender cylinders on relatively large bases (Fig. 21). They contrast with the pegs of the arm, that are slightly conical, short, relatively much stouter, and occupy a much larger portion of the polygonal process (Fig. 19). Each process and peg is covered with small holes, those on the surface of a peg being approximately 0·07 μm in diameter.

The epithelial cells underlying the infundibulum of the sucker are shown in Fig. 22. Many of these cells are long, some being 32 μm. The processes with projecting pegs overlie these cells and the most peripheral processes partly cover the shorter epithelial cells at the edge.

Discussion

Although *A. subulata* eats crustaceans and fish as well as its fellows (Bidder, 1950) we do not know how it feeds. Both the arms and the tentacles are muscular. The suckers of the tentacle have conical teeth and long pegs on the infundibular processes and could all act in helping to retain food. They contrast with the arm suckers which have blunt teeth and stout pegs and resemble the suckers of *Sepia officinalis*. Again like the sepioids the infundibular processes and the inner ring are covered with very small holes, perhaps serving for adhesion.

Oegopsida, ***Mastigoteuthis*** *sp.*

Distribution and depth

This squid is found in all oceans of the world between latitude 50°N and 50°S (Clarke, 1966). With opening-closing nets *Mastigoteuthis* has been captured between 200 and 1500 m (Clarke, 1969; Clarke & Lu, 1974, 1975; Lu & Clarke, 1975a,b). It is normally considered to be part of the bathypelagic fauna sharing features present in other cephalopods that live below 700 m (Voss, 1967). Roper &

Young (1975) in their study conclude that members of the genus are deep-living, having an upper daytime limit of 600 to 700 m while at night some may spread upwards and are found between 200 and 700 m.

Habits and food

Small crustaceans were found in the stomach by Verrill (1882) while Rancurel (1971) found antennae from crustaceans which must have been larger than the predator, as well as crustacean eggs.

Physiology and behaviour

Mastigoteuthis is neutrally buoyant owing to the presence of ammonium chloride in its tissues (Denton & Gilpin-Brown, 1973), many of which are extensively vacuolated (Dilly, Nixon & Young, 1977). This animal lies vertically in the sea with its vacuolated arms held upwards (Denton & Gilpin-Brown, 1973).

The arms and tentacles

Figure 23 shows part of the extremely long whip-like tentacles covered with minute suckers on long stalks (Verrill, 1882). The very large ventral pair of arms are almost twice the length of the longest of the remaining three pairs (Chun, 1910, 1914). Each arm has two longitudinal rows of small suckers quite closely arranged, except on the ventral arms where they are more widely separated (Dilly, Nixon & Young, 1977). The arms of *Mastigoteuthis* are highly vacuolated and the muscle much reduced (Fig. 24). In marked contrast the tentacles have dense musculature including both longitudinal and transverse fibres (Fig. 25).

The arm suckers

The inner ring of the sucker has conical teeth pointing towards the mouth on one half, and blunt, wedge-shaped ones on the remaining part (Fig. 26). The conical teeth are on the part of the ring furthest from the mouth and so point towards it. The inner ring and its teeth are formed by the aggregation of rods between which are spaces opening as pores approximately 0·16 μm in diameter (Fig. 27). The rods forming these pores are rather stouter than those found in the polygonal processes of the infundibulum (see Fig. 29).

The infundibulum has two or three rings of polygonal processes with pegs projecting from the surface (Fig. 26). The shape and form of these pegs varies with their position on the infundibulum. Those closest to the central conical teeth of the inner ring are

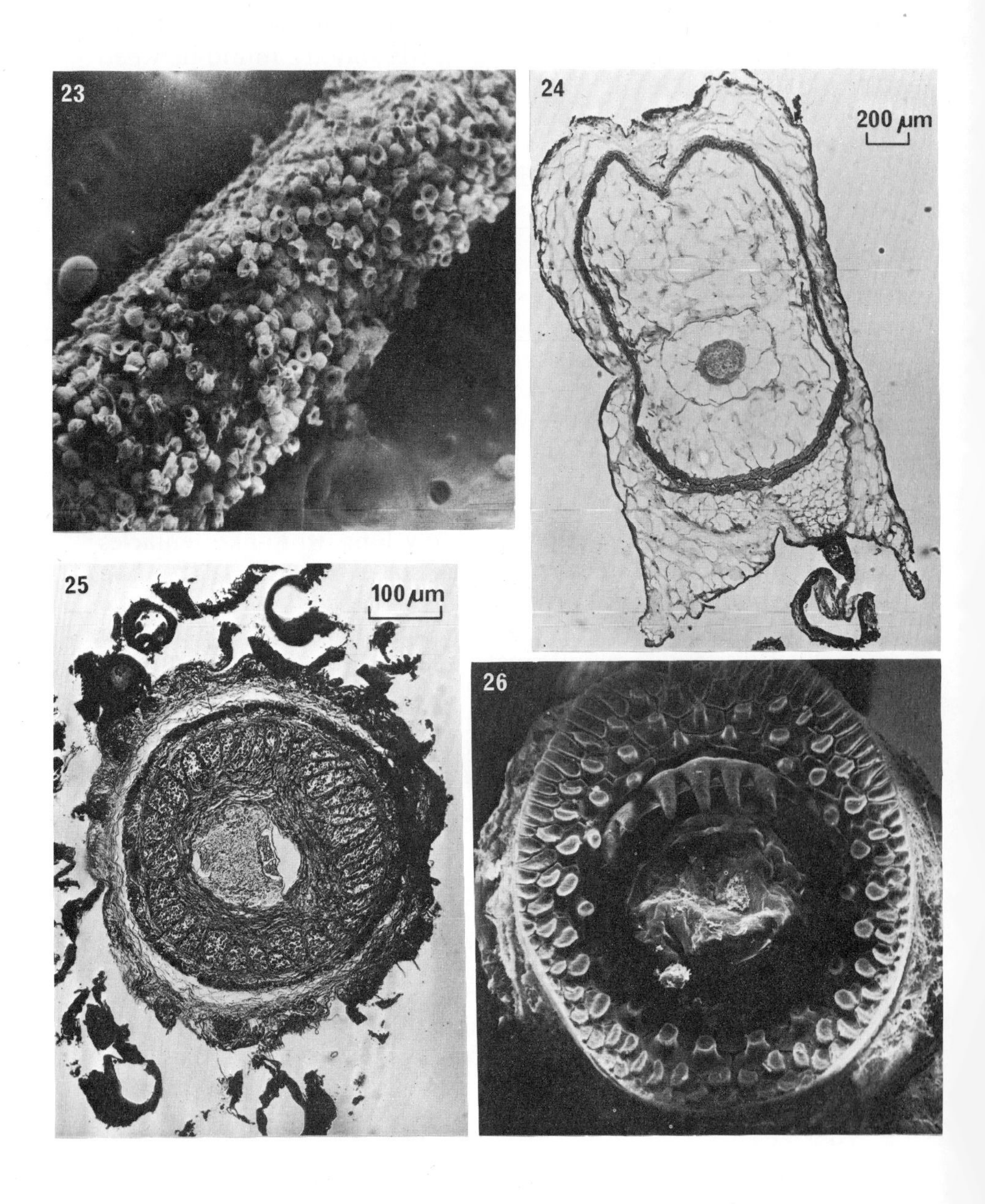
23
24
200 µm
25
100 µm
26

slender and taller than the remaining ones. On the opposite side of the infundibulum the pegs are much stouter and occupy a larger portion of the base of the process. The processes adjacent to the inner ring have taller pegs than do the more peripheral ones. At the edge of the infundibulum the flat processes are radially arranged. Again, as in *Sepia officinalis*, there is evidence of an inward migration of these processes together with development of the peg from the distal part of the process. Small flat processes come to occupy the space left by the developing polygonal process. The transformation is well seen in Fig. 28 where the infundibulum has broken. The total height of the pegs and the depth of the polygonal process can also be seen. The polygonal processes and their pegs are shown at high magnification in Fig. 29; the pores on the surface are approximately 0·08 μm in diameter.

The tentacular suckers

These are minute and very numerous and appear to cover the tentacle along most of its length (Fig. 23).

The inner ring is almost smooth and has no teeth (Fig. 30). The infundibulum has relatively large pegs which are quite close to each other and their surfaces are nearly flat. The size of the pegs and the polygonal processes increases from the periphery inwards. The number of polygonal processes decreases from 39 on the peripheral ring, to 27 on the middle one and 13 on the innermost one (Fig. 30). The surfaces of the polygonal processes and pegs have pores of approximately 0·05 μm. At the periphery of the infundibulum the flat processes are arranged radially. The transformation already seen in *Sepia* from an elongated process to a polygonal one with a peg can also be seen.

Mastigoteuthis sp.

FIG. 23. Part of one of the long tentacles to show its numerous minute suckers. (Field width 1482 μm.)

FIG. 24. A transverse section through one of the arms to show the extensive vacuolation of the tissues. (Cajal.)

FIG. 25. A transverse section through one of the tentacles. This has dense and complex musculature and is in marked contrast with the arms. The section shows many suckers cut in various planes. (Cajal.)

FIG. 26. The surface of a sucker from an arm. One part of the inner ring has long, conical teeth, while the remaining part has blunt, wedge-shaped ones. The pegs on the infundibulum show considerable differentiation, those close to the central conical teeth of the inner ring being tall and slender on a relatively large base, while those on the opposite side are large and stout and occupy much of the base of the process. The pegs diminish in size towards the periphery. (Field width 123 μm.)

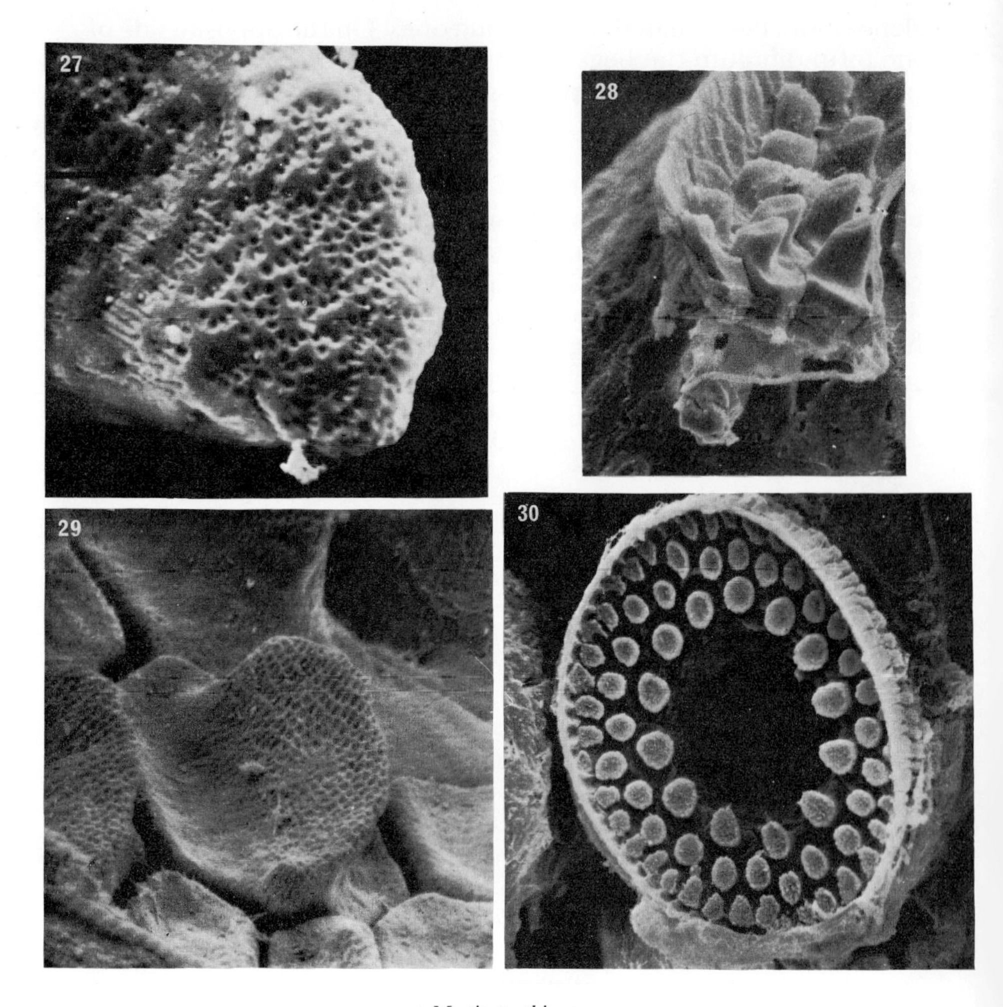

Mastigoteuthis sp.

FIG. 27. The upper surface of one of the teeth of the inner ring to show its formation by the aggregation of fine rods or perhaps tubes. The appearance of the surface suggests that at their surface at least, the pores have no contents after preparation for the SEM. (Field width 8 μm.)

FIG. 28. The infundibulum of one sucker had broken to reveal the depth of cuticle of the polygonal processes. The developmental stages from a flat peripheral process to one with a peg can be seen, changes occurring in the underlying part of the process as well as at its surface. (Field width 43 μm.)

FIG. 29. The polygonal process and the projecting peg is formed by an aggregation of rods or tubes; the pores do not appear to have any contents. The walls around the pores are thinner than those of the tooth shown in Fig. 27. (Field width 8 μm.)

FIG. 30. The surface features of a tentacular sucker. The inner ring has no teeth and the infundibular pegs are large relative to their polygonal bases. (Field width 83 μm.)

Discussion

There are only a few, small suckers along the arms, and the conical teeth on the inner ring could, perhaps, act to help to retain food particles. The tentacles, in marked contrast, have numerous, minute suckers. In spite of the size difference they resemble the suckers of *Sepia officinalis* and, by analogy, probably act by adhering to prey. The number and closeness of these suckers and their adhesive properties would make each tentacle resemble a rod of fly paper. Indeed the tentacles alone of *Mastigoteuthis* sp. are found attached to nets and lines although the rest of the animal is missing (M. R. Clarke, pers. comm.). The long tentacles with their adhesive suckers perhaps stick to planktonic organisms.

If primary receptors are present in the suckers of these tentacles as in *Sepia officinalis* (Graziadei, 1964a) and *Lolliguncula brevis* (Santi, 1975), then they could obtain tactile, chemical and mechanical information from a large region of the sea around the animal.

Oegopsida, ***Taonius megalops***

Distribution and depth

This is a cranchiid squid whose young stages have been caught in the North Atlantic (Muus, 1956; Lu & Clarke, 1975b). Using opening-closing nets the majority have been taken above 200 m, the remaining ones being captured down to 1000 m. The information obtained indicates an ontogenetic vertical spreading after an early ascent of the very young (Lu & Clarke, 1975b).

Habits and food

Small members of this species have been captured in quite large numbers (Lu & Clarke, 1975b). These animals have an escape mechanism in which they take on the form of a ball (Dilly, 1972), as has also been seen in another cranchiid, *Cranchia scabra* (R. E. Young, 1972; Angel, 1974).

Nothing is known of the prey as no food remains have been found in the specimens examined (Dilly & Nixon, 1976).

Physiology and behaviour

Members of the Cranchiidae exhibit several distinctive features. They are usually transparent (Chun, 1910, 1914; Dilly, 1973), and have a relatively very large coelom filled with a solution largely of ammonium chloride which, by its low density, makes the animal neutrally buoyant (Denton, Gilpin-Brown & Shaw, 1969). Clarke (1966) described them as "... passive, balloon-like drifters ...".

The arms and tentacles

The larval form of *Taonius megalops* has very short arms and exceedingly long tentacles. Suckers extend all along the length of the tentacles; the club develops during the post-larval stage when the arms are a little longer. In the juvenile the tentacles are only slightly longer than the longest arms. The musculature of the arms and tentacles is dense and complex at all stages examined from larva to juvenile (Dilly & Nixon, 1976).

The arm suckers

In very small animals the arms are barely present. A sucker from an animal of 34 mm dorsal mantle length is shown in Fig. 31. It has blunt, rather wedge-shaped teeth on some two-thirds of the ring, and the rest of the ring is almost smooth. The infundibulum has two to three rings of polygonal processes with large, short pegs all quite close together. Pores are present on the surface of the polygonal processes and the pegs, approximately 0·08 μm in diameter. The development of pegs from the flat peripheral processes takes place as has already been described for *Sepia officinalis* and *Alloteuthis subulata*.

Around the outside of the sucker are isolated pores from which cilia protrude. This is better seen in Fig. 32 (arrowed) of a sucker from a larger animal, of 46 mm dorsal mantle length.

Taonius megalops

FIG. 31. The inner ring of a sucker from an arm has some blunt, wedge-shaped teeth while the remainder of the ring is almost smooth. The infundibulum has stout, blunt pegs with little space between them. (Field width 227 μm.)

FIG. 32. An arm sucker from another animal at higher magnification to show the tufts of cilia (arrowed) emerging from pores at isolated intervals around the outside. (Field width 74 μm.)

FIG. 33. A tentacular sucker from a very small animal. The inner ring has rather poorly developed conical teeth on one half while the remaining half is almost smooth. Although a young animal, the polygonal processes are already of different sizes and shapes. Some tufts of cilia (arrowed) can be seen. (Field width 41 μm.)

FIG. 34. One of the large suckers from the manus of a tentacular club, of an animal of 180 mm dorsal mantle length, showing the quite sharp conical teeth on one half of the inner ring and blunt ones on the remaining part. The infundibulum of this sucker is notably different in that it lacks the polygonal processes and pegs. Instead the processes appear elongated and finger-like; they are radially arranged and may be free at their central end. (Field width 1600 μm.)

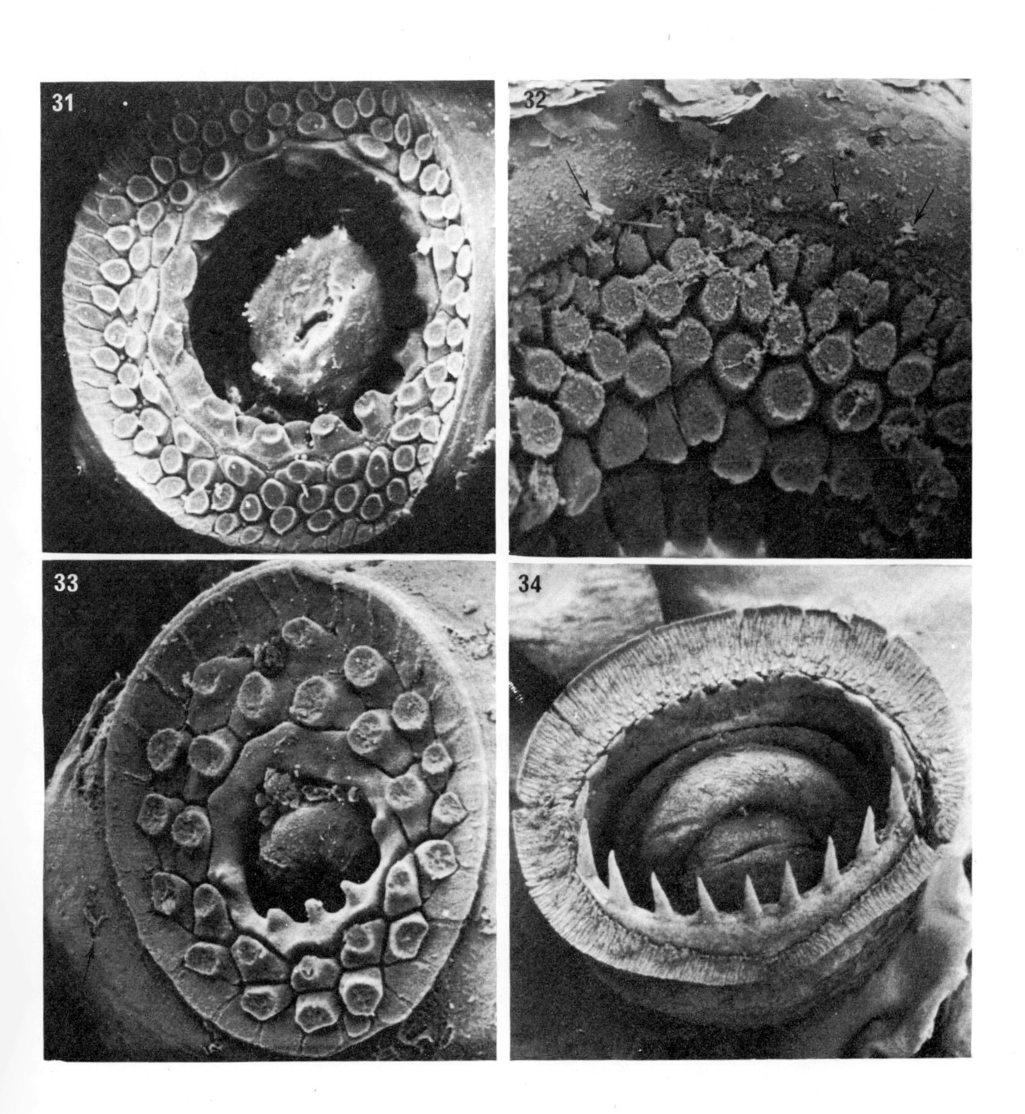
31
32
33
34

The general changes with growth have been followed in the sixth sucker of right arm 1 in animals of 34 to 180 mm dorsal mantle length, the results of which are reported in detail elsewhere (Dilly & Nixon, 1976). The main differences seen are the change from the sucker shown in Fig. 31 with an inner ring with blunt teeth to one with an almost smooth and very prominent ring in the large animal, in which the infundibulum is extremely difficult to discern (Dilly & Nixon, 1976).

The tentacular suckers

Figure 33 shows a sucker from one of the very long tentacles of a larval specimen, 4 mm dorsal mantle length. There is no differentiation of the suckers at this stage and they show a considerable resemblance to those of the arm seen in Fig. 31. The inner ring has five, slightly blunt, conical teeth, pointing towards the mouth, on one half while the remaining part is smooth. There are two rings of polygonal processes on the infundibulum and differences in the shape, size and form of the processes and their pegs can be seen (Fig. 33). The pegs close to the conical teeth of the ring are smaller than those on the opposite side, where they are very much larger. There is a peripheral ring of flat, radially arranged processes. The outer surface of the sucker, just beyond the infundibulum, has isolated pores from which cilia protrude (arrowed in Fig. 33).

The club develops during the post-larval stage and differentiation of the suckers begins. In a late juvenile there are three forms of sucker on the club (Dilly & Nixon, 1976). The toothed suckers of the manus are quite large, and one is shown in Fig. 34. The inner ring has sharp, conical teeth which point towards the mouth. The teeth on the remaining part of the ring are short, blunt and wedge-shaped. The infundibulum differs from the younger stages and all the others examined here. It is covered with elongated, radially arranged processes (Fig. 34). However, the small suckers on the carpus and dactylus regions of the club have polygonal processes with pegs (Dilly & Nixon, 1976).

Discussion

In the larval form the tentacular suckers must be of great importance in prey capture since there are no others. The surface of these suckers resembles those of *Sepia officinalis* and presumably adheres to the prey. After the club develops the large suckers of the manus have quite sharp teeth, which would act in retaining prey. The suckers of the arms of older animals also resemble somewhat those of *Sepia* and perhaps are used for adhering to food retained by the tentacular suckers.

A notable difference in the toothed sucker of the manus was the lack of polygonal processes and pegs on the infundibulum, and instead there were radially arranged, elongated, finger-like processes. Nothing is known of their internal structure or function. However, the infundibulum of a tentacular sucker of *Bathothauma lyromma*, another member of the Taoniinae, figured by Aldred (1974), had similar processes.

VAMPYROMORPHA

Vampyroteuthis infernalis

Distribution and depth

V. infernalis, the only species of the order, lives only in the great depths of all tropical and subtropical oceans (Pickford, 1940). Its vertical distribution is between 300 and 3000 m, with the majority between 1500 and 2000 m, usually in regions where the oxygen content of the water is low (Pickford, 1949). Using opening-closing nets two animals were caught during the day between 910 and 1500 m, and three at night between 900 and 1250 m (Clarke & Lu, 1974). Roper & Young (1975) captured some 185 specimens and found the majority at depths of between 700–800 m and 900–1100 m. Larval forms were found below 900 m while the adults were above this depth. No diel migration was found to occur.

Habits and food

It has been suggested that *Vampyroteuthis* feeds on plankton, perhaps catching the prey with the well-developed interbrachial web (Voss, 1967). One animal has been found with a relatively large prawn in its stomach while the remains found in the stomach of another were identified as coelenterates by the late W. J. Rees of the British Museum (R. E. Young, pers. comm.).

The web can be used in an escape reaction in rapid swimming when, with the arms, a medusoid action is adopted (R. E. Young in Roper & Brundage, 1972). This animal is at least partially, or neutrally, buoyant by reduction of sulphate (J. B. Gilpin-Brown, pers. comm.).

The arms

Each of the eight arms has a single row of suckers alternating with paired cirri (Pickford, 1949) (Fig. 35). The suckers near the base of the arm are usually sessile but the small, distal, recently formed ones have a peduncle. Beyond the last sucker is a series of sucker-forming papillae on the filamentous extremity of the arms (Pickford, 1949). The sessile suckers are like an urn with a straight neck surrounding the opening. The acetabulum is wide and the distal suckers are similar but less compressed.

There may be up to ten pairs of primary cirri along the base of the arm before the first sucker, beyond which they alternate with the suckers (Pickford, 1949). The suckers and the tips of the cirri are tinged with crimson, the pigment being in the epithelial cells of the epidermis (Pickford, 1940). Figure 36 shows extensive vacuolation of the arm, cirri and even in the large peduncle of the sucker. In contrast the sucker has quite dense radial muscles around the acetabulum (and see Fig. 2). There is a slight demarcation between the muscles that perhaps indicates a division between the acetabulum and the infundibulum, by analogy with the suckers of octopods (see Figs 2 and 3).

The suckers

The suckers are simpler than any of the others studied here. The rim of the sucker has a radial arrangement with complex folds (Fig. 37). Along each sector are small hillocks from some of which cilia project (Fig. 38). These are about 0·1 μm in diameter and 5 μm in length (Fig. 39).

The straight "neck" region of the sucker, leading to the acetabulum, has a fairly smooth surface (Figs 35 and 37) with a

---

Vampyroteuthis infernalis

FIG. 35. Two suckers and the cirri on one side; the corresponding ones on the other side are not shown. (Field width 2777 μm.)

FIG. 36. A transverse section through an arm, a sucker and two cirri. Most of the tissue is vacuolated, the exceptions being the sucker, a narrow band of muscle in the arm and the axial nerve cord. (Cajal.)

FIG. 37. The rim is the most noticeable part of the sucker. The surface of the rim while being arranged radially has quite deep folds superimposed on it. Some of the apices of these folds have tufts of cilia protruding. (Field width 952 μm.)

FIG. 38. A small part of the rim at higher magnification to show the occurrence of the tufts of cilia (arrowed). (Field width 95 μm.)

FIG. 39. One of the apical tufts of cilia on the rim of the sucker. (Field width 11 μm.)

FIG. 40. The tip of a cirrus on which the cells can be seen outlined by their elevated junctions. At the apices of the junctions of some of the cells, tufts of cilia emerge through pores (arrowed). The surface of the cirrus facing the sucker is deeply folded and appears as a densely stained area in Fig. 36. (Field width 190 μm.)

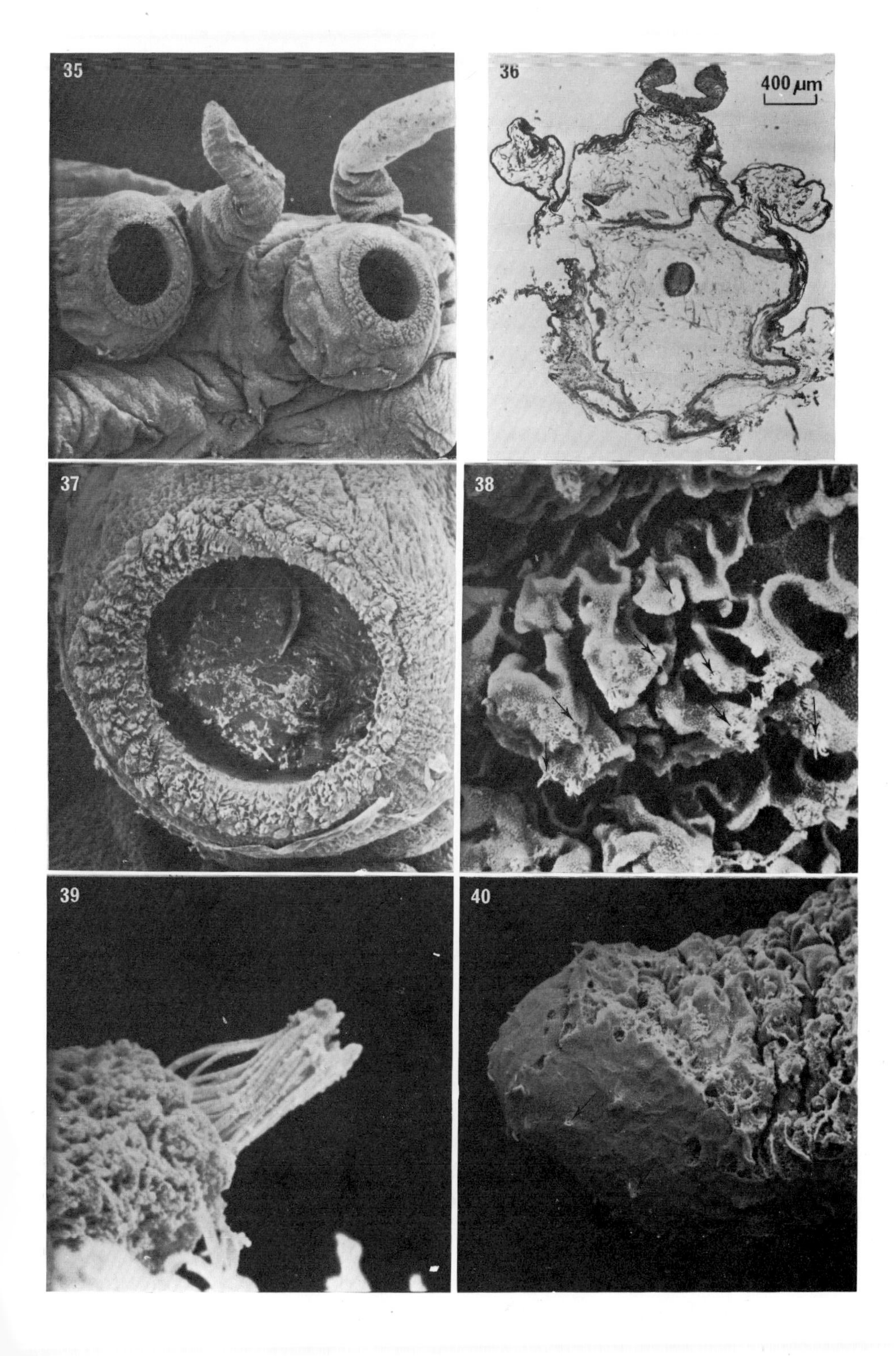
35
36
400 µm
37
38
39
40

cubical epithelium and appears to lack the cuticle found in the octopod suckers examined (see p. 481). There does not appear to be any sphincter present (see Fig. 3) and so presumably the chamber may be regarded as the acetabulum.

The cirri

The inner tip of the cirrus is much folded (Figs 36 and 40) and light microscopy has revealed the presence of abundant secretory cells as well as a few "olfactory-like" cells (R. E. Young, pers. comm.).

The remaining smoother region of the tip has elevated junctions delineating the underlying cells (Fig. 40). Between some of the apical borders of the cells are pores from which a few cilia protrude.

Discussion

These suckers are unique, at least amongst those seen in this survey, in the absence of a cuticle with pegs on the infundibulum. They are also the only ones that have cilia on the rim; it is unlikely that the cilia are motile, but they are well-sited to sample the surrounding sea for possible food organisms. As the arms are highly vacuolated these animals may depend upon the detection and capture of passing prey with their outstretched arms, suckers and cirri, both the latter being apparently well-endowed with sensory mechanisms. The apparent absence of the infundibulum and its associated cuticle could mean that its habits are different or that the conditions encountered at the depths inhabited by this animal do not necessitate such an elaborate means of mechanical attachment to its prey.

OCTOPODA

Cirrata, ***Cirroteuthis*** *sp.*

Distribution and depth

Cirrate octopods are rare in collections but have recently been found to be rather common at great depths. They have been photographed from a deep-sea search vehicle (Roper & Brundage, 1972). They were seen at depths of 2300 to 2500 m over a bottom which varied from sediment to coarse gravel.

Habits and food

The animals observed by Roper & Brundage (1972) did not feed but were often hovering, always some 2–4 m from the bottom, with the aid of the fins, looking like an open umbrella. When hovering in this way the cirri were erect on either side of the single row of suckers, the mouth being at the centre of the open arms, and it was suggested that the animals were seeking food. Indeed, if the cirri are sensory, then in this position they would obtain the maximum tactile and chemical information available in their vicinity.

Amongst the animals found associated with these cirrate octopods were several species of fish, decapod crustaceans, and sea cucumbers, the last two groups being present in greatest numbers (Roper & Brundage 1972). In another cirrate, *Grimpoteuthis*, various deep-living benthic crustaceans, 5–10 mm long, have been found in the stomach (Scott, 1910).

The cirrates seen from the search vehicle were able to jet, taking on an elongated, cone-like form with arms and web closed together as a tapering tail (Roper & Brundage, 1972).

The arms

The arms are subequal and the longest are some two-thirds of the total length of the animal. The suckers are firm and muscular and are embedded in the soft tissues of the arms (Hoyle, 1885). The cirri are long and are present from between the second and third suckers to the edge of the web. The erect cirri are well seen in one of the photographs taken from the search vehicle (Roper & Brundage, 1972: fig. 20).

The internal structure of the arm, cirri and peduncle is shown in Fig. 12 of J. Z. Young (this volume, p. 392). The tissue is somewhat vacuolated, the arm muscle being reduced to a peripheral band (Robson, 1925). The vacuolation is presumably a mechanism for buoyancy as in *Japetella diaphana* (see p. 495). The sucker has a stout peduncle, and very deep muscles of the acetabulum (Fig. 41). There is a projecting ridge which demarcates the acetabulum and the infundibulum, and associated with this ridge is a sphincter seen as a lighter staining area.

The suckers

The rim of the sucker is divided radially (Fig. 42) and each sector is thrown into complex folds (Fig. 43) but no tufts of cilia could be found as in *Vampyroteuthis infernalis* (Figs 38 and 39).

If the sphincter in the sucker of incirrate octopods (Fig. 3) is taken as the demarcation between the infundibulum and acetabulum, then the infundibulum is extremely small in *Cirroteuthis* (Fig. 41). It occupies the region between the rim and the projection associated with the sphincter (Fig. 44). The epithelium

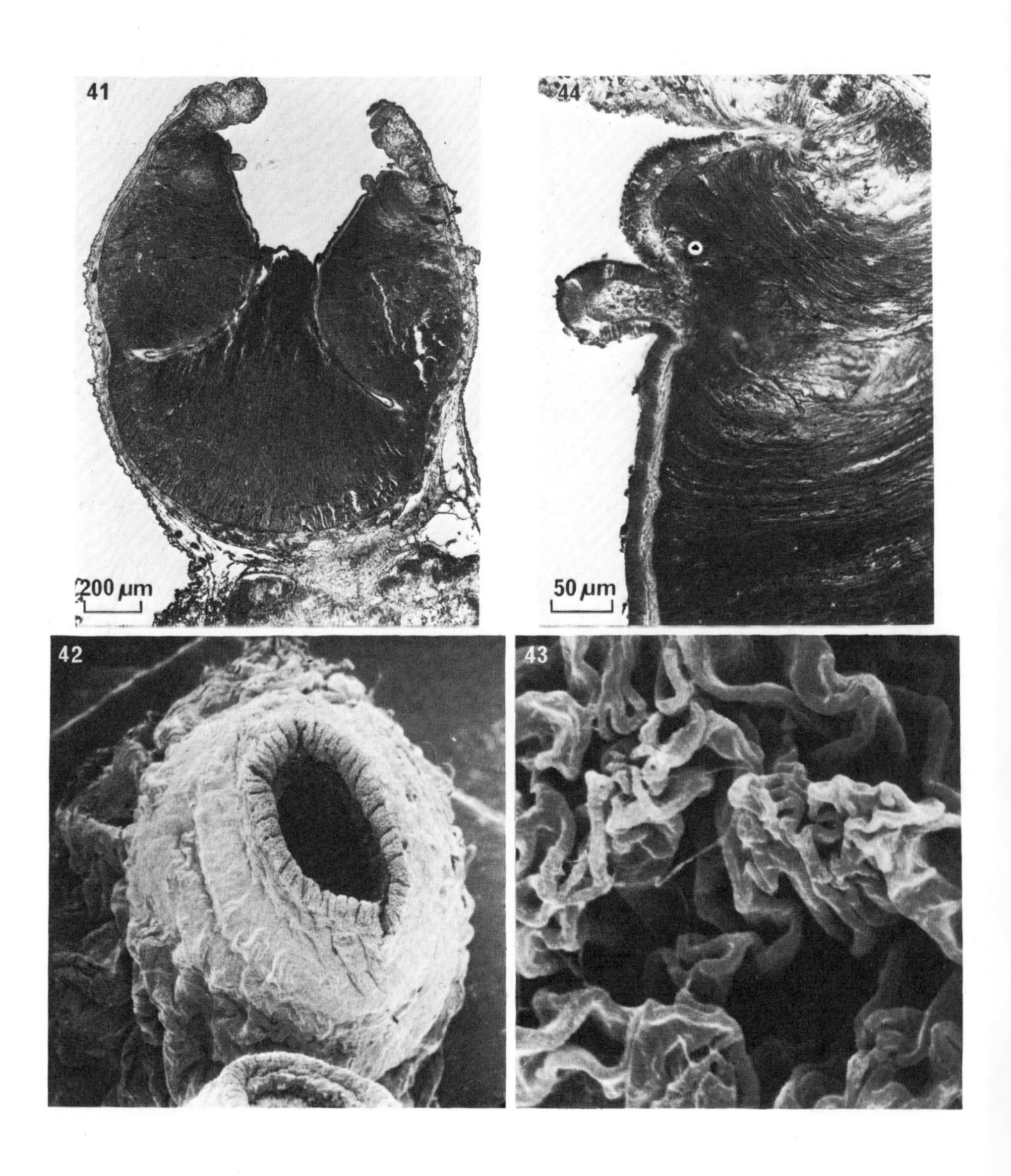
41
200 μm
44
o
50 μm
42
43

of the infundibulum is cubical in contrast with the incirrate octopods where it is of tall columnar cells (Fig. 6). The epithelium of the acetabulum is also cubical (Fig. 44) and is like that of *Octopus vulgaris* (Fig. 3).

Discussion

The musculature of the suckers suggests they would be efficient suction mechanisms and are in contrast with the vacuolated arms. The much-folded rim is reminiscent of that of *Vampyroteuthis infernalis* although *Cirroteuthis* lacks the cilia. The infundibulum is notably small and does not appear to have the pegs found on that of the incirrate octopods. The adhesion mechanism of the suckers is thus much reduced and may have a different action.

Incirrata, ***Octopus vulgaris***

Distribution and depth

Octopus is a very widely distributed genus in warm and temperate waters, particularly between latitudes 33°N and 40°S (Robson, 1929; Rees, 1951; Mangold-Wirz, 1963). A coastal animal, it lives between 0 and 100 m, but may be found to depths of 250 m.

Habits and food

O. vulgaris inhabits crevices amongst rocks or hollow objects on the sea bed. Two studies by aqualung divers have shown that an animal will occupy the same home for quite long periods (Altman, 1967; Kayes, 1974). The majority of animals under observation left their home at dusk and were found in the open, presumably hunting during the hours of darkness.

O. vulgaris feeds mainly on crustaceans and molluscs (Taki, 1941; Altman, 1967; Nixon, 1968). It will capture a crab and paralyse it with cephalotoxin (Ghiretti, 1960) in less than 30 sec.

Cirroteuthis sp.

FIG. 41. A sagittal section showing the deep radial muscles of the sucker. The infundibulum is small and its separation from the acetabulum is marked by the sphincter and associated ridge projecting into the sucker chamber. (Cajal.)

FIG. 42. When seen at low magnification the sucker resembles that of *Vampyroteuthis infernalis* (Fig. 37). The rim is radially arranged with complex folds superimposed upon it. (Field width 1904 μm.)

FIG. 43. At higher magnification the folds of the rim show no indication of cilia. (Field width 19 μm.)

FIG. 44. At higher magnification the small infundibulus has a cubical epithelium which may support small pegs. (Cajal.)

The crab is then separated at the joints, entirely without breaking the exoskeleton, and the tissue removed leaving it quite clean (Altman & Nixon, 1970). Manipulation of the crab during this process is presumably by the suckers, although this cannot be readily ascertained. Shelled molluscs are bored with radula and a paralysing substance introduced (Pilson & Taylor, 1961; Arnold & Arnold, 1969; Wodinsky, 1969; J. Z. Young, pers. comm.).

Physiology and behaviour

Behavioural studies have already revealed the capacity of the octopus to discriminate between live and wax-filled lamellibranchs (Wells & Wells, 1956), and various chemical substances in very dilute solutions (Giersberg, 1926; Wells, 1963a; Wells, Freeman & Ashburner, 1965).

It appears that physical differences between objects are distinguished in terms of the distortion they impose on the suckers with which they are in contact. Features like shape, size, surface patterning or weight may still be discriminated because of the differences in the degree of distortion. The implication from a variety of experiments is that proprioception plays no part in tactile discrimination (Wells, 1960).

Electrical stimulation of the rim of the sucker indicated tactile units with limited areas of representation in the arm nerve cords (Rowell, 1963). Tactile signals are carried to the brain, after passing further relays in the cord, by a relatively small number of fibres (Graziadei, 1965; Rowell, 1966). Rowell (1966) was unable to find proprioceptor inputs in the brachial nerve cords, thus confirming the training experiments of Wells (1963b, 1964).

Another important function of the suckers of these bottom-living animals is to explore, hold and retain objects. The suction power of isolated suckers of *O. bimaculatus* Verrill was examined by Parker (1921) who found a relationship between the breaking force and the area of the sucker; there was a rather higher efficiency for the smaller suckers than the larger ones. The force exerted by *Octopus vulgaris* was tested experimentally and animals of 1·3–2·5 kg were able to exert a total pull of 18 kg (Dilly, Nixon & Packard, 1964).

The arms

Each of the eight arms has a complex musculature and two longitudinal rows of suckers each attached by a large muscular peduncle (Guérin, 1908). The suckers at the base of the arms are small and gradually increase in size to the seventh or eighth sucker. There is a dramatic increase in the diameter of the next sucker, of arms two and three, in the males (Robson, 1929; Packard, 1961). The suckers along the rest of the arm gradually diminish in size.

Primary receptors are present on the surface of the arms but it is in the epithelium of the suckers that they are most numerous (Graziadei, 1971). The receptors are present in the order of tens of thousands in one sucker, and an estimate of the total number in all the suckers of the arms is in the region of 4 000 000 (Graziadei, 1971). There appear to be three types: (a) cells with a round body, (b) irregular multipolar cells and (c) tapered or flask-shaped cells with a crown of cilia, each with an accessory cell with internal cilia and a "rootlet reservoir" (Graziadei, 1971; Graziadei & Gagne, 1973). Ciliated cells form about 80–90% of the receptors of the suckers. The round and multipolar cells may be mechanoreceptors and the ciliated ones chemoreceptors, but physiological studies are needed to be certain of the function of these morphologically different cells. Information from all of these receptors passes to the sucker ganglion and then through the axial ganglia and cord to the complex central nervous system (J. Z. Young, 1971).

The suckers

The sucker has an encircling rim with a pattern of complex radial folds. Within the rim is the infundibulum with radially arranged cushions, as in all the octopods (d'Orbigny, 1845) and grooves (Niemec, 1885). The infundibulum has a cuticle which is covered by a large number of tiny pegs. The cuticle is shed at intervals by animals kept in aquaria, and floats as a disc to the surface (Girod, 1884) and is renewed continuously (Naef, 1921, 1923). The infundibular chamber is separated from the underlying suction chamber, the acetabulum, by a circular sphincter muscle.

The general features of the sucker are shown in Figs 3, 5 and 45. The relative proportions of the rim and infundibulum change as the suckers increase in size; the former is the most prominent feature of small ones (Fig. 5) and the latter in the large suckers (Fig. 45). The rim has quite deep radiating folds in a small sucker of 0·4 mm diameter (Figs 5 and 46), but it gets progressively more complex and more peripheral as the sucker becomes larger (Fig. 45, and see fig 3.1 in Graziadei, 1971).

In small suckers the surface of each of the sectors of the rim has a number of hillocks with only a slight indication of alternation (Fig. 46). The apices of these hillocks have pores, often in clusters

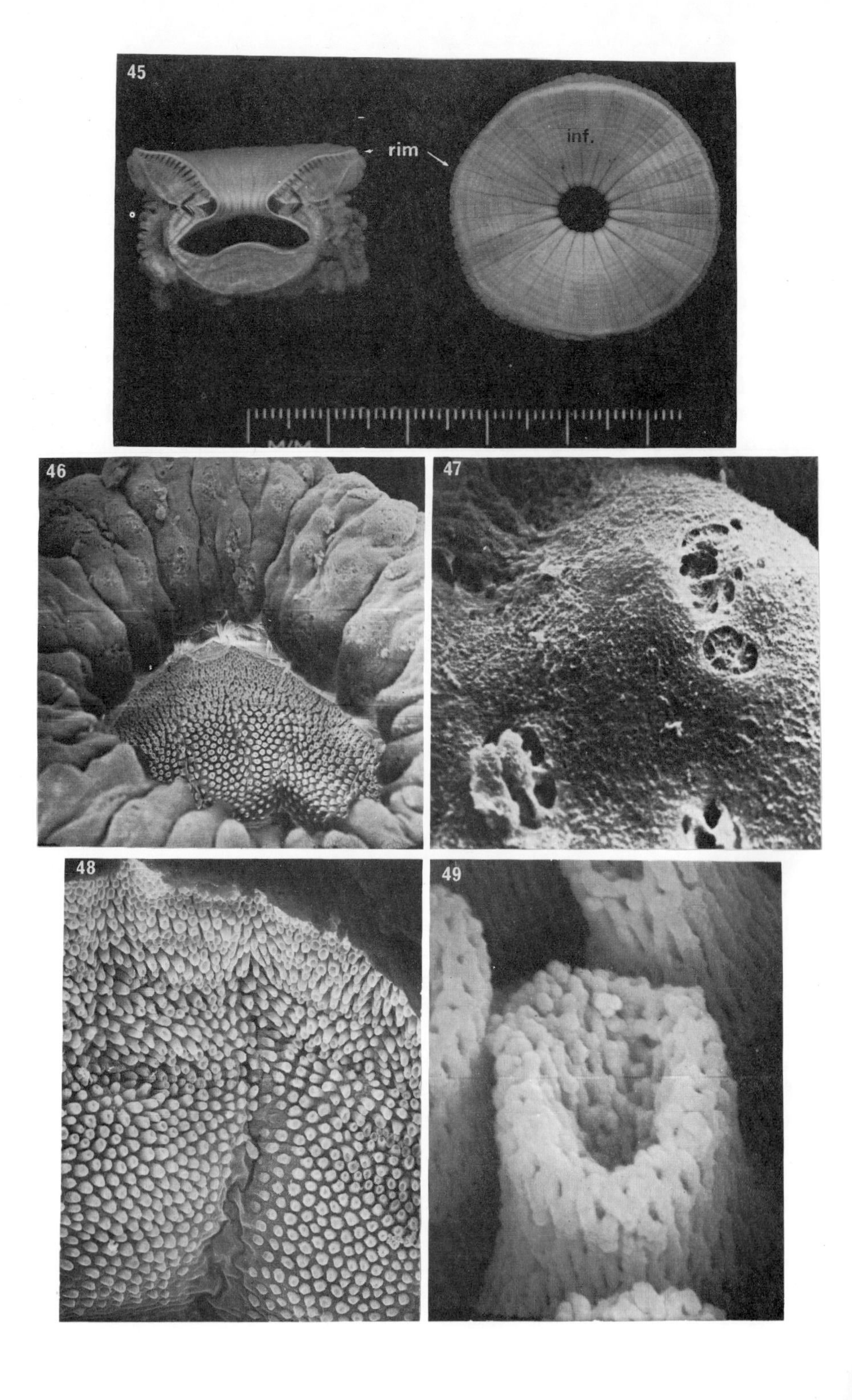
45
rim
inf.
46
47
48
49

(Figs 46 and 47), some of which have structures within their openings. These may be the cilia of the receptor cells described by Graziadei (1964b).

The narrow region between the rim and the infundibulum appears as a thin layer of much-folded tissue (Fig. 46). The infundibulum, like the rim, is radially arranged (Figs 45 and 46) and the cuticle of each sector is covered with pegs. On the upper part of each sector the pegs are quite tall and rod-like, with a deep concavity on the upper surface (Fig. 48) while lower down, near to the sphincter, the pegs are less densely packed, shorter and rather conical in form with only a shallow concavity on the surface. Each peg is 1–3 μm in height and the surface punctuated by very small pores. Examination of a sucker at a higher magnification indicates that each peg is formed by an aggregation of rods, packed in such a way as to leave fairly regular spaces of approximately 0·08 μm in diameter (Fig. 49). The rods may be secreted at fairly regular intervals, as there are narrow regions along each one, by the columnar epithelial cells that lie below. The formation of the cuticle of the suckers was said to be a cellular secretion (Kölliker, 1858) from the underlying epithelial cells (Figs 6 and 50). In Fig. 51 (of a microdissected sucker) the radial muscles, the tall columnar epithelium and the infundibular cuticle can be seen. One of the pegs has broken and appears hollow. This would give space for the formation of a new cuticular peg below, as seen in the photomicrographs where old cuticular discs are in the process of being shed (Figs 62 and 65).

Octopus vulgaris

FIG. 45. One of the large suckers from an adult male animal. The rim lies around the edge of the now relatively very large infundibulum (inf.) which forms an almost flat disc. The radial divisions of the infundibulum are readily seen. The hemisection shows the width of the acetabular chamber. (Compare with the small sucker in Fig. 5.)

FIG. 46. A small sucker, 0·4 mm diameter, in which the rim is relatively very large. The rim is radially arranged and each of the sectors forms a cushion on which there are hillocks superimposed in a fairly random manner. Between the rim and the infundibulum is a thin but folded epithelium. The infundibulum also has a radial arrangement and has numerous small pegs covering its surface. (Field width 190 μm.)

FIG. 47. The hillocks on the rim have pores which appear to be structural features and not artefacts. The pores do not appear to have any contents, at least after preparation for the SEM. (Field width 38 μm.)

FIG. 48. Two radial sectors of the infundibulum to show the small pegs densely packed on the upper region but somewhat sparse on the lower region. The pegs on the upper part are stout and have a deep concavity on their upper surfaces; in contrast those at the bottom are conical with only a shallow concavity on their surfaces. (Field width 100 μm.)

FIG. 49. The pegs at higher magnification to show the pores covering their surface. (Field width 1.5 μm.)

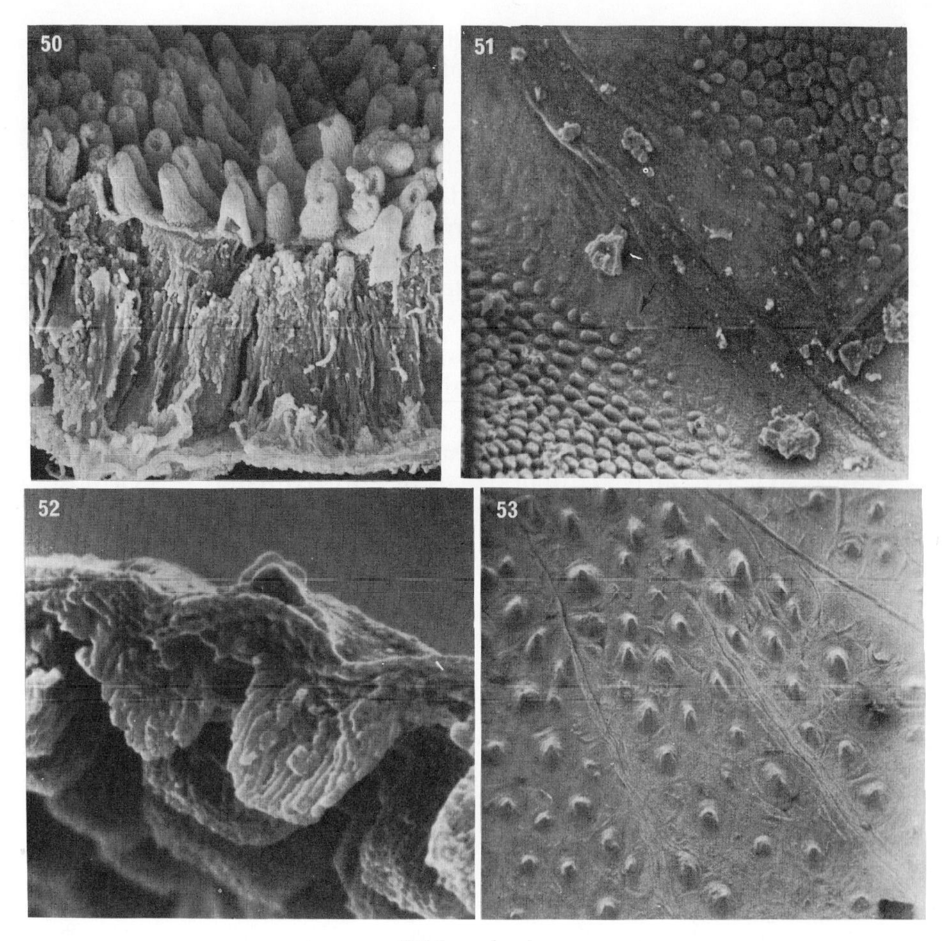

Octopus vulgaris

FIG. 50. A sucker was microdissected to show the infundibular cuticle and epithelium. The heights of the columnar cells are well shown in relation to the cuticular pegs. Some of these pegs have broken and appear as hollow cones. (Field width 33 μm.)

FIG. 51. A shed disc was spread over the flat surface of a rivet and this exposed the cuticle otherwise hidden in the radial folds. Here the pegs are absent or small only becoming prominent beyond the region of the fold. The outline of the underlying cells can be seen by their raised border (arrow) and one of the underlying cells seen in Figs 6 and 50 secretes one peg. (Field width 80 μm.)

FIG. 52. Part of the cuticular disc had broken to show each peg formed by an aggregation of rods. The pegs appear hollow at their centre. (Field width 8 μm.)

FIG. 53. The shed disc was 6 mm in diameter, a much larger sucker than that in Fig. 5. On the disc there were, in addition to the pegs over the surface, other quite large protuberances which increase both in size and number towards the centre of the disc. (Field width 1333 μm.)

The whole infundibular cuticle is often to be seen floating as a disc in an aquarium after being shed by an octopus (Le Souef & Allan, 1933; Taki, 1941). Some discs from an animal of about 400 g were examined with the SEM after being allowed to dry flat on a stub. The area between two sectors was exposed, and shown to be without pegs. The cells, about 6 by 3 μm, could be seen outlined by their raised junctions (Fig. 51), and this photograph confirms that one peg is supported by one cell. One part of the disc had broken and the pegs were seen to be formed by an aggregation of rods (Fig. 52). The pores, formed by the space between the rods, may be open at each end, although this cannot be so when an old cuticle is being shed. The shed cuticular disc was approximately 6 mm in diameter and each sector had protuberances of a different order of magnitude to the pegs (Fig. 53). It is possible to see these at low magnifications (see fig. 3.1 of Graziadei, 1971). The number and size of these protuberances increase from the periphery to the centre.

To act as the efficient suction organ we know it to be (Parker, 1921; Dilly, Nixon & Packard, 1964), the sucker must first form a seal. The soft, much-folded rim and the infundibular cuticle make the seal while the infundibulum becomes attached to the foreign surface. This could be by adhesion followed by suction, perhaps by each of the tiny pores on the surface of the pegs on the infundibulum. Once the sucker and the foreign surface are intimately connected then the muscles of the sucker can act to produce suction within the acetabular chamber.

In order to ascertain whether the fine pores on the pegs of the infundibulum could act in such a manner the approximate number on a sucker of 0·4 mm diameter has been estimated. This sucker was taken from an animal of 120 g, and 84 mm mantle length, and had about 250 pegs on each of some 12 sectors making a total of approximately 3000 pegs on the sucker. The average number of pores, each almost 0·0001 mm diameter, on the upper surface of ten pegs was 40, thus making a total of 120 000 (say 10^5) pores on this very small sucker.

If each pore is the surface representation of a capillary and acts as a miniature suction chamber then it may be the mechanism whereby the sucker is attached to a foreign surface. From the capillary size we have calculated the theoretical number needed to support 10 g weight at depths of 10 and 100 m (see Macdonald, 1975).

To find the number of capillaries needed on the infundibulum of one sucker to support a mass of 10 g at different depths some assumptions have been made. These are:

(1) that 10 m of water is equivalent to 1 atmosphere, both of which produce a pressure of

$$h\rho g = 10 \times 10^3 \times 10 = 10^5 Nm^{-2}$$

[the acceleration due to gravity (g) is taken as 10 msec^{-2}]

(2) that the diameter of the capillary 10^{-4} mm is 10^{-7} m, and the radius is $\frac{10^{-7}}{2}$

(3) that the pressure in the suction chamber is $= 0$.

Then from:

$$\text{Force } (F) = \text{Pressure } (P) \times \text{Area } (A)$$

$$A = \pi r^2 = \pi \times \left(\frac{10^{-7}}{2}\right)^2 = \frac{\pi}{4} \times 10^{-14}\ \text{m}^2$$

At a depth of 10 m then:

$$P = \text{pressure due to 10 m water} + \text{atmospheric pressure}$$
$$= 10^5 + 10^5$$
$$= 2 \times 10^5\ Nm^{-2}$$

$$F = P \times A = 2 \times 10^5 \times \frac{\pi}{4} \times 10^{-14}$$

$$= \frac{\pi}{2} \times 10^{-9}\ N$$

Mass to be supported (10 g) has weight

$$= 10^{-2} \times gN = 10^{-1} N$$

So number of capillaries required is

$$= \frac{10^{-1}}{\pi/2 \times 10^{-9}} = \frac{2 \times 10^8}{\pi}$$

and this is approximately:

$$\simeq 2.\pi \times 10^7 \simeq 6 \times 10^7$$

At a depth of 100 m then:

$$P = \text{pressure due to 100 m water} + \text{atmospheric pressure}$$

$$= 10\times10^5 + 1\times10^5 = 11\times10^5 Nm^{-2}$$

$$F = \frac{11}{2}\times\frac{\pi}{2}\times10^{-9}\,N$$

So number of capillaries required to support $10^{-1}N$

$$= \frac{10^{-1}}{\frac{11}{2}\times\frac{\pi}{2}\times10^{-9}} = \frac{2}{11}\times\frac{2}{\pi}\times10^8$$

and this is approximately:

$$= \frac{2}{11}\times2.\pi\times10^7 \simeq 10^7.$$

We cannot at present make any direct comparisons with the observations of the breaking force of a single sucker made by Parker (1921), as the suckers were larger than those examined here.

Discussion

The arms of the octopus present, to the unaided eye, a multiple array of superficially simple suction cups. The scanning electron microscope has, however, shown the cuticular surface of the infundibulum to have numerous pegs with minute pores on their surface. This cuticle is presumably pliant since it is essential for intimate contact to be made with the very variable contours encountered. If this is so then initial contact with a foreign surface would compress the pegs and any fluid in the capillaries would be extruded on impact. Each capillary could adhere, make a seal and then act as a suction chamber. The muscles of the sucker would then react to reduce pressure in the chamber. All this must remain speculative until critical experiments can be carried out to test whether the capillaries function in this manner.

If this proposition is correct then it suggests a possible reason for Parker's (1921) findings. Of the suckers tested he found the smaller ones to be more efficient than the larger ones and it is the larger suckers of *Octopus vulgaris* that have the additional, rather large, protuberances. These may have some value when the animal is "walking". The large protuberances are most numerous at the centre of the infundibulum and least at the periphery; this is in contrast with the small pegs which are most numerous at the periphery, where the new ones develop, and least numerous at the centre of the infundibulum. The effect, however, of these large protuberances would be to reduce the number of small pegs in contact with a foreign surface.

Graziadei (1964b, 1965) found the largest number of primary receptors to be in the rim. This is a very prominent feature of the small suckers, although less so in the larger ones where the rim has become much more folded. It is the arm tips, with their small suckers, that explore the surroundings thus obtaining an abundance of information through the many primary receptors of the relatively large rim (Fig. 5). The larger suckers nearer the middle of the arms are used in walking. In them the rim is peripheral to the almost flat infundibulum (Fig. 45), and the primary receptors in it are still well placed to gather information about the substrate.

Besides the change in the relative proportions of thc rim and infundibulum there are also others. The rim is initially fairly simple, radially arranged, cushions with deep folds. In the larger suckers it has many superimposed complex folds. It is in these latter suckers that the rim appears quite small round the edge of the infundibulum, which now has quite large protuberances in addition to the small pegs. These changes occur during growth but may also represent functional changes, the protuberances being present on the larger "walking" suckers. A detailed examination of the suckers along the length of an arm should resolve these problems.

We have found the possible external representation of some receptors in the rim of the suckers, but have not seen any cilia protruding beyond the surface. However, we have found cilia around the sucker of both *Sepia officinalis* and *Taonius megalops*; but notably they were in young specimens. This certainly does not preclude their presence from larger specimens nor from *Octopus vulgaris*, but it must be noted that the fixatives used are not ideal, although they do appear to preserve the tissue of young specimens better.

When an octopus attacks a test stimulus in the laboratory the suckers often retain their hold for some time. Indeed a constantly moving stimulus on an automatic feeder had to be kept stationary after an attack otherwise it was held by the animal for up to three minutes (Nixon, 1968).

J. Z. Young (pers. comm.) has observed the phenomenon of the so-called "sticky suckers" that develop after lesions in the central nervous system. He found it to correlate with damage to the inferior frontal system thus demonstrating a neural basis for the release of the suckers. The suckers make the initial contact with food and by a combination of movements, of both suckers and arms, convey it to the mouth; the opposite action is one of rejection and has been found to be controlled by the posterior buccal lobe (Altman, 1971). These reactions of the suckers are, at least finally, controlled by the central nervous system (see J. Z. Young, 1971).

Incirrata, **Eledone moschata**

Distribution and depth

Eledone moschata is found in the Mediterranean and the adjoining regions of the North Atlantic, usually at depths of between 10 and 100 m, but occasionally to 300 m in some places (Rees, 1956; Mangold-Wirz, 1963).

Habits and food

This species is generally said to prefer sites where the bottom is of mud, but it is also found in sand or gravel and, rarely, amongst rocks (Robson, 1932; Mangold-Wirz, 1963).

The arms

The arms are almost equal in length and are about 75% of the total length. Each has a single longitudinal row of up to 100 suckers. The largest usually reach a diameter of 11% of the mantle length, but in males may reach 22%. The suckers are like those of *Octopus vulgaris* (Robson, 1932). The arms and the suckers have dense and complex musculature, much like *O. vulgaris*.

The suckers

A small sucker has a radially arranged rim with small hillocks, as seen in a sucker of similar size in *O. vulgaris*. The pegs on the infundibulum are also very similar to those of *O. vulgaris*, having a rod-like form with a slight concavity on the upper surface (Fig. 54). The pores on the surface of the pegs are about 0·07 μm in diameter. There is a reduction in the number of pegs passing from the edge down the infundibulum.

The thin epithelial tissue between the rim and the infundibulum is folded (Fig. 54).

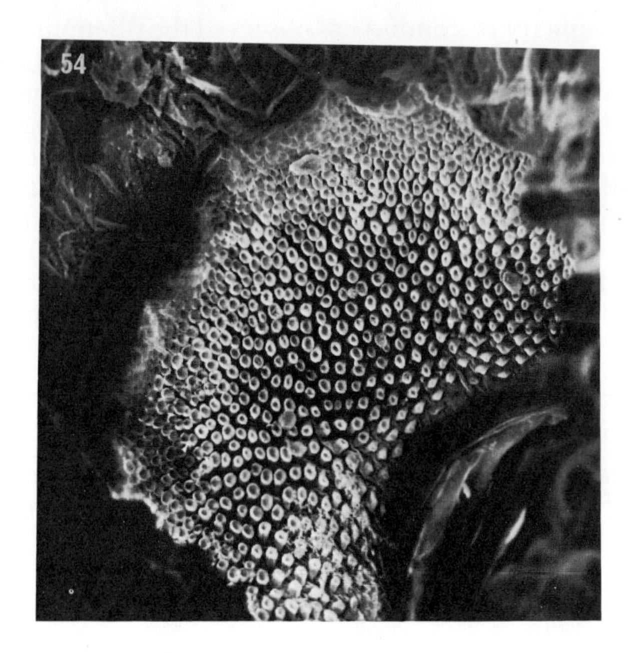

Eledone moschata

FIG. 54. To show the similarity in the surface of the infundibulum with that of *Octopus vulgaris* in Fig. 48. (Field width 147 μm.)

The action of the suckers

The suckers of this octopod, like those of *Octopus vulgaris*, held firmly to the wood on which they rested when the arms were severed from the animal. However, it was much more difficult to get these suckers to adhere to a glass slide than those of *Sepia officinalis* (p. 456).

Again the rim formed a seal, and ink pipetted between the suckers did not penetrate the infundibulum which appeared closely in contact with the glass. The diameter of the infundibulum changed from 8 to 7 mm when the peduncle was pulled. The suckers of *Eledone moschata* were more readily pulled from the glass than were those of *Sepia officinalis*.

Discussion

The shape, form and arrangement of the pegs on the infundibulum is very like that in *Octopus vulgaris*. Both animals are found in similar depths of coastal waters and probably take much the same sort of prey.

Incirrata, ***Octopus salutii***

Distribution and depth

In the Mediterranean it is found from the Bay of Naples to the Catalan Sea (Naef, 1923; Robson, 1929; Mangold-Wirz, 1963). It lives where the bottom is of mud, or mud and gravel, at depths of 70 to 400 m, the majority being found between 120 and 250 m (Mangold-Wirz, 1963).

The arms

The arms are some 75% of the total length (Robson, 1929) and the third pair are the longest (Naef, 1921, 1923). The suckers attain a maximum diameter of 5% of the mantle length in the female, while the 10–12th suckers are strikingly enlarged in the male (Naef, 1921, 1923; Robson, 1929). Figure 55 shows the complex musculature of part of the arm and of the sucker.

The suckers

The rim of the sucker is arranged radially and the cushions have deep folds (Figs 56–58a), and on their surface the borders of the underlying cells can be discerned by their elevated junctions (Fig. 56). At the apices of some of these borders there are openings about 9 μm in diameter, but nothing was seen to protrude from them. The depth of the folds is best seen in a section of the sucker (Fig. 57).

The infundibulum was also radially arranged and the pegs restricted to longitudinal cushions separated by regions without pegs (Fig. 58a), thus effectively reducing the number of pegs and so the number of capillaries. The pegs do not have such a uniform appearance as those of *Octopus vulgaris* and their upper surface is rounded (Fig. 58b) and without the well-defined concavity seen on the pegs of *O. vulgaris*. The size of the pores of the surface of the pegs is about 0·06 μm in diameter.

Discussion

Unfortunately little is known of the habits of *Octopus salutii* except of its reproduction (Mangold-Wirz, 1963).

We know from the calculation for *O. vulgaris* that, at least theoretically, fewer capillaries are needed at greater depths to produce the same breaking force. This arrangement of spaces between the cushions of pegs on the infundibulum provides some confirmation of the suggestion that the pores on the pegs act as suction devices.

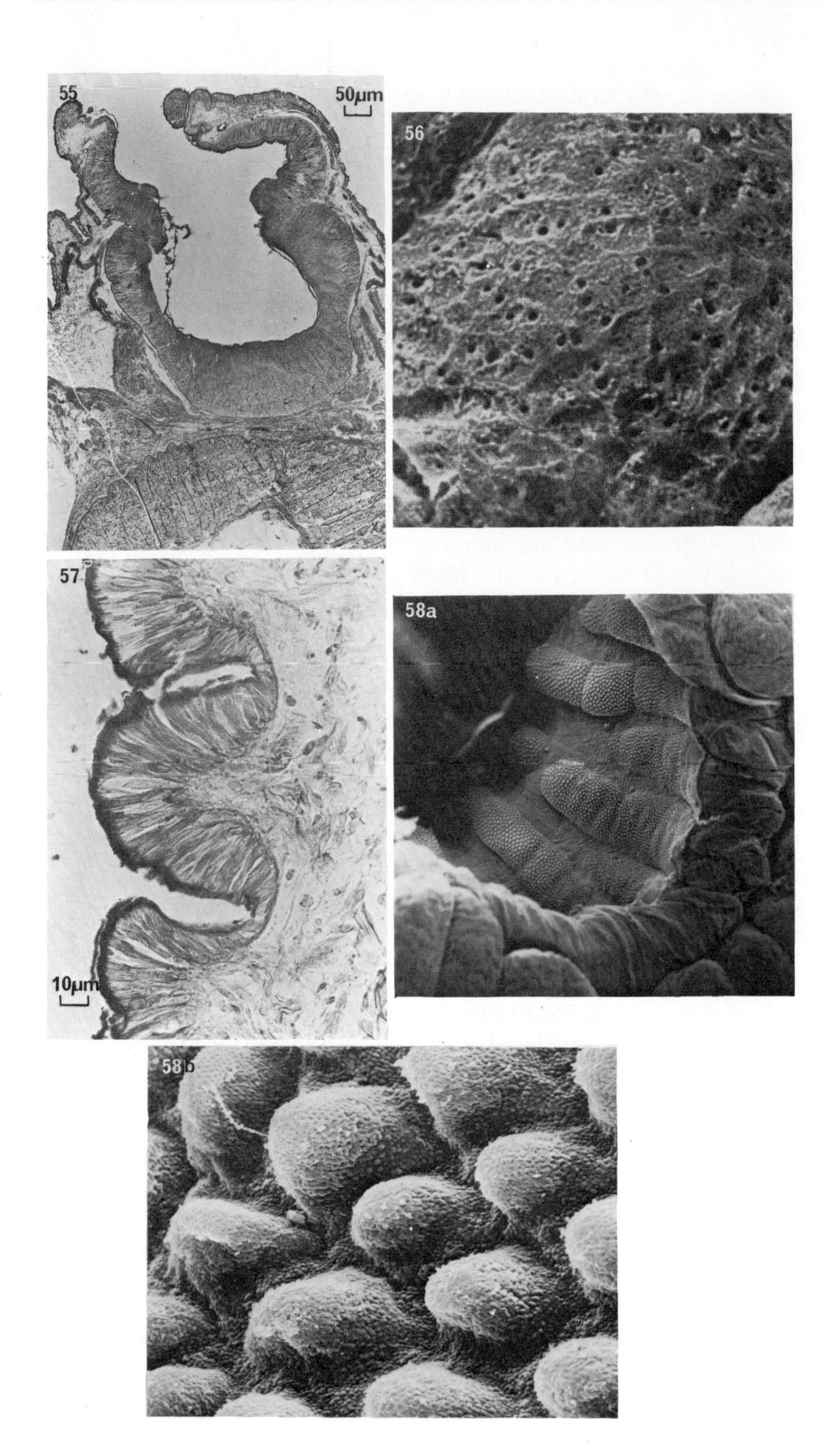

55
50μm
56
57
10μm
58a
58b

The openings on the rim of the sucker are quite large but as yet we know nothing of their function.

Incirrata, ***Scaeurgus unicirrhus***

Distribution and depth

This is a cosmopolitan species recorded in the warm waters of all oceans, and in the Mediterranean (Rees, 1954; Mangold-Wirz, 1963). The adults are benthic, living on the continental shelf at depths of 60–450 m (Mangold-Wirz, 1963). The larval phase is oceanic, and the so-called macrotritopus larvae have been caught in the Atlantic Ocean far from land (Clarke, 1969). The majority have been taken between 0 and 100 m, but a small number were caught below 600 m (Rees, 1954; Clarke, 1969; Lu & Clarke, 1975b).

The arms

The larval stage is striking in that the third pair of arms are greatly elongated, being some 50% of the total length of the animal (Rees, 1954; Clarke, 1969). This is in contrast with the adult condition where all the arms are similar in length although the third pair remain the longest (Mangold & Portmann, 1964). Figure 59 shows the musculature of part of the arm and a sucker of an adult.

The suckers

The sucker in Fig. 60 has been distorted, but the rim is radially arranged. The pegs are cone-shaped and closely packed on long cushions arranged radially on the infundibulum, alternating with areas devoid of pegs (Fig. 61). The old cuticle is in the process of being shed, like that of *Octopus vulgaris*, and the new one lies below

---

Octopus salutii

FIG. 55. A sagittal section to show the general features of the suckers. The rim has quite deep folds and the thin epithelial tissue between the rim and the infundibulum can be seen. The slight obliqueness of the section reveals the cushions on the infundibular surface overlying the tall columnar cells. The radial muscles are deep and the sphincter quite large. (Cajal.)

FIG. 56. The rim has clearly demarcated radial cushions and at high magnification the surface has pores, of up to 9 μm diameter, whose contents, if any, were below the surface or lost during preparation. (Field width 454 μm.)

FIG. 57. The rim of a sucker at high magnification to show the depth of the folds. (Cajal.)

FIG. 58. (a) The infundibulum has a well defined radial arrangement in which the pegs are confined to elongated cushions. (Field width 654 μm.) (b) The infundibulum at higher magnification to show the rather rounded pegs and the small pores. (Field width 8 μm.)

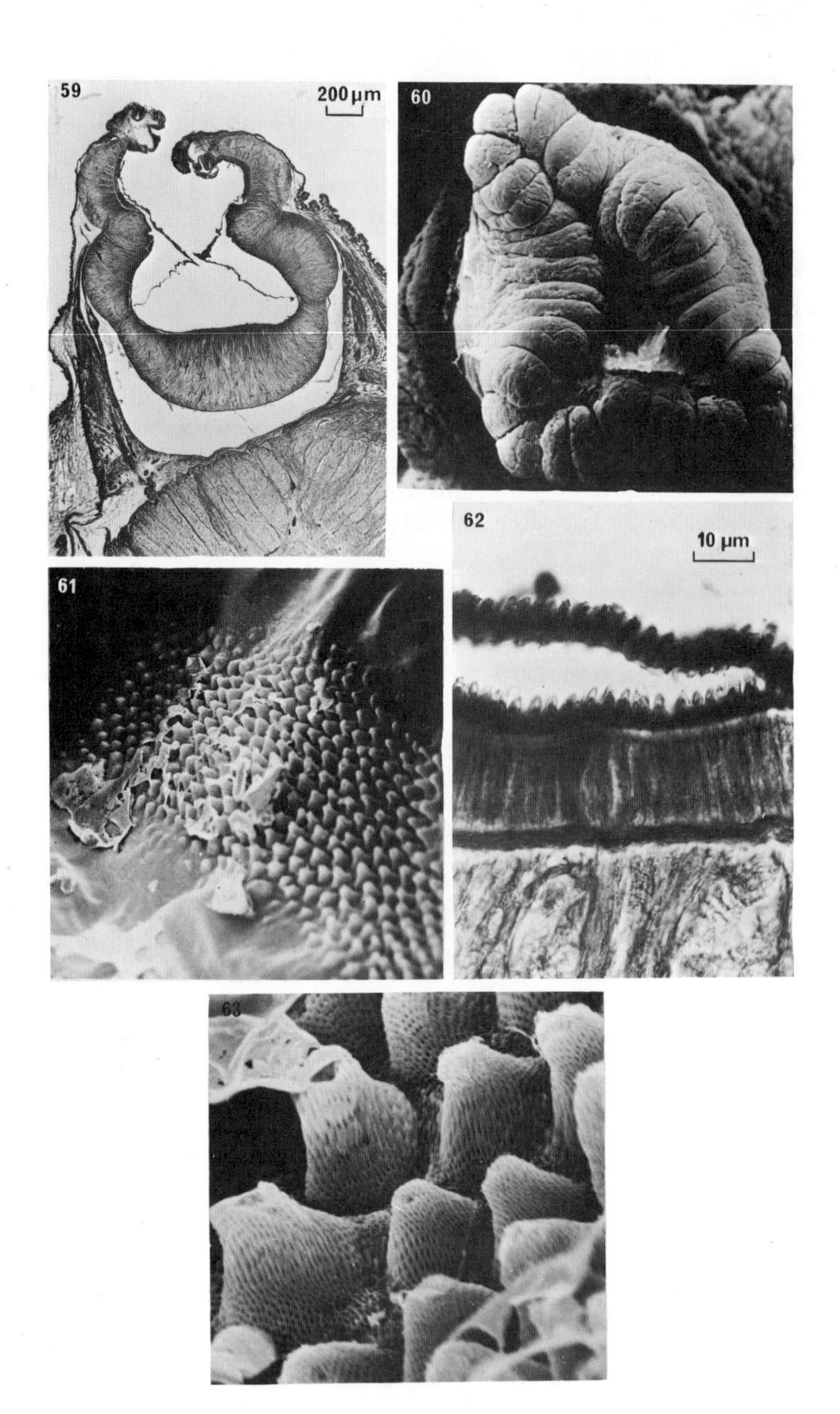
59
200 µm
60
61
62
10 µm
63

attached to the tall columnar cells (Fig. 62). Most of the pores on the surface of the pegs are rather elongated, but those at the base are more nearly circular; they are about 0·06 μm in diameter. The apex of some pegs is slightly concave (Fig. 63).

Discussion

It is notable that both *Octopus salutii* and *Scaeurgus unicirrhus*, rather deeper water octopods than *Octopus vulgaris* or *Eledone moschata*, have reduced the number of pegs on the infundibulum by limiting them to cushions separated by regions without pegs. In *Scaeurgus unicirrhus* the pegs are cone-chaped so that it is only the capillaries on or near the apices of the pegs that would make contact with a foreign surface. Thus a smaller number of capillaries would make contact than in *Octopus vulgaris* or *Eledone moschata*, where the pegs are rod-like with a larger area at their surface of contact, and hence a larger number of capillaries. This goes some way to confirm the suggestion that fewer capillaries are needed to form the close mechanical attachment of sucker infundibulum to the prey at a greater depth.

Incirrata, ***Japetella diaphana***

Distribution and depth

This is a pelagic octopod and has a wide distribution around the world between latitudes 40°N and 50°S (Robson, 1932). It has been taken in opening-closing nets from 0 to 1500 m depth, with some indication of an ontogenetic descent (Clarke & Lu, 1974, 1975; Lu & Clarke, 1975b).

Physiology and behaviour

This octopod is gelatinous. Owing to the reduction in concentration of sulphate, it is half-way to being neutrally buoyant (Denton & Gilpin-Brown, 1973).

Scaeurgus unicirrhus

FIG. 59. A photomicrograph showing the general features of arm and suckers. (Cajal.)

FIG. 60. This sucker was distorted during fixation, as were the others, nevertheless the radial arrangement of the rim is readily seen. (Field width 1250 μm.)

FIG. 61. The infundibulum, like that of *Octopus salutii*, has pegs on radially arranged longitudinal cushions separated by areas without pegs. (Field width 129 μm.)

FIG. 62. The infundibulum to show one cuticle in the process of being shed with the new cuticle below, attached to tall columnar cells. (Cajal.)

FIG. 63. The pegs are conical with pores over their surface and the apex of the cone may have a slight concavity. (Field width 30 μm.)

The arms

The tissues are extensively vacuolated (J. Z. Young, his chapter, this volume) including the arms, but the suckers are muscular (Fig. 64). The arms are unequal in length, the third pair being almost as long as the body and nearly twice as long as the others (Hoyle, 1885). The suckers are small, close together and arranged in a single row.

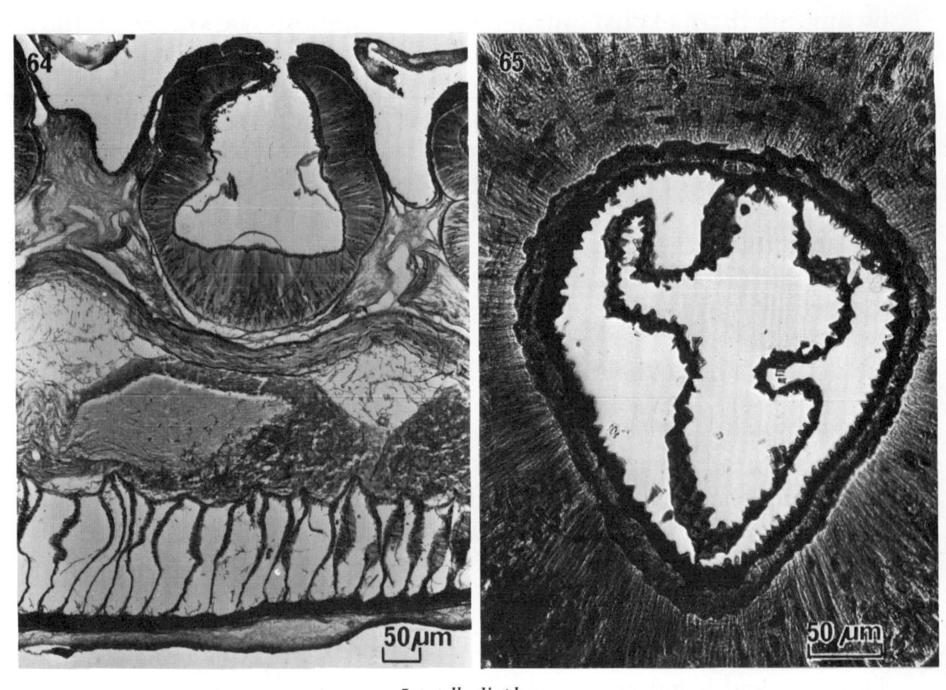

Japetella diaphana

FIG. 64. A longitudinal section of an arm showing the organization of the vacuolated tissues in compartments. The suckers by contrast have dense musculature. (Cajal.)

FIG. 65. A transverse section across a sucker to show the conical form of the infundibular pegs. One disc of cuticle is in the process of being shed. (Cajal.)

The suckers

Figure 65 is a transverse section through the infundibulum and the cone-shaped pegs are quite prominent. It is of note that the old cuticle is in the process of being shed. The shape of the pegs suggests they resemble those of *Scaeurgus unicirrhus* (see p. 494) although we were unable to obtain material for scanning electron microscopy.

Discussion

We know nothing of the habits of this animal but it has been found at considerable depths, and the shape of the pegs on the infundibulum may well be similar to those of *S. unicirrhus.*

Incirrata, ***Argonauta argo***

Distribution and depth

A. argo, the paper nautilus, is a cosmopolitan species of subtropical waters (Robson, 1932; Mangold-Wirz, 1963). It lives near the surface, and has been caught in opening-closing nets at 50–20 m depth but may be capable of some vertical migration (Roper, 1972). *Argonauta* sp. were captured in opening-closing nets in the North Atlantic, where the majority were between 0 and 50 m depth, and one female at 350–300 m (Lu & Clarke, 1975b).

Habits and food

The paper nautilus is said to eat small fish and crustaceans (see Robson, 1932). It has occasionally been observed in aquaria (J. Z. Young, 1960; Zeiller, 1970). When feeding an adult female, definite reactions were elicited as pieces of dead fish touched the extended web (a thin flattened extension of the first pair of arms that almost covers the shell). The ventral arm of the ipsilateral side swept upwards over the web until the food was seized. The suckers and the edge of the web were found to be the most sensitive regions (J. Z. Young, 1960). Another animal kept in an aquarium (shell width 120 mm) was given live brine shrimps which, although ignored in daylight, were eaten during the night (Zeiller, 1970). A small fish, 20 mm long, gently touched the web without reaction from *Argonauta,* but when it touched the beak, was quickly bitten in two with one clean snap. The posterior half of the fish was held while the front half was eaten, and the whole fish was soon gone.

The arms

Niemec (1885) examined the suckers histologically and found marginal hooks between the rim and the infundibulum. The small tooth-like processes present on the infundibular surface were also seen by Niemec who suggested that they augmented the adhesive power of the sucker.

Unlike the bottom-living octopods, the arms of the paper nautilus are partly muscular and partly vacuolated (Fig. 66). The suckers are muscular and of the usual octopod form, with dense musculature in contrast with the peduncle and the arm.

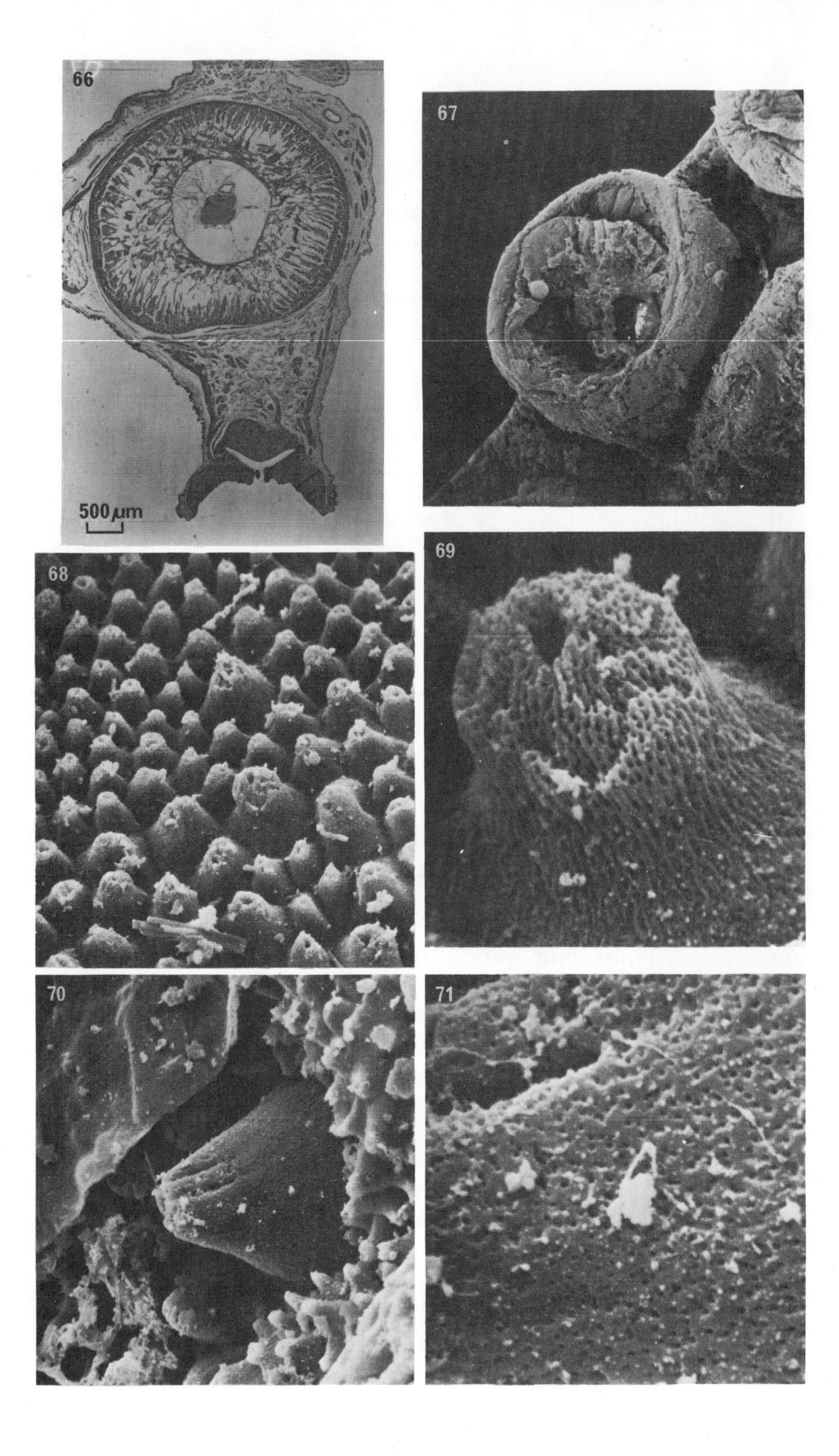
66
67
500 μm
68
69
70
71

The suckers

The infundibulum is divided radially (Fig. 67) and each sector has quite closely packed pegs (Fig. 68). They are mainly conical but there is considerable variation in both size and form. Many of the pegs in the field appear not only to have minute pores over the surface but in addition to have a rather larger aperture near the centre of the slight concavity on the apex of the cone. This is best seen on one of the larger pegs at high magnification (Fig. 69). The large central hole, approximately 0·6 μm in diameter, appears to be a structural feature as the wall is smooth. The small pores are approximately 0·08 μm in diameter.

In addition to these pegs there are others, some five times taller, to be seen chiefly near the rim (Fig. 70). These are stout conical forms whose surface is studded with holes some 0·09 μm in diameter, the same order of magnitude as those on the pegs (Fig. 71). The matrix of these large cones appears on the surface to be denser than that of the small cuticular pegs.

Discussion

The infundibular surface of the sucker of *Argonauta argo* differs from that of the other octopods examined here in having the relatively much larger, cone-shaped pegs. They are a special feature perhaps for feeding or capture of prey by these largely surface-living octopods.

DISCUSSION

Structural features of the sucker surface

There is little doubt that the suckers on the tentacular clubs of *Sepia officinalis* are powerful organs of attachment. This was clearly

---

Argonauta argo

FIG. 66. A transverse section through an arm to show the extent of the vacuolation in both the arm and the peduncle of the sucker, whereas the sucker has dense muscles. (Cajal.)

FIG. 67. A sucker showing the general external features although it is not well preserved. (Field width 1904 μm.)

FIG. 68. The infundibulum is covered with densely packed, conical pegs of rather variable size. The majority have quite a large pore at their apices in addition to the fine pores all over the cuticle. (Field width 43 μm.)

FIG. 69. One of the pegs to show the small pores over the surface and the larger apical pore. This appears to be a structural feature as its walls are smooth and without distortion. (Field width 9 μm.)

FIG. 70. In addition to the pegs seen in Figs 68 and 69 there are other much larger protuberances. (Field width 46 μm.)

FIG. 71. The large protuberance in Fig. 70 at higher magnification to show its surface covered with small pores. The matrix appears much denser than that of the pegs in Fig. 69 and the pores are smaller. (Field width 9 μm.)

demonstrated by the ability of the animal to catch prawns after the removal of one tentacle (Messenger, 1968). It seems likely that these properties are largely due to the rim forming a seal initially, then to the form of the infundibular surface, followed by the action of the muscles. Thus several stages in the action of the suckers can be tentatively recognized. The initial impact forms the seal and is followed by compression of the pegs on the infundibulum, then perhaps by adhesion of the pores on the pegs acting like small suction chambers. At the same time the infundibular surface must retain its shape and hold the seal so that the acetabular muscles can act to produce suction in the chamber. Finally the sucker must be released.

An experimental study of the function of the sucker is needed to obtain proof, or refutation, of the hypothesis proposed in this paper. Nothing is known either of the chemical or physical properties of the infundibular cuticle, or the inner rings of the decapod sucker, except that the inner ring contains no chitin (Rudall, 1955). An analysis of the chemical properties of these structures is in progress (S. Hunt & Nixon).

That a seal is obtained is certain in *Octopus bimaculatus* (Parker, 1921) and *O. vulgaris* (Dilly, Nixon & Packard, 1964), otherwise the breaking forces obtained from these animals would not be possible. Parker (1921) using isolated suckers of *Octopus bimaculatus* found ones of 2·3 and 6 mm diameter to have breaking forces of 68 and 50% respectively of the theoretical maximum. If the larger suckers of *O. bimaculatus* develop protuberances, as do those of *O. vulgaris*, then these would reduce their efficiency by preventing the small pegs from making such intimate contact as in the smaller suckers where the protuberances are absent.

The hypothesis given on p. 486, for the mechanical attachment of a sucker to a foreign surface by the minute pores on the pegs of the infundibulum acting as individual suction chambers, indicates that fewer pores are needed at greater depth. This is lent support by the examination of suckers of other incirrate octopods. It is the more shallow-living *Octopus vulgaris* and *Eledone moschata* that have the infundibulum covered with pegs, the only exception being in the depth of the fold between the cushions, whereas in the two deeper-living species examined, *Octopus salutii* and *Scaeurgus unicirrhus*, the pegs are restricted to longitudinal cushions separated by areas entirely without pegs, so that the number of pores is reduced. Moreover, *S. unicirrhus*, the deeper-living of these two species, has cone-shaped pegs, where presumably only the pores close to the apex would act as suction chambers, while in *O. salutii* the pegs are blunt, similar to those of *O. vulgaris* and *Eledone moschata*. Once the infundibular surface is in intimate contact with a foreign surface then the muscles of the acetabulum can act to produce suction.

Further support for the hypothesis comes from the structure of the suckers of the very deep-living cephalopods. In *Cirroteuthis* sp. the infundibulum occupies only a very small region of the sucker and does not appear to have the cuticular pegs seen in other octopods and decapods. The major part of the sucker is the chamber surrounded by very deep radial muscles with a cubical epithelium suggesting that it is analogous with the acetabulum of the incirrate octopods. The sucker of *Vampyroteuthis infernalis* appears to lack an infundibulum; the straight, smooth region of the neck, although without pegs, may be considered as the infundibulum. The rim of this animal also shows some resemblance to that of *Cirroteuthis*.

It is notable that the size of the pores on the surface of the pegs on the infundibulum of the suckers of decapods and octopods, of various sizes, is very constant between 0·04 and 0·10 μm (Table II).

TABLE II

A summary of the approximate diameter of the pores of the infundibular pegs of the animals investigated

	Arm (μm)	Tentacle (μm)
Spirula spirula	0·07	0·10
Sepia officinalis	0·07	0·08
Alloteuthis subulata	0·10	0·07
Mastigoteuthis sp.	0·08	0·05
Taonius megalops	0·08–0·10	0·04
Scaeurgus unicirrhus	0·06	
Octopus salutii	0·06	
Octopus vulgaris	0·08	
Eledone moschata	0·07	
Argonauta argo	0·08	

This is perhaps the best support for the suggestion of a similar function of the infundibulum throughout these two major groups of cephalopods. Some other animals which attach themselves to foreign surfaces make use of fine tubular hairs and setae (see Nachtigall, 1974). The climbing organ of *Rhodnius prolixus*, a blood-sucking bug, has 5000 tubular hairs, each about 1μm in diameter. These tubes are also the outlets for a glandular secretion (Gillet & Wigglesworth, 1932). The digital setae of lizards have slightly spatulate ends of about 0·2–0·7 μm diameter that allow these animals to walk up vertical walls and even across ceilings. There are a total of about one million setae on the digital lamellar surface of a gecko (Ruibal & Ernst, 1965). There is, however, some difference since in the cephalopods the pores on the surface of the pegs of the infundibulum are almost certainly the spaces between a series of rods and are not tubules as in the insect *Rhodnius*, but until a transmission electron microscope study is made we cannot be absolutely certain of this.

The release mechanism appears to be slow-acting under some conditions, but rapid under others. *Octopus vulgaris* will hold test stimuli for as long as 2 to 3 min (Dews, 1959; Nixon, 1968). The tentacular club of *Sepia officinalis* remains attached to a glass jar for some seconds when trying to attack mysids within it (M. J. Wells, pers. comm.). *Loligo vulgaris*, when attempting to take prawns from the bottom of a polythene tank, were unable to release their tentacular suckers for a minute or more (J. Barlow, pers. comm.). This suggests that one release mechanism is similar in these three animals belonging to three groups of coleoids. It is presumably mechanical, muscular and neural. In those decapods where there are sharp, retaining, teeth on the inner ring, these would penetrate the prey and the release mechanism may differ. On some occasions, however, *Octopus vulgaris* can release its suckers rapidly: for example, on encountering a noxious object such as a hermit crab in a shell with an anemone attached to it (Boycott, 1954).

The suckers and prey capture

We know that *Sepia officinalis* feeds mainly on prawns and crabs, usually captured with the suckers of the tentacles, but sometimes with those of the arms (Messenger, 1968). The surface architecture of the suckers of both the tentacles and the arms is very similar. The inner ring has blunt projections and the pegs of the infundibulum are rod-like with a concavity on the upper surface which is rather rough. Their features resemble those of both arms and tentacles of *Spirula spirula*, the tentacles of the larval *Taonius megalops*, and those of the arms of *Alloteuthis subulata*. All these animals prey at least some of the time upon crustaceans.

Where the inner ring of the sucker has sharp, pointed teeth they are presumably used in some way to retain prey or food particles. Such toothed suckers may be on arms or tentacles or both (see for example Chun, 1910, 1914). The small suckers of the arms of *Mastigoteuthis* have teeth which could help in holding small particles. In many squids the inner ring has sharp teeth, and in a few there are only large hooks. Such structures are perhaps most important in the capture of gelatinous animals, or to help to retain prey or food particles.

It is of note that while the arms of some of the octopods and decapods were vacuolated this was never found in the tentacles; thus while the arms may become modified to function as organs of buoyancy the suckers and tentacles are not so sacrificed.

Sensory structures

A prerequisite of a chemoreceptor is to be in a site where it can monitor small changes in the chemicals of the environment rapidly. In invertebrates both cilia and microvilli have been proposed as chemosensory (Laverack, 1974). In this brief survey we have found tufts of cilia protruding from pores on the outside of the suckers of larval and post-larval *Taonius megalops*, on the suckers and the protective margin of a 12-day-old *Sepia officinalis* as well as on the rim of the sucker and on the tip of the cirrus of an adult *Vampyroteuthis infernalis*. The pores on the hillocks of the rim of *Octopus vulgaris* and the numerous pores on the rim of *O. salutii* are well sited for the ciliated endings of receptor cells described in this region of *O. vulgaris* (Graziadei, 1962, 1965), an octopod (Graziadei & Gagne, 1973), in the external epithelium of *Sepia officinalis* (Graziadei, 1964a), of *Lolliguncula brevis* (Santi, 1975; Santi & Graziadei, 1975) and on the ocular tentacles of *Nautilus* (Barber & Wright, 1969). The cilia are present in very large numbers and closely packed on the ocular tentacles of *Nautilus* (Dilly & Nixon, unpublished). On the other sites found so far the cilia emerge in small tufts from relatively well separated pores, suggesting that these latter are not motile. Although we have only found cilia on the suckers of young animals, this in no way precludes their presence in larger animals, but they may be more widely separated with growth.

Graziadei (1964b) found sensory cells in the infundibular epithelium of *Octopus vulgaris* and suggested that the distal poles of these cells reached the surface, through the cuticle, to be in contact with the sea-water. The free surface of some sensory cells have about 30 microvilli, about 0·1 μm in diameter and 1·5 μm long. These microvilli might be able to enter the capillaries, 0·06–

0·10 μm in diameter, of the pegs of the infundibulum (Table II); they are less likely to reach the upper surface of the pegs since these were 3 μm in height in one specimen of *O. vulgaris*. However, the cilia of the receptor cells may barely extend beyond the surface of the cell as in the olfactory organ of *Lolliguncula brevis* (Emery, 1975). They may perhaps enter the hollow part of the infundibular pegs, but this needs further examination with the transmission electron microscope to elucidate the exact sites of the endings of these receptors. Graziadei (1964a) demonstrated multipolar neurons between the epithelial cells of the infundibulum of the sucker of *Sepia officinalis* and suggested that they were mechanoreceptors. In a squid, *Lolliguncula brevis*, multipolar cells were shown to be in a discrete ring below the outer ring of pegs of the infundibulum (Santi, 1975; Santi & Graziadei, 1975). There was also evidence of inhibition by efferent fibres which synapse on these receptors.

We know from behavioural and neurological studies that the suckers of *Octopus vulgaris* have a much greater repertoire of responses to chemical and tactile stimuli than do those of *Sepia officinalis*. *Octopus vulgaris* has some tens of thousands of primary receptors in each of some 2400 suckers (Graziadei, 1971), *Sepia officinalis* about 1200 receptors (Graziadei, 1965) in each of some 1000 suckers (estimated from Tompsett, 1939) including those of the tentacles, while *Lolliguncula brevis* has about 465 receptors (Santi, 1975) in each of about 600 suckers (estimated from Dillon & Dial, 1962), again including tentacular ones (Table III). There

TABLE III

A very crude estimate of the total number of receptors present on one sucker from a cuttlefish, squid and octopus

		Total number suckers (including tentacles) approx.	Primary receptors estimated for one small sucker	Approximate estimate of total number of receptors in the suckers
Sepioidea	*Sepia officinalis*	1000	1120–1420	1 300 000
Teuthoidea	*Lolliguncula brevis*	600	465	279 000
Octopoda	*Octopus vulgaris*	2400	tens of thousands	24 000 000

must therefore be very marked differences in the amount of information that these animals would be able to obtain from their environment, it being most restricted in the squid. These animals thus form a series and the numbers of suckers and receptors in them are a reflection of their habits. *Octopus vulgaris* lives mostly in contact with, or close to, the sea floor, swimming mainly at dawn and dusk, *Sepia officinalis* lies buried in sand or gravel during the day but swims at night, while *Lolliguncula brevis,* like the majority of squids, probably spends most of its life swimming. The value of chemotactile information thus diminishes as less time is spent on the sea bottom.

Growth of the infundibular surface

In the decapods the size of the infundibulum is increased by the transformation of the flat peripheral processes into polygonal ones with pegs. The underlying epithelial cells are shortest at the periphery of the infundibulum and longest when adjacent to the inner ring.

The octopod infundibulum also increases in size by addition to its periphery. The numbers of pegs over a given surface area is much greater at the periphery than it is close to the sphincter muscle; there the pegs are sparser and much shorter. The octopods also shed the infundibular cuticle at intervals.

Other changes were found to occur in the suckers of *Taonius megalops* with the growth of the animal (see Dilly & Nixon, 1976). There is, however, much need of further studies on the development and changes in shape and form of the suckers, particularly in species where the feeding habits are already known, such as *Sepia officinalis* (Messenger & Nixon, in progress) and *Octopus vulgaris.*

Phylogenetic considerations

The suckers of *Vampyroteuthis* may well show an early stage in coleoid evolution, consisting as they do of simple cups (Fig. 2). However, this simplification may be a secondary result of their deep-sea habitat. The suckers of decapods and octopods all have a common plan, with division into an inner acetabulum and an outer infundibulum. In both groups the infundibulum carries a system of minute pores, probably serving for adhesion. These similarities suggest that sepioids, teuthoids and octopods all come from a common stock, as is concluded also by J. Z. Young (his chapter, this volume) from his study of the nervous system.

The suckers of cirroteuthid octopods show some similarities with those of *Vampyroteuthis* with a very muscular acetabulum and poorly developed infundibulum. These similarities may be the result of the shared deep-sea habitat, where the pressure perhaps reduces the need for close adhesion, being quite efficient with a seal and the use of muscles for suction. But it may be that cirroteuthids represent an early stage of evolution of octopods and indeed of all coleoids; their brains are also very different from those of other octopods (J. Z. Young, his chapter, this volume).

Conclusion

From the examination of the suckers of *Sepia officinalis* and knowledge of its habits in prey capture we can now make some predictions about the food of other cephalopods by looking at the surface of their suckers. The octopods showed considerable uniformity in their sucker architecture while there was more variation amongst the decapods.

This study has exposed a host of further problems. Not least of these is the need for a transmission electron microscope study of the formation of the infundibular cuticle and its intimate connection with the underlying epithelial cells. This structure is able to adhere to a foreign surface during the periods when considerable forces of suction are exerted. The cuticle appears to have somewhat unusual properties and an analysis of its constituents may provide some useful clues. Critical experiments are needed to ascertain whether the capillaries of the infundibular cuticle can act individually as suction chambers as has been postulated. Further study of these organs may provide valuable information for making adhesive materials.

That the suckers are efficient organs of attachment cannot be doubted by anyone, particularly those of us who have watched or handled these animals.

Acknowledgements

We are deeply indebted to Drs E. A. Bradley, M. R. Clarke, D. O'Dell, P. J. Herring, K. Mangold and C. F. E. Roper, and to the Staff of The Laboratory, Plymouth; Laboratoire Arago, Banyuls; Stazione Zoologica, Naples; Institute of Oceanographic Sciences, Surrey, for both facilities and help in obtaining material. To Dr. A. Boyde for considerable help in providing facilities for the use of the scanning electron microscope, and also helpful discussion of the manuscript. To Professors R. McN. Alexander and E. J. Denton FRS, Dr Q. Bone and Dr J. Gilpin-Brown, Mr P. Howells, Mr A. P. Kahn and Ms M. Lord for much helpful discussion; Dr J. B. Messenger and Dr M. Wells for reading the manuscript and giving valuable critical advice and discussion. And to Professor J. Z. Young FRS, for reading the manuscript and for all his enthusiastic help and advice throughout this project, as in so many others in the past, and we hope in the future.

To Ms P. R. Stephens for her skill in preparing the histological material and helpful advice during the study; to Ms E. Bailey for much skill and assistance with the SEM and photography.

References

Adam, W. & Rees, W. J. (1966). A review of the cephalopod family Sepiidae. *Scient. Rep. John Murray Exped. 1933–34.* 11: 1–165.

Aldred, R. G. (1974). Structure, growth and distribution of the squid *Bathothauma lyromma* Chun. *J. mar. biol. Ass. U.K.* **54**: 995–1006.

Altman, J. S. (1967). The behaviour of *Octopus vulgaris* Lam. in its natural habitat: a pilot study. *Underwat. Ass. Rep.* **1966–67**: 77–83.

Altman, J. S. (1971). Control of accept and reject reflexes in the octopus. *Nature, Lond.* **229**: 204–206.

Altman, J. S. & Nixon, M. (1970). Use of the beaks and radula by *Octopus vulgaris* in feeding. *J. Zool., Lond.* **161**: 25–38.

Angel, M. & Angel, H. (1974). *Ocean life.* London: Octopus Books.

Arnold, J. M. & Arnold, K. O. (1969). Some aspects of hole-boring predation by *Octopus vulgaris. Am. Zool.* **9**: 991–996.

Barber, V. C. & Wright, D. E. (1969). The fine structure of the sense organs of the cephalopod mollusc *Nautilus. Z. Zellforsch. microsk. Anat.* **102**: 293–312.

Bidder, A. M. (1950). The digestive mechanism of the European squids *Loligo vulgaris, Loligo forbesii, Alloteuthis media* and *Alloteuthis subulata. Q. Jl microsc. Sci.* **91**: 1–43.

Boletzky, S. v. (1974). Élevage de céphalopodes en aquarium. *Vie Milie* **24**(A): 309–340.

Boycott, B. B. (1954). Learning in *Octopus vulgaris* and other cephalopods. *Pubbl. Staz. zool. Napoli* **25**: 67–93.

Boycott, B. B. (1961). The functional organization of the brain of the cuttlefish *Sepia officinalis. Proc. R. Soc.* (B) **153**: 503–534.

Boyde, A. (1974). Histological and cytological methods for the SEM in biology and medicine. In *Scanning electron microscopy.* Wells, O. C. (ed.). New York: McGraw Hill.

Bruun, A. Fr. (1943). The biology of *Spirula spirula* (L). *Dana Rep.* No. 24: 1–46.

Chun, C. (1910, 1914). Die Cephalopoden. *Wiss. Ergebn. dt. Tiefsee-Exp. "Valdivia"* **13**: 1–401; **18**: 403–552.

Clarke, M. R. (1966). A review of the systematics and ecology of oceanic squids. *Adv. mar. Biol.* **4**: 91–300.

Clarke, M. R. (1969). Cephalopoda collected on the SOND cruise. *J. mar. biol. Ass. U.K.* **49**: 961–976.

Clarke, M. R. & Lu, C. C. (1974). Vertical distribution of cephalopods at 30°N 23°W in the North Atlantic. *J. mar. biol. Ass. U.K.* **54**: 969–984.
Clarke, M. R. & Lu, C. C. (1975). Vertical distribution of cephalopods at 18°N 25°W in the North Atlantic. *J. mar. biol. Ass. U.K.* **55**: 165–182.
Colman, J. S. (1954). Gear, narrative and station list. *Bull. Br. Mus. nat. Hist.* (Zool.) **2**: 119–130.
Denton, E. J. & Gilpin-Brown, J. B. (1960). Daily changes in the buoyancy of the cuttlefish. *J. Physiol., Lond.* **151**: 36–37 *P.*
Denton, E. J. & Gilpin-Brown, J. B. (1961). The effect of the light on the buoyancy of the cuttlefish. *J. mar. biol. Ass. U.K.* **41**: 343–350.
Denton, E. J. & Gilpin-Brown, J. B. (1973). Flotation mechanisms in modern and fossil cephalopods. *Adv. mar. Biol.* **11**: 197–268.
Denton, E. J., Gilpin-Brown, J. B. & Howarth, J. V. (1967). On the buoyancy of *Spirula spirula*. *J. mar. biol. Ass. U.K.* **47**: 181–191.
Denton, E. J., Gilpin-Brown, J. B. & Shaw, T. I. (1969). A buoyancy mechanism found in a cranchid squid. *Proc. R. Soc.* (B) **174**: 271–279.
Dews, P. B. (1959). Some observations on an operant in the octopus. *J. exp. anal. Behav.* **2**: 57–63.
Dillon, L. S. & Dial, R. S. (1962). Notes on the morphology of the common Gulf squid *Lolliguncula brevis* (Blainville). *Texas J. Sci.* **14**: 156–166.
Dilly, P. N. (1972). *Taonius megalops*, a squid that rolls up into a ball. *Nature, Lond.* **237**: 403–404.
Dilly, P. N. (1973). The enigma of colouration and light emission in deep-sea animals. *Endeavour* **32**: 25–29.
Dilly, P. N. & Nixon, M. (1976). Growth and development of *Taonius megalops* (Mollusca, Cephalopoda) and some phases of its life cycle. *J. Zool., Lond.* **179**: 19–83.
Dilly, N., Nixon, M. & Packard, A. (1964). Forces exerted by *Octopus vulgaris*. *Pubbl. staz. zool. Napoli* **34**: 86–97.
Dilly, P. N., Nixon, M. & Young, J. Z. (1977). *Mastigoteuthis*—the whip-lash squid. *J. Zool., Lond.* **181**: 527–559.
Emery, D. G. (1975). The history and fine structure of the olfactory organ of the squid *Lolliguncula brevis* Blainville. *Tiss. Cell* **7**: 357–367.
Ghiretti, F. (1960). Toxicity of *Octopus* saliva against Crustacea. *Ann. N.Y. Acad. Sci.* **90**: 726–741.
Giersberg, H. (1926). Ueber den chemischen Sinn von *Octopus vulgaris*. *Z. vergl. Physiol.* **3**: 827–838.
Gillet, J. D. & Wigglesworth, V. B. (1932). The climbing organ of an insect, *Rhodnius prolixus* (Hemiptera; Reduviidae). *Proc. R. Soc.* (B) **111**: 364–376.
Girod, P. (1884). Recherches sur la peau des Céphalopodes. La ventouse. *Archs zool. exp. gén.* (2) **2**: 379–401.
Gray, J. E. (1845). On the animal of *Spirula*. *Ann. Mag. nat. Hist.* **15**: 257–260.
Graziadei, P. (1959). Sulla presenza di elementi nervosi negli epiteli di rivestimento della ventosa di *Sepia officinalis* *Z. Anat. EntwGesch.* **121**: 103–115.
Graziadei, P. (1962). Receptors in the suckers of *Octopus*. *Nature, Lond.* **195**: 57–59.
Graziadei, P. (1964a). Receptors in the sucker of the cuttlefish. *Nature, Lond.* **203**: 384–386.
Graziadei, P. (1964b). Electron microscopy of some primary receptors in the sucker of *Octopus vulgaris*. *Z. Zellforsch. mikrosk. Anat.* **64**: 510–522.

Graziadei, P. (1965). Sensory receptor cells and related neurons in cephalopods. *Cold Spring Harb. Symp. quant. Biol.* **30**: 45–57.

Graziadei, P. (1971). The nervous system of the arms. In *The anatomy of the nervous system of* Octopus vulgaris. Young, J. Z. Oxford: Clarendon Press.

Graziadei, P. P. C. & Gagne, H. T. (1973). Neural components in the octopus suckers. (Paper 242 in Abstracts of papers presented at the thirteenth annual meeting of the American Society for cell biology.) *J. Cell Biol.* **59**: 121a.

Guerin, J. (1908). Contribution à l'étude des systèmes cutanés musculaire et nerveux de l'appareil tentaculaire de Céphalopodes. *Archs zool. exp. gén* (4) **8**: 1–178.

Holmes, W. (1940). The colour changes and colour patterns of *Sepia officinalis* L. *Proc. zool. Soc. Lond.* (A) **110**: 17–36.

Hoyle, W. E. (1885). Diagnosis of new species of Cephalopoda collected during the cruise of H.M.S. *Challenger*—Part I. The Octopoda. *Ann. Mag. nat. Hist.* (5) **15**: 222–236.

Kayes, R. J. (1974). The daily activity pattern of *Octopus vulgaris* in a natural habitat. *Mar. Behav. Physiol.* **2**: 337–343.

Kerr, J. G. (1931). Notes upon the Dana specimens of *Spirula* and upon certain problems of cephalopod morphology. *Dana Rep.* No. 8: 1–34.

Kölliker, A. (1858). Untersuchungen zur vergleichenden Gewebelehre angestellt in Nizza im Herbste 1856. *Verh. Phys.-Med. ges.* Würzburg. **8**: 1–128.

Laverack, M. S. (1974). The structure and function of chemoreceptor cells. In *Chemoreception in marine organisms*. Grant, P. T. & Mackie, A. M. (eds). London and New York: Academic Press.

Le Souef, A. S. & Allan, J. K. (1933). Habits of the Sydney octopus (*Octopus cyaneus*) in captivity. *Aust. Zool.* **7**: 373–376.

Lu, C. C. & Clarke, M. R. (1975a). Vertical distribution of cephalopods at 11°N, 20°W in the North Atlantic. *J. mar. biol. U.K.* **55**: 369–389.

Lu, C. C. & Clarke, M. R. (1975b). Vertical distribution of cephalopods at 40°N, 53°N and 60°N at 20°W in the North Atlantic. *J. mar. biol. Ass. U.K.* **55**: 143–163.

Macdonald, A. G. (1975). *Physiological aspects of deep sea biology*. Monograph of the Physiological Society No. 31. Cambridge: University Press.

Mangold-Wirz, K. (1963). Biologie des céphalopodes benthiques et nectoniques de la Mer Catalane. *Vie Milieu* Suppl. No. 13: 1–285.

Mangold, K. & Portmann, A. (1964). Dimensions et croissance relatives des octopodidés Méditerranéens. *Vie* Milieu Suppl. No. 17: 213–233.

Messenger, J. B. (1968). The visual attack of the cuttlefish, *Sepia officinalis. Anim. Behav.* **16**: 342–357.

Messenger, J. B. (1973). Learning in the cuttlefish, *Sepia. Anim. Behav.* **21**: 801–826.

Muus, B. J. (1956). Development and distribution of a North Atlantic pelagic squid, Family Cranchiidae. *Meddr Danm. Fisk-og. Havunders,* (N.S.) **1**: 3–13.

Muus, B. J. (1963). Cephalopoda. *Fich. Ident. Zooplancton.* Nos 94–98.

Nachtigall, W. (1974). *Biological mechanisms of attachment. The comparative morphology and bioengineering of organs for linkage, suction and adhesion.* Berlin: Springer-Verlag.

Naef, A. (1921, 1923). Cephalopoda. *Fauna Flora Golfo Napoli* **35** (1) **1**: Fasc. I and II. (Translated from German; Israel Program for Scientific Translations, Jerusalem 1972.)

Niemec, J. (1885). Recherches morphologiques sur les ventouses dansle règne animal. *Rech. zool. Suisse* **2**: 1–147.
Nixon, M. (1968). *Feeding mechanisms and growth in* Octopus vulgaris. Ph.D. Thesis, Univeristy of London.
d'Orbigny, A. (1845). *Mollusques vivants et fossils ou description de toutes les espèces de Coquilles et de Mollusques.* **I**. *Geologique et Geographique.* Paris: Gide.
Owen, R. (1879). Observations on the anatomy of *Spirula australis*, Lamarck. *Ann. Mag. nat. Hist.* (5) **3**: 1–16.
Packard, A. (1961). Sucker display of *Octopus. Nature, Lond.* **190**: 736–737.
Parker, G. H. (1921). The power of adhesion in the suckers of *Octopus bimaculatus* Verrill. *J. exp. Zool.* **33**: 391–394.
Pickford, G. E. (1940). The Vampyromorpha, living-fossil Cephalopoda. *Trans. N.Y. Acad. Sci* (2) **2**: 169–181.
Pickford, G. E. (1949). *Vampyroteuthis infernalis* Chun, an archaic dibranchiate cephalopod. II. External anatomy. *Dana Rep.* No. 82, 1–132.
Pilson, M. E. Q. & Taylor, P. B. (1961). Hole drilling by *Octopus. Science, N.Y.* **134**: 1366–1368.
Rancurel, P. (1971). *Mastigoteuthis grimaldi* (Joubin, 1895). Chiroteuthidae peu connu de l'Atlantique tropical (Cephalopoda-Oegopsida). *Cah. ORSTOM* (Océanogr.) **9**: 125–145.
Rees, W. J. (1951). The distribution of *Octopus vulgaris* in British waters. *J. mar. biol. Ass. U.K.* **29**: 361–378.
Rees, W. J. (1954). The *Macrotritopus* problem. *Bull. Br. Mus.* (*nat. Hist.*) (*Zool.*) **2**: 67–100.
Rees, W. J. (1956). Notes on the European species of *Eledone* with special reference to eggs and larvae. *Bull. Br. Mus.* (*nat. Hist.*) (*Zool.*) **3**: 283–293.
Robson, G. C. (1925). The deep-sea Octopoda. *Proc. zool. Soc. Lond.* **1926**: 1323–1356.
Robson, G. C. (1929 and 1932). *A monograph of the recent Cephalopoda.* Parts I and II. Printed by order of the Trustees of the British Museum, London.
Roper, C. F. E. (1972). Ecology and vertical distribution of Mediterranean pelagic cephalopods. In *Mediterranean biological studies.* Goodyear, R. H., Gibbs, Jr., R. H., Roper, C. F. E., Kleckner, R. C., Sweeney, M. J. & Zahuranee, B. J. (eds). Final report Vol. I. Smithsonian Institution, Washington.
Roper, C. F. E. & Brundage, W. L. (1972). Cirrate octopods with associated deep-sea organisms: new biological data based on deep benthic photographs (Cephalopoda). *Smithson. Contrib. Zool.* No. 121: 1–46.
Roper, C. F. E. & Young, R. E. (1975). Vertical distribution of pelagic cephalopods. *Smithson. Contrib. Zool.* No. 209: 1–51.
Roper, C. F. E., Young, R. E. & Voss, G. L. (1969). An illustrated key to the families of the order Teuthoidea (Cephalopoda). *Smithson. Contrib. Zool.* No. 13: 1–32.
Rowell, C. H. F. (1963). Excitatory and inhibitory pathways in the arm of *Octopus. J. exp. Biol.* **40**: 257–270.
Rowell, C. H. F. (1966). Activity of interneurones in the arm of *Octopus* in response to tactile stimulation. *J. exp. Biol.* **44**: 589–605.
Rudall, K. M. (1955). The distribution of collagen and chitin. *Symp. Soc. exp. Biol.* **9**: 49–71.
Ruibal, R. & Ernst, V. (1965). The structure of the digital setae of lizards. *J. Morph., Lond.* **117**: 271–294.

Sanders, F. K. & Young, J. Z. (1940). Learning and other functions of the higher nervous centres of *Sepia*. *J. Neurophysiol.* **3**: 501–526.
Santi, P. A. (1975). *The morphology of putative receptor neurons and other neural components in the suckers of the squid,* Lolliguncula brevis. Doctoral dissertation, Florida State Univeristy, Tallahassee.
Santi, P. A. & Graziadei, P. P. C. (1975). A light and electron microscope study of intra-epithelial putative mechanoreceptors in squid suckers. *Tiss. Cell* **7**: 689–702.
Schmidt, J. (1922). Live specimens of *Spirula*. *Nature, Lond.* **110**: 788–790.
Scott, T. (1910). Notes on crustacea found in the gizzard of a deep-sea cephalopod. *Ann. Mag. nat. Hist.* (8) **5**: 51–54.
Stephens, P. R. (1971). Histological methods. In *The anatomy of the nervous system of* Octopus vulgaris. Young, J. Z. Oxford: Clarendon Press.
Taki, I. (1941). On keeping octopods in an aquarium for physiological experiments, with remarks on some operative techniques. *Venus, Kyoto* **10**: 140–156.
Thompson, D'Arcy W. (1910). *The works of Aristotle.* **4**. *Historia Animalum.* Oxford: Clarendon Press.
Tompsett, D. H. (1939). *Sepia. L.M.B.C. Mem. Typ. Br. mar. Pl. Anim.* **32**: 1–184.
Verrill, A. E. (1882). Report on the cephalopods of the north eastern coast of America, *Rep. U.S. Commnr Fish.* **1879**: 211–450.
Voss, G. L. (1967). The biology and bathymetric distribution of deep-sea cephalopods. *Stud. trop. Oceanogr.* **5**: 511–535.
Wells, M. J. (1958). Factors affecting reactions to *Mysis* by newly hatched *Sepia. Behaviour* **13**: 96–111.
Wells, M. J. (1960). Proprioception and visual discrimination of orientation in *Octopus. J. exp. Biol.* **37**: 489–499.
Wells, M. J. (1962). Early learning in *Sepia. Symp. zool. Soc. Lond.* No. 8: 149–169.
Wells, M. J. (1963a). Taste by touch; some experiments with *Octopus. J. exp. Biol.* **40**: 187–193.
Wells, M. J. (1963b). The orientation of *Octopus. Ergeb. Biol.* **26**: 40–54.
Wells, M. J. (1964). Tactile discrimination of surface curvature and shape by the octopus. *J. exp. Biol.* **41**: 433–445.
Wells, M. J., Freeman, N. H. & Ashburner, M. (1965). Some experiments on the chemotactile senses of octopuses. *J. exp. Biol.* **43**: 553–563.
Wells, M. J. & Wells, J. (1956). Tactile discrimination and the behaviour of blind *Octopus. Pubbl. Staz. zool. Napoli* **28**: 94–126.
Wodinsky, J. (1969). Penetration of the shell and feeding on gastropods by *Octopus. Am. Zool.* **9**: 997–1010.
Young, J. Z. (1960). Observations on *Argonauta* and especially its method of feeding. *Proc. zool. Soc. Lond.* **133**: 471–479.
Young, J. Z. (1971). *The anatomy of the nervous system of* Octopus vulgaris. Oxford: Clarendon Press.
Young, R. E. (1972). The systematics and areal distribution of pelagic cephalopods from the seas off Southern California. *Smithson. Contrib. Zool.* No. 97: 1–159.
Zeiller, W. (1970). Rare gifts from the sea. *Sea Front.* **16**: 322–327.

Symp. zool. Soc. Lond. (1977) No. 38, 513 523.

THE EVOLUTION OF HAEMOCYANIN

ANNA GHIRETTI-MAGALDI, F. GHIRETTI
and B. SALVATO

Istituto di Biologia Animale, Università di Padova, Padova, Italy

SYNOPSIS

In this paper a brief review is given of the current knowledge about the structure of haemocyanin. The available data of the quaternary structure and the chemical composition of the haemocyanins from Mollusca and Arthropoda indicate that both classes of cuproproteins, which are currently considered isologous and not homologous, may have a common origin. This hypothesis is based on the amino acid composition of haemocyanins from different phyla as well as on the type of polypeptide chains that build up the functional subunit.

INTRODUCTION

In his classical paper published in 1878 Léon Fredericq first described and identified the extracellular pigment present in the blood of *Octopus*, a pigment which contained copper and had the remarkable property of becoming blue in the presence of oxygen.

> "Je fus frappé de la teinte bleu foncé que prenait le sang artériel au sortir de l'organe respiratoire. Il me vint á l'esprit que ce changement de couleur pouvait avoir une signification analogue à celle de la transformation du sang veineux en sang artériel dans notre poumon. . . . Je constatai que le sang du Poulpe se décolorait quand on lui enlevait l'oxygène, qu'il redevenait bleu quand on lui permettait de réabsorber ce gaz, que les changements de couleur avaient pour support une matière albuminoide de constitution analogue à notre hémoglobine. . . . J'y cherchai vainement le fer, mais au lieu de ce métal j'en trouvai un autre, le cuivre." (Bacq & Florkin, 1936.)

Fredericq named the pigment Haemocyanin.

> "Je propose de l'appeler Hémocyanine de ἇιμα = sang et κύανος = bleu, terme rappelant la parenté étroite avec l'hémoglobine du sang des Vertébrés. La combinaison avec l'oxygène serait naturellement l'oxyhémocyanine." (Fredericq, 1878.)

We wanted to recall this pioneering work to the participants of this symposium because it bears some similarity to the scientific discoveries of Professor J. Z. Young to whom the symposium is dedicated, a similarity which goes beyond the preference for the same experimental animal. In fact, what the work of Fredericq did for the physiology of the respiratory pigments, the pioneering work of J. Z. Young on the brain of *Octopus* has done for the physiology of the nervous system, opening new horizons.

Almost a hundred years have elapsed since the discovery of haemocyanin. Since then the respiratory function of the pigment has been established for *Octopus* as well as for all the animals that carry it; the pigment was found to be a protein, and its chemical composition and molecular weight were investigated in several groups of animals. It was also demonstrated that the name given to the pigment does not imply the presence of a haem group: the "strict relationship" with haemoglobin claimed by Fredericq refers only to its physiological role, i.e. to the property of haemocyanin to combine reversibly with oxygen and to function as an oxygen-carrying pigment.

Haemocyanins, therefore, are usually described among the respiratory proteins. This function is certainly the most peculiar property of the protein; amongst others are the catalase and phenolase activities, which are considered secondary and of less physiological significance. From the chemical point of view, however, haemocyanins are better considered as members of the large group of copper proteins which are present in all living organisms, plants and animals, from bacteria to man: tyrosinase, laccase, ascorbic acid oxidase, ceruloplasmin, and so on. It is quite possible that haemocyanins have an ancestor in common with one of these copper proteins. Unfortunately little is known about the primary structure either of haemocyanins or of any other copper protein and only limited information has been gained about the active site of all of them.

It is possible, however, to derive some conclusions from the data recently obtained on the chemical composition of these proteins in order to find out whether haemocyanins have a polyphyletic or monophyletic origin. We shall summarize the most relevant features of the haemocyanin molecule, namely: the quaternary structure, the smallest functional subunit, and the polypeptide chain.

THE QUATERNARY STRUCTURE

Haemocyanins are present in two phyla: Mollusca and Arthropoda. In molluscs these respiratory pigments have been found in the classes of Cephalopoda, Gasteropoda and Amphineura; among the arthropods, in the classes of Merostomata and Crustacea. Haemocyanins have also been identified in some species of scorpions, spiders and centipedes (van Holde & van Bruggen, 1971; Lontie & Witters, 1973) (Fig. 1).

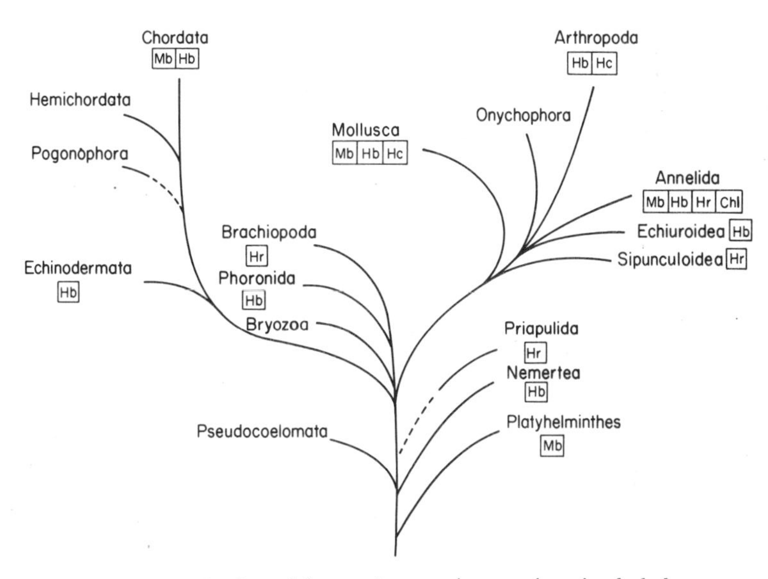

FIG. 1. Distribution of the respiratory pigments in animal phyla.

In the blood of both molluscs and arthropods circulating haemocyanin is present in large aggregates of high molecular weight. This has been demonstrated by sedimentation and with the transmission electron microscope. In Arthropoda haemocyanin shows sedimentation coefficients of 16, 24, 33 and 57 S (Svedberg unit) according to the zoological groups (Fig. 2). The molecular weight ranges from 400 000 to 3 300 000 daltons (Eriksson-Quensel & Svedberg, 1936). These aggregates have been interpreted as polymers (dimers, tetramers and octamers) of the 16 S subunit; after negative staining the molecules are seen as squares, rectangles and hexagons (van Bruggen, 1968). The monomer (16 S, m.w. 400 000) is found in Caridea, Palinura and Anomura; the dimer (24 S m.w. 800 000) in Astacura and Brachiura; the tetramer (33 S, m.w. 1 300 000) in Scorpionidea and Arachnida; the octamer (m.w. 3 300 000) together with the other components, in *Limulus polyphemus*.

The haemocyanins of Mollusca, as seen with the transmission electron microscope, appear as circles and rectangles which have

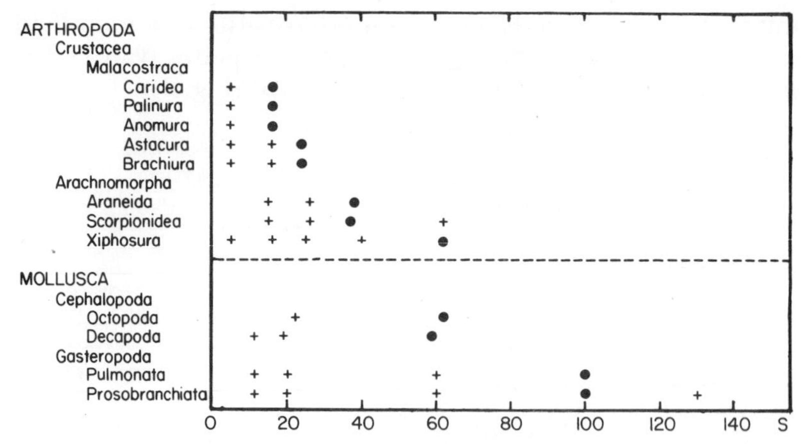

FIG. 2. Molecular aggregates of molluscan and arthropod haemocyanins. Circles: main components present in the blood. Crosses: minor components.

been interpreted as cylinders with a fivefold or tenfold symmetry (Fernandez-Moran, van Bruggen & Ohtsuki, 1966). In the ultracentrifuge two aggregates are found with sedimentation coefficients of 59 S, and 105 S, corresponding to monomers and dimers respectively. Only the monomer is found in Cephalopoda (Eriksson-Quensel & Svedberg, 1936). At the isoelectric point the dimer found in Gasteropoda shows a tendency to aggregate in long linear polymers with sedimentation values up to 178 S (Condie & Langer, 1964) (Fig. 3).

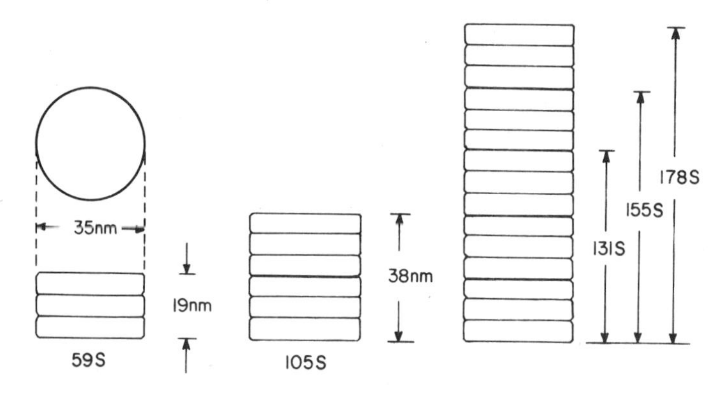

FIG. 3. The quaternary structure of molluscan haemocyanin. (From: Lontie & Vanquickenborne, 1974.)

THE SMALLEST FUNCTIONAL SUBUNIT

The copper content of haemocyanin is characteristic for each phylum: on average 0·25% for Mollusca and 0·17% for Arthropoda. Assuming that the metal is evenly distributed in the protein, these percentages suggest a minimum molecular weight of about 25 000 and 37 400 daltons respectively. Since in the oxygenation reaction one oxygen molecule is bound by two atoms of copper, the smallest functional subunit should have an average molecular weight of 50 000 for molluscan haemocyanin and 75 000 for the protein from arthropods, if all copper atoms are effective (Redfield, Coolidge & Schott, 1928; Redfield, Coolidge & Montgomery, 1928; Guillemet & Gosselin, 1932).

Accordingly, it is generally accepted that haemocyanins fall into two classes, which differ basically for the molecular weight of the smallest functional subunit. The monomer of molluscan haemocyanin is made up of about 150 subunits containing 75 oxygen binding sites. As long ago as 1936 Eriksson-Quensel and Svedberg found that, under different experimental conditions, the largest molecular aggregates undergo a reversible dissociation down to components of lower molecular weight. This has been demonstrated for all haemocyanin species (Brosteaux, 1937; van Holde & Cohen, 1964; Lontie & Witters, 1966; Konings, Dijk *et al.*, 1969). It would be expected, therefore, that by further dissociation the functional subunits of 50 000 and 75 000 daltons could also be obtained.

From arthropod haemocyanin the functional subunit (5 S, m.w. 75 000) has been obtained in several laboratories under mild conditions. At pH 9·3 in the absence of calcium functional components ranging from 4·5 to 5·5 S have been found for the haemocyanins of *Homarus americanus*, of *Limulus polyphemus* and, more recently, of the spider *Dugesiella* (Printz, 1963; Pickett, Riggs & Larimer, 1966; Loewe & Linzen, 1973).

Under similar conditions molluscan haemocyanin dissociates to a fraction of the original aggregate, which still has a considerable size and a high molecular weight.

A systematic study of the effect of pH, ionic strength, temperature, ion concentration and oxygen binding on the dissociation curve of *Helix pomatia* haemocyanin has been carried out by Lontie and his collaborators and by van Bruggen and his school. The 105 S haemocyanin (m.w. 9×10^6) dissociates easily into half-size molecules (65 S, m.w. $4{\cdot}5 \times 10^6$), one-tenth size molecule (20 S, m.w. 900 000) and finally into one-twentieth-size molecules (12 S, m.w. 450 000). According to these authors, one-twentieth is the smallest functional subunit obtainable under mild conditions, i.e. without covalent cleavage (Konings, Siezen & Gruber, 1969; Konings, Dijk *et al.*, 1969; Cox, Witters & Lontie, 1972).

Further dissociation by strong dissociating agents brings about complete loss of the property of reacting with oxygen (Brouwer & Kuiper, 1973). It has been recently reported that, by enzymatic digestion with trypsin or subtilisin, functional subunits of decreasing size down to 50 000 daltons may be obtained from *Helix pomatia* haemocyanin (Brauns & Lontie, 1960; Gielens, Préaux & Lontie, 1973; Lontie, De Ley *et al.*, 1973; Brouwer, 1975). These results, however, involve the cleavage of peptide bonds, so that the problem of the number and size of the smallest functional subunits that make up the quaternary structure remains unsolved.

While studying the action of urea on the haemocyanin from *Octopus vulgaris,* we observed that the protein retains its ability to combine with oxygen in the presence of this denaturing agent. It was demonstrated that, up to an urea concentration of 3M, no gross conformational changes occur, as can be judged by the reaction with oxygen, by the circular dichroic and the fluorescence emission spectra. The extent of dissociation of the haemocyanin molecule under these conditions has been studied by molecular sieving chromatography, by electrophoresis on acrylamide gel and by ultracentrifugation. The results indicate that *Octopus vulgaris* haemocyanin can be dissociated down to components of approximately 250 000 daltons without affecting the tertiary structure of the protein and even less without breaking covalent bonds. This 250 000-dalton subunit, which contains five oxygen binding sites, is the smallest functional unbroken subunit obtainable from molluscan haemocyanin (Salvato, in prep. a).

Three questions arise from these results.

(1) What is the structure of this 250 000 component?
(2) Why does further dissociation of this component require intensive chemical treatment and covalent cleavage?
(3) Why does arthropod haemocyanin under mild experimental conditions easily dissociate down to the smallest functional subunit of 75 000?

To these questions we can give only a tentative answer. The chemical composition of molluscan and arthropod haemocyanins is very similar except for the sugar content. Whereas *Octopus* haemocyanin has about 5% of carbohydrates comprising mannose, fucose and galactosamine (Albergoni, Cassini & Salvato, 1972), no sugar can be detected with conventional analytical methods in the haemocyanins of Crustacea and of *Limulus*. This difference could have an important structural significance. In fact the 50 000 subunit containing one oxygen binding site could be covalently bound by sugar moieties in the 250 000-dalton component. Nothing is known about how sugars are linked to the polypeptide chains. When the haemocyanin from *Octopus vulgaris* is treated with acetic anhydride for extensive acetylation, a 50 000-dalton subunit is obtained by gel filtration in SDS. Acetic anhydride is a good reagent for breaking sugar–protein bonds, when threonine or serine are involved. Therefore the formation of a 50 000-dalton subunit is a strong indirect support that polysaccharide bridges may play a role in the quaternary structure of this protein.

If this hypothesis is correct and if it is proved by further experiments, the structural differences between molluscan and arthropod haemocyanins which are seen at the quaternary level of organization can be explained on the basis of the chemical data. These are essentially two: the absence of sugar moieties and the low number of disulphide bridges in arthropod haemocyanin (Ghiretti-Magaldi & Nuzzolo, 1965). Both of them may strongly affect the quaternary structure of the protein.

THE POLYPEPTIDE CHAIN

It is generally accepted that the smallest functional subunit of molluscan haemocyanin is made up of two polypeptide chains of 25 000 daltons each (Dijk *et al.*, 1970; Cox *et al.*, 1972; Salvato *et al.*, 1972). On the other hand, the molecular weight of the polypeptide chain(s) that build up the active subunit of arthropod haemocyanin is a matter of controversy. Pickett *et al.* (1966) found (by succinylation) the smallest inactive subunits of 37 000 daltons from the haemocyanin of *Homarus americanus*. Other authors, however, believe that the functional unit of arthropod haemocyanin is made by a single polypeptide chain of about 75 000 daltons binding two copper atoms (Loher & Mason, 1973; Carpenter & van Holde, 1973).

In 1973 we obtained from the haemocyanin of *Carcinus maenas*, after exhaustive acetylation and gel filtration in sodium dodecyl sulphate (SDS), components of 75 000, 50 000 and 25 000 daltons. (In the same conditions the complete dissociation to 25 000 components of haemocyanin from *Octopus vulgaris* is achieved only when the disulphide bridges are broken by reduction and the SH groups are blocked; Salvato *et al.*, 1972.) This was the first experimental demonstration that the functional subunits of molluscan and arthropod haemocyanin were composed of polypeptide chains of the same size and that the minimal subunit has a molecular weight of approximately 25 000 daltons in both classes of haemocyanins. Since then, these results have been confirmed and extended to other haemocyanin species.

More recently, the functional subunit of *Carcinus maenas* has been dissociated in 8M urea into components that have been separated by ion exchange chromatography and analysed for the amino acid composition (Salvato, in prep. b).

The results obtained until now can be interpreted by assuming that in arthropod haemocyanin the 75 000 functional subunit containing two atoms of copper is formed by three polypeptide chains of approximately 25 000, of which two are "functional" and one is "structural" i.e. without copper (Table I).

TABLE I

Levels of organization in haemocyanins

Subunits	Mollusca	Arthropoda
Polypeptide chain	25 000	25 000
↓		
Smallest functional	$(25\,000 \times 2) \times 5$	$25\,000 \times 3$
↓		
Further association	$3{\cdot}7 \times 10^6 \rightarrow$	$4 \times 10^5 \rightarrow$

In spite of the large differences found in the quaternary structure there seems to be a great similarity in molluscan and arthropod haemocyanins. This similarity appears more striking when the amino acid composition of the two haemocyanins is compared.

In 1973 Harris & Teller demonstrated a close relationship between the amino acid composition and the amount of sequence homology in proteins. These authors defined a function, called "the composition divergence", which allows us to predict the sequence homology of unsequenced proteins on the basis of their amino acid composition. They applied this method to the haemocyanin species that had been analysed in our laboratory (Ghiretti-Magaldi, Nuzzolo & Ghiretti, 1966) and established that, regardless of species and phyla, these proteins show a level of homology ranging from 54 to 96%. The degree of homology calculated for α and β chains of human haemoglobin is about 46%.

This high homology indicates that the two classes of haemocyanin have a monophyletic origin. Although sequence work is necessary for establishing the evolutionary history of these proteins, it can be argued that generally speaking, homologous proteins have tended to maintain the same length (Acher, 1974).

Parallel gene duplication for producing two new distinct genes is a very well known event in evolution; a good number of ascertained cases are reported in Dayhoff (1972). It can therefore be supposed that also in haemocyanin, gene duplication may have produced the two chains of the active subunit of molluscan haemocyanin and the three chains of the arthropod protein.

If arthropod haemocyanin is made of a single chain of 75 000 or of two chains of 37 000 daltons, a more complicated assumption is required to explain its history. Mollusca have diverged before Arthropoda from their common ancestor. The presence of two polypeptide chains in their haemocyanin may be considered as originated by duplication of a single gene. A 75 000-molecular weight polypeptide chain may derive from the common ancestor only by a further duplication followed by splicing of the genes; two 37 000 dalton chains would require duplication, splicing and chain shortening of half a chain length.

As long as the sequence of these proteins remains unknown, it is probably more fruitful to accept as a working hypothesis that the haemocyanins of Mollusca and Arthropoda, in spite of the differences observed in their chemical composition and their quaternary structure, have evolved from a common ancestor by a process of gene duplication followed by divergence.

In his book *An introduction to the study of man*, Professor J. Z. Young, while considering similarities hidden beneath the immense diversities of living things, says that perhaps somebody from another planet would notice more than us the surprising chemical and structural uniformity of terrestrial forms of life. This observer, from another planet, he says [I am translating from the Italian translation (1974)], would describe the structure and composition of terrestrial life in terms of different hierarchic levels. Discovering unity amongst diversity is one of the aims of scientific research and we would predict that the haemocyanins, of such diverse groups as cephalopods and arthropods, are related members of a single biochemical family, despite their more obvious differences.

References

Acher, R. (1974). L'évolution moléculaire au niveau des protéines. *Biochimie* **56**: 1–19.

Albergoni, V., Cassini, A. & Salvato, B. (1972). The carbohydrate portions of hemocyanin from *Octopus vulgaris*. *Comp. Biochem. Physiol.* **41B**: 445–451.

Bacq, Z. M. & Florkin, M. (1936). Léon Fredericq. *Archs Int. Pharmacodyn. Thér.* **52**: 245–280.

Brauns, G. & Lontie, R. (1960). Etude de l'hydrolyse partielle de l'hémocyanine *d'Helix pomatia*. *Arch. Intern. Physiol. Biochim.* **68**: 211–212.

Brosteaux, I. (1937). Ueber den einfluss del Ca- und Mg-ionen auf die stabilität des Hämocyanins. *Naturwissen.* **25**: 249.

Brouwer, M. (1975). *Structural domains in* Helix pomatia *α-hemocyanin.* Ph.D. Thesis: Groningen University.

Brouwer, M. & Kuiper, H. A. (1973). Molecular weight analysis of *Helix pomatia* α-hemocyanin in guanidine hydrochloride, urea and sodium dodecylsulfate. *Eur. J. Biochem.* **35**: 428–435.

Carpenter, D. E. & van Holde, K. E. (1973). Amino acid composition, amino terminal analysis and subunit structure of *Cancer magister* hemocyanin. *Biochemistry* **12**: 2231–2238.

Condie, R. M. & Langer, R. B. (1964). Linear polymerization of a gastropod hemocyanin. *Science, N.Y.* **144**: 1138–1140.

Cox, J., Witters, R. & Lontie, R. (1972). The quaternary structure of *Helix pomatia* hemocyanins as determined by alkali treatment and succinylation. *Int. J. Biochem.* **3**: 283–293.

Dayhoff, M. O. (1972). *Atlas of protein sequence and structure*, **5**. Washington, D.C.: National Biomedical Research Foundation.

Dijk, J., Brouwer, M., Coert, A. & Gruber, M. (1970). The smallest subunits of α- and β-hemocyanin of *Helix pomatia*: size, composition, N- and C-terminal amino acids. *Biochim. Biophys. Acta* **221**: 467–479.

Eriksson-Quensel, I. B. & Svedberg, T. (1936). The molecular weights and pH stability regions of the Hemocyanins. *Biol. Bull. mar. biol. Lab., Woods Hole* **71**: 498–574.

Fernandez-Moran, H., van Bruggen, E. F. J. & Ohtsuki, M. (1966). Macromolecular organization of hemocyanins and apohemocyanins as revealed by electron microscopy. *J. Molec. Biol.* **16**: 191–207.

Fredericq, L. (1878). Recherches sur la physiologie du poulpe commun. (*Octopus vulgaris*). *Arch Zool. exp.* **7**: 535–583.

Ghiretti-Magaldi, A. & Nuzzolo, C. (1965). Thiol groups in hemocyanin. *Comp. Biochem. Physiol.* **16**: 249–252.

Ghiretti-Magaldi, A., Nuzzolo, C. & Ghiretti, F. (1966). Chemical studies on hemocyanins. I. Amino acid composition. *Biochemistry* **5**: 1943–1951.

Gielens, C., Préaux, G. & Lontie, R. (1973). Isolation of the smallest functional subunit of *Helix pomatia* hemocyanin. *Arch. Intern. Physiol. Biochim.* **81**: 182.

Guillemet, R. & Gosselin, G. (1932). Sur les rapports entre le cuivre et la capacité respiratoire dans les sangs hémocyaniques. *C.r. Séanc. Soc. Biol.* **111**: 733–735.

Harris, C. E. & Teller, D. C. (1973). Estimation of primary sequence homology from amino acid composition of evolutionary related proteins. *J. theor. Biol.* **38**: 347–362.

Konings, W. N., Dijk, J., Wichertjes, T., Beuvery, E. C. & Gruber, M. (1969). Structure and properties of hemocyanins. IV. Dissociation of *Helix pomatia* hemocyanin by succinylation into functional subunits. *Bioch. Biophys. Acta* **188**: 43–54.

Konings, W. N., Siezen, R. J. & Gruber, M. (1969). Structure and properties of hemocyanins. VI. Association-dissociation behaviour of *Helix pomatia* hemocyanin. *Biochim. Biophys. Acta* **194**: 376–385.

Loewe, R. & Linzen, B. (1973). Subunits and stability region of *Dugesiella californica* hemocyanin. *Hoppe Seyler's Z. Physiol. Chem.* **354**: 182–188.

Loher, J. S. & Mason, H. S. (1973). Dimorphism of *Cancer magister* hemocyanin subunits. *Biochem. Biophys. Res. Comm.* **51**: 741–745.

Lontie, R., De Ley, M., Robberecht, H. & Witters, R. (1973). Isolation of small functional subunits of *Helix pomatia* hemocyanin after subtilisin treatment. *Nature, New Biology, Lond.* **242**: 180–182.

Lontie, R. & Vanquickenborne, L. (1974). The role of copper in hemocyanins. In *Ions in biological systems.* **3**. *High molecular complexes*: 183–200. Sigel, H. (ed.). New York: Marcel Decker.

Lontie, R. & Witters, R. (1966). *Helix pomatia* hemocyanins. In *The biochemistry of copper*: 455–463. Peisach, J., Aisen, P. & Blumberg, W. E. (eds). London and New York: Academic Press.

Lontie, R. & Witters, R. (1973). Hemocyanins. In *Inorganic biochemistry* **1**: 344–358. Eichhorn, G. L. (ed.). Amsterdam: Elsevier Publ. Co.

Pickett, S. M., Riggs, A. F. & Larimer, J. L. (1966). Lobster hemocyanin: properties of the minimum functional subunit and of aggregates. *Science, N.Y.* **151**: 1005–1007.

Printz, M. P. (1963). An investigation of *Limulus* hemocyanin and the formation of its active monomer. *Fedn Am. Socs exp. Biol. Proc.* **22**: 291.

Redfield, A. C., Coolidge, T. & Montgomery, H. (1928). The respiratory proteins of the blood. II. The combining ratio of oxygen and copper in some bloods containing hemocyanins. *J. biol. Chem.* **76**: 197–205.

Redfield, A. C., Coolidge, T. & Schott, M. A. (1928). The respiratory proteins of the blood. I. The copper content and the minimal molecular weight of the hemocyanin of *Limulus polyphemus*. *J. biol. Chem.* **76**: 185–195.

Salvato, B. (in prep. a). *Functional subunits of* Octopus vulgaris *hemocyanin.*

Salvato, B. (in prep. b). *The minimal functional subunit of arthropod hemocyanins.*

Salvato, B., Sartore, S., Rizzotti, M. & Ghiretti-Magaldi, A. (1972). Molecular weight determination of polypeptide chains of molluscan and arthropod hemocyanins. *F.E.B.S. Letters* **22**: 5–7.

van Bruggen, E. F. J. (1968). Electron microscopy of hemocyanins. In *Physiology and biochemistry of hemocyanins*: 37–48. Ghiretti, F. (ed.). London and New York: Academic Press.

van Holde, K. E. & Cohen, L. B. (1964). The dissociation and reassociation of *Loligo pealei* hemocyanin. *Brookhaven Symp. Biol.* No. 17: 184–193.

van Holde, K. E. & van Bruggen, E. F. J. (1971). The hemocyanins. In *Subunits in biological systems*: 1–53. Timasheff, N. S. & Fashman, G. D. (eds). New York: Marcel Decker.

Young, J. Z. (1974). *La scienza dell'uomo. Biologia, evoluzione e cultura.* Milano: Boringhieri.

Symp. zool. Soc. Lond. (1977) No. 38, 525–540.

OPTIC GLANDS AND THE ENDOCRINOLOGY OF REPRODUCTION

M. J. WELLS and J. WELLS

Department of Zoology, Cambridge University, Cambridge, England

SYNOPSIS

Sexual maturity in *Octopus* is always associated with enlargement of the optic glands. Precocious maturity can be induced by cutting the nerve supply from the subpedunculate lobe to the glands, by blinding the animals, or by implanting glands into immature recipients. Implants are effective regardless of the sex of the donor, or the state of the glands at the time of implantation; interspecific and intergeneric implants from other octopods are as effective as those derived from the same species. The optic gland secretion appears to affect the condition of a wide range of structures, including the sex ducts. There is no evidence for a gonadal hormone; in castrates the ducts remain unchanged, behaviour appears to be unaffected, and there are no indications of a feedback regulation of optic gland secretion. In at least one process the hormone is known to act directly: *in vitro* experiments show that fragments of ovary will synthesize yolk protein from ^{14}C-leucine present in a nutritive medium, and both uptake and synthesis are greatly enhanced when optic gland extracts are added. Electron micrographs indicate that the sites of synthesis are the follicle cells that surround the egg as it approaches maturity. With the development of *in vitro* systems it will soon be possible to characterize this important hormone, which appears to have widespread effects upon cephalopod metabolism quite apart from its specific role in regulating the onset of sexual maturity.

INTRODUCTION

Quite a lot is now known about the breeding habits of cephalopods; most, and perhaps all of them, normally breed once and then die. Experimental work on the endocrine control of this terminal sexual maturation has been limited to *Octopus vulgaris* and *Sepia officinalis*, with a few observations on *Eledone cirrosa*. Although corresponding experiments have yet to be made on squids, there is every reason to believe that what we understand from work on *Octopus* will apply with little or no modification to the whole range of dibranchiate cephalopods.

Octopus vulgaris in the Mediterranean normally breeds in its second year, the females laying eggs when they have reached a weight of 1 kg or more (Mangold-Wirz, 1963; Mangold & Boletzky, 1973). Precocious maturation can be induced in animals weighing as little as 50 g by cutting the optic nerves, by removal of the subpedunculate lobe from the hind part of the supraoesophageal brain, or by cutting the nerve supply from this lobe to the optic glands, two small pale yellowish spheres on the stalks joining the large optic lobes to the rest of the central nervous system. Maturation, whether induced experimentally or occurring naturally in the sea, is always associated with gross enlargement of the optic glands, which swell up and become bright orange in colour (Wells & Wells, 1959).

The structure and innervation of the optic glands has been studied by a number of workers (Björkman, 1963; Boycott & Young, 1956; Defretin & Richard, 1967; Froesch, 1974; Nishioka, Bern & Golding, 1970; Wells & Wells, 1959). There appears to be a single sort of gland cell, producing a single product that is released into the bloodstream. The gland cells are innervated by axons originating in the subpedunculate lobe.

The hormone produced by the optic glands controls the onset of sexual maturity. If the glands are removed neither blinding nor brain lesions will induce sexual maturity, and the gonads of animals that have already begun to develop regress (Wells, O'Dor & Buckley, 1975; Wells & Wells, 1972a). Organ culture experiments with *Sepia* show that the cells of the germinal epithelium cease to divide if optic gland hormone is absent (Durchon & Richard, 1967; Richard, 1970).

Besides these effects upon the gonads, secretion by the optic glands appears to regulate the condition of the sex ducts, which also enlarge or regress if the glands are activated or removed. Castration is not followed by regression of the male or female ducts, which indicates that the effect of the hormone is direct, rather than via the gonad; there is at present no evidence for the existence of a gonadal hormone in cephalopods (Callan, 1940; Wells, 1960; Wells & Wells, 1972a,b; although Taki, 1944, found oedema followed by degeneration in male ducts after castration, as did Wells & Wells, 1972a, in a few animals; the most likely explanation of this is interference with the blood supply to the ducts).

IMPLANTATION EXPERIMENTS

Implants derived from ***Octopus vulgaris***

If, as it appears, the optic glands are controlled by an inhibitory nerve supply, transplanted glands should become active, regardless of their condition when removed from the donor. This proves to be the case; optic glands taken from mature or immature octopuses and placed in the blood sinus enclosing the "white body" of immature animals regularly induce precocious maturity (Table

TABLE I

A summary of experiments in which optic glands were implanted into Octopus vulgaris

Sex of recipient	Sex of donor	Condition of host's gonad at death		Condition of the gland when implanted	
				Enlarged	Not enlarged
♀	♂	Enlarged	6	2	4
		Normal	10	4	6
	♀	Enlarged	11	3	8
		Normal	8	4	4
♂	♂	Enlarged	8	1	7
		Normal	16	2	14
	♀	Enlarged	2	2	—
		Normal	9	6	3

"Normal" means that the host's gonad was smaller than the largest found in controls.

I). In this situation the glands often adhere to the walls of the sinus where they attract a capillary blood supply and begin to secrete. The gonads and ducts of the recipients can enlarge as quickly as if their own optic glands had been denervated. Usually, enlargement is slower; the results show a great deal of variability as one might expect from the somewhat unpredictable fate of implants, which may or may not settle upon a membrane where they can readily develop a blood supply. Figures 1a and b compare the effect of implantation with the effect of blinding and brain lesions in female octopuses. Similar results were obtained with males. Implantation has no externally visible effect on the optic glands of the recipient (Wells & Wells, 1975).

Secretion by implanted glands shows that the gland cells can operate without a nerve supply. Froesch (1974) has shown both axo-glandular and axo-axonal synapses in the optic glands, and points out that this could mean presynaptic inhibition superimposed on an excitatory innervation, normally responsible for regulating secretion. This finding is not incompatible with the results

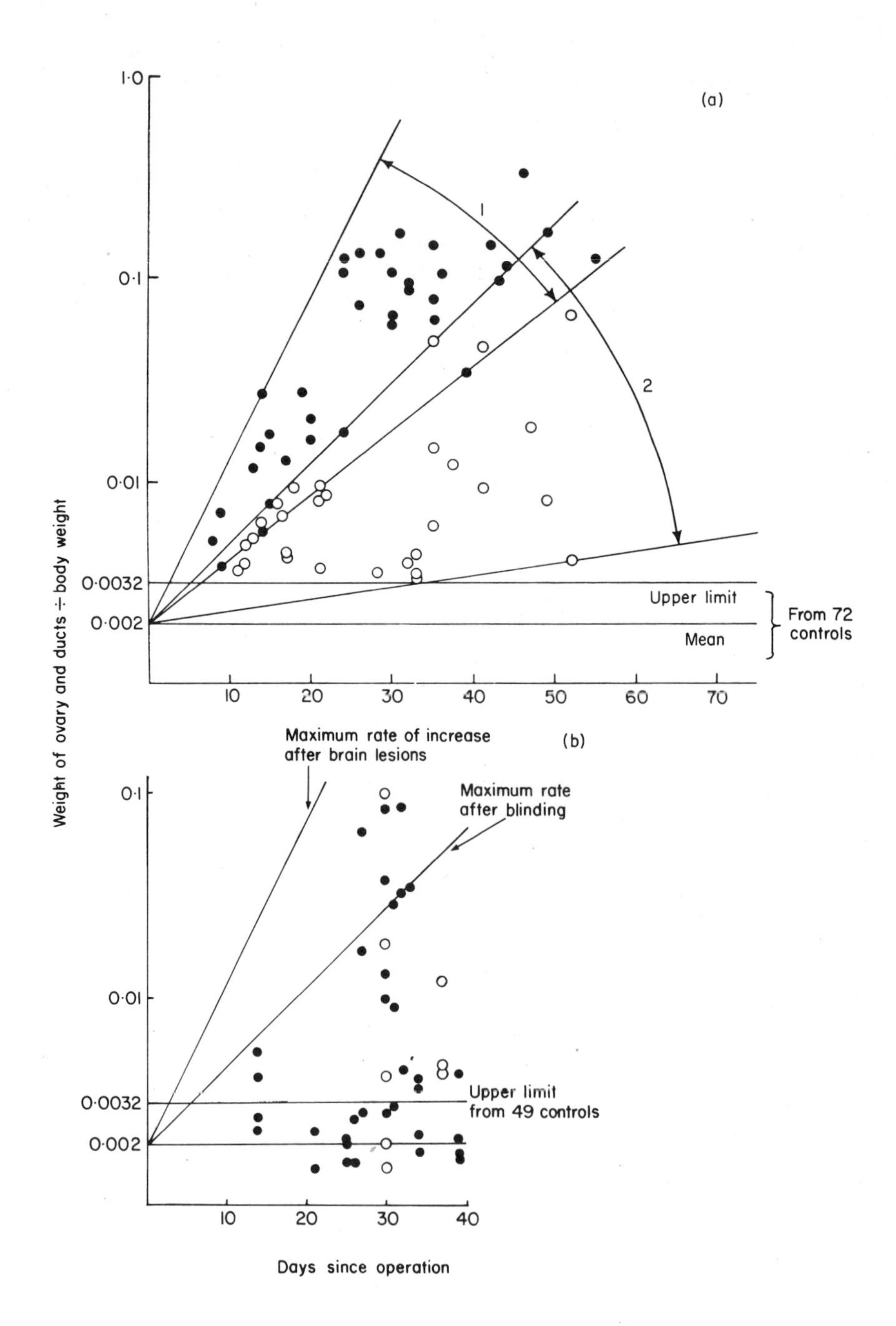
1·0
(a)
1
2
0·1
0·01
0·0032
0·002
Upper limit
Mean
From 72 controls
10
20
30
40
50
60
70
Weight of ovary and ducts ÷ body weight
Maximum rate of increase after brain lesions
(b)
Maximum rate after blinding
Upper limit from 49 controls
Days since operation

of nerve-cutting or implantation experiments, both of which produce a very (? perhaps unnaturally) rapid response. We know, from the experiments of Durchon & Richard (1967), Richard (1970) and Wells & Wells (1972a) that the optic glands must produce a small quantity of secretion even when they are not visibly enlarged, and the innervation discovered by Froesch could represent a more refined control system that exists in the intact animal.

Implants derived from the opposite sex and/or other species and genera

The secretory state of the implant at the time of transplantation is unimportant since the gland cells soon begin to secrete in the absence of their nerve supply. It seems that the sex of the donor may also be unimportant; male glands can cause ovarian enlargement in females and vice versa (Table I). However, there is some indication both from implantation experiments and from the type of *in vitro* experiments discussed below that glands of the same sex may be marginally more effective than those derived from the opposite sex. The differences shown in Table I are in this direction though not statistically significant at any acceptable level. The *in vitro* results summarized in Fig. 3 were obtained using ovaries from different animals and different dosages of optic glands, so they are not strictly comparable with one another. But they too indicate that extracts derived from female optic glands or from mixtures of male and female glands seem to be more effective in inducing protein synthesis by eggs than extracts made from male glands alone.

Extension of the implantation experiments to include transplants derived from other species and genera have shown that octopod to octopod implants at least can survive and cause enlargement of the gonad of recipients. Three out of five *Octopus macropus* implants caused ovarian enlargements in *O. vulgaris* (they were, incidentally, the three out of the five derived from female animals) as did all three glands derived from *Eledone moschata* (all females).

FIG. 1. (a) The effect of (1) removal of the subpedunculate lobe, source of the inhibitory innervation to the optic gland and (2) optic nerve section, on the growth of the ovaries of *Octopus vulgaris*. Both operations cause the gonads of immature animals to enlarge precociously (from Wells & Wells, 1959). (b) The effect of implanting optic glands into immature *Octopus vulgaris*. The subsequent rate of ovarian enlargement is more variable, but comparable at its most rapid with that following subpedunculate lobe removal. In this graph ● shows the results of implanting glands taken from other members of the recipient's species and ○ the effect of implanting glands derived from *O. macropus* or *Eledone*.

Eighteen decapod (*Sepia* or *Loligo*) implants were ineffective. Similar results were obtained with male recipients (Wells & Wells, 1975).

We do not know why decapod implants fail. It could mean that the decapod optic gland hormone is different from that of octopods. Alternatively, and in our present view more likely, the tissue is recognized as "foreign" and attacked by the host amoebocytes. Out of 23 decapod to octopod implants, only four were recovered in recognizable condition on dissection of the host; nine out of 15 octopod to octopod implants were still visible as distinct round bodies adhering to their host's sinus membranes, or embedded in the host's white body.

IN VITRO EXPERIMENTS

Implanted glands could have their effect indirectly, by exciting some further control system in the host. The likelihood of a multi-stage control system is considerably reduced by two sorts of *in vitro* experiment, both of which suggest that the optic gland hormone acts directly on its target tissues.

Fragments of *Sepia* testes and ovaries will survive well in organ culture; the spermatocytes and oocytes continue to develop but are not replaced by further germ cell divisions unless a few cells of optic gland material are included. The gland cells secrete under these conditions and presumably act directly upon the germ cells (Durchon & Richard, 1967; Richard, 1970).

Organ culture experiments are not well suited for investigation of the later stages of gonadal development, since the metabolic demands of developing ovaries (in particular) make an intact circulation virtually essential. A different type of experiment is necessary, in which the tissue to be studied spends only a few hours *in vitro*; nutrition and the accumulation of metabolic wastes are then less of a problem, and there is no need to use antibiotics to ensure sterility. One possible approach is that adopted by O'Dor & Wells (1973, 1975). They examined the effect of optic gland hormone on the deposition of yolk in the maturing ovary of *Octopus*. ^{14}C-leucine, injected into the branchial hearts of octopuses maturing precociously as a result of brain lesions, is rapidly taken up by the ovary, which eventually accumulates about 40% of all the counts injected. The amino acid is made into protein in the ovary itself, a rather unusual situation, since vertebrates and arthropods synthesize most of their yolk proteins at a distance and carry them in the bloodstream to their final site of deposition. Removal of the active optic glands stops uptake and synthesis in the ovary within about five days (Fig. 2). If eggs are removed from the ovary two or

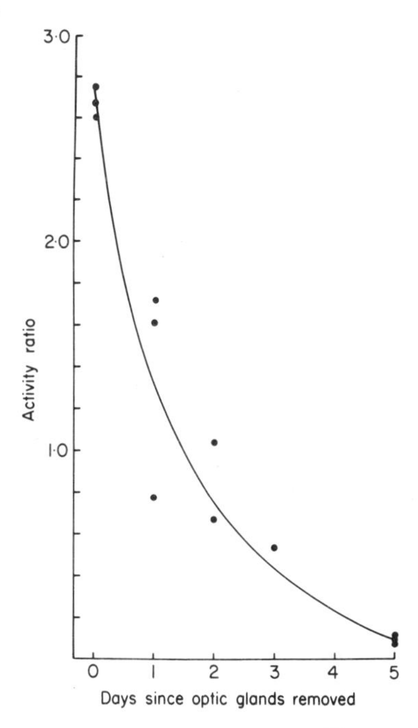

FIG. 2. Protein synthesis by the ovaries of *Octopus*. ^{14}C-leucine was injected into the bloodstream of precociously maturing animals. The octopuses were killed 5 hr later and the amount of labelled protein in the ovaries was measured. The ordinate shows the "activity ratio", being the DPM in protein/g of ovary divided by the DPM amino acid injected/g body weight.

three days after removal of the optic glands and incubated in a nutrient medium they will still take up ^{14}C-leucine and synthesize a little protein. Synthesis is much enhanced if an extract of fresh or frozen optic glands is added (Fig. 3).

Amino acid uptake and yolk protein synthesis by the ovary

When eggs are incubated in synthetic media for 3–6 hours, there is a steady uptake of amino acids. Only a fraction of this material is converted into protein within the period of the experiments. Both uptake and synthesis are increased by the addition of optic gland hormone but there is no clear relation between the two as might be expected if either were the single rate-limiting process.

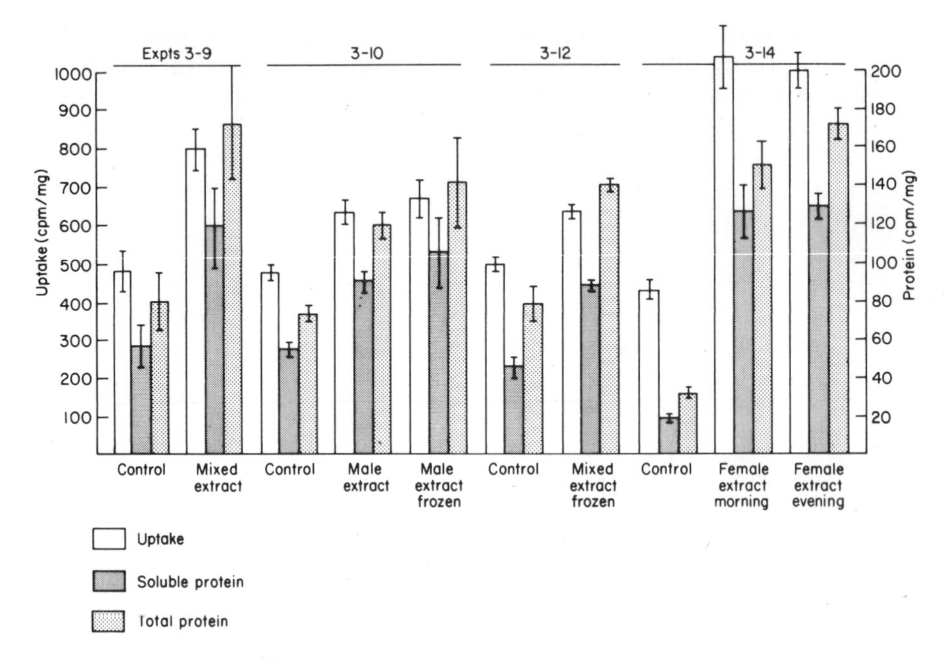

FIG. 3. Uptake of ^{14}C-leucine and synthesis of protein by eggs *in vitro*. Eggs were taken from maturing octopuses, pre-incubated for 3 hr in sea-water and then incubated for a further 3 hr in a nutrient medium including the radioactive amino acid. Some of the samples had optic gland extracts added, as indicated. The egg donor's own optic glands had been removed two days before removal of the eggs in order to reduce the amount of hormone present in controls (see Fig. 2). Results with eggs taken from four different ovaries are shown here. For each series, the left-hand column shows uptake, the centre column soluble protein, and the right-hand column total protein made: mean and s.e. mean values derived in each experiment from six replicate samples.

The passage of labelled amino acid into the eggs depends upon both uptake and exchange. The latter should reach equilibrium after a period that will depend upon the permeability of the cells concerned and the concentration of the "cold" amino acid inside them. The exchange component should be increased by a period of pre-incubation in a leucine-rich medium, reduced by pre-incubation in sea-water. Experimentally, it has been found that pre-incubation for three hours in sea-water is sufficient to run down the free leucine available for exchange to the point where exchange has no detectable effect upon the rate of uptake over a further three hours in a medium containing ^{14}C-leucine. Figure 4 shows the uptake of ^{14}C-leucine by two samples of eggs pre-incubated in sea-water, one of which had optic gland extract added.

In both cases uptake was linear when plotted against time; it was approximately doubled by the addition of the hormone.

The increased uptake could have been due to an increase in cell permeability. If this were the result of adding hormone, one would expect the treated cells to be comparatively "leaky" when transferred to sea-water or a "cold" medium. They were not. In the second part of the experiment summarized in Fig. 4 the medium containing ^{14}C-leucine was replaced by a similar medium containing only non-radioactive leucine. In this situation, exchange should be proportional to concentration inside the cells at the start of the run. The concentration of ^{14}C-leucine inside the eggs is known, from the "uptake" stage of the experiment. The expected escape, occurring simultaneously with continued uptake, can be calculated, using an equation (see caption to Fig. 4) developed by Sheppard & Beyl (1951) to describe the uptake of ^{24}Na into red blood cells which also show a steady uptake superimposed upon exchange. In Fig. 4 the solid lines are calculated, the points show observations. Theory and experiment agree remarkably well; there is even some indication of a reversal of the outflow of label at about $0+9$ hours, when uptake is once again beginning to exceed exchange. Escape from the eggs throughout this part of the experiment is proportional to concentration in both control and experimental groups. The additional uptake associated with the presence of optic gland hormone cannot therefore be attributed to an increase in cell permeability; it must be due to the acceleration of an active process.

The quantity of protein made was estimated from samples taken at the beginning and the end of the "cold" medium period. Protein synthesis is clearly not sufficient to account for the uptake found. As the experiments summarized in Fig. 3 show, there is no consistent relationship between uptake and protein synthesis. The implication is that active uptake and protein synthesis are independently controlled by the optic gland hormone. Either could provide a measure by which glandular extracts could be assayed (Wells, O'Dor & Buckley, 1975).

The site of yolk synthesis and the nature of the yolk protein

The "egg" samples used in the *in vitro* experiments consisted of bunches of 50 to 100 ova each surrounded by a coating of follicle cells. The eggs were at various stages of development, and differed so very greatly in size that the total volume of each sample was almost entirely contained in the 20 or 30 largest eggs.

Early oocytes have a thin skin of flattened follicle cells. Electron micrographs show that the oocytes contain mitochondria, Golgi

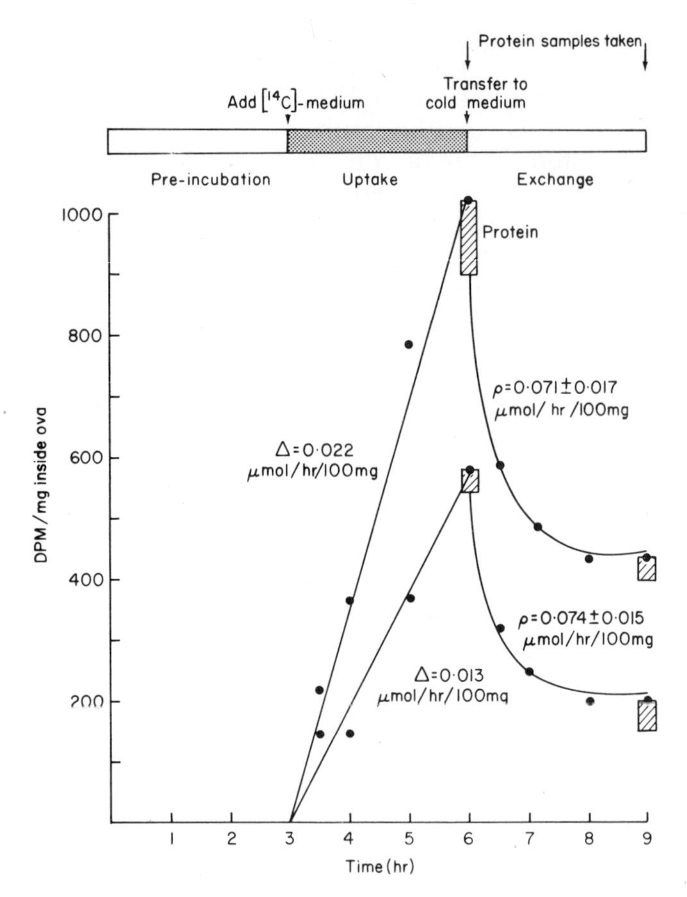

FIG. 4. The action of optic gland extracts on eggs *in vitro*. Samples of eggs from maturing animals deprived of their optic glands two days previously were pre-incubated in sea-water. One sample had optic gland extract added. Three hours later a nutrient medium including ^{14}C-leucine was added. Uptake by the eggs was estimated by measuring the disappearance of counts from the medium. After 3 hr the eggs were transferred to a "cold" medium, samples taken to estimate the protein synthesized, and the escape of label into the medium followed. After a further 3 hr a final protein sample was taken. Cross-hatched columns show labelled protein made during the 3 hr-uptake period and additional labelled protein made during the final incubation in "cold" medium. Optic gland extract increases the net inward movement (Δ) of ^{14}C-leucine without affecting exchange (ρ) which depends simply on the concentration present in the cells; see text (from Wells, O'Dor & Buckley, 1975).

and other cell organelles, while histochemical tests indicate that they accumulate some carbohydrate and lipid materials at this stage (Buckley, unpublished). As the egg grows the follicle cells expand to form a columnar epithelium served by an extensive

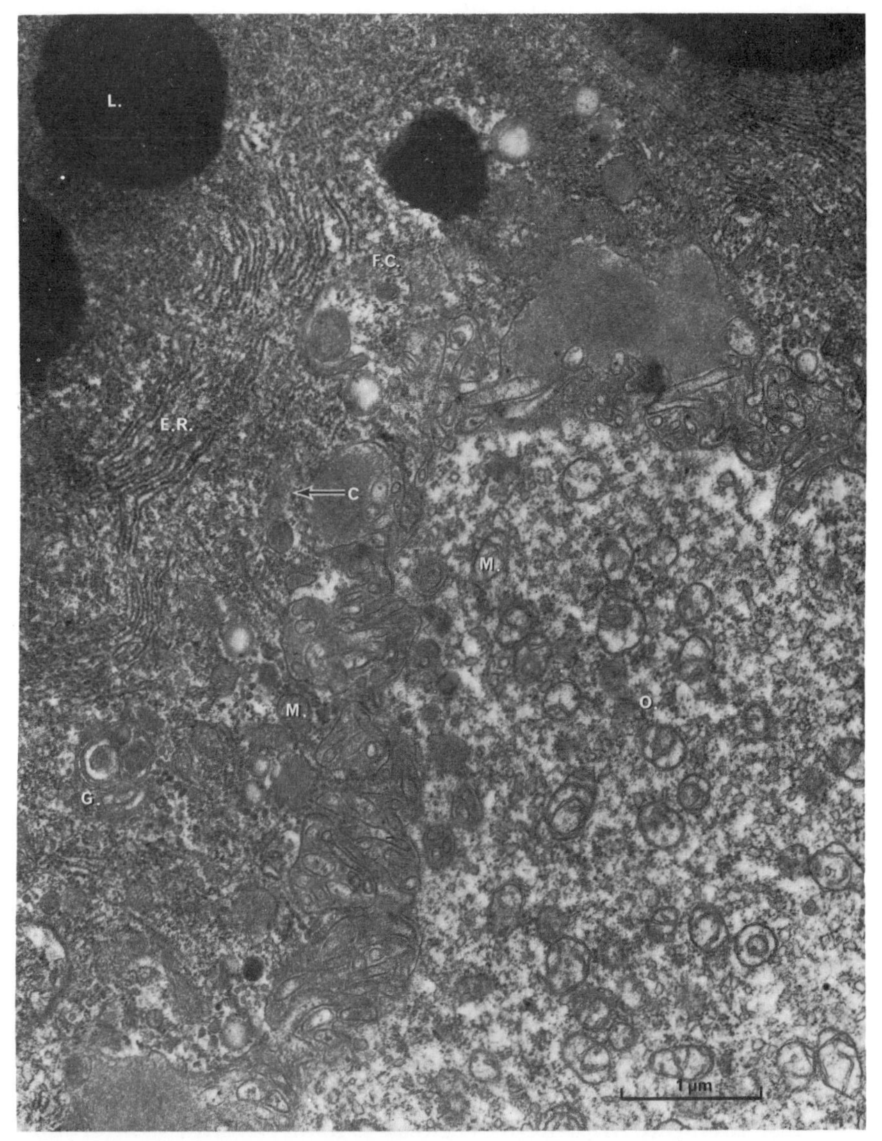

FIG. 5. Section through a follicle cell and the ovum of an *Octopus* that is beginning to lay down proteinaceous yolk. This section is taken close to the animal pole, and nucleus, of the egg of a naturally maturing animal of 2090 g; most of the ovum organelles are concentrated in this region. Nevertheless, the difference between the egg cell, which has many mitochondria but few other organelles, and the follicle cell, with its abundant rough endoplasmic reticulum, Golgi fields and mitochondria is obvious (Buckley, unpublished). Abbreviations: C., chorion; E.R., endoplasmic reticulum; F.C., follicle cell; G., golgi; L., lipid droplet; M., mitochondria; O., ova.

blood supply. The follicle cell layer invaginates the egg cell in a series of longitudinal folds, and a massive accumulation of proteinaceous yolk begins. This was the stage studied in the present series of *in vitro* experiments. The follicle cells at this stage in development are crammed with organelles, including a great quantity of rough endoplasmic reticulum that has no parallel in the ovum at this or at any other stage in its development. The follicle cells are linked to the ovum through finger-like extensions that penetrate the chorion, which now begins to separate the two. It appears that the follicle cells are the principal site of protein yolk production, assembling and exporting materials to the ovum (Buckley, unpublished). The rather sparse organelles in the ovum itself—mainly mitochondria— are pressed to the outermost edges and around the nucleus of the cell. In Fig. 5, which is of a comparatively early stage in this process, the chorion is only just beginning to develop; notice the contrast between the two sorts of cell. It is taken from the ovary of a naturally-maturing animal of 2090 g.

Removal of the optic glands from an animal with eggs at this stage in development is soon followed by disintegration of the organelles in the follicle cells (Wells, O'Dor & Buckley, 1975). The rough endoplasmic reticulum vesiculates, and the Golgi saccules collapse and shrink within 2 to 5 days of the operation. The structural appearance can be correlated with the progressive loss of the capacity for protein synthesis noted in Fig. 2.

Cephalopod yolk protein is unusual in that it will dissolve in solutions of low ionic strength (Fugii, 1960). In the *in vitro* experiments, somewhat more than two-thirds of the protein made and subsequently extracted from the ovum-follicle cell complex was soluble in distilled water (Fig. 3; see also O'Dor & Wells, 1975). This sort of material hardly seems likely to form cell organelles or the chorion surrounding the eggs, so it is, presumably, yolk protein. The remaining insoluble protein fraction is very probably material either destined for the chorion or egg stalk, or used to form organelles within the egg and its coating of follicular cells.

The total rate of protein synthesis is in any event considerable; the follicle cells cannot comprise more than about 5% of the total weight of the ovary, yet even on this basis they are producing yolk protein at a rate of 200 mg per g of cells per day during the fortnight or so before egg-laying. This is comparable with the rate of milk protein production in the mammary glands of lactating rats (O'Dor & Wells, 1973).

DISCUSSION

The gross appearance of the optic glands can be correlated with the sexual condition of *Octopus* and other cephalopods, and there is every reason to believe that a secretion produced by the glands is responsible for the onset of sexual maturity. If the glands are removed, operations such as removal of the subpedunculate lobe, or blinding, which would otherwise cause precocious maturity, are without effect. The implantation of glands into immature animals induces precocious maturation. Since the only change made in these experiments is removal or addition of optic glands, these must be the source of a product that dictates the state of the gonad.

In all cases the state of the sex ducts can be correlated with the condition of the gonad, yet there is rather clearly no gonadal hormone involved; castration has no detectable effect upon the ducts of either sex and no visible effect upon the condition of the optic glands of the castrate, which might be expected to hypertrophy if there were any feedback system involved. The implication is that the optic glands control the state of the sex ducts as well as that of the testes or ovary.

Figure 6 is a summary of the known effects of optic gland hormone on the sex organs of *Octopus*. There could also be widespread effects on other systems. We know, for example, that octopuses cease to feed as they become fully mature. In females a failure to feed could be the result of compression of the gut by the expanding ovary, but this cannot be the explanation for males, which also stop eating and eventually die (van Heukelem, 1973). It is known, besides, that the proteolytic activity of the secretions of the posterior salivary glands and hepatopancreas falls dramatically (to about 5% of its former value) in octopuses that have laid their eggs (Sakaguchi, 1968) so that even if the animals did feed they might be unable to digest their prey. The animals become shrunken in appearance, wound healing declines, and there are progressive degenerative changes in the capacity to control skin colour and texture (van Heukelem, 1973).

It is possible that the optic gland hormone could sometimes be acting through a further control stage. We do not know, for example, whether the state of the male or female ducts is determined directly, or indirectly through a further and as yet unidentified nervous or hormonal link; all we do know is that this link does *not* operate through the gonad. The only effects that we can be certain are direct are those that it is possible to demonstrate *in vitro*.

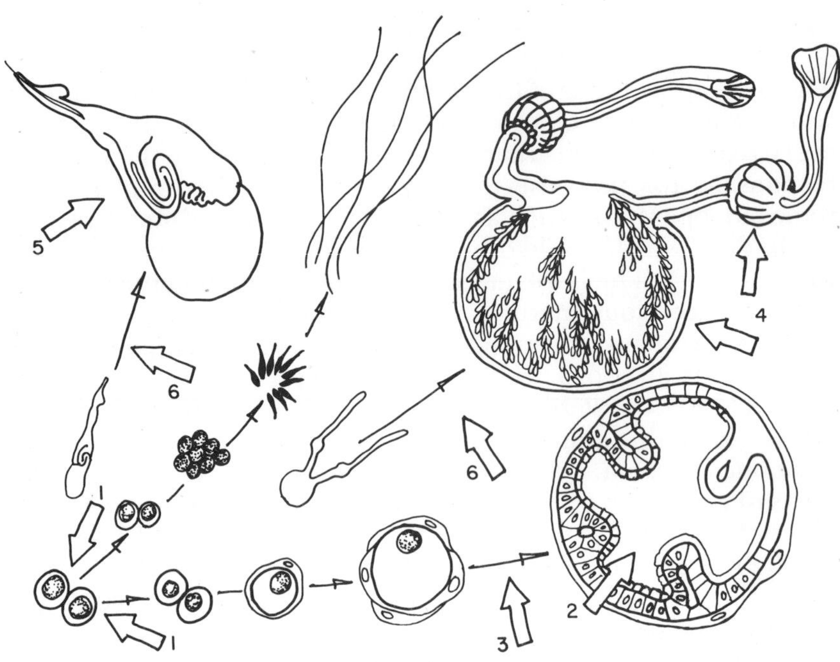

FIG. 6. Diagram summarizing the known effects of the optic gland hormone. (1) Stimulation of germ cell divisions (Richard, 1970). (2) Uptake of amino acids and yolk protein synthesis by the follicle cells (Wells, O'Dor & Buckley, 1975). (3) Final stages of oocyte and follicle cell development (Wells & Wells, 1959, 1975). (4) Maturation of ovisac, oviducts and oviducal glands independent of the ovary (Wells, 1960). (5) Growth and maintenance of the male ducts independently of the testis (Wells & Wells, 1972b). (6) Maturation of the male and female gonad and ducts (Wells & Wells, 1959, 1975).

The early stages of egg and sperm formation (Richard, 1970) and the control of protein yolk deposition (reviewed above) must both depend directly upon the optic gland hormone since its effects are apparent in the absence of the whole of the rest of the octopus. In the absence of any evidence to the contrary, the simplest hypothesis is that all the other varied and widespread effects of optic gland hormone are mediated directly, a variety of target tissues responding variously to a single product of the one sort of gland cell present. Since we now have *in vitro* assay systems available, it should soon be possible to characterize this important material.

REFERENCES

Björkman, N. (1963). On the ultrastructure of the optic gland in *Octopus*. *J. Ultrastruct. Res.* **8**: 195.

Boycott, B. B. & Young, J. Z. (1956). The subpedunculate nerve and other organs associated with the optic tract of cephalopods. In *Bertil Hanström, Zoological Papers in Honour of his Sixty-fifth Birthday*: 76–165. Wingstrand, K. G. (ed.). Lund: Zool. Inst.

Callan, H. G. (1940). The absence of a sex hormone controlling regeneration of the hectocotylus in *Octopus vulgaris* Lam. *Pubbl. Staz. zool. Napoli* **18**: 15–19.

Defretin, R. & Richard, A. (1967). Ultrastructure de la glande optique de *Sepia officinalis* (Mollusque, Céphalopode). Mise en évidence de la sécrétion et de son controle photopériodique. *C.r. hebd. Séanc. Acad. Sci., Paris* **265**: 1415–1418.

Durchon, M. & Richard, A. (1967). Étude, en culture organotypique, du rôle endocrine de la glande optique dans la maturation ovarienne chez *Sepia officinalis* L. (Mollusque, Céphalopode). *C.r. hebd. Séanc Acad. Sci., Paris* **264**: 1497–1500.

Froesch, D. (1974). The subpedunculate lobe of the octopus brain: evidence for dual function. *Brain Res.* **75**: 277–285.

Fugii, T. (1960). Comparative biochemical studies on the egg-yolk proteins of various animal species. *Acta Embryol. Morph. exp.* **3**: 260–285.

Mangold-Wirz, K. (1963). Biologie des céphalopodes benthiques et nectoniques de la Mer Catalane. *Vie Milieu* **13** (suppl): 1–285.

Mangold, K. & Boletzky, S. v. (1973). New data on reproductive biology and growth of *Octopus vulgaris. Mar. Biol.* **19**: 7–12.

Nishioka, R. S., Bern, H. A. & Golding, D. W. (1970). Innervation of the cephalopod optic gland. In *Aspects of neuroendocrinology*: 47–54. Bargman, W. & Scharrer, B. (eds) (5th International symposium on Neurosecretion). Berlin: Springer Verlag.

O'Dor, R. K. & Wells, M. J. (1973). Yolk protein synthesis in the ovary of *Octopus vulgaris* and its control by the optic gland gonadotropin. *J. exp. Biol.* **59**: 655–674.

O'Dor, R. K. & Wells, M. J. (1975). Control of yolk protein synthesis by *Octopus* gonadotropin *in vivo* and *in vitro*. *Gen. comp. Endocr.* **27**: 129–135.

Richard, A. (1970). Différenciation sexuelle des céphalopodes en culture *in vitro*. *Année biol.* **9**: 409–415.

Sakaguchi, H. (1968). Studies on digestive enzymes of devil fish. *Bull. Jap. Soc. scient. Fish.* **34**: 716–729.

Sheppard, C. W. & Beyl, G. E. (1951). Cation exchange in mammalian erythrocytes: 3. The prolytic effect of X-rays on human cells. *J. gen. physiol.* **34**: 691–704.

Taki, I. (1944). [Studies on Octopus (2). Sex and the genital organ.] *Jap. J. Malac.* (*Venus*) **13**: 267–310. [In Japanese.]

van Heukelem, W. F. (1973). Growth and lifespan of *Octopus cyanea* (Mollusca: Cephalopoda). *J. Zool., Lond.* **169**: 299–315.

Wells, M. J. (1960). Optic glands and the ovary of *Octopus*. *Symp. zool. Soc. Lond.* No. 2: 87–107.

Wells, M. J., O'Dor, R. K. & Buckley, S. K. L. (1975). An *in vitro* bioassay for a molluscan gonadotropin. *J. exp. Biol.* **62**: 433–446.

Wells, M. J. & Wells, J. (1959). Hormonal control of sexual maturity in *Octopus*. *J. exp. Biol.* **36**: 1–33.

Wells, M. J. & Wells, J. (1972a). Optic glands and the state of the testis in *Octopus*. *Mar. Behav. Physiol.* **1**: 71–83.

Wells, M. J. & Wells, J. (1972b). Sexual displays and mating of *Octopus vulgaris* Cuvier and *O. cyanea* Gray and attempts to alter performance by manipulating the glandular condition of the animals. *Anim. Behav.* **20**: 293–308.
Wells, M. J. & Wells, J. (1975). Optic gland implants and their effects on the gonads of *Octopus*. *J. exp. Biol.* **62**: 579–588.

Symp. zool. Soc. Lond. (1977) No. 38, 541–555.

A RECONSIDERATION OF FACTORS ASSOCIATED WITH SEXUAL MATURATION

KATHARINA MANGOLD and D. FROESCH

Laboratoire Arago, Banyuls-sur-mer, France

SYNOPSIS

Until now, it has not been possible to describe unequivocally any external factor that would induce sexual maturation in cephalopods in general.

The present investigation is based on (a) a gonad weight/body weight analysis of *Octopus vulgaris* captured during the last seven years in the Catalonian Sea off Banyuls-sur-mer (France) and (b) a gonad weight/body weight analysis and an ultrastructural study of optic glands of different octopods kept under various experimental conditions in the laboratory. Our data from Mediterranean octopods suggest that the gonads develop relatively independently of external factors. At the ultrastructural level we have found no evidence for hormone synthesis and release in the optic gland at any moment in the life of an octopus. The abundance of lipofuscin and the presence of haemocyanin in the cells of the optic gland suggest that it may also be involved in other functions than the control of sexual maturation.

INTRODUCTION

In the search for external factors that might influence gonad maturation in cephalopods biologists have hitherto depended almost entirely on field data by examining freshly captured animals. Consequently, the habitat and the period of spawning of a great number of species is known. In addition, the approximate lifespan of a few species reared in the laboratory is known. In Fig. 1 the biological data of some randomly selected cephalopod species are given (Boletzky, 1975; Boletzky & Boletzky, 1969; Boletzky, Boletzky, Froesch & Gaetzi, 1971; Choe, 1966; Fields, 1965; Mangold-Wirz, 1963; Mangold & Boletzky, 1973; Van Heukelem, 1973); this figure comprises species with different periods of spawning, though living in similar habitats, and vice versa. We do not know the factors which account for these differences.

It has been shown by the experiments of various authors that daylength (Richard, 1967; Wells & Wells, 1959, 1969, 1972; Laubier-Bonichon & Mangold, 1975), temperature (Mangold-Wirz, 1963), and feeding habits (Mangold & Boucher-Rodoni, 1973; Rowe & Mangold, 1975) may have some influence on maturation. However, a hypothesis of how these factors act upon the gonad is still lacking.

The present investigation is based on (a) a gonad weight/body weight analysis of *Octopus vulgaris* captured in the Catalonian Sea

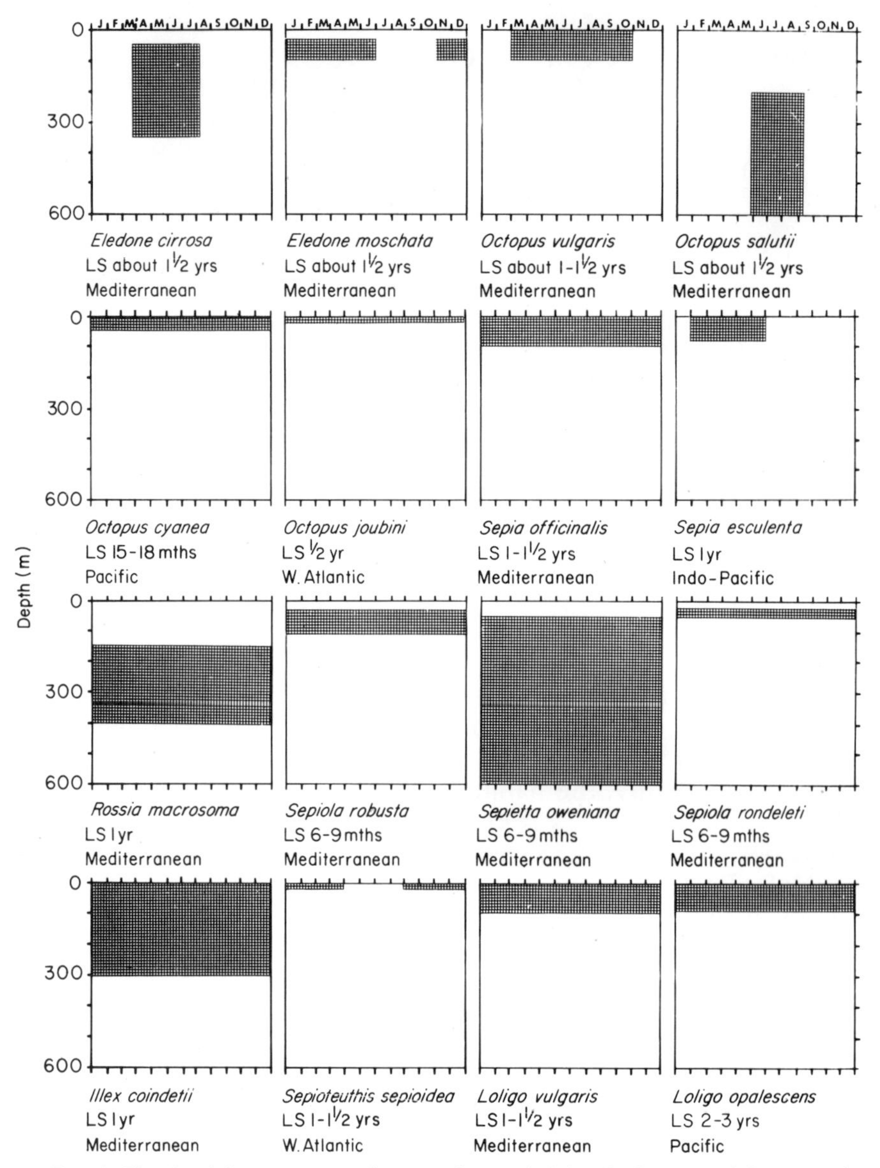

FIG. 1. The depth in metres (m), the spawning period (hatched area) and the approximate lifespan (LS) of mature females of 16 species of cephalopods.

off Banyuls-sur-mer (France) during the last seven years and (b) a gonad weight/body weight analysis and an ultrastructural study of optic glands of three octopod species kept under different experimental conditions.

RESULTS

The gonads of several hundred freshly caught animals were weighed and related to their total body mass. There is an important difference between the growth of the gonads in male and in female cephalopods: there is a permanent production and release of spermatophores in mature males whereas in females there is a continuous accumulation of material in the ovary up to the time of spawning. One might therefore expect a higher variability in the male gonad than in the female one, but the contrary is true: the

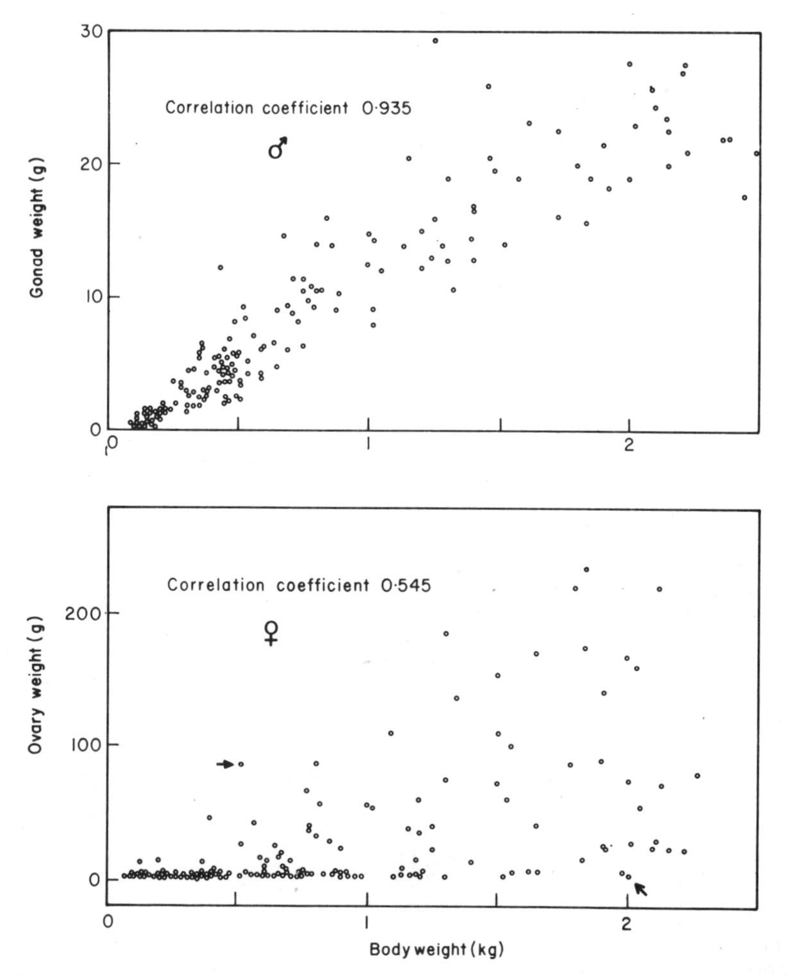

FIG. 2. The gonad weight/body weight correlation of freshly captured *Octopus vulgaris*, 155 ♀ and 161 ♂, comprising animals caught throughout the year.

correlation between gonad weight and total body weight is considerably better in males ($r = 0{\cdot}935$) than in females ($r = 0{\cdot}545$); Fig. 2. In females we find a substantial proportion of immature individuals at weights of over 2 kg or even 4 kg, but we also find females of 500 g that are already mature (i.e. their eggs are freed from the follicular sheath). This can hardly be due to seasonal variations, since the two females indicated by arrows in Fig. 2—512 g/85 g (mature) and 2013 g/3 g (immature)—were captured in the same week of October 1974 and at the same depth, 50 m approximately. This difference in the developmental pattern of growth of the gonads in males and females is remarkable and possibly an important fact, although we cannot understand its significance yet.

We have tried to ascertain whether the different patterns of growth of the male and the female gonad could be correlated with that of the optic gland, which is considered a gonadotropic gland (Wells, 1960; Wells & Wells, 1959, 1969, 1972, 1975; Bonichon, 1967; Richard, 1971). Figure 3 is a section through the optic gland of an immature female of *Eledone cirrosa*. It shows a large number

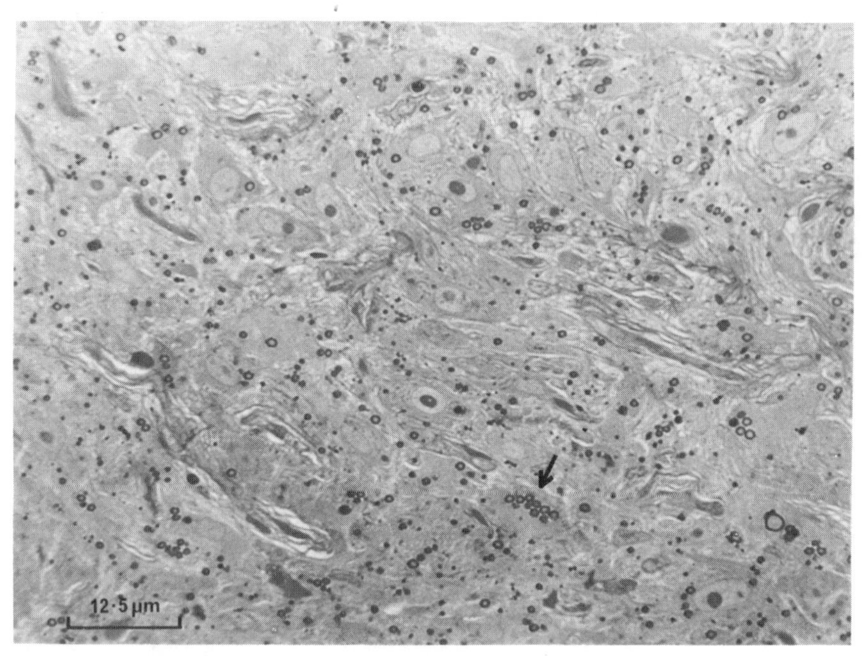

FIG. 3. A section of the optic gland of an immature female of *Eledone cirrosa*. The gland was fixed by immersion in an isotonic solution of 2% osmic acid and embedded in Epon. The section is stained with methylene blue and shows a large number of yellow inclusions.

of yellow inclusions surrounded by a dark halo. These inclusions are usually described as secretory grana of the optic gland or vacuoles of secretory material (Nishioka, Bern & Golding, 1970; Wells, 1960; Wells & Wells, 1959). We examined the relationship of the secretory grana to the body and gonad weight in 52 specimens of *Octopus vulgaris*, 24 females and 28 males. The number of grana showed good correlation with the body weight in both sexes (Fig. 4: in males r was 0·899, in females r was 0·932) and it did not reflect the high variability of the female gonad weight shown in Fig. 2. The secretory grana are already present in octopods of less than 100 g (Fig. 6b).

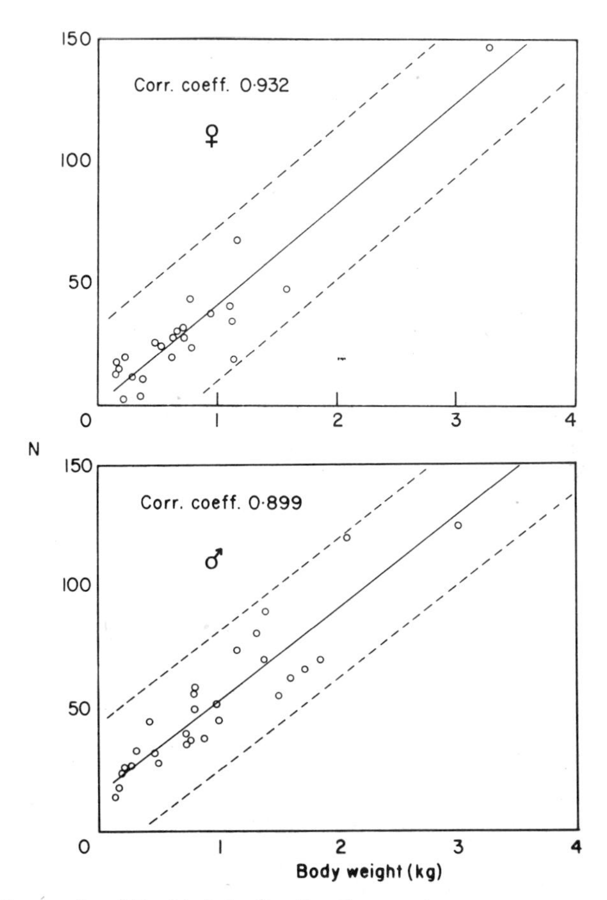

FIG. 4. The number (N) of lytic bodies (i.e. the secretory grana or vacuoles of other authors, referred to in the text) counted per 100 sq. μm of optic gland sections, plotted against the body weight, of 24 ♀ and 28♂ octopuses. Each dot is the mean value of 10 counts.

Previous experiments with octopods and sepiolids have failed to demonstrate unequivocally that varied photoperiods would cause significant differences in gonad development (Boletzky, 1975; Mangold, Froesch *et al.*, 1975). However, one could still expect differences at the level of the optic gland. We therefore isolated 18 specimens of *Octopus vulgaris*, nine males and nine females, and kept them under the following daylength conditions: 0, 1, 4, 8, 12, 16, 20, 23 and 24 hours. The animals were given crabs (*Carcinus maenas*) to eat. At the beginning of the experiment the animals weighed between 63 and 220 g and at the end, after four months, between 290 and 1052 g. We could find neither a correlation between daylength and gonad weight nor daylength and the number of yellow inclusions. Rather, we noticed that the number of inclusions was considerably smaller in all the isolated animals than it was in the 52 control animals just caught from the sea (Fig. 5). Thus, the optic gland contained fewer secretory grana in all the isolated animals, whatever daylength they were kept under.

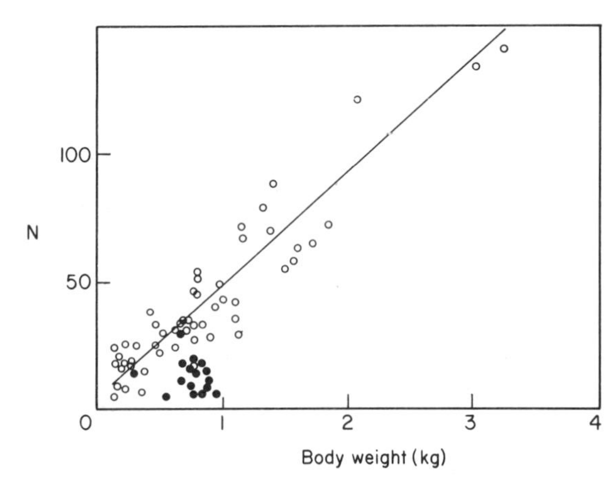

FIG. 5. The number (N) of lytic bodies per 100 sq. μm of optic gland sections, freshly captured (○) and of laboratory kept (●) animals. The latter were isolated for four months and kept at different daylengths, from 0 to 24 hr. The isolated animals accumulated fewer lytic bodies than their controls, regardless of daylength. (Two out of 18 animals died.) Each dot is the mean value of 10 counts.

Since we had hoped to find in the number of secretory grana of the optic gland a useful parameter for the reproductive biology of octopods, these two results were rather disappointing. We had expected the grana to be correlated somehow with the state of maturation and eventually reflect the length of the artificial photoperiods.

There remained the possibility that differences concerning the stage of maturation or various photoperiods could be found at the ultrastructural level of the gland and we therefore set out to find the site of hormone release in the gland and other features related to its secretory activity. The yellow secretory grana visible in light microscopic sections are present in the cell body and the cell processes of the optic gland chief or stellate cells (Nishioka *et al.*, 1970; Wells & Wells, 1959). They measure between one and several microns, and occasionally are nearly as large as the nucleus of their cell. The electron micrographs of Figs 6 and 7 show some of these inclusions: they resemble the residual or lytic bodies which are present in almost every animal cell, in varying numbers but particularly abundant in non-dividing cells (Beck, Lloyd & Squier, 1972; Daems, Wisse & Brederoo, 1972). They are lysosome-derived structures involved in the catabolism of endogenous and exogenous material.

Lytic bodies are found, for instance, in the pituitary gland where they remove excess endogenous hormone (Smith & Farquhar, 1966), and in a number of other vertebrate organs such as the liver, the spleen, the lymph nodules and the kidney where they are charged with the degradation of exogenous material like bacteria or haemoglobin.

A histochemical examination of Epon-embedded glands suggests that the core of the residual bodies contains a lipoprotein pigment; the following methods were positive for lipofuscin:

(a) Hueck's method
(b) Lillie's method
(c) Haematoxylin
(d) Sudan black

(methods and references according to Ganter & Jollès, 1970). The yellow autofluorescence and the acid-fast nature of the substance also support the claim that it is lipofuscin, a "waste pigment", that is an orange-coloured, fluorescent, insoluble and biologically inactive protein.

The yellow inclusions in the optic gland being residual bodies and not secretory grana, we had to look for other signs of endocrine function such as glandular epithelia bordering the blood, secretory vesicles, granular endoplasmic reticulum (GER), exocytosis, and so forth. Yet, whatever the state of gonad maturation, the features required for secretory activity are sparse (GER and Golgi-associated vesicles) or even absent (glandular cell/blood contact and exocytosis). At any moment of the life of an octopod the optic gland chief cells contain considerable amounts of

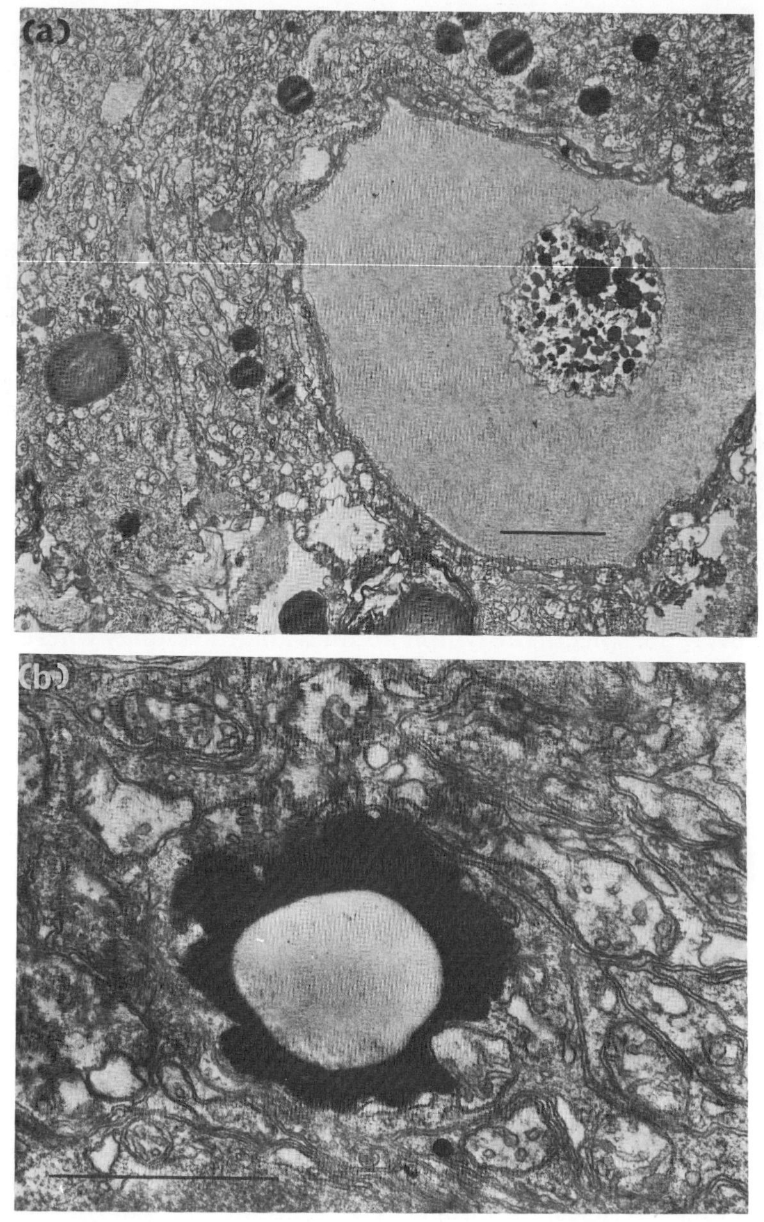

FIG. 6. (a) Section of the optic gland of an *Eledone moschata* (body weight = 800 g) after six months of breeding. A capillary with an amoebocyte is cross cut. The gland cells contain characteristically dark inclusions and many mitochondria. The bar is 3 μm. (b) A higher magnification from a gland cell of a newly hatched *Eledone moschata*. It shows a lytic body, with pale core and dark halo, surrounded by mitochondria. The bar is 1 μm.

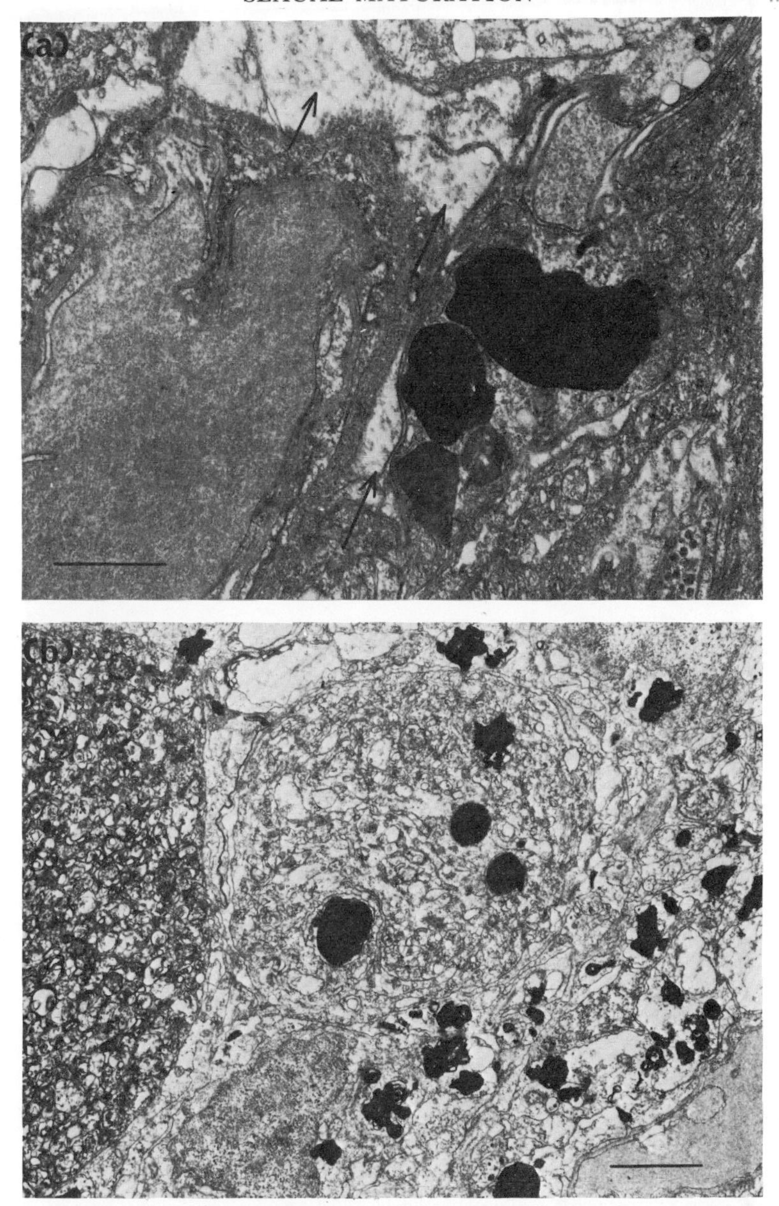

FIG. 7. (a) Section of the optic gland of a female *Octopus vulgaris* (body weight = 518 g, ovary weight = 0·6 g). Four dark inclusions lie in a process of an optic gland cell. The capillary is surrounded by a considerable external basement membrane. Haemocyanin can be seen in the blood vessels as well as inside the cells (arrows) bordering the outer basement membrane. The bar is 1 μm. (b) Section of the optic gland of an *Eledone moschata* ready to spawn. The cells bear numerous dark inclusions and mitochondria. The debris to the left of the micrograph consists mainly of degenerating mitochondria. The bar is 2 μm.

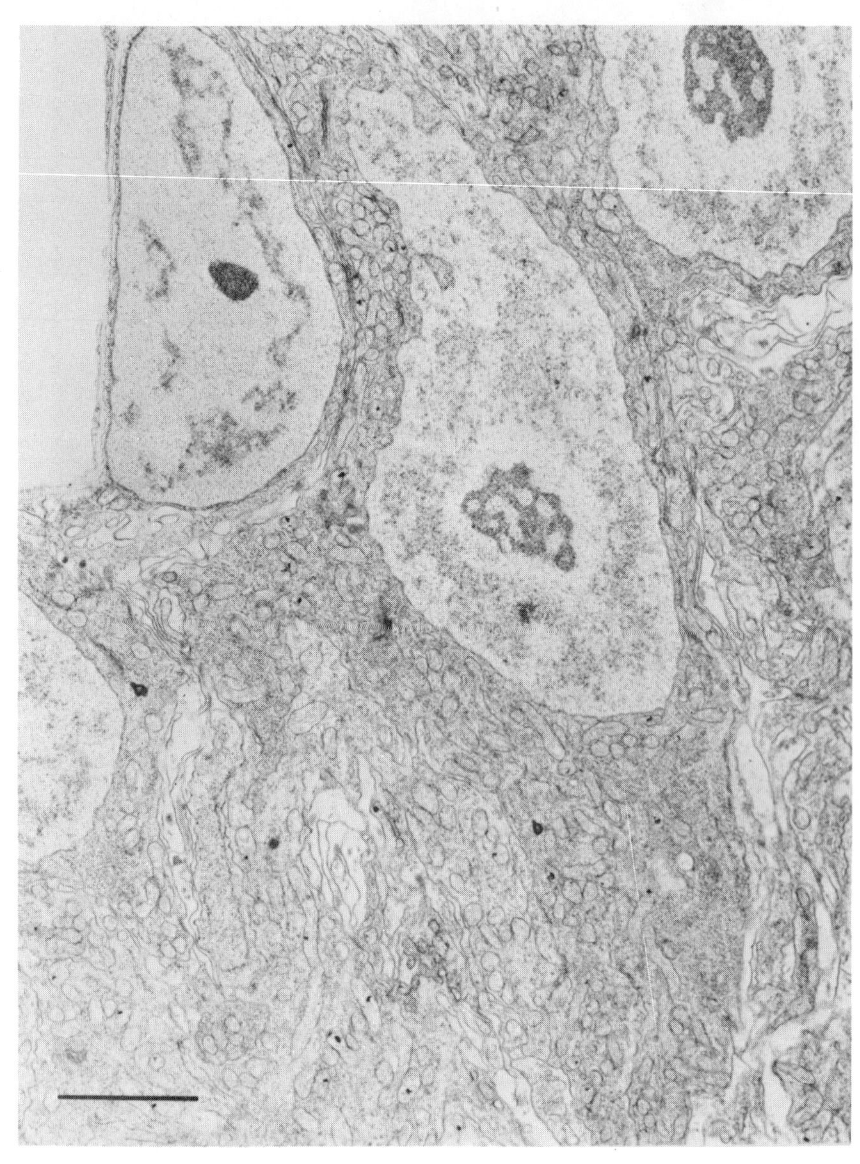

FIG. 8. Section of the optic gland of a female *Octopus vulgaris* (body weight = 778 g, ovary weight = 4·8 g, egg length = 1·0 mm). The nucleus bordering the perfused capillary belongs to a pericyte. The gland cells contain large numbers of mitochondria and few GER. The bar is 2 μm.

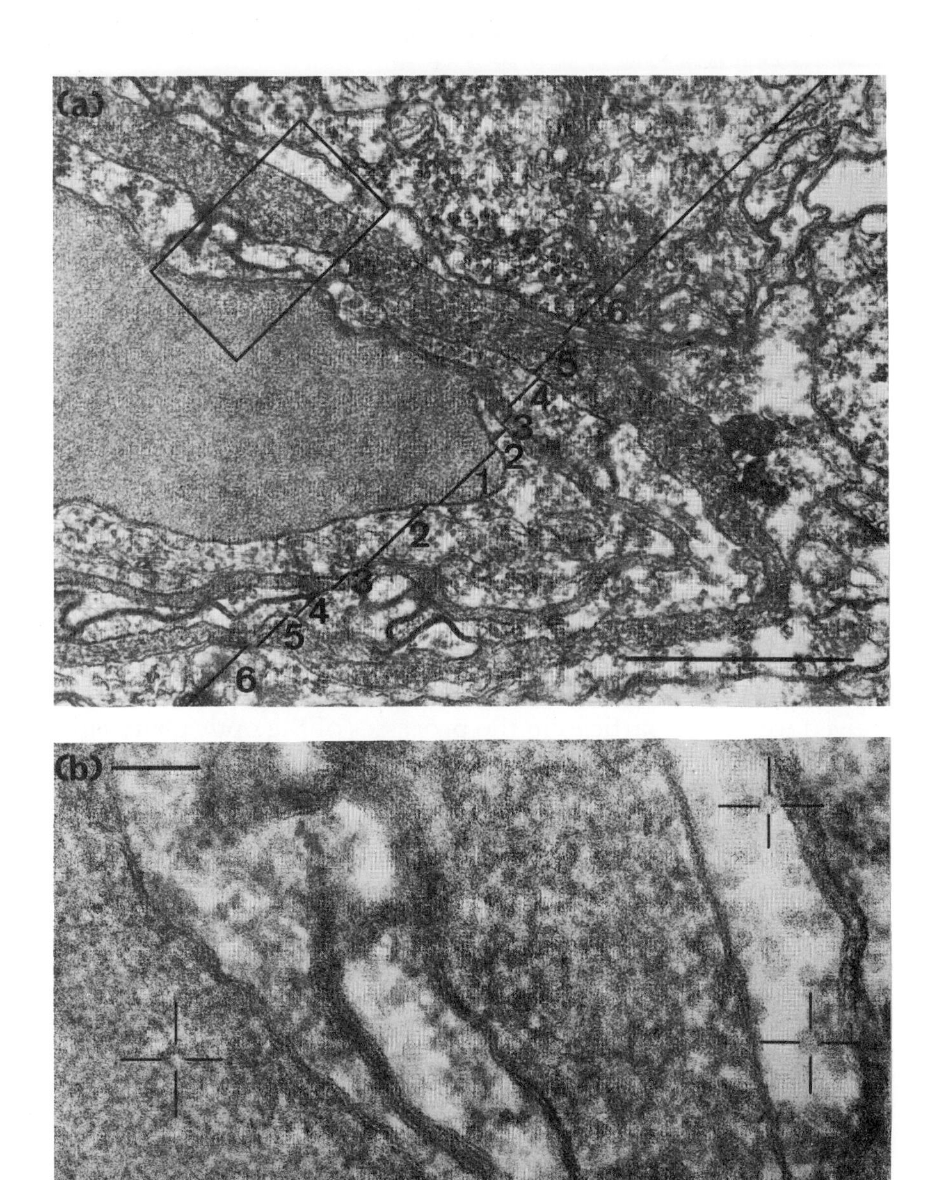

FIG. 9. (a) Section of the optic gland of a female *Octopus vulgaris* (body weight = 160 g, ovary weight less than 1 g), demonstrating the basic organization of the gland. (1) Vessel light with haemocyanin; (2) endothelial cell; (3) internal basement membrane; (4) pericyte; (5) external basement membrane; (6) optic gland cells. The bar is 1 μm. (b) Inset of (a). Haemocyanin is found in the blood vessels and inside the optic gland cell processes (crosses). The bar is 0·1 μm.

mitochondria and lysosomes as the predominant organelles (Figs 6, 7 and 8). They never face the blood directly as is usual in an endocrine gland. They are separated from the blood not only by the wall of the capillary (i.e. the pericyte), but also by an enormous external basement membrane, comparable with the reticulin coat of spleen capillaries, Fig. 9. Other blood vessels investigated in *Octopus*, such as capillaries in many lobes of the brain, in mantle muscle, salivary gland, oviducal gland, statocysts, iris and stellate nerves as well as the blood vessels in cephalopods described by Barber & Graziadei (1965, 1967a, 1967b) do not have this specialization.

The absence of an endocrine fine structure in the chief cells (see also Björkman, 1963 and Barber, 1967), the abundance of mitochondria and lytic bodies and finally the similarity of optic gland and spleen capillaries, led us to the hypothesis that the organ could be involved in a catabolic function: removal of endogenous hormone like gonadotropin or degradation of exogenous material like bacteria or haemocyanin. Soon after the hypothesis was conceived we found particles in the processes of the optic gland chief cells identical in size and structure with the haemocyanin molecules circulating in the capillaries (Figs 7a and 9). This finding was later confirmed by Ghiretti-Magaldi (pers. comm.).

The rate of accumulation of residual bodies was then determined in animals that had undergone an operation inducing

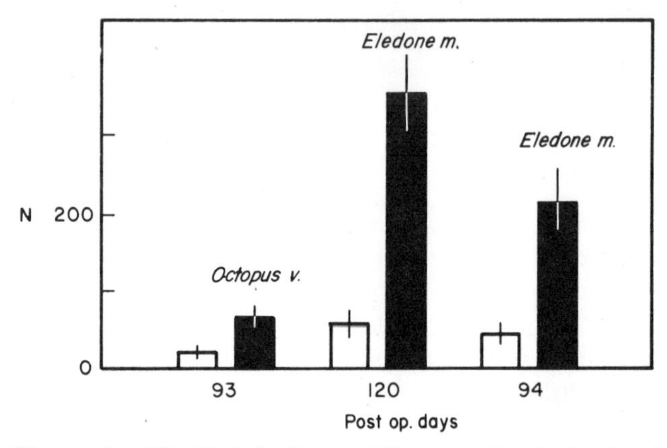

FIG. 10. The number (N) of lytic bodies per 100 sq. μm of optic gland sections of three animals which had their right optic stalk sectioned. The denervated glands accumulate about three to six times more lytic bodies within 93–120 post-operative days. The white columns represent the unoperated gland, the black ones the operated gland of the same individuals.

precocious maturation (Wells & Wells, 1959). In three animals the nerve supply to the optic gland was interrupted by severing the optic stalk unilaterally. Separated from the brain an optic gland produced within about 100 days three to six times more lytic bodies than did the unoperated contralateral gland (Fig. 10). It even became orange with the high content of lipofuscin. The same effect—acute lipofuscinosis—can be obtained by two other operations that have been performed by Wells & Wells (1959): destruction of the subpedunculate lobe, site of the origin of the optic gland nerve and the subpedunculate nerve, and section of the optic nerves, which join the retina to the optic lobe. An elevated number of residual bodies was the only structural difference between operated and control glands in our three animals.

DISCUSSION

The evidence of O'Dor & Wells (1973), that animals without optic glands accumulate more radioactively-labelled haemocyanin in the blood than unoperated controls, and the occurrence of haemocyanin in the chief cells of the optic gland, suggest that the organ may play a role in the metabolism of the blood pigment—in addition to its gonadotropic function that has been demonstrated clearly and repeatedly by Wells & Wells (1959, 1969, 1972, 1975), Durchon & Richard (1967), O'Dor & Wells (1973) and Wells, O'Dor & Buckley (1975).

The optic gland gonadotropin is the only internal factor so far described that controls sexual maturation. Our investigation of optic gland structure in octopods does not support the great bulk of behavioural evidence for an endocrine function of the organ. Indeed, so far there is no ultrastructural evidence for synthesis of a gonadotropin, nor has its chemical structure been elucidated.

Richard gave a careful and detailed account of *Sepia* optic gland structure in his Ph.D. Thesis (1971). In the search for secretory grana he described a number of Golgi-associated vesicles that we have found also in octopods. These structures were no more prominent than in normal cells nor could the authors find evidence for their accumulation or release.

It remains to be clarified whether the haemocyanin in the gland cells is of physiological relevance and, if so, how it could be linked with the gonadotropic activity of the organ.

The most important external factor, on the other hand, that has been reported by various authors to induce gonad maturation, is that of light. Our field data and experiments with Mediterranean octopods do not confirm this hypothesis; they rather suggest that the gonads develop relatively independently of external factors.

References

Barber, V. C. (1967). A neurosecretory tissue in *Octopus*. *Nature, Lond.* **213**: 1042–1043.

Barber, V. C. & Graziadei, P. (1965). The fine structure of cephalopod blood vessels. I. Some smaller peripheral vessels. *Z. Zellforsch. mikrosk. Anat.* **66**: 765–781.

Barber, V. C. & Graziadei, P. (1967a). II. The vessels of the nervous system. *Z. Zellforsch. mikrosk. Anat.* **77**: 147–161.

Barber, V. C. & Graziadei, P. (1967b). III. Vessel innervation. *Z. Zellforsch. mikrosk. Anat.* **77**: 162–174.

Beck, F., Lloyd, J. B. & Squier, C. A. (1972). Histochemistry. In *Lysosomes, a laboratory handbook*: 200–239. Dingle, J. T. (ed.). Amsterdam: North-Holland Publishing Company.

Björkman, N. (1963). On the ultrastructure of the optic gland in *Octopus*. *J. ultrastruct. Res.* **8**: 195.

Boletzky, S. von. (1975). The reproductive cycle of the Sepiolidae. *Pubbl. Staz. zool. Napoli* **39** (suppl. 1): 84–95.

Boletzky, S. von & Boletzky, M. V. von. (1969). First results in rearing *Octopus joubini*. *Verh. naturf. Ges. Basel* **80**: 56–61.

Boletzky, S. von, Boletzky, M. V. von, Froesch, D. & Gaetzi, V. (1971). Laboratory rearing of Sepiolinae (Mollusca: Cephalopods). *Mar. Biol.* **8**: 82–87.

Bonichon, A. (1967). Contribution à l'étude de la neurosécrétion et de l'endocrinologie chez les Céphalopodes. I. *Octopus vulgaris*. *Vie Milieu* **18** (2A): 228–252.

Choe, S. (1966). On the eggs, rearing, habits of the fry and growth of some Cephalopoda. *Bull. Mar. Sci. Gulf Caribb.* **16**: 330–348.

Daems, W. T., Wisse, E. & Brederoo, P. (1972). Electron microscopy of the vacuolar apparatus. In *Lysosomes, a laboratory handbook*: 150–199. Dingle, J. T. (ed.). Amsterdam: North-Holland Publishing Company.

Durchon, M. & Richard, A. (1967). Étude, en culture organotypique, du rôle endocrine de la glande optique dans la maturation ovarienne chez *Sepia officinalis* L. (Mollusque Céphalopode). *C.r. hebd. Séanc. Acad. Sci. Paris* (D) **264**: 1497–1500.

Fields, W. G. (1965). The structure, development, food relations, reproduction and life history of the squid *Loligo opalescens*. *Fish. Bull., Calif.* **131**: 1–108.

Ganter, P. & Jollès, G. (1970). *Histochimie, normale et pathologique*. Paris: Gauthier-Villars.

Laubier-Bonichon, A. & Mangold, K. (1975). La maturation sexuelle chez les mâles *d'Octopus vulgaris* (Céphalopode, Octopode) en relation avec le réflexe photo-sexuel. *Mar. Biol.* **29**: 45–52.

Mangold, K. & von Boletzky, S. (1973). New data on reproductive biology and growth in *Octopus vulgaris*. *Mar. Biol.* **19**: 7–12.

Mangold, K. & Boucher-Rodoni, R. (1973). Rôle du jeûne dans l'induction de la maturation génitale chez les femelles d'*Eledone cirrosa* (Cephalopoda, Octopoda). *C.r. hebd. Séanc. Acad. Sci. Paris* (D) **276**: 2007–2010.

Mangold, K., Froesch, D., Boucher-Rodoni, R. & Rowe, V. L. (1975). Factors affecting sexual maturation in cephalopods. *Pubbl. Staz. zool. Napoli* **39** (Suppl. 1): 259–266.
Mangold-Wirz, K. (1963). Biologie des Céphalopodes benthiques et nectonique de la Mer Catalane. *Vie Milieu* Suppl. **13**: 1–285.
Nishioka, R. S., Bern, H. A. & Golding, D. W. (1970). Innervation of the Cephalopod optic gland. In *Aspects of neuroendocrinology*. Bargmann, W. & Scharrer, B. (eds). Heidelberg: Springer Verlag.
O'Dor, R. K. & Wells, M. J. (1973). Yolk protein synthesis in the ovary of *Octopus vulgaris* and its control by the optic gland gonadotropin. *J. exp. Biol.* **59**: 665–674.
Richard, A. (1967). Rôle de la photopériode dans le déterminisme de la maturation génitale femelle du Céphalopode *Sepia officinalis* L. *C.r. hebd. Séanc. Acad. Sci. Paris* (D.) **264**: 1315–1318.
Richard, A. (1971). *Contribution à l'étude expérimentale de la croissance et de la maturation sexuelle de* Sepia officinalis *L.* (*Mollusque, Céphalopode*). Ph.D. Thesis, University of Lille, France: 264 pp.
Rowe, V. L. & Mangold, K. (1975). The effect of starvation on sexual maturation in *Illex illecebrosus* (Cephalopoda: Teuthoidea). *J. exp. mar. Biol. Ecol.* **17**: 157–163.
Smith, R. E. & Farquhar, M. (1966). Lysosome function in the regulation of the secretory process in cells of the anterior pituitary gland. *J. Cell Biol.* **31**: 319–347.
Van Heukelem, W. F. (1973). Growth and life-span in *Octopus cyanea* (Mollusca: Cephalopoda). *J. Zool., Lond.* **169**: 299–315.
Wells, M. J. (1960). Optic glands and the ovary of *Octopus*. *Symp. zool. Soc. Lond.* No. 2: 87–107.
Wells, M. J., O'Dor, R. K. & Buckley, S. K. L. (1975). An *in vitro* bioassay for a molluscan gonadotropin. *J. exp. Biol.* **62**: 433–446.
Wells, M. J. & Wells, J. (1959). Hormonal control of sexual maturity in *Octopus*. *J. exp. Biol.* **36**: 1–33.
Wells, M. J. & Wells, J. (1969). Pituitary analogue in the *Octopus*. *Nature, Lond.* **222**: 293–294.
Wells, M. J. & Wells, J. (1972). Optic glands and the state of the testis in *Octopus*. *Mar. Behav. Physiol.* **1**: 71–81.
Wells, M. J. & Wells, J. (1975). Optic gland implants and their effects on the gonads of *Octopus*. *J. exp. Biol.* **62**: 579–588.

Note Added in Proof

In the meantime, we have been able to produce further evidence supporting a catabolic function of the optic gland (Froesch, D. & Mangold, K. (1976) *Cell Tiss. Res.* **170**: 549–551). The main cells are capable of rapidly taking up and accumulating exogeneous ferritin.

Symp. zool. Soc. Lond. (1977) No. 38, 557–567.

POST-HATCHING BEHAVIOUR AND MODE OF LIFE IN CEPHALOPODS

S. v. BOLETZKY

Laboratoire Arago, Banyuls-sur-Mer, France

SYNOPSIS

Only in two families of cephalopods, both benthic, are there species in which the newly hatched animals differ radically in their mode of life from the adult. Observation on young animals in one of these families, the Octopodidae, show that adaptations to the adult, benthic mode of life can be detected in the planktonic young.

INTRODUCTION

Cephalopods fall into two categories, depending on whether they are natatory or sedentary. Although generally one of these modes clearly prevails in a given group, virtually all benthic cephalopods have a "nektonic potential", because they can swim by jet propulsion.

In view of the various ecological adaptations enabling cephalopods to live in very different habitats, this division may appear extremely crude. Yet it is the only useful one for the comparison of all newly hatched cephalopods.

In an earlier paper the problem of what can be considered an "adult-like" mode of life in young cephalopods has been discussed (Boletzky, 1974b). We have stressed that a fundamental difference between the early planktonic and the adult nektonic mode does not exist in pelagic cephalopods, so that the mode of life of most young cephalopods largely corresponds to that of the adult. A distinct difference between the post-hatching and the adult mode of life exists, as far as we know, only in two families: in the Idiosepiidae and in many species among the Octopodidae.

In the present paper, we briefly summarize the characteristics of young cephalopods, report recent observations on young octopodids, and discuss them with regard to the problem of behavioural changes during post-embryonic development.

Our present very limited knowledge of the behaviour of young cephalopods is almost entirely based on laboratory observations. In recent years, the techniques for maintenance and rearing of cephalopods, hatched in the laboratory, have improved (Boletzky, 1974a) and facilities for studies of their behaviour are thus improved. However, our knowledge is still extremely limited as far as the study of pelagic species living in off-shore waters is concerned.

SOME CHARACTERISTICS OF YOUNG CEPHALOPODS

Nektonic (oceanic and neritic) cephalopods

As indicated by Table I, a nektonic adult mode of life is achieved at hatching. The young animals are very small and swim actively by jet propulsion. They are probably all carnivorous predators. Despite their small size relative to the propulsive mantle-funnel complex, the arms although often very small, as in some of the pelagic squids, are equipped with suckers for seizing comparatively large prey (Boletzky, 1974b). The young usually have relatively large eyes, a functional ink sac and chromatophores (see Packard, 1972, for literature).

TABLE I

Post-hatching and adult mode of life in cephalopods

	Adult mode	Post-hatching mode
Sepioidea	(Nekto-)benthic	(Nekto-)benthic
		Idiosepiidae: planktonic
	Spirula: macroplanktonic	Planktonic
	Heteroteuthinae: nektonic	Planktonic
Teuthoidea	Nektonic	Planktonic
Octopoda (Incirrata)	Nektonic	Planktonic
	Octopodidae: benthic	Benthic
		Young, very small relative to the adult: planktonic

Benthic sepioids

Benthic (or nekto-benthic) sepioids show the adult mode of life from the beginning of post-embryonic development. Their body proportions are similar to those of the adults and they have well developed arms and long tentacles, which are ejected for prey capture (see Messenger, 1968). Chromatophore patterns are complex in young *Sepia*. But in sepiolids the simpler colour patterns seem to allow mimicry of certain backgrounds, from hatching onward. Arm displays, which probably serve as a camouflage in a similar way to the "flamboyant posture" of benthic young octopuses (Packard & Sanders, 1971), are common in young and adult Sepiolinae (Boletzky, 1974b).

Most benthic sepioids bury themselves in soft bottoms during the day. The behavioural pattern of this sand-covering is well established at hatching. It is comparatively simple in *Sepia*: the animal settles on the bottom and blows the sand from underneath its body by jets of water from the funnel. The sand particles thus whirl up and descend upon the animal, which gradually settles deeper until it is entirely covered.

In the sepiolids, there is an additional process. It consists of a series of sweeping movements of the dorso-lateral arms which gather sand particles to completely cover the animal (Boletzky & Boletzky, 1970).

Young *Sepia* settled on a hard substrate show a response unique amongst the benthic cephalopods. They attach themselves with the ventral integument of the mantle and ventral arms, whose muscles act as a set of "suckers". This ability gradually disappears when the animals grow larger (Boletzky, 1974b).

Among benthic sepioids, only the newly hatched animals of the pygmy cuttlefish, *Idiosepius pygmaeus*, are known to be planktonic (Natsukari, 1970, and pers. comm.). The most striking feature of these young is the absence of tentacles; they are formed during post-embryonic development. The general aspect of the arm-crown thus reminds one of planktonic young octopodids. How and when the young pygmy cuttlefish take up an adult mode of life is unknown.

The benthic octopodids

In the benthic octopodids, there are numerous species in which the young live for some time in the plankton. Common characters of these young are the small size relative to the adult [dorsal mantle length (ML) = 2–5% of adult], the small arm-length (35–50% of ML) and few suckers (3–10 per arm) (Fig. 1).

The newly hatched octopodids that are benthic are larger relative to the adult (ML = 6–20% of adult ML) and already have long arms (arm-length $\geqslant$ ML) and a large number of suckers (20 or more per arm). Their body proportions thus correspond to those

FIG. 1. Newly hatched benthic *Eledone moschata*, on the left (dorsal view), and newly hatched *Eledone cirrosa*, on the right (ventral view). Both animals are anaesthetized.

of young *Octopus vulgaris* when they become benthic (Itami *et al.*, 1963).

As the adult sizes differ greatly among octopodids, newly hatched animals of similar size may show different modes of life: for example, planktonic young *Eledone cirrosa* and benthic young *Octopus joubini*, both with a mantle length of 4·5 mm at hatching, which is 3·5 and 18% respectively of the adult mantle length.

Among the planktonic young octopodids, the mantle length of newly hatched animals varies from about 2 mm to about 4–5 mm. One would expect the number of suckers, which varies among species, to be related to the absolute size of the newly hatched young. This is not so, however.

Table II indicates, for newly hatched planktonic young, the mantle length, the relative size of the animal, and the number of suckers in five species. It shows that the number of suckers increases, although irregularly, with the relative size of the newly hatched animals. Thus amongst the planktonic young, those larger

TABLE II

Body size and number of suckers in planktonic young octopodids

Newly hatched animal	ML (mm)	ML as a percentage of adult ML	No. of suckers per arm
Octopus vulgaris	2	2%	3
Scaeurgus unicirrhus	2	2·2%	4
Octopus salutii	3·5	3·5%	4–5
Eledone cirrosa	4·5	3·5%	8
Hapalochlaena lunulata	2·3	4%	10

relative to the adult are also more like the adults in having a greater number of suckers. This raises the question of whether they are also closer to settling than the other planktonic young that have a more rudimentary sucker equipment. This problem will be considered later.

A character that is apparently not related to the post-hatching mode of life is the number of chromatophores. According to our observations, it is merely related to body size, so that large planktonic and small benthic young have similar colour patterns.

A peculiarity of all young octopods so far studied, with the exception of the benthic young of *Octopus briareus* and *O. maya*, is the presence of very numerous tufts of chitinous setae all over the skin (Brocco, O'Clair & Cloney, 1974). The size of these Kölliker tufts does not vary greatly among different species, so that they are much larger relative to the body size in planktonic young compared with the benthic ones; in the latter, their distribution is also less dense. These tufts are associated with the integumental musculature, which allows their repeated evagination and spreading (Boletzky, 1973). This spreading has been observed under experimental conditions only. The supposed "spreading response" of intact animals has not yet been seen but the small size of the tufts, which have a diameter of less than 100 μm when spread, makes their observation under natural conditions practically impossible. If such a response does exist, it may function, for example, to slow down passive sinking. Have the Kölliker tufts any function in benthic young? They are shed after a few weeks of benthic life, but it is not yet known whether they are retained in pelagic octopods (but see Chun, 1902).

SIGNS OF ADULT HABITS IN PLANKTONIC YOUNG OCTOPODIDS

For the planktonic young octopodid, settling means giving up an exclusive swimming existence and taking up an adult mode of life that comprises, in addition to occasional swimming, certain responses concerned with the substrate. These responses must be innate, but it remains to be established whether the central nervous system of the newly hatched animals has only a preliminary programme for a later establishment of the nervous control of these responses, or whether these are functional already at hatching. If the latter is true, the so-called adult responses to a substrate must remain inhibited throughout planktonic life. The following observations suggest they are mature, but their inhibition may be raised very early.

We have observed young *Eledone cirrosa* for several days after hatching. They mostly swam about (Fig. 2), but at intervals would

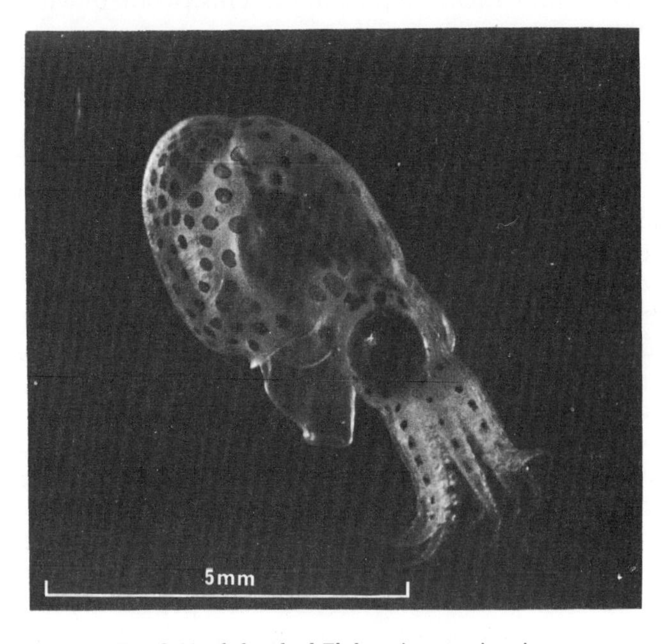

FIG. 2. Newly hatched *Eledone cirrosa*, swimming.

settle on the bottom or on the wall of the tank to which they became attached with their suckers (Fig. 3). In other words although the tendency to swim clearly dominated over the tendency to settle, the settling behaviour can, and does, emerge on occasion.

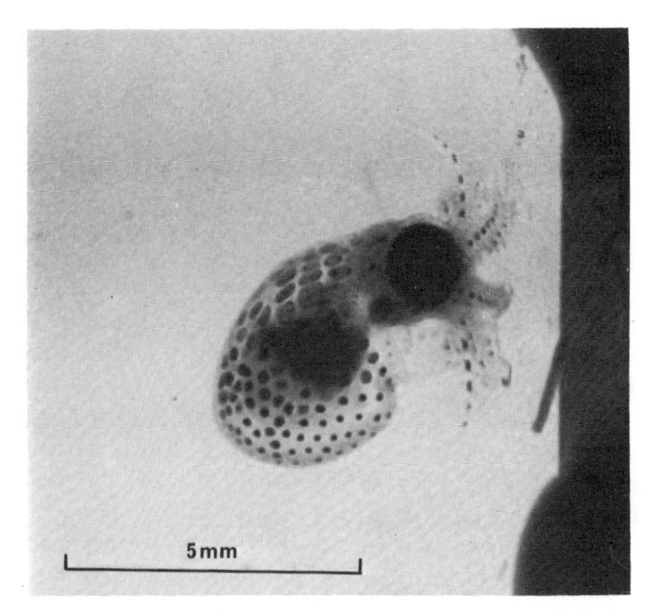

FIG. 3. Newly hatched *Eledone cirrosa*, settled on the wall of the aquarium.

In young animals that have only three or four suckers on (the proximal part of) each arm, we have never seen this mode of settling under normal conditions. Animals may accidentally attach themselves with their suckers to the wall of the tank when they miss prey, because their attack is a very rapid forward thrust. Otherwise they do not attach themselves to a substrate. The same seems to be true, however, for young *Hapalochlaena lunulata*, which have a comparatively large number of suckers (see below). Thus more adult-like sucker equipment is not necessarily a sign of imminent behavioural changes resulting in settling.

Certainly the settling response is the prerequisite of the typical adult mode of life of benthic octopodids. However, its release is but the first step towards a change. Benthic life does not consist of clinging to the substrate, interspersed only with occasional swimming. Octopuses also crawl along the ground. This crawling or walking comprises a well co-ordinated action of the arms and suckers. As newly settled animals do not depend on this for their locomotion, it would seem likely that the co-ordination of sucker movements would only be established with the formation of more suckers on the distal part of the arms. To our surprise, we found that even small planktonic young that have few suckers may already show the typical crawling movement.

In fact newly hatched individuals of *Scaeurgus unicirrhus* placed in a drop of sea-water on a hollow slide so that free swimming was impossible, immediately attached themselves with their suckers to the slide and crawled forward with perfectly co-ordinated alternate movements of the arms, including the alternate attachment and release of the suckers.

Under similar conditions, young *Hapalochlaena lunulata* attached themselves to the substrate, but they were not observed to crawl (Overath & Boletzky, 1974).

Assuming that the newly hatched *Scaeurgus unicirrhus* are not an exception to the rule, we may conclude that the fundamental responses of the benthic adult life are ready for release in all young octopodids, but that they are generally inhibited in those that are very small relative to the adult. The periodical settling under normal conditions, as observed in young *Eledone cirrosa*, suggests that the change of mode of life is a gradual process. This is also indicated by the observations on settling of Itami and his collaborators (1963). These authors overcame the problem of food supply and were thus able to raise young *Octopus vulgaris* to the benthic stage. The young octopuses were fed prawn larvae raised parallel with them. What is particularly interesting is that this prey was supplemented by other food only after the octopuses had settled. This suggests that the feeding behaviour does not change to induce settling. But does it change after settling?

From several observations (Messenger, 1963; Boletzky & Boletzky, 1969; Wells & Wells, 1970; Thomas & Opresko, 1973) it appears that benthic young octopuses feed on benthic prey. In the laboratory, they also accept dead food. They attack living prey either by pouncing upon it, propelled by a jet of water from the funnel, and seizing it with all arms, or by remaining on the ground and extending an arm towards the prey to seize it (Wells, 1962).

This latter method of attack is necessarily associated with benthic life. The pouncing attack, however, largely corresponds to the mode observed in the planktonic young. Thus the change from the planktonic to the benthic mode of life does not imply a complete change of predatory behaviour, but only the addition of a new, purely "benthic" alternative mode.

In the benthic young of *Eledone moschata*, on the other hand, we have observed chase and attack off the bottom. Young animals that had been starved for several days were presented with newly hatched *Sepia officinalis*, similar in size to the young *Eledone*. The latter immediately attacked the cuttlefish, which escaped by jetting away. The young *Eledone* pursued them, swimming forwards at high speed; their body took on a very slender form, and the long arms were tightly held together and stretched out toward the prey. These slender squid-like animals shooting through the water looked very different indeed from the stout little octopuses they had been a few seconds earlier. Swimming in this squid-like posture has also been observed in the benthic young *Octopus joubini*, where it was not, however, associated with the attack of prey (Boletzky & Boletzky, 1969).

Finally it remains to establish whether fixation of the prey is monocular or binocular. In the planktonic young octopodids we have observed feeding, and in the young *Eledone* mentioned above, fixation appeared to be binocular as in *Sepia* (Messenger, 1968). In young *Eledone moschata* we have observed rapid pouncing attacks on prawns on the bottom, and as the animals moved forward fixation of the prey may have been binocular.

To summarize, planktonic young octopodids do not give up their mode of prey capture when they become benthic, and animals that are benthic immediately after hatching may still feed in open water. Only in those species with well developed arms, such as the common octopus, will the mode of prey capture that is typical of the planktonic young be later restricted to the exclusive pouncing attack on the bottom.

DISCUSSION

The benthic habit is well established in all octopodids, although there are differences of swimming ability among them. For instance, *Eledone* is a far better swimmer than *Octopus*, and *Octopus* is a better swimmer than *Bathypolypus* (personal observation). Yet on the whole, the sedentary mode of life is clearly dominant even in *Eledone*.

We have suggested that the nervous control of the so-called "benthic responses" is functional even in very small planktonic young, so one may wonder why there is a planktonic phase in so many species.

It has often been stressed that benthic animals need a planktonic stage in order to maintain a wide distribution of the species. However, as some benthic octopodids are good swimmers, a planktonic post-hatching mode of life would not seem to be important. All we know is that the most widely distributed octopodids are in fact among those species which have small planktonic young (Mangold, 1972).

What is important is that these species doubtless represent a primitive situation from which the exclusively benthic octopodids which have large eggs, hence large young, are derived. The question then is: how has selection acted to maintain in so many species this inherited strategy of dispersal, which seems to be opposed by the pressure of a benthic adult?

Although a great advantage of the benthic mode of life is low energy consumption, young animals need large quantities of food for growth. The plankton may constitute a far richer source of suitable prey for very small, young, octopodids.

Clearly a planktonic post-hatching phase has proved successful in evolution. It may have resulted in certain adaptations such as the slow growth of some (Rees, 1954) or all of the arms. Such reduction must be adaptive because unless large arms are neutrally buoyant they will constitute a mechanical disadvantage. This reduction of the arms for locomotory reasons does not prevent them from functioning, at least on occasion, as they do in the adult. Apparently the juvenile central nervous system contains at least some of the adult programmes for controlling the behaviour of the arms.

References

Boletzky, S. v. (1973). Structure et fonctionnement des organes de Kölliker chez les jeunes octopodes (Mollusca, Cephalopoda). *Z. Morph. Tiere* **75**: 315–327.

Boletzky, S. v. (1974a). Elevage de céphalopodes en aquarium. *Vie Milieu* **24** (2A): 309–340.

Boletzky, S. v. (1974b). The "larvae" of Cephalopoda—a review. *Thalassia jugosl.* **10**: 45–76.

Boletzky, S. v. & Boletzky, M. V. v. (1969). First results in rearing *Octopus joubini* Robson 1929. *Verh. naturf. Ges. Basel* **80**: 56–61.

Boletzky, S. v. & Boletzky, M. V. v. (1970). Das Eingraben in Sand bei *Sepiola* und *Sepietta* (Mollusca, Cephalopoda). *Rev. suisse Zool.* **77**: 536–548.

Brocco, S. L., O'Clair, R. M. & Cloney, R. A. (1974). Cephalopod integument: The ultrastructure of Kölliker's organs and their relationship to setae. *Cell Tiss. Res.* **151**: 293–308.

Chun, C. (1902). Ueber die Natur und die Entwicklung der Chromatophoren bei den Cephalopoden. *Verh. dtsch. zool. Ges.* **12**: 162–182.

Itami, K., Izawa, Y., Maeda, S. & Nakai, K. (1963). Notes on the laboratory culture of the octopus larvae. *Bull. Jap. Soc. scient. Fish.* **29**: 514–520.

Mangold, K. (1972). Le potentiel évolutif des céphalopodes récents. In *Fifth European Marine Biology Symposium*: 307–316. Battaglia, B. (ed.). Padova: Piccin Editore.

Messenger, J. B. (1963). Behaviour of young *Octopus briareus* Robson. *Nature, Lond.* **197**: 1186–1187.

Messenger, J. B. (1968). The visual attack of the cuttlefish, *Sepia officinalis*. *Anim. Behav.* **16**: 342–357.

Natsukari, Y. (1970). Egg-laying behaviour, embryonic development and hatched larva of the pygmy cuttlefish, *Idiosepius pygmaeus paradoxus* Ortmann. *Bull. Fac. Fish. Nagasaki Univ.* **30**: 15–29.

Overath, H. & Boletzky, S. v. (1974). Laboratory observations on spawning and embryonic development of a blue-ringed octopus. *Mar. Biol.* **27**: 333–337.

Packard, A. (1972). Cephalopods and fish: the limits of convergence. *Biol. Rev.* **47**: 241–307.

Packard, A. & Sanders, G. D. (1971). Body patterns of *Octopus vulgaris* and maturation of the response to disturbance. *Anim. Behav.* **19**: 780–790.

Rees, W. J. (1954). The *Macrotritopus* problem. *Bull. Br. Mus. nat. Hist. (Zool.)* **2**: 67–100.

Thomas, R. F. & Opresko, L. (1973). Observations on *Octopus joubini*: four laboratory reared generations. *Nautilus* **87**: 61–65.

Wells, M. J. (1962). *Brain and behaviour in cephalopods.* London: Heinemann.

Wells, M. J. & Wells, J. (1970). Observations on the feeding, growth rate and habits of newly settled *Octopus cyanea. J. Zool., Lond.* **161**: 65–74.

APPENDIX I

Studies on Cephalopods by J. Z. Young

1929

- Fenomeni istologici consecutivi alla sezione dei nervi nei cefalopodi. *Boll. Soc. ital. Biol. sper.* **4**: 741–745.
- Sopra un nuovo organo dei cefalopodi. *Boll. Soc. ital. Biol. sper.* **4**: 1022–1024.

1932

- On the cytology of the neurons of cephalopods. *Q. Jl microsc. Sci.* **75**: 1–47.
- with E. SERENI. Nervous degeneration and regeneration in cephalopods. *Pubbl. Staz. zool. Napoli* **12**: 173–208.

1934

- Structure of nerve fibres in *Sepia*. *J. Physiol., Lond.* **83**: 27*P*–28*P*.

1935

- with J. Y. BOGUE and H. ROSENBERG. Electrotonus in the fin nerve of *Sepia*. *J. Physiol., Lond.* **86**: 6*P*.

1936

- Structure of nerve fibres and synapses in some invertebrates. *Cold Spring Harb. Symp. quant. Biol.* **4**: 1–6.
- The structure of nerve fibres in cephalopods and Crustacea. *Proc. R. Soc.* (B) **121**: 319–337.
- The giant nerve fibres and epistellar body of cephalopods. *Q. Jl microsc. Sci.* **78**: 367–386.

1937

- The structure and functioning of the higher nervous centres of cephalopods. *Rep. Br. Ass. Advmt Sci.* **107**: 366–367.
- with R. S. BEAR and F. O. SCHMITT. The sheath components of the giant nerve fibres of the squid. *Proc. R. Soc.* (B) **123**: 496–504.
- with R. S. BEAR and F. O. SCHMITT. The ultrastructure of nerve axoplasm. *Proc. R. Soc.* (B) **123**: 505–519.
- with R. S. BEAR and F. O. SCHMITT. Investigations on the protein constituents of nerve axoplasm. *Proc. R. Soc.* (B) **123**: 520–529.
- with C. L. PROSSER. Responses of muscles of the squid to repetitive stimulation of the giant nerve fibres. *Biol. Bull. mar. biol. Lab., Woods Hole* **73**: 237–241.

1938

- The functioning of the giant nerve fibres of the squid. *J. exp. Biol.* **15**: 170–185.
- Synaptic transmission in the absence of nerve cell bodies. *J. Physiol., Lond.* **93**: 43*P*–45*P*.
- with R. J. PUMPHREY. The rates of conduction of nerve fibres of various diameters in cephalopods. *J. exp. Biol.* **15**: 453–466.

1939

- Fused neurons and synaptic contacts in the giant nerve fibres of cephalopods. *Phil. Trans. R. Soc.* (B) **229**: 465–503.

1940

- with R. J. PUMPHREY and O. H. SCHMITT. Correlation of local excitability with local physiological response in the giant axon of the squid (*Loligo*). *J. Physiol., Lond.* **98**: 47–72.
- with F. K. SANDERS. Learning and other functions of the higher nervous centres of *Sepia. J. Neurophysiol.* **3**: 501–526.
- with D. A. WEBB. Electrolyte content and action potential of the giant nerve fibres of *Loligo. J. Physiol., Lond.* **98**: 299–313.

1944

- Giant nerve-fibres. *Endeavour* **3** (11): 108–113.

1950

- with B. B. BOYCOTT. The comparative study of learning. *Symp. Soc. exp. Biol.* No. 4: 432–453.

1951

- Ferrier Lecture. Growth and plasticity in the nervous system. *Proc. R. Soc.* (B) **139**: 18–37.

1955

- with B. B. BOYCOTT. A memory system in *Octopus vulgaris* Lamarck. *Proc. R. Soc.* (B) **143**: 449–480.
- with B. B. BOYCOTT. Memories controlling attacks on food objects by *Octopus vulgaris* Lamarck. *Pubbl. Staz. zool. Napoli* **27**: 232–249.

1956

- Visual responses by *Octopus* to crabs and other figures before and after training. *J. exp. Biol.* **33**: 709–729.
- with B. B. BOYCOTT. Reactions to shape in *Octopus vulgaris* Lamarck. *Proc. zool. Soc. Lond.* **126**: 491–547.
- with B. B. BOYCOTT. The subpedunculate body and nerve and other organs associated with the optic tract of cephalopods. In *Bertil Hanström, zoological papers in honour of his sixty-fifth birthday, November 20th, 1956*: 76–105. Wingstrand, K. G. (ed.). Lund: Zool. Inst.

1957

- with B. B. BOYCOTT. Effects of interference with the vertical lobe on visual discriminations in *Octopus vulgaris* Lamarck. *Proc. R. Soc.* (B) **146**: 439–459.

1958

- Effect of removal of various amounts of the vertical lobes on visual discrimination by *Octopus. Proc. R. Soc.* (B) **149**: 441–462.
- Responses of untrained octopuses to various figures and the effect of removal of the vertical lobe. *Proc. R. Soc.* (B) **149**: 463–483.
- with B. B. BOYCOTT. Reversal of learned responses in *Octopus vulgaris* Lamarck. *Anim. Behav.* **6**: 45–52.

1959

- Squids, cuttlefishes and octopuses. *Proc. R. Instn Gt Br.* **37**: 394–411.
- Extinction of unrewarded responses in *Octopus. Pubbl. Staz. zool. Napoli* **31**: 225–247.

- with J. E. AMOORE and K. RODGERS. Sodium and potassium in the endolymph and perilymph of the statocyst and in the eye of *Octopus*. *J. exp. Biol.* **36**: 709–714.

1960

- Observations on *Argonauta* and especially its method of feeding. *Proc. zool. Soc. Lond.* **133**: 471–479.
- The statocysts of *Octopus vulgaris*. *Proc. R. Soc.* (B) **152**: 3–29.
- Unit processes in the formation of representations in the memory of *Octopus*. *Proc. R. Soc.* (B) **153**: 1–17.
- The failures of discrimination learning following the removal of the vertical lobes in *Octopus*. *Proc. R. Soc.* (B) **153**: 18–46.
- The visual system of *Octopus*. (1) Regularities in the retina and optic lobes of *Octopus* in relation to form discrimination. *Nature, Lond.* **186**: 836–839.

1961

- Learning and discrimination in the octopus. *Biol. Rev.* **36**: 32–96.
- Eyes, colours and shapes. *Proc. R. Instn Gt Br.* **38**: 401–413.
- Rates of establishment of representations in the memory of octopuses with and without vertical lobes. *J. exp. Biol.* **38**: 43–60.

1962

- The retina of cephalopods and its degeneration after optic nerve section. *Phil. Trans. R. Soc.* (B) **245**: 1–18.
- The optic lobes of *Octopus vulgaris*. *Phil. Trans. R. Soc.* (B) **245**: 19–58.
- Courtship and mating by a coral reef octopus (*O. horridus*). *Proc. zool. Soc. Lond.* **138**: 157–162.
- Reversal of learning in *Octopus* and the effect of removal of the vertical lobe. *Q. Jl exp. Psychol.* **14**: 193–205.
- Repeated reversal of training in *Octopus*. *Q. Jl exp. Psychol.* **14**: 206–222.
- The thirty-sixth Maudsley Lecture: Memory mechanisms of the brain. *J. ment. Sci.* **108**: 119–133.
- with W. R. A. MUNTZ and N. S. SUTHERLAND. Simultaneous shape discrimination in *Octopus* after removal of the vertical lobe. *J. exp. Biol.* **39**: 557–566.
- How can the memory of the nervous system be studied? pp. 275–292 in *Il metodo sperimentale in biologia da Vallisneri ad oggi*. Simposio nel III centenario della nascita di Antonio Vallisneri, 29–30 Settembre–1 Ottobre 1961. Padova Universita degli studi Accademia Patavina di Scienze Lettere ed Arti.
- with J. R. PARRISS. The limits of transfer of a learned discrimination to figures of larger and smaller sizes. *Z. vergl. Physiol.* **45**: 618–635.

1963

- The number and sizes of nerve cells in *Octopus*. *Proc. zool. Soc. Lond.* **140**: 229–254.
- Light- and dark-adaptation in the eyes of some cephalopods. *Proc. zool. Soc. Lond.* **140**: 255–272.
- Some essentials of neural memory systems. Paired centres that regulate and address the signals of the results of action. *Nature, Lond.* **198**: 626–630.
- with P. N. DILLY and E. G. GRAY. Electron microscopy of optic nerves and optic lobes of *Octopus* and *Eledone*. *Proc. R. Soc.* (B) **158**: 446–456.

1964

- *A model of the brain*. Oxford: Clarendon Press.
- Paired centres for the control of attack by *Octopus*. *Proc. R. Soc.* (B) **159**: 565–588.
- with E. G. GRAY. Electron microscopy of synaptic structure of *Octopus* brain. *J. Cell. Biol.* **21**: 87–103.

1965

- The central nervous system of *Nautilus*. *Phil. Trans. R. Soc.* (B) **249**: 1–25.
- The diameters of the fibres of the peripheral nerves of *Octopus*. *Proc. R. Soc.* (B) **162**: 47–79.
- The buccal nervous system of *Octopus*. *Phil. Trans. R. Soc.* (B) **249**: 27–44.
- The centres for touch discrimination in *Octopus*. *Phil. Trans. R. Soc.* (B) **249**: 45–67.
- The nervous pathways for poisoning, eating and learning in *Octopus*. *J. exp. Biol.* **43**: 581–593.
- Two memory stores in one brain. *Endeavour* **24**: 13–20.
- The Croonian Lecture, 1965: The organization of a memory system. *Proc. R. Soc.* (B) **163**: 285–320.
- Influence of previous preferences on the memory of *Octopus vulgaris* after removal of the vertical lobe. *J. exp. Biol.* **43**: 595–603.
- with T. M. O. BURROWS, I. A. CAMPBELL and E. HOWE. Conduction velocity and diameter of nerve fibres of cephalopods. *J. Physiol., Lond.* **179**: 39*P*–40*P*.
- with M. J. WELLS. Split-brain preparations and touch learning in the octopus. *J. exp. Biol.* **43**: 565–579.

1966

- *The memory system of the brain*. Berkeley and Los Angeles: University of California Press. Oxford: Clarendon Press.
- with R. S. NISHIOKA, I. YASUMASU, A. PACKARD and H. A. BERN. Nature of vesicles associated with the nervous system of cephalopods. *Z. Zellforsch. mikrosk. Anat.* **75**: 301–316.
- with M. NIXON. Levels of responsiveness to food or its absence and the vertical lobe circuit of *Octopus vulgaris* Lamarck. *Z. vergl. Physiol.* **53**: 165–184.
- with M. J. WELLS. Lateral interaction and transfer in the tactile memory of the octopus. *J. exp. Biol.* **45**: 383–400.

1967

- The visceral nerves of *Octopus*. *Phil. Trans. R. Soc.* (B) **253**: 1–22.

1968

- Reversal of visual preference in *Octopus* after removal of the vertical lobe. *J. exp. Biol.* **49**: 413–419.
- with M. J. WELLS. Changes in textural preferences in *Octopus* after lesions. *J. exp. Biol.* **49**: 401–412.
- with M. J. WELLS. Learning with delayed rewards in *Octopus*. *Z. vergl. Physiol.* **61**: 103–128.

1969

- with P. R. STEPHENS. The glio-vascular system of cephalopods. *Phil. Trans. R. Soc.* (B) **255**: 1–12.

- with M. J. WELLS. The effect of splitting part of the brain or removal of the median inferior frontal lobe on touch learning in *Octopus*. *J. exp. Biol.* **50**: 515–526.
- with M. J. WELLS. Learning at different rates of training in the Octopus. *Anim. Behav.* **17**: 406–415.

1970

- Short and long memories in *Octopus* and the influence of the vertical lobe system. *J. exp. Biol.* **52**: 385–393.
- The stalked eyes of *Bathothauma* (Mollusca, Cephalopoda). *J. Zool., Lond.* **162**: 437–447.
- Neurovenous tissues in cephalopods. *Phil. Trans. R. Soc.* (B) **257**: 309–321.
- with F. BAUMANN, A. MAURO, R. MILECCHIA and S. NIGHTINGALE. The extra-ocular light receptors of the squids *Todarodes* and *Illex*. *Brain Res.* **21**: 275–279.
- with M. J. WELLS. Single-session learning by octopuses. *J. exp. Biol.* **53**: 779–788.
- with M. J. WELLS. Stimulus generalization in the tactile system of *Octopus*. *J. Neurobiol.* **2**: 31–46.

1971

- *The anatomy of the nervous system of* Octopus vulgaris. Oxford: Clarendon Press.
- with J. BARLOW and E. G. GRAY. Microneurons and their synapses in the stellate ganglion of *Octopus*. *J. Anat., Lond.* **109**: 337–338.

1972

- The organization of cephalopod ganglion. *Phil. Trans. R. Soc.* (B) **263**: 409–429.
- with N. M. CASE and E. G. GRAY. Ultrastructure and synaptic relations in the optic lobe of the brain of *Eledone* and *Octopus*. *J. Ultrastr. Res.* **39**: 115–123.
- with M. J. WELLS. The median inferior frontal lobe and touch learning in the *Octopus*. *J. exp. Biol.* **56**: 381–402.

1973

- Receptive fields of the visual system of the squid. *Nature, Lond.* **241**: 469–471.
- The giant fibre synapse of *Loligo*. *Brain Res.* **57**: 457–460.
- with M. J. HOBBS. A cephalopod cerebellum. *Brain Res.* **55**: 424–430.

1974

- The central nervous system of *Loligo*. I. The optic lobe. *Phil. Trans. R. Soc.* (B) **267**: 263–302.
- with G. D. SANDERS. Reappearance of specific colour patterns after nerve regeneration in *Octopus*. *Proc. R. Soc.* (B) **186**: 1–11.

1975

- Systems connecting endoplasmic reticulum with axon and dendrites. *J. Physiol., Lond.* **248**: 17–19*P*.
- Sources of discovery in neuroscience. In *The neurosciences*: *paths of discovery*: 15–46. Worden, F. G., Swazey, J. P. & Adelman, G. (eds.). London: MIT Press.
- with E. A. BRADLEY. Are there circadian rhythms in learning by *Octopus*? *Behav. Biol.* **13**: 527–531.
- with E. A. BRADLEY. Comparison of visual and tactile learning in *Octopus* after lesions to one of the two memory systems. *J. Neurosci. Res.* **1**: 185–205.

- with P. N. DILLY and P. R. STEPHENS. Receptors in the statocyst of squids. *J. Physiol., Lond.* **249**: 59–61*P*.
- with P. R. STEPHENS. The statocysts of various cephalopods. *J. Physiol., Lond.* **249**: 1*P*.
- with M. J. WELLS. The subfrontal lobe and touch learning in the octopus. *Brain Res.* **92**: 103–121.

1976

- The nervous system of Loligo. II. Suboesophageal centres. *Phil. Trans. R. Soc.* (B), **274**: 101–167.
- The "cerebellum" and the control of eye movements in cephalopods. *Nature, Lond.* **264**: 572–574.
- with P. R. STEPHENS. The statocyst of *Vampyroteuthis infernalis* (Mollusca: Cephalopoda). *J. Zool., Lond.* **180**: 565–588.

1977

- The central nervous system of *Loligo* III. Higher motor centres. The basal supraoesophageal lobes. *Phil. Trans. R. Soc.* (B) **276**: 351–398.
- with P. N. DILLY and M. NIXON. *Mastigoteuthis*—the whip-lash squid. J. Zool., Lond. **181**: 527–559.

APPENDIX II

Classification of Recent Cephalopods

Class **CEPHALOPODA** Cuvier, 1798

- *Subclass* **NAUTILOIDEA** Agassiz, 1847
  - *Family* **Nautilidae** Blainville, 1825
    - *Nautilus* Linné, 1758
- *Subclass* **COLEOIDEA** Bather, 1888
  - *Order* **Sepioidea** Naef, 1916
    - *Family* **Spirulidae** Owen, 1836
      - *Spirula* Lamarck, 1801
    - *Family* **Sepiidae** Keferstein, 1866
      - *Sepia* Linné, 1758
      - *Sepiella* Gray, 1849
    - *Family* **Sepiadariidae** Naef, 1912
      - *Sepiadarium* Steenstrup, 1881
      - *Sepioloidea* Orbigny, 1840
    - *Family* **Sepiolidae**
      - *Subfamily* **Rossiinae** Appellöf, 1898
        - *Rossia* Owen, 1834
        - *Semirossia* Steenstrup, 1887
        - *Neorossia* Boletzky, 1971
      - *Subfamily* **Heteroteuthinae**
        - *Heteroteuthis* Gray, 1849
        - *Nectoteuthis* Verrill, 1883
        - *Iridoteuthis* Naef, 1912
        - *Stoloteuthis* Verrill, 1881
        - *Sepiolina* Naef, 1912
      - *Subfamily* **Sepiolinae** Appellöf, 1898
        - *Sepiola* Leach, 1817
        - *Euprymna* Steenstrup, 1887
        - *Rondeletiola* Naef, 1921
        - *Sepietta* Naef, 1912
        - *Inioteuthis* Verrill, 1881
    - *Family* **Idiosepiidae** Appellöf, 1898
      - *Idiosepius* Steenstrup, 1881
  - *Order* **Teuthoidea** Naef, 1916
    - *Suborder* **Myopsida** Orbigny, 1845
      - *Family* **Pickfordiateuthidae** Voss, 1953
        - *Pickfordiateuthis* Voss, 1953
      - *Family* **Loliginidae** Steenstrup, 1861
        - *Loligo* Schneider, 1784
        - *Doryteuthis* Naef, 1912
        - *Lolliguncula* Steenstrup, 1881
        - *Sepioteuthis* Blainville, 1824
        - *Alloteuthis* Wülker, 1920
        - *Uroteuthis* Rehder, 1945
        - *Loliolus* Steenstrup, 1856
        - *Loliolopsis* Berry, 1929

Suborder **Oegopsida** Orbigny, 1845
Family **Lycoteuthidae** Pfeffer, 1908
Subfamily **Lycoteuthinae** Pfeffer, 1908
Lycoteuthis Pfeffer, 1908
Oregoniateuthis Voss, 1956
Selenoteuthis Voss, 1958
Nematolampas Berry, 1913
Subfamily **Lampadioteuthinae** Berry, 1916
Lampadioteuthis Berry, 1916
Family **Enoploteuthidae** Pfeffer, 1900
Subfamily **Enoploteuthinae** Pfeffer, 1912
Enoploteuthis Orbigny, 1839
Abralia Gray, 1849
Abraliopsis Joubin, 1896
Watasenia Ishikawa, 1913
Subfamily **Ancistrocheirinae** Pfeffer, 1912
Ancistrocheirus Gray, 1849
Thelidioteuthis Pfeffer, 1900
Subfamily **Pyroteuthinae** Pfeffer, 1912
Pyroteuthis Hoyle, 1904
Pterygioteuthis Fischer, 1896
Family **Octopoteuthidae** Berry, 1912
Octopoteuthis Rüppell, 1844
Taningia Joubin, 1931
Family **Onychoteuthidae** Gray, 1849
Onychoteuthis Lichtenstein, 1818
Onykia Lesueur, 1821
Chaunoteuthis Appellöf, 1890
Ancistroteuthis Gray, 1849
Moroteuthis Verrill, 1881
Family **Cycloteuthidae** Naef, 1923
Cycloteuthis Joubin, 1919
Discoteuthis Young & Roper, 1969
Family **Gonatidae** Hoyle, 1886
Gonatus Gray, 1849
Gonatopsis Sasaki, 1920
Berryteuthis Naef, 1921
Family **Psychroteuthidae** Thiele, 1921
Psychroteuthis Thiele, 1921
Family **Lepidoteuthidae** Naef, 1912
Lepidoteuthis Joubin, 1895
Pholidoteuthis Adam, 1950
Tetronychoteuthis Pfeffer, 1900
Family **Architeuthidae** Pfeffer, 1900
Architeuthis Steenstrup, 1857
Family **Histioteuthidae** Verrill, 1881
Histioteuthis Orbigny, 1841
Family **Alluroteuthidae** Odhner, 1923
Alluroteuthis Odhner, 1923
Neoteuthis Naef, 1921

Family **Bathyteuthidae** Pfeffer, 1900
Bathyteuthis Hoyle, 1885
Family **Ctenopterygidae** Grimpe, 1922
Ctenopteryx Appellöf, 1899
Family **Brachioteuthidae** Pfeffer, 1908
Brachioteuthis Verrill, 1881
Family **Batoteuthidae** Young & Roper, 1968
Batoteuthis Young & Roper, 1968
Family **Ommastrephidae** Steenstrup, 1857
Subfamily **Illicinae**
Illex Steenstrup, 1880
Todaropsis Girard, 1890
Subfamily **Todarodinae**
Todarodes Steenstrup, 1880
Nototodarus Pfeffer, 1912
Martialia Rochebrune & Mabille, 1889
Subfamily **Ommastrephinae**
Ommastrephes Orbigny, 1835
Dosidicus Steenstrup, 1857
Symplectoteuthis Pfeffer, 1900
Ornithoteuthis Okada, 1927
Hyaloteuthis Gray, 1849
Family **Thysanoteuthidae** Keferstein, 1866
Thysanoteuthis Troschel, 1857
Cirrobrachium Hoyle, 1904
Family **Chiroteuthidae** Gray, 1849
Chiroteuthis Orbigny, 1839
Chiropsis Joubin, 1933
Valbyteuthis Joubin, 1931
Family **Mastigoteuthidae** Verrill, 1881
Mastigoteuthis Verrill, 1881
Echinoteuthis Joubin, 1933
Family **Promachoteuthidae** Naef, 1912
Promachoteuthis Hoyle, 1885
Family **Grimalditeuthidae** Pfeffer, 1900
Grimalditeuthis Joubin, 1898
Family **Joubiniteuthidae** Naef, 1922
Joubiniteuthis Berry, 1920
Family **Cranchiidae** Prosch, 1849
Subfamily **Cranchiinae** Prosch, 1849
Cranchia Leach, 1817
Leachia Lesueur, 1821
Drechselia Joubin, 1931
Liocranchia Pfeffer, 1884
Liguriella Issel, 1908
Pyrgopsis Rochebrune, 1884
Subfamily **Taoniinae** Pfeffer, 1912
Taonius Steenstrup, 1861
Phasmatopsis Rochebrune, 1884
Egea Joubin, 1933

Belonella Lane, 1957
Megalocranchia Pfeffer, 1884
Helicocranchia Massy, 1907
Ascocranchia Voss, 1962
Galiteuthis Joubin, 1898
Corynomma Chun, 1906
Bathothauma Chun, 1906
Mesonychoteuthis Robson, 1925

Order **Octopoda** Leach, 1818
Suborder **Cirrata** Grimpe, 1916
Family **Cirroteuthidae** Keferstein, 1866
Cirroteuthis Eschricht, 1838
Cirrothauma Chun, 1911
Family **Stauroteuthidae** Grimpe, 1916
Stauroteuthis Verrill, 1879
Grimpoteuthis Robson, 1932
Chunioteuthis Grimpe, 1916
Froekenia Hoyle, 1904
Family **Opisthoteuthidae** Verrill, 1896
Opisthoteuthis Verrill, 1883
Suborder **Incirrata** Grimpe, 1916
Family **Bolitaenidae** Chun, 1911
Bolitaena Streenstrup, 1859
Japetella Hoyle, 1885
Eledonella Verrill, 1884
Dorsopsis Thore, 1949
Family **Amphitretidae** Hoyle, 1886
Amphitretus Hoyle, 1885
Family **Idioctopodidae** Taki, 1962
Idioctopus Taki, 1962
Family **Vitreledonellidae** Robson
Vitreledonella Joubin, 1918
Family **Octopodidae** Orbigny, 1845
Subfamily **Octopodinae** Grimpe, 1921
Octopus Lamarck, 1798
Enteroctopus Rochebrune & Mabille, 1889
Danoctopus Joubin, 1933
Cistopus Gray, 1849
Pinnoctopus Orbigny, 1845
Robsonella Adam, 1938
Scaeurgus Troschel, 1857
Macrochlaena Robson, 1929
Pteroctopus Fisher, 1882
Hapalochlaena Robson, 1929
Berrya Adam, 1939
Sasakinella Taki, 1964
Euaxoctopus Voss, 1971
Subfamily **Eledoninae** Gray, 1849
Eledone Leach, 1817

Pareledone Robson, 1932
Thaumeledone Robson, 1930
Bentheledone Robson, 1932
Velodona Chun, 1915
Subfamily **Bathypolypodinae** Robson
Bathypolypus Grimpe, 1921
Graneledone Joubin, 1918
Benthoctopus Grimpe, 1921
Teretoctopus Robson, 1929
Grimpella Robson, 1928
Haptochlaena Grimpe, 1922
Tetracheledone Voss, 1955
Family **Tremoctopodidae** Brock, 1882
Tremoctopus Delle Chiaje, 1829
Family **Ocythoidae** Gray, 1849
Ocythoe Rafinesque, 1814
Family **Argonautidae** Naef, 1912
Argonauta Linné, 1758
Family **Allopsidae** Verrill, 1882
Alloposus Verrill, 1880
Order **Vampyromorpha** Pickford, 1939
Family **Vampyroteuthidae** Thiele, 1915
Vampyroteuthis Chun, 1903

We are greatly indebted to Dr G. L. Voss for this classification of the cephalopods; it will meet a great need as few have been published in recent times. There is also an annotated classification in the Encyclopaedia Britannica (15th edition, 1974) by Dr Voss.

AUTHOR INDEX

Numbers in italics refer to pages in the References at the end of each article.

Numbers in bold refer to the first page of the author's chapter in this volume.

Y

Z

SUBJECT INDEX

B

D

E

Q

R